KU-571-152

*Science and
Technology in
World History*

Science and Technology in World History

AN INTRODUCTION

Third Edition

James E. McClellan III
and Harold Dorn

Johns Hopkins University Press
BALTIMORE

© 2015 Johns Hopkins University Press
All rights reserved. Published 2015
Printed in the United States of America on acid-free paper
9 8 7 6 5 4 3 2

Johns Hopkins University Press
2715 North Charles Street
Baltimore, Maryland 21218-4363
www.press.jhu.edu

Library of Congress Cataloging-in-Publication Data
McClellan, James E., III (James Edward), 1946–
    Science and technology in world history : an introduction / James E. McClellan
III and Harold Dorn. — Third edition.
        pages cm
    Includes bibliographical references and index.
    ISBN 978-1-4214-1774-5 (hardcover : alk. paper) — ISBN 978-1-4214-
1775-2 (pbk. : alk. paper) — ISBN 978-1-4214-1776-9 (electronic) — ISBN
1-4214-1774-X (hardcover : alk. paper) — ISBN 1-4214-1775-8 (pbk. :
alk. paper) — ISBN 1-4214-1776-6 (electronic) 1. Science—History. 2.
Technology—History. 3. Tool and die makers—History. I. Dorn, Harold, 1928–
2011 II. Title.
Q125.M414 2015
509—dc23   2014047474

A catalog record for this book is available from the British Library.

*Special discounts are available for bulk purchases of this book. For more
information, please contact Special Sales at 410-516-6936 or specialsales@press
.jhu.edu.*

Johns Hopkins University Press uses environmentally friendly book materials,
including recycled text paper that is composed of at least 30 percent post-
consumer waste, whenever possible.

Contents

Preface

This book was initially written as an introduction for general readers and students to provide the "big picture" that an educated person might wish to have of the history of science and technology. It was not written for scholars or experts, and its character as a textbook is self-evident. The style and format grew out of our extensive experience in engaging undergraduates in these matters, and the hard knocks of the classroom have suggested both the essential lessons and what materials and examples work well in conveying those lessons.

The first edition of this book appeared in 1999, the second in 2006. In approaching this work for the third time, we think that perhaps we wrote a more serious book than we realized. The book at hand examines the history of science and technology from prehistory to the present, and from this overarching point of view it has fundamental things to say about the history of science and technology across that sweep of time that escape attention or are harder to see in more fine-grained scholarly works. In the introduction that follows ("The Guiding Themes"), we enumerate these things and the thematics of science and technology in society as we see them playing out over the millennia of history. Because of this reappraisal the introduction has been substantially revised from previous editions.

The success of the earlier editions of this work exceeded our expectations and hopes. The book has been widely adopted at the college level in history of science and technology courses, and also in courses devoted to world civilization and modernization. To judge from correspondence sent to us, this book has been well received by a lay public beyond the walls of the university, evidently attracted by its broad subject. And, a surprise to us, it has also been translated into Chinese, German, Turkish, and Korean. Undoubtedly, what appeals to foreign publishers and readers was foreshadowed in the title and in our vision for this book: "Science and Technology in *World* History," and we still take pride that the first edition of this work earned laurels from the World History Association.

We are gratified by the reception our work has received, and we are glad for the opportunity to craft an improved-upon third edition. We have corrected several small errors that crept into earlier editions, and we have introduced a few stylistic changes that we hope improve the clarity of our presentation and prose. We have likewise sought to provide updated facts and figures where called for in our later chapters.

In addition to reworking the above-mentioned thematic introduction, the major change to this edition concerns chapter 17 ("Toolmakers Take Command"). Already in the second edition we rewrote this chapter to reflect a more encompassing view of the major technological systems that make up industrial civilization today. We might have otherwise just tweaked the chapter for this edition, except for one horrible particular: Superstorm Sandy and the 19-foot storm surge that slammed into the New York–New Jersey region on the night of October 29–30, 2012. Across the Hudson River from midtown Manhattan, Superstorm Sandy flooded Hoboken, New Jersey, where one of us (McClellan) lives, cutting off electricity, cell phone and Internet service, and so much else, just as it cut off whole communities in the New York Metropolitan region from the rest of civilization. We were on TV, but we couldn't get TV. McClellan was fortunate that at Stevens Institute of Technology, where he teaches and which was likewise cut off in Hoboken, a vigorous discussion took place regarding the effects of Superstorm Sandy and what the event has to tell us about technology and the modern world. In the ten dark and cold days before power was restored and in conferences and panels afterward, a stricken community of flood victims and those otherwise affected grappled intellectually, not just physically, with what was occurring around them: residents, involved students, Stevens professors and research scientists, other professionals, community activists, and local politicians struggled to make sense of what had happened. McClellan's good fortune extended to having Professors Andrew L. Russell, Lee Jared Vinsel, and John Horgan as his specialist colleagues in STS (science and technology studies). Their discussions converged on the notion of a technological supersystem undergirding industrial civilization today. The fragility and vulnerabilities of this supersystem and our utter dependence on it became abundantly plain in Superstorm Sandy. The idea that not all technological systems are equal became clear, too, with electricity being the foundational system that makes industrial civilization possible. How this thinking works itself out can be seen further in the revised chapter 17.

Previous editions acknowledged the people and institutions that aided us, and it is unnecessary to repeat those thanks at this remove. For this edition we have especially and as warmly as possible to thank Professor Joseph November, historian of science at the University of South Carolina. Joe November and his colleagues have taught using previous editions of this book, and to help us in preparing this edition Joe took the time to reread and annotate the second edition carefully. We owe Joe November a great deal for his willingness to do this and for his many critical suggestions for improvements. We can only hope that we have responded adequately to his input. We owe an additional word of thanks to the aforementioned Stevens colleagues, Andrew Russell, Lee Vinsel, and John Horgan for having read and commented on drafts of the new material presented here. Our colleague in the Physics Department at Stevens, Professor Edward A. Whittaker, saved us from more than one gaffe in dealing with modern physics.

Our students at Stevens Institute of Technology were once again helpful, and we acknowledge their many suggestions for enhancing the presentation. Bits and pieces of several student research projects have found their way into the present work, and in particular, we thank Jovanna Manzari, Muhammad Abd Malik, and Emanuel Rodriguez. We thank Alex McClellan and Julian McClellan for their insights into smartphones and social media. Tamar Boodaghians did good work and graciously aided us in updating the Web resources presented at the end of this book. Our Stevens colleague Professor Andrew Rubenfeld prepared the index, and we thank Andrew warmly for his friendship, effort, and skills. Jackie McClellan again earns heartfelt praise and thanks for her lifelong companionship and her editorial expertise.

We are especially grateful to Dr. Robert J. Brugger and his associates at Johns Hopkins University Press. Bob initiated this whole enterprise over a lunch with Dorn and McClellan in Baltimore over two decades ago, and in many ways this is his book as well as ours. The professionalism and effectiveness of everyone at the Press are once again manifest in the book at hand. We have particularly to thank Juliana McCarthy, Kathryn Marguy, Hilary Jacqmin, and especially Mary Lou Kenney whose many touches as a real editor substantially improved this work.

Regrettably, Harold Dorn passed away in 2011. His imprint on this book, particularly through his chef-d'œuvre, *The Geography of Science* (1991), remains substantial, and this third edition is no less a complete collaboration as before.

*Science and
Technology in
World History*

The Guiding Themes

This book surveys a history of science and technology from the Old Stone Age—or Paleolithic era, stretching back tens of thousands and hundreds of thousands of years—to the present. One of its special concerns is with the relations between science and technology over this long period, but before we can broach our subject we first need to address what we mean by science and technology.

Defining science and technology is not a straightforward matter. Simple dictionary definitions will not do. Part of the problem is that the way we use the terms today is relatively new. The word *science* derives from the Latin *scientia*, meaning knowledge. It is used in English from the Middle Ages onward to cover a variety of meanings broadly concerned with knowledge, often bordering on philosophy, or a formalized skill, and only in the nineteenth century did the word science come to have its more restricted modern meaning of or pertaining to the natural or physical sciences, such as physics, chemistry, biology, and so forth. Famously and tellingly in this context, the English word *scientist* was coined only in 1840. Is that to say that scientists did not exist before 1840? Obviously not, but caution is clearly called for in using the labels science or scientists in reference to ancient Greece or medieval China, for example. Similarly, the word *technology* derives from the Greek *techné* (τέχνη) having to do with the practical arts, craftsmanship, and technique. Much of that sense is preserved in our word technology, but the word in English does not appear before the seventeenth century, and the more modern meaning of technology as the scientific study and systematic pursuit of the practical arts and their improvements does not enter our vocabularies again until the middle of the nineteenth century.

Knowledge and know-how have a history, but given how their meanings have shifted over time, it becomes harder to use the terms science or technology uncritically, especially in diverse historical contexts. The danger of anachronism or projecting our notions back onto the past looms large in such a work as this one. In always placing ourselves in the context of the times and in trying to think as contemporaries thought, however, we can identify institutions, individuals, and activities that touch on the investigation of the natural world (science) or that sought practical mastery of that world (technology). For simplicity's sake, in what follows we do speak about science and technology in ancient Egypt, Greece, China, and elsewhere. Occasionally we use the unavoidable term scientist, but always keeping in mind the precise context.

Our definitional problem is compounded because science and technology clearly aren't single, unitary things easily identified as such, even today, much less in changing historical contexts. Are we to conflate Aristotle's science with what happens at CERN (the European Center for Nuclear Research) or the National Science Foundation today? Is the technology of the Roman chariot to be thought of on par with a sleek new car? More than that, what we know as science and technology have grown over time and have accreted meanings as the fields have developed, so any static notion of the nature of science or technology has to be abandoned at the outset. But this dilemma offers its own solutions, as we can define science and technology—the subjects of this history—by how the activities we find variously associated with each (inquiry into nature/craft activities) have developed and accrued over time. Once we have done that we will ask the further question about applied science and the historical connections of science and technology.

What Is Science?

The long view offered by the sweep of history incorporated into this book can help us articulate what we mean by science. Plainly, "science" has never been a single thing. That's easy to say, but at least three different kinds of considerations are wrapped up in such a statement: the first concerns science as a body of knowledge about the natural world around us; another concerns science as a social institution and the place of science in society; and the third concerns the scientific enterprise itself and what its practitioners do. Let's take these up in turn.

In terms of content, science has considerable cultural and intellectual prestige in providing our modern scientific worldview. Science tells amazing and amazingly detailed stories about the cosmos and the world around us. The details are esoteric and reserved for experts; who gets to define the boundaries of science is a question unto itself, and there are competing bodies of knowledge, as in certain fundamentalisms or many New Age beliefs. That said, basic scientific understandings still generally frame people's view of the world: That we live on planet Earth that spins on its axis and circles the Sun, that things fall down due to gravity, that salt on the table consists of molecules of sodium and chlorine, that egg and sperm make babies, not to mention the further reaches of Big Bang cosmology, black holes, and evolutionary perspectives on life. Along with priests and theologians since antiquity, scientists and natural philosophers have long provided and still provide basic ideas about nature and the physical world around us. Where are we? What is the world made up of? Who are we? What is going on? Our ideas about nature have a dynamic history of their own, of course, and one of the purposes of this book is to provide an overview of how it is that we arrive at the stories we tell about nature today. In so doing we quickly realize that an intellectual history of science is punctuated by radical shifts in worldviews and reinterpretations of fundamental features of nature, such as in moving from a geocentric to a heliocentric system. We want to be attuned to scientific revolutions in general and sensitive to how worldviews have radically shifted over time.

The second consideration concerns the place of science in society, and it shows up in the different social settings where we can identify activities that seem scientific in nature. It's one thing, for example, to think about a state-supported astrologer or astronomer contemplating the heavens; it's another to consider a freelance Greek natural philosopher perambulating with adepts, or yet again a modern scientist in a lab coat involved in large-scale, complex research. "Science" is found in varied social and institutional contexts, not to mention in diverse social roles exercised by individuals. This insight into the varied and changing social circumstances of science holds true for scientific practice today and is visible, say, in the different loci where we find science, such as in universities, government, or corporations. But this diversity of social circumstances is even more apparent in larger historical and cultural frames, as we move from the pharaohs of ancient Egypt, through the city-states of ancient Greece, or courts in the Islamic world or in India, China, or Renaissance Italy. A more narrowly focused sociology of science likewise fits into seeing science as an enterprise of one sort or another unfolding in a larger social context. And so, asking what science is, our chapters trace out this dynamic social history down to the advent of the modern scientist and the scientific enterprise today.

But, if science is not a single or static entity either as an intellectual enterprise or social institution, science is likewise not a unitary activity in terms of what its practitioners (scientists) do. At its heart science has to do with understanding the natural world around us, but what individuals and communities have done to pursue that end has varied considerably over time and varies considerably today. That is, scientists do different things in doing science: empirical work, theoretical ruminating, puzzle-solving, philosophical inquiry, experimental research using instruments, more solitary research versus being part of a large team of scientists doing Big Science work, applied research and development, secret/military/"homeland security" work, entrepreneurial activity, and so forth, not to mention teaching. It makes a difference whether you're a graduate student, postdoc, professor or research scientist of one sort or another, or hunkered down in a start-up somewhere. And here, the sociology of science illuminates the practice of science. There's also wide variety between and among the different scientific disciplines, their different traditions, and how they are practiced (think physics versus botany, for example).

The long sweep of history allows us to see how these different activities and methodologies arose to give us science as we know it today. That is, different notions emerged over time of how and what it means to pursue inquiry into nature. Thus, subsequent chapters outline novelties as they arose from simple observation of nature to recordkeeping, mastering the regularities of the heavens, research guided—or not—by theory, mathematics in natural philosophy, changing methods for producing knowledge, and so forth. Scientific instruments, like the telescope or the barometer, for example, appeared only in the seventeenth century as an agreed-upon element of scientific research. This growing complexity regarding the scientific enterprise and its tools and methodologies is especially clear in historical context, as more and more becomes included in various scientific traditions over time. And so, in asking what science is, we can answer by pointing to a history of how science came to be understood and practiced at different times and in different contexts including down to the scientific enterprise today. But we also need to emphasize here that if, at its most basic, we think of science as simply knowledge of nature, then we are led back to the first humans and the knowledge they had of the natural world around them. And so, against the grain of most histories of science, we begin with the Paleolithic era and place the origins of science, as well as human technology, there.

A Word about Technology

Take a look around you. Look at the various things you see. They are largely all manufactured, aren't they? Think about where you live. Think about houses generally, our offices, our roads, and our GPS- and otherwise equipped vehicles we and things move around in. Think about the Internet and your phone. Think about TV, streaming video, and social media. Think about the thousands of airplanes in the air and ships at sea right now as you read this. Think about flipping a switch and all kinds of things getting powered, including the lights. Think about the food you'll eat today or the least thing in your environment . . . a toothpick . . . the penny in your pocket. Stopping to think about these things reveals unequivocally our utter dependence on our supporting technologies and technological systems of industrial civilization and just what a complicated, amazing world we live in today.

All these things we see and experience are not natural or part of nature, but are manmade, artificial technologies that we have created to sustain ourselves. Humankind has always sheltered itself in a cocoon of technologies, as we will see, but today especially the cascade of interlocking technologies that make up contemporary industrial civilization underscores the centrality and overarching importance of technology in structuring today's world. How did we come to live this way? This is the largest question this book seeks to answer.

There are different ways to think about technologies. We most often think of technology as artifacts: The light bulb in the socket, the flat tire on my car, an X-ray, your computer or smartphone, and so forth. However, historians and students of technology for decades now have conceptualized technology more in terms of technological systems, that is, technologies as more complex assemblages of components that function to some end, such as lighting the room. This style of systems thinking informs our treatment of technology and its history here, and we will have a richer and clearer understanding of technologies only if we conceptualize them as systems. As we will see in chapter 17 ("Toolmakers Take Command"), electric lighting, for example, incorporates a complex and interacting set of parts that runs from the generating plant and includes the light bulb and so much more, from the electrons that run through the wires to the stockholders invested in the power company. A corollary of this point is that even the earliest and simplest technologies, such as stone tools or the control of fire, also need to be conceived of in terms of technological systems. For making a fire, for example, we might ask where the flint comes from and how it gets to the user. What is the tinder, and how is it collected, processed, and stored? How does one acquire the skill to knap stone or spark a fire? And what about fuel? No, a simple, but integrated system of elements allowed primitive man to organize the production of fire.

We also have to recognize that intangible technique and know-how are to be considered part of what we mean by technology. The blacksmith and the computer programmer have sophisticated skill sets; they know how to do things, and the practical—even nonverbal or motor—techniques of *doing* have also to be folded into our understanding of technology. Beyond all that, we need to remember that technologies do not exist on some abstract plane but are always situated in historical and cultural context. Cultures and societies always developed technologies that seemed appropriate for their economies, lifestyles, and circumstances, and it is from this vantage point that particular technologies—as artifacts or as systems—need to be considered.

Technology has several other features that distinguish it from science in particular. One is geographical locale. Technology is found everywhere: Not only in bustling cities or ports, but more tellingly in rural and even remote areas because humans can't survive without their technologies. Science, on the other hand, is not similarly or so universally diffused. There have been minor exceptions and although research is sometimes carried out in faraway places, overwhelmingly organized science resides not in the countryside but in urban settings. Another distinction is that traditional technologies are in a sense more egalitarian because at least until comparatively recently untold legions of artisans and craft specialists plied their trades in decentralized, local venues. Throughout most of history the cobbler, the mason, the butcher, and their like served and sustained local communities everywhere. By contrast, only a comparatively few people have ever become involved in scientific pursuits, and they have pursued their science in restricted centers. Yet at the same time, unlike traditional artisans, those interested in science formed part of extended communities spread across space and time. One might contrast restricted local and even private knowledge (in guilds, say) of traditional technologies with the universalist claims of extended communities in science. And finally in this connection, how technologies have typically been transmitted from one generation to another also differs in important ways from how the scientific enterprise reproduces itself. Literacy and schooling, as mentioned, are the major differentiators here. Historically, technological practitioners were at least largely unlettered and learned their trades through apprenticeships and hands-on experience, their practice largely bereft of theory. Entry into the world of science requires its own apprenticeship, but seemingly of a different sort because of schooling, literacy, and the role of theory and research in science.

Social divisions and occupational gradations fragment the world of technology. In this connection one thinks of class or caste distinctions and a scale separating the peasant farmer, a highly specialized artisan who might craft an inlaid table, a master builder employed by a town to build a cathedral, or the engineers and project managers who built the pyramids for the pharaoh. And at this last level the world of technology meets up with science in history and the application of expert knowledge to practical problems. At the top of this sociological pyramid stands an exclusive set of literate engineers and architects, such as the Roman engineer and architect Vitruvius (who died in 15 CE) or the Roman Senator Frontinus (40–130 CE), who wrote expertly on aqueducts and water supply. These men approached and wrote about architecture and civil and military engineering in much the same manner and social circumstances as contemporary natural philosophers. These literary engineers are certainly noteworthy, but they remain exceptions compared to the great mass of individuals engaged in traditional trades and craft practice that sustained humanity for so many centuries prior to the nineteenth century.

The coming of industrial civilization reshaped the sociology and practice in the world of technology and forged new connections between science and technology. Today, engineers, architects, and technological innovators are trained in a manner akin to scientists, and they are not sociologically that distinct from scientific or other professionals. New technologies come into being quickly; they usually result from complicated planning and team efforts; they are much more closely connected today to industry and centers of money and power than ever before; and in one way or another they often entail the application of scientific knowledge. A sense of traditional technologies and how they became transformed in the modern world is an underlying theme that finds expression throughout.

Is Technology Applied Science?

Science is generally considered to be the fount from which today's technological marvels spring. Not to mention the classical examples of the atomic bomb or the computer chip, many of the most exciting technologies of our time—the Internet, advances in medicine, or the latest hot app or electronic gadget—seem to connect to science somehow, and so we commonly think that technology is (simply) applied science. Indeed, science has become so identified with practical benefits that the dependence of technology on science is commonly assumed to be a timeless relationship and a single enterprise. Science and technology, research and development—these are assumed to be almost inseparable twins, and they rank among the sacred phrases of our time. The belief in the coupling of science and technology is now enshrined in the dictionary definition of technology as applied science, and journalistic reports under the rubric of "science news" are, in fact, often accounts of engineering rather than scientific achievements. One need hardly mention the scientific training that engineers and others at the cutting edge of new technologies receive to reinforce the sense that technology is indeed applied science. Certainly in today's world, surrounded and supported as we are by a host of amazing technologies that seem self-evidently to involve science in some way or another, it's not hard to think that, yes, technology has to be applied science.

The predominant guiding theme of this work indeed concerns the historical relations between science and technology. Our initial motivation in writing this book, in fact, was to break down the cliché that "technology is applied science." We wanted to show this belief to be an artifact of today's cultural attitudes superimposed without warrant on the historical record. The historical record does show that in the earliest civilizations under the patronage of pharaohs and kings, and in general whenever centralized states arose, knowledge of nature was applied in some fashion or another and exploited for useful purposes. But it cannot be said that science and technology were systemically and closely related throughout most of history. In ancient Greece (where theoretical science had its beginning), among the scholastics of the Middle Ages, in the time of Galileo and Newton, and even for Darwin and his contemporaries in the nineteenth century, science constituted a learned calling whose results were recorded in scientific publications, while technology was understood as the crafts practiced by unschooled artisans. Until the second half of the nineteenth century few artisans or engineers attended a university or, in many cases, received any formal schooling at all. Conversely, the science curriculum of the university centered largely on pure mathematics (not to mention Greek and Latin) and what was often termed natural philosophy—the philosophy of nature—and was written in technical terms (and often a language) foreign to artisans and engineers.

Particularly since the nineteenth century, science has undoubtedly bestowed genuine benefits on humankind, and it has fostered the hope that research can be channeled in the direction of improving the human condition. But a more secure understanding of science, one less bound by the cultural biases of our time, can be gained by viewing it through the lens of history. Seen thus, with its splendid achievements but also with its blemishes and social constraints, the enterprise of science becomes a multidimensional reality rather than a culture-bound misconception. At the same time, a more accurate historical appreciation of technology places proper emphasis on independent traditions of skilled artisans whose talents crafted everyday necessities and amenities throughout the millennia of human existence. Such a historical reappraisal will also show that in many instances technology directed the development of science, rather than the other way around—the telescope and the steam engine providing classic examples.

In order to develop the argument that the relationship between enterprises we might label science and technology has been a historical process and not an inherent identity, we trace the joint and separate histories of science and technology from the prehistoric era to the present. We intend first and foremost to review the common assumption that technology is applied science and show, instead, that in most historical situations prior to the nineteenth century, science and technology have progressed in either partial or full isolation from each other—both intellectually and sociologically. In the end, an understanding of the historical process will shed light on the circumstances under which science and technology have indeed merged over the past two hundred years.

Even if technology has been or is in some fashion applied science in many cases today, the question remains about how science actually becomes applied in various technologies. Might there be varieties of applied science? Scientific knowledge taken from the cutting edge of new research is one thing, for example; an engineer consulting a table is another. How and in what precise contexts did and does science become "applied"? This issue is addressed more particularly in chapter 19.

The Babylonian, Hellenic, and Hellenistic in Science History

A key to understanding the history of science (and technology as applied science) lies in the recognition that science has not always been situated or has not functioned the same way in societies. We mentioned above the social history of science and the necessity of keeping in mind local circumstances in following science and technology in history, but we have something more in mind in the present observation: we can identify at least three and now possibly four *general* ways in which science has been structured socially and has functioned in particular societies. Identifying these patterns helps us analytically as it opens the door to cross-cultural comparisons and provides deeper insight into the Western tradition in science. That's why they deserve attention here at the outset. The first of these types we are labeling the Babylonian model.

If the enterprises of science and technology were largely separate throughout most of history, that truth does not mean that they were (or were always) entirely separate. We shouldn't go overboard and think that technology becomes applied science in the nineteenth century and not at all before. Thus, in what follows we want to show what the connection between science and technology *was not*, but also to show *what it was* over time. The real story of the historical connections between social, technical, and intellectual enterprises we call science and technology is more nuanced, more interesting, and more revealing. The general historical separation of science and technology must not obscure key instances where science or expert knowledge was applied in a technology to serve some practical end. This crucial strand in the historical intersection of science and technology goes back much further than the nineteenth century. It goes back to the third millennium BCE (before the common era), in fact, and the dawn of civilization, and it arises coincident with the coming of the state and the requirements of reckoning, recordkeeping, and governance. The pattern we repeatedly see for this kind of applied science dates from ancient Babylonia and Egypt and involves state patronage for and institutionalization of expert knowledge for the purposes of governance. Literacy is a defining trait here, and the object of such state patronage was to produce experts of all sorts, including ones that we would see as scientist-like: scribes, tax collectors and accountants, physicians and healers, astronomers/astrologers, alchemists and chemists, engineers, bureaucrats and administrators of one sort or another, lawyers, priests, and so on. In this sense scientific expertise (whatever that might mean in a particular context) is a subset of a larger panoply of expertise needed to run a state. The pharaoh could not rule Egypt alone, so to speak, and ever after in history across a variety of cultural contexts we find subsidized experts who were literate, schooled, and employed in institutions of one sort or another, often anonymously, applying their specialized knowledge to support the ends and interests of governing and patronizing authority. This "Babylonian model" of state-supported applied science appears repeatedly in this account whenever we find centralized authority in need of experts to run governments and armies, keep track of the calendar, and otherwise manage polities. Civilizations that arose in the ancient Near East, India, China, the Americas, and elsewhere incorporated this kind of applied science again and again into ruling regimes. This key strand in the history of science, and technology as applied science, weaves its way through this account. But the Babylonian model aside, until recently the wider world of technology was under the masterful control of artisans and the unlettered.

In addition to the "Babylonian," two other revealing modes for the organization and pursuit of science present themselves: the Hellenic and the Hellenistic. The term Hellenic, of course, harks back to the apogee of classical Greece, the "Greek miracle" of the fifth and fourth centuries BCE, and the birth of philosophy and scientific theory. The era of the Presocratics, as well as Plato or Aristotle, as we will see further in chapter 4, is famous for the invention of what we usually call pure science or the disinterested pursuit of knowledge for the sake of knowledge itself. More than that, Hellenic natural philosophers repudiated the practical application of knowledge and proposed that science should have nothing to do with technology, the crafts, or the improvement of the human condition. Helping to run the affairs of state, as in Babylonia, similarly was not part of the ideology of scientific inquiry in the Hellenic era. Whereas the Babylonian model for science arose independently many times over, the Hellenic model of knowledge for its own sake unsupported by institutions or patrons seems to have been a unique historical event that emerged coincident with city-states and economies of classical fifth- and fourth-century Greece down to the death of Alexander the Great in 323 BCE. The phenomenon of pure science pursued by disinterested inquirers into nature untrammeled by any considerations of applying knowledge for some practical end only spread among scientific intellectuals in periods following the Hellenic. But the Hellenic tradition continued mostly on an individual basis and largely privately among like-minded but usually scattered scientific intellectuals forming spatially and temporally extended communities. Except for isolated instances otherwise, this disinterested "pure science," theory-oriented tradition floated in non-institutionalized and nonpatronized settings, sociologically detached from the heart of society or the supporting economy.

Our third model, the Hellenistic, combines features of both the Babylonian model and the Hellenic model for the organization and pursuit of science. The Hellenistic era immediately followed the Hellenic and the conquests of Alexander the Great (356–23 BCE). As seen best in the Museum and Library at Alexandria, detailed in chapter 5, the Hellenistic model for science combines Babylonian state support for the practical *and* subsidies for Hellenic-style, pure science pursuits. In this way the disinterested, theoretical pursuit of inquiry into nature for the sake of knowledge and insight alone found a niche alongside the practical and applied science in societies that succeeded the Hellenistic in the Near East.

These three different historical styles for the organization, practice, and patronage of science play themselves out in different and revealing ways over time. At the least, these distinctions provide useful entrées for thinking about science or science-like activities in a variety of social and chronological contexts. More than that, separating out these three alternatives this way, however crudely, enables a comparative history of science in world cultures. That is, again, we repeatedly find the Babylonian model worldwide whenever state- or near-state-level societies appear. We'll see this in our chapters devoted to the earliest civilizations, and further concerning medieval China, India, and the Americas. We focus on the marginal but telling case of Anasazi Indian settlements in and around Chaco Canyon in the American Southwest as a test of this very proposition.

Of these three archetypes, the Hellenistic combination of patronage for nonutilitarian as well as useful knowledge is especially valuable analytically. Unlike the Hellenic, whose influence was vague and which was largely detached from society, the Hellenistic model did formally repeat itself in a limited set of societies that followed in the ancient world, as we will see in chapter 6. What we usually label as Western civilization and Western scientific traditions trace themselves back to Egypt and Mesopotamia, not to mention the birth of theory in the Hellenic. Nevertheless, the Hellenistic tradition of external support for pure and applied science came to characterize the history of science in the West. That is, with its dual roots in theory and utility, the Hellenistic-style science passed to subsequent Near-Eastern civilizations, Islam, and eventually the European West, whereas the pattern in civilizations not deeply impacted by Hellenistic-style science remained largely Babylonian in nature. Following our Hellenistic model in this way helps clarify what we mean when we speak of the Western tradition and the history of Western science. Observing the continuations of the Hellenistic model over time similarly helps us better understand the multifaceted character of science and applied science today. The fourth possibility emerges only recently, where the Hellenistic model centered on *government* support for science and its experts has evolved into a drastically new mode of science embedded in industry and production as well as married to government.

The Problem of Europe

From the point of view of an overarching world history, historians have had to grapple with two related, sometimes uncomfortable questions: Why European powers became so impactful on a world scale from the fifteenth and sixteenth centuries, and why the Scientific Revolution and the modern heliocentric worldview originated in Europe. Metaphorically, how do we explain Columbus (1492) and Copernicus (1543)?

This "problem of Europe" is sometimes paired with the related "problem of China," which asks why the Scientific Revolution did not arise in China, which was more developed than Europe prior to the sixteenth century. The latter is a false question, however, as we take up in chapter 7, because science and technology functioned well in medieval China. Historians are called on to explain what happened in Europe, not what didn't happen in China. The discomfort stems from not wanting to somehow favor Europe, yet at the same time needing to explain these two great world-historical phenomena of the coming of modern science and European expansion. That is, during a thousand years of a medieval era lasting on a world-scale from roughly 500 to 1500 CE, a jostling, but relative balance maintained itself among most of the world's cultures. (We examine several of the most important of these and their scientific traditions in part II of this volume.) Then, dramatically, after 1500 Spain, Portugal, Holland, France, and England transformed the direction of world history through European colonization and later imperial expansion. Not coincidentally, at the same time Copernicus proposed a sun-centered universe and the Scientific Revolution and modern science followed through Newton's *Principia* (1687) and beyond.

Part III of this book ("Europe and the Solar System") focuses on the Scientific Revolution and the technological wherewithal of European expansion. Chapter 10 provides an account, although not a wholly original one, of how Europe could move from being a cultural and technological backwater to the forefront of the world historical stage, and how a select group of nations in Europe with their ships and cannons became poised to break out on a world scale and so shift the course of history. The other chapters in part III offer an account of the Scientific Revolution of the sixteenth and seventeenth centuries and the coming into being of our modern worldview. We also examine the Scientific Revolution as the proto-typical revolution in science and as a model for how science changes in general. In considering these developments we try to provide explanations that do not somehow privilege

Europe in any intrinsic fashion, but only as the lucky beneficiary of geography, technology, and specific historical circumstances.

By the same token, as much as European expansion and the origination of modern science and the modern scientific enterprise in Europe were key features of modern world history since 1500, we argue at the end of this work that we can no longer think of science as particularly or distinctively European or Western, except for its historical roots, because science and the scientific enterprise have institutionally expanded to the world level today. World science has succeeded Western science.

Industrial Civilization and Capitalism

From a global perspective three great socio-technological revolutions in their turn utterly transformed how humans lived: the Neolithic revolution (beginning circa 10,000 BCE), the Urban Bronze Age revolution (circa 3500 BCE), and the Industrial Revolution (from circa 1750 or so of our era). As we will see in subsequent chapters, each of these revolutions radically shifted the course of human history and ushered in fundamental changes in lifestyle and economy. The Neolithic revolution separates the eons of migratory Paleolithic food-collecting (hunting and gathering) from settled Neolithic villages and food-producing (horticulture and animal husbandry with domesticated plants and animals). The Urban Bronze Age revolution brought cities and civilization, true agriculture, a significant surplus, specialization and social stratification, monumental building, and the advent of kings and rulers, state violence, and governing power. We are all more familiar with the Industrial Revolution and industrial civilization of much more recent days, with the mechanization of production in factories, new classes of workers to work in them, new transportation systems, increased urbanization, changing market structures, and extraordinary new technologies (think the telegraph, railroad, telephone, or radio) leading to industrial civilization full blown as we know it today with its smartphones, social media, and instantaneous global communications and connections.

Two important conclusions are to be drawn from this bird's-eye view of human history. One, the revolutions that gave rise to the Neolithic, Bronze Age, or our industrial age were all fundamentally technological revolutions. They involved radically new technologies of production (horticulture, field agriculture, or industrial production, for example), and they led to radically new forms of social organization (villages, cities and polities, the modern nation-state or corporation). How and why these three fundamental socio-technological transformations occurred call out for explanation by historians, and we suggest possibilities and parameters as we proceed. But indisputably, the three historical turning points of the Neolithic, Urban Bronze Age, and contemporary industrial civilization are the three most important chapters in any history of technology.

The other observation to be made in this connection is that, comparatively speaking, the Industrial Revolution occurred just yesterday, beginning in the middle of the eighteenth century. Two hundred or three hundred years might seem like a long time ago, given the history that is usually taught, but not compared to the 5,000 years that separates us from the beginning of the Urban Bronze Age, or 12,000 years since the origins of the Neolithic, or the tens of thousands of years that separate us from people living in the Paleolithic era. It's a blink of the eye to get from eighteenth-century England to today. Thus, industrialization is presented here as a relatively recent and still unfolding process that has a unitary history and is not a series of distinct revolutions (the first, second, third Industrial Revolution, etc.) leading to the computer revolution and the new digital world of the Internet and Information Age. Accordingly, we devote the last part of this volume (part IV: "Science, Technology, and Industrial Civilization") to the Industrial Revolution and the unfolding of industrial civilization today. It seems we have entered a new phase in human history where nations and peoples are knitted together into one global system with science and technology essential in providing the infrastructure. A related theme in earlier editions and continued here is globalization with modern science and technology presented as distinctive elements of our connected world.

In earlier editions of this book, we did not make an explicit link between industrial civilization and capitalism. In retrospect we should have woven the history of capitalism and business history more closely into our text. We now see more clearly that lurking behind our anodyne term, industrial civilization, is the story of capitalism and of economic systems and ideologies that arose to invest in science and technology—research and development (R&D) and profit. Here and there in what follows we have altered our wording to be more suggestive of the links between our story of the history of science and technology and the history of capitalism. We address this point a little in our section in chapter 20, "Science as a Mode of Production," but otherwise we leave it to readers to fill in the blanks. Linking industrial civilization to capitalism aids in understanding why industrial civilization is likely not sustainable.

The Way Ahead

So, we end up with a textbook for students and the general reader, after all. Part I ("Origins to the End of Antiquity") traces the history of science and technology from the Paleolithic era though 500 CE and the end of antiquity. Part II ("Thinking and Doing among the World's Peoples") offers a world tour of science and technology in major world civilizations across a medieval millennium lasting from 500 to

1500. Part III ("Europe and the Solar System") focuses on the above-mentioned "Problem of Europe" and the Scientific Revolution of the sixteenth and seventeenth centuries. Finally, part IV ("Science, Technology, and Industrial Civilization") examines the Industrial Revolution and the further evolution of science, technology, and industrial civilization that has given us the amazing but challenging world we live in today. Nevertheless, as outlined here, we hope this volume offers something more substantial than a mere survey, a mini-encyclopedia, or a watered-down guide to a more extensive literature.

Part I

Origins to the
End of Antiquity

The origin of technology is rooted in biology. Not to mention tools and techniques observed in the animal world, technology in the form of stone tools originated literally hand in hand with humankind. Two million years ago a species of primate evolved which anthropologists have labeled Homo habilis, *or "handy man," in recognition of its ability, far beyond that of any other primate to that point, to fashion tools. Over the next two million years our ancestors were food collectors—hunters and gathers—and they used a toolkit that slowly became more elaborate and complex. A conservative estimate would have people like us—modern* Homo sapiens sapiens—*on the scene by 100,000 years ago using these advanced toolkits. We lived by our technologies then and now.*

History is often said to begin with the first written records at the dawn of the first civilizations roughly 5,000 years ago. The long eons prior to that date thus are characterized as human prehistory, long more the province of the archeologist and paleontologist than the historian. The history/pre-history division is too artificial, but it helps orient us chronologically, and plainly the scale of human prehistory is enormous, approximately 95 percent of human existence counting back only to 100,000 BCE (before the common era). At the end of that long old stone age epoch human technologies and human societies underwent two revolutionary transformations in prehistory and the cusp of the historical era and those written records in the fourth millennium BCE. In the Neolithic Revolution a few communities gave up the foraging way of life, around 12,000 years ago, in favor of horticulture and herding and developed radically new tools and techniques for earning their literal bread. The Urban Bronze Age Revolution began to unfold circa 3500 BCE in the ancient Near East and wrote a new chapter in the history of technology with field agriculture, a large surplus, and the advent of state-level societies with kings and rulers. Many other important technological novelties arose in the ancient world:

domestication of the horse, the invention of the potter's wheel, the use of cement, and a myriad of other technological developments in specialized trades and crafts. But the Urban Bronze Age Revolution of 3500 BCE set an enduring pattern of human social, economic, and technical existence with by far most people being peasants directly involved in agricultural production and with a surplus siphoned off to support a thin slice of the population that governed or served as military or priests. This pattern remained in place for roughly five thousand years until the third great technological revolution, the Industrial Revolution, began to bring about its changes in more recent times.

The history of science through the end of antiquity is clearly not congruent with the history of technology. Only toward the end of the long prehistoric era did peoples begin to observe and record the natural world systematically in ways that appear akin to science. Even when they established settled Neolithic societies only limited evidence is seen of investigating nature or patronizing research. Only with the onset of the historical era, when civilized — city-based — empires emerged in the ancient Near East did monarchs come to value higher learning for its applications in the management of complex societies and found institutions for those ends — what we are calling the Babylonian model for organization and pursuit of science. The ancient Greeks — the Hellenic Greeks of the fifth and fourth centuries BCE — then added natural philosophy; abstract theoretical science thereby took its place as a component of knowledge. This different pattern for the social organization of science we are labeling the Hellenic model. In the period following the Hellenic — the Hellenistic and Greco-Roman eras — we find a blending of state patronage for pure and applied science, a historical novelty we lump under the aegis of a Hellenistic model for how science can be organized and practiced. These models help us better grasp the history of science in antiquity and thereafter.

By the end of antiquity and the ancient world at roughly 500 of the common era (CE), both technology and science had enjoyed significant, but remarkably different, histories. Little overlap unites these two stories or their historical trajectories, but what little overlap there is was nontrivial and of further historical importance. An account of science and technology from prehistory through the end of antiquity forms the subject matter of part I.

Humankind Emerges:
Tools and Toolmakers

Scholars customarily draw a sharp distinction between prehistory and history. Prehistory is taken to be the long era from the biological beginnings of humankind over 2 million years ago to the origins of civilization about 5,000 years ago in the first urban centers of the Near East. The transition to civilization and written records at that point traditionally mark the commencement of history proper. History begins at Sumer. Everything before that is prehistory.

Prehistory, because of the exclusively material nature of its artifacts, mainly in the form of stone, bone, or ceramic products, has inescapably become the province of the archaeologist, while the historical era, with its documentary records, is the commonly assigned domain of the historian. The pre-history/history distinction has its problems, but it helps situate us in time as we start out. More than that, however, the single label "prehistory" obscures two distinctly different substages in prehistory: the Paleolithic, or Old Stone Age, which held sway for around 2 million years, is marked by rudimentary stone tools designed for collecting and processing wild food sources, while the succeeding Neolithic, or New Stone Age, which first took hold in the Near East around 12,000 years ago, entailed substantially more complex stone implements adapted to the requirements of an economy of low-intensity food production in the form of gardening or herding.

The immense scale of prehistory and especially the Paleolithic era needs emphasis. The two million years since the genus *Homo* appeared on earth is 400 times longer than the mere five thousand years of history that have elapsed since the Egyptian pyramids and written records. Using the narrower scale of 100,000 years that anatomically modern humans roamed the world, the ratio of prehistory to history as we know it is more than twenty to one. Even on this scale, reduced to a 24-hour day, history with its written records begins at 10:45 PM. One can only be struck by the immensity of the thousands and tens of

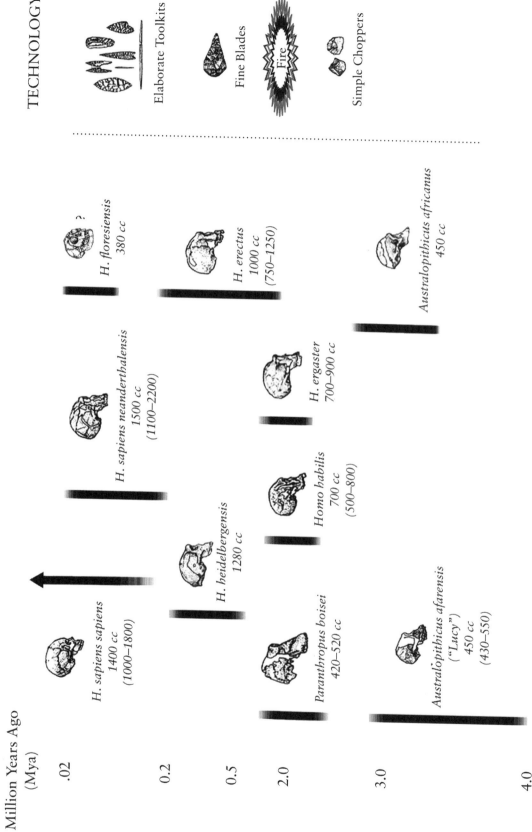

TECHNOLOGY

Elaborate Toolkits

Fine Blades

Fire

Simple Choppers

Million Years Ago
(Mya)

.02

0.2

0.5

2.0

3.0

4.0

H. sapiens sapiens
1400 cc
(1000–1800)

H. floresiensis
380 cc

H. sapiens neanderthalensis
1500 cc
(1100–2200)

H. erectus
1000 cc
(750–1250)

Australopithecus africanus
450 cc

H. heidelbergensis
1280 cc

H. ergaster
700–900 cc

Homo habilis
700 cc
(500–800)

Paranthropus boisei
420–520 cc

Australopithecus afarensis
("Lucy")
450 cc
(430–550)

thousands of years of the Paleolithic era in particular. Such unimaginable continuity over such unfathomable stretches of time, one wonders what our Paleolithic ancestors may have imagined of themselves and their own place in time.

The technologies of both the Paleolithic and Neolithic eras have left a rich legacy of material artifacts. In contrast, only a feeble record exists of any scientific interests in these preliterate societies, mainly in the form of astronomically oriented structures. Thus, at the very outset, the evidence indicates that science and technology followed separate trajectories during the eons of prehistory. Technology—the crafts— formed an essential element of both the nomadic food-collecting economy of Paleolithic societies and the food-producing activities in Neolithic villages, while science, as an abstract and systematic interest in nature, was essentially nonexistent, or, at any rate, has left little trace.

The Arrival of Handy Man

By most accounts human beings appeared on Earth only recently, as measured on the scales of cosmic, geologic, or evolutionary time. As scientists now believe, the cosmos itself originated with the "Big Bang" some 13.8 billion years ago. Around 4.5 billion years ago the earth took shape as the third in a string of companion planets to an ordinary star near the edge of an ordinary galaxy; soon the self-replicating chemistry of life began. Biological evolution then unfolded over the next millions and billions of years. In the popular imagination the age of the dinosaurs exemplifies the fantastic history of life in past ages, and the catastrophic event—probably a comet or an asteroid colliding with the earth—that ended the dinosaur age 65 million years ago illustrates the vicissitudes life suffered in its tortuous evolution. The period that followed is known as the age of mammals because these animals flourished and diversified in the niches vacated by the dinosaurian reptiles. By about 4 million years ago a line of "ape-men" arose in Africa—the australopithecines—our now-extinct ancestral stock.

Figure 1.1 depicts the several sorts of human and prehuman species that have arisen over the last 4 million years. Experts debate the precise evolutionary paths that join them, and each new fossil discovery readjusts the details of the story. Yet its broad outlines are not in dispute.

The figure shows that anatomically modern humans, *Homo sapiens sapiens*, or the "wise" variety of "wise Man," evolved from a series of human and prehuman ancestors. Archaic versions of modern humans made their appearance around 200,000 years ago. They migrated out of Africa in several waves, the latest occurring at around 60,000 years ago, and by that time anatomically modern humans—people physically like you and me—wandered the world.

The Neanderthals were an extinct race of humans that existed mainly in the cold of Europe from about 200,000 to 40,000 years ago. Experts

Fig. 1.1. (opposite) Human evolution. Modern humans (*Homo sapiens sapiens*) evolved from earlier, now extinct, human and prehuman ancestors. (Plants and animals are classified according to the binomial nomenclature of genus and species: genus being general groups of related species, and species being specific interbreeding populations of individuals. Thus, *Homo* is the genus, and *sapiens* the species; the third name indicates a subspecies.) In general, brain size and technological sophistication increased over time, but there is no strict correlation between species and technologies. For example, *Paranthropus* and *Homo habilis* may both have used simple choppers; *H. erectus* and archaic *H. sapiens* cannot be distinguished by their respective fine-blade toolkits. Aspects of this picture are matters of debate, notably the relationship of Neanderthals to modern humans. New findings regularly shed light on the details of human biological and cultural evolution.

differ over the modernity of Neanderthals and whether one would or would not stand out in a crowd or in a supermarket. Many scientists look upon them as so similar to ourselves as to form only an extinct variety or race of our own species, and so label them, as we do here *Homo sapiens neanderthalensis*. Others thought Neanderthals more "brutish" than anatomically modern humans and therefore regarded them as a separate species, *Homo neanderthalensis*. Neanderthals and modern humans interacted culturally and interbred. Recent DNA evidence indicates that about 5 percent of genes non-African people carry are inherited from Neanderthal ancestors.

Preceding *Homo sapiens*, the highly successful species known as *Homo erectus* arose around 2 million years ago and spread throughout the Old World (the continents of Africa, Europe, and Asia). Before that, the first species of human being, *Homo habilis*, coexisted with at least two other species of upright hominids, the robust and the gracile forms of the genus *Paranthropus*. At the beginning of the sequence stood the ancestral genus *Australopithecus* (or "Southern Ape") that includes *Australopithecus afarensis*—represented by the fossil "Lucy."

This sequence highlights several points of note. First is the fact of human evolution—that we arose from more primitive forebears. Among the more significant indicators of this evolution is a progression in brain size, from around 450 cubic centimeters (cc) in the case of pre-human Lucy, only slightly larger than the brain of a modern chimpanzee, through an average of 750 cc for *Homo habilis*, 1000 cc for *Homo erectus*, to around 1400 cc for humanity today. An as-yet-unexplained irony of this "progression" is that Neanderthals had slightly larger brains than today's humans.

Bipedality—or upright walking on two feet—represents another defining feature of this evolutionary sequence. Experts debate whether Lucy and her kin were fully bipedal, but her successors certainly were. An upright stance allows the hand and arm to become a multipurpose utensil for grasping and carrying items. Lucy and her type had probably adopted male-female cooperation, at least temporary pair-bonding, and a "family" structure for raising offspring.

From the point of view of the history of technology, however, the most important lesson to be drawn from figure 1.1 concerns tool use among our ancestors. It used to be thought that tool use—technology—is an exclusively human characteristic; the oldest fossil of the human genus, *Homo habilis*, received its name ("handy man") both because of its "human" skeletal features and because it was discovered along with simple stone choppers. However, the older notion can no longer be maintained. Indeed, the origin of technology is rooted in biology. Some nonhuman animals create and use tools, and technology as a cultural process transmitted from generation to generation arises occasionally among monkey and ape communities. Chimpanzees in the wild sometimes "fish" for termites by carefully preparing a twig, inserting it into a termite nest, and licking off the insects that cling to it. Since the activity is not instinctive but is instead taught to juveniles by their mothers, it must be regarded as cultural, unlike, say, the instinct of bees to build hives. Reportedly, chimpanzees have also culturally transmitted knowledge of medicinal plants, so it may be possible to identify the origins of medical technology outside of the human genus, too. Perhaps the best documented feats of technical innovation and cultural transmission in the animal world concern a single female, Imo, the "monkey genius" of a colony of Japanese macaques. Incredibly, Imo made two separate technical discoveries. First she discovered that to remove sand from potatoes thrown on the beach she could wash them in the sea rather than pick off the sand with her fingers. Then, in an even more remarkable display of ingenuity, Imo found that to separate rice from sand she did not have to pick out the individual grains; the mixture can be dropped into water where the sand will sink, and the rice will float and can be easily recovered. Both techniques were adopted by younger members of the troop as well as by older females and passed on to the next generation.

Recent discoveries have shown that stone tools antedate *Homo habilis* by 700,000 years, and a species of *Paranthropus* may have used fire. Furthermore, little correlation exists between species type and different types of tool-kits. For example, Neanderthal tools varied little from the precedents set by *Homo erectus*. The record reveals only a weak correlation between biological species and the toolkit used.

That said, however, making and using tools and the cultural transmission of technology became essential to the human mode of existence and was practiced in all human societies. Moreover, humans seem to be the only creatures who fashion tools to make other tools. Without tools humans are a fairly frail species, and no human society has ever existed without technology. Humankind owes its evolutionary success in large measure to mastery and transmission of toolmaking and -using, and thus human evolutionary history is grounded in the history of technology.

Control of fire represented a key new technology for humankind. Fire provided warmth. Fire made human migration into colder climes possible, opening up huge and otherwise inhospitable areas of the globe for human habitation. The technology of fire also supplied artificial light, thus extending human activity after dark and into dark places, such as caves. Fire offered protection against wild animals. Fire permitted foods to be cooked, which lessened the time and effort required to eat and digest meals. Fire-hardened wooden tools became possible. And fire no doubt served as a hearth and a hub for human social and cultural relations for a million years. Their practical knowledge of fire gave early humans a greater degree of control over nature. *Homo erectus* was an exceptionally successful animal, at least as measured by its spread across the Old World from Africa to Europe, Asia,

Fig. 1.2. "*H. erectus*
Utilizing a Prairie Fire,"
by Jay H. Matternes.
Control of fire became a
fundamental technology
in the human odyssey.
Undoubtedly, members of
the genus *Homo* first used
wildfires before learning
to control them.

Southeast Asia, and archipelagoes beyond. That success in large measure depended on mastering fire.

The grasping hand constitutes one human "tool" that evolved through natural selection; speech is another. Speech seems to be a relatively recent acquisition, although paleontologists have not yet reached agreement on how or when it first appeared. Speech may have evolved from animal songs or calls; novel brain wiring may have been involved. But, once acquired, the ability to convey information and communicate in words and sentences must have been an empowering technology that produced dramatic social and cultural consequences for humanity.

A turning point occurred around 40,000 years ago. Previously, Neanderthals and anatomically modern humans had coexisted for tens of thousands of years in the Middle East and in Europe. Around 35,000 years ago Neanderthals became extinct, possibly exterminated through conflict with a new population, or they may have interbred and become absorbed into the modern human gene pool. A cultural discontinuity manifested itself around the same time. Whereas Neanderthals had produced simple, generalized, multipurpose tools from local materials, we—*Homo sapiens sapiens*—began to produce a great assortment of tools, many of which were specialized, from stone, bone, and antler: needles and sewn clothing, rope and nets, lamps, musical instruments, barbed weapons, bows and arrows, fish hooks, spear throwers, and more elaborate houses and shelters with fireplaces. Humans began to conduct long-distance trade of shells and flints through exchange over hundreds of miles, and they produced art, tracked the moon, and buried their dead. And yet, in terms of their basic social and economic way of life, they continued along the same path—they remained nomadic food-collectors.

Foraging for a Living

Prehistorians classify the period from 2 million years ago to the end of the last Ice Age at about 12,000 years ago as a single era. They label it the Paleolithic (from the Greek, paleo, "ancient"; lithos, "stone") or Old Stone Age. Food-collecting is its essential attribute, codified in the term hunter-gatherer society. Paleolithic tools aided in hunting or scavenging animals and for collecting and processing plant and animal food, and it is now understood that Paleolithic technology developed in the service of a basic food-collecting economy.

Paleolithic food-collecting bespeaks a subsistence economy and a communal society. Seasonal and migratory food-collecting produced little surplus and thus permitted little social ranking or dominance and no coercive institutions (or, indeed, any institutions) of the kind needed in stratified societies to store, tax, and redistribute surplus food. The record indicates that Paleolithic societies were essentially egalitarian, although grades of power and status may have existed within groups. People lived in small bands or groups of families, generally numbering fewer than 100. Much circumstantial evidence suggests that a division of labor based on gender governed the pattern of food collection. Although one has to allow for sexually ambiguous roles and individual exceptions, males generally attended to hunting and scavenging animals, while females most likely went about gleaning plants, seeds, and eggs as food and medicines. Men and women together contributed to the survival of the group, with women's work often providing the majority of calories. *Homo sapiens sapiens* lived longer than Neanderthals, it would seem; more true elders thus added experience and knowledge in those groups. Paleolithic bands may have converged seasonally into larger clans or macrobands for celebrations, acquiring mates, or other collective activities, and they probably ingested hallucinatory plants. Except as located in a handful of favored spots where year-round hunting or fishing might have been possible, Paleolithic food-collectors were nomadic, following the migrations of animals and the seasonal growth of plants. In some instances Paleolithic groups engaged in great seasonal moves to the sea or mountains. In the Upper Paleolithic (around 30,000 years ago) spear-throwers and the bow and arrow entered the weapons arsenal, and the dog (wolf) became domesticated, the first domesticated animal possibly as an aid in hunting.

Ice Age art is the most heralded example of the cultural flowering produced after anatomically modern humans appeared on the scene. Earlier human groups may have made beautified objects of perishable materials, but several late Upper Paleolithic cultures in Europe (30,000 to 10,000 years ago) produced enduring and justly renowned paintings and sculptures in hundreds of sites, often in hard-to-reach galleries and recesses of caves. Artists and artisans also created jewelry and portable adornments, and decorated small objects with animal motifs and other embellishments. No one has yet fully decoded what purposes cave paintings fulfilled; anthropologists have suggested hunting rituals, initiation rites, magical beliefs, and sexual symbolism. The many "Venus" statuettes with exaggerated feminine features, characteristic of the Paleolithic, have been interpreted in terms of fertility rituals and divination of one sort or another. By the same token, they may represent ideals of feminine beauty. Beyond the many uncertainties remaining, Ice Age art shows that human communities evoked symbolic representation in their dealings with the world. And then, we should not overlook the technical dimension of Ice Age art, from pigments and painting techniques to ladders and scaffolding. The great cave paintings of Europe are the better known, but literally and figuratively Paleolithic peoples the world over left their artistic handprints.

Neanderthals had already begun to care for their old and invalid, and by 100,000 years ago they ceremonially buried some of their dead. Centers of mortuary and burial activity may have existed, and one can speak of a "cult of the dead" beginning in the Middle Paleolithic (100,000 to 50,000 years ago). Intentionally burying the dead is a distinctly human activity, and burials represent a major cultural landmark in human prehistory. They bespeak self-consciousness and effective social and group cohesion, and they suggest the beginning of symbolic thought.

It may be enlightening to speculate about the mental or spiritual world of Paleolithic peoples. What we have already seen and said of Paleolithic burials and cave art strongly suggests that Paleolithic populations, at least toward the end of the era, developed what we would call religious or spiritual attitudes. They may well have believed the natural world was filled with various gods or deities or that objects and places, such as stones or groves, were themselves alive. Religious beliefs and practices—however we might conceive them—formed a social technology, as it were, that knitted communities together and strengthened their effectiveness.

For anatomically modern humans the Paleolithic way of life continued unabated and essentially unchanged for 30,000 years, a phenomenally long and stable cultural era, especially compared to the

Fig. 1.3. Paleolithic art. In the late Paleolithic era food-collecting populations of *Homo sapiens* began to create art in many parts of the world. In southwestern Europe they adorned the walls of caves with naturalistic representations of animals.

rapid pace of change in the periods that followed. Paleolithic peoples doubtless lived relatively unchanging lives involving great continuity with their own past. Well fed on a varied diet that included significant amounts of meat, not having to work too hard, cozy in fur and hide, comfortable by a warm fire, who can deny that our Paleolithic ancestors often enjoyed the good life?

Over the entire 2 million years of the Paleolithic, beginning with the first species of *Homo*, population density remained astonishingly low, perhaps no more than one person per square mile, and the rate of population increase, even in the late (or Upper) Paleolithic, may have been only one five-hundredth of what it has been for modern populations over the past few centuries. The very low rate of population increase derives from several factors acting singly or in combination to restrict fertility rates: late weaning of infants (since nursing has somewhat of a contraceptive effect), low body fat, a mobile lifestyle, and infanticide. Nevertheless, humankind slowly but surely fanned out over the earth and, as long as suitable food-collecting habitats could be found, humanity had no need to alter its basic lifestyle. Food-collecting groups simply budded off from parent populations and founded new communities. Paleolithic peoples spread through Africa, Asia, Europe, and Australia, while waves of hunters and gatherers reached North America by at least 12,000 years ago, if not well before, ultimately spreading the Paleolithic mode of existence to the southernmost tip of South America. After many millennia of slow expansion, Paleolithic humans "filled up" the world with food-collectors. Only then, it seems, did population pressure against collectible resources trigger a revolutionary change from food-collecting to food-producing in the form of horticulture or herding.

Is Knowledge Science?

The extraordinary endurance of Paleolithic society and mode of existence depended on human mastery of an interlocked set of technologies and practices. It is sometimes said that Paleolithic peoples needed and possessed "science" as a source of the knowledge that underpinned their practical activities. It is all too easy to assume that in making and using fire, for example, Stone Age peoples practiced at least a rude form of "chemistry." In fact, however, while both science and technology involve "knowledge systems," the knowledge possessed by food-collectors cannot reasonably be considered theoretical or derivative of science or theories of nature. Although evidence of something akin to science appears in late Paleolithic "astronomy," there was no call for systematic experimentation or directed inquiry into nature beyond accumulated experience in the practice of Paleolithic crafts. To discover the origins and character of that science we need to understand why it did not impact technology.

Knowledge comes in many forms. Practical knowledge embodied in the crafts is different from knowledge deriving from some abstract understanding of a phenomenon. To change a car tire, for example, one needs direct instruction or hands-on experience, not any special knowledge of mechanics or the strength of materials. By rubbing sticks together or sparking flint into dry kindling, a scout can build a fire without knowing the oxygen theory (or any other theory) of combustion. And conversely, knowledge of theory alone does not enable one to change a tire or make a fire. It seems fair to say that Paleolithic peoples applied practical skills and practical knowledge rather than any theoretical or systematized knowledge to practice their crafts. More than that, Paleolithic peoples may have had explanations for fire without it being meaningful to speak about Paleolithic "chemistry"—for example, if they somehow thought they were invoking a fire god or a spirit of fire in their actions. A major conclusion about Paleolithic technology follows from all this: to whatever small extent we may be able to speak about "science" in the Paleolithic, Paleolithic technologies clearly were prior to and independent of any such knowledge.

The record (or rather the absence of one) indicates that Paleolithic peoples did not self-consciously pursue "science" or deliberate inquiries into nature. Does the Paleolithic period nevertheless offer anything of note for the history of science? On the most rudimentary level we must recognize the extensive "knowledge of nature" possessed by Paleolithic peoples and gained directly from experience. They had to be keen observers since their very existence depended on what they knew of the plant and animal worlds around them. And, like surviving food-collectors observed by anthropologists, they probably developed taxonomies and natural histories to categorize and rationalize what they found in nature. The history of science plainly finds its origins here.

Fig. 1.4. Paleolithic lunar observations. (a) An engraved mammoth tusk from Gontzi, Ukraine, that has been interpreted as a record of lunar cycles. Thousands of these artifacts have been found stretching back 30,000 years. This one dates from approximately 15,000 years ago. (b) A diagrammatic rendition of the artifact showing cycles of four lunar months aligned with the engraved markings.

Even more noteworthy, the archaeological record for the late Paleolithic era, beginning around 40,000 years ago, offers striking evidence of activities that look a lot like science. That evidence appears in the form of thousands of engraved fragments of reindeer and mammoth bones that seem to have recorded observations of the moon. An "unbroken line" of such artifacts stretches over tens of thousands of years. The engraved mammoth tusk from Gontzi in Ukraine is an example of such lunar records, which may have been kept at all major habitation sites. Pictured in figure 1.4, it dates from around 15,000 years ago.

We can only speculate, of course, but as Paleolithic peoples lived close to nature, the waxing and waning moon would naturally present itself as a significant object of interest with its obvious rhythms and periods. One can easily imagine our intelligent forebears following those rhythms and beginning to record in one fashion or another the sequence and intervals of full and new moon. Moreover, the Gontzi bone and others like it could have served as a means of reckoning time. Although we cannot go so far as to say that Paleolithic peoples possessed a calendar, we can surmise that knowledge of the moon's periods would be useful in time-reckoning. For example, dispersed groups might have come together seasonally and would have needed to keep track of the intervening months. We need not envision a continuous tradition of such lunar records, for the process may have been invented and reinvented hundreds of times over: a simple counter fashioned over the course of a few months and discarded. The artifacts in question evidence the active observation and recording of natural phenomena over time. That activity indicates only a rudimentary approach to theoretical knowledge, but its results seem more abstract than knowledge gained from direct experience and different from what Paleolithic peoples otherwise embodied in their crafts.

Leaving the Garden

This picture of life in the Paleolithic period, which has emerged from the research of archaeologists, paleoanthropologists, and prehistorians, raises several puzzling questions about the dynamics of social change. How can we explain the steadfast durability of a food-collecting social system for 2 million years including more than 200,000 years populated by our own species? How can the relative lack of technological innovation be accounted for? Why, after anatomically modern humans flourished culturally in the Paleolithic 40,000 to 30,000 years ago, did they continue to live as food-collectors, making stone tools and following a nomadic way of life? And why did the pace of change accelerate 12,000 years ago, as food-collecting finally gave way to food-producing, first in the form of gardening (horticulture) and animal husbandry in the Neolithic era and later, after another technological revolution in the form of intensified farming (agriculture) under the control and management of the political state?

Different accounts have been offered to explain the social and economic transformations that occurred at the end of the Paleolithic. Changes may have been set in motion by the retreat of the glaciers at the end of the last Ice Age about 10,000–12,000 years ago. The extinction of many large-bodied animals occurred then, restricting the food supply, and other animal-migration patterns shifted northward, probably leaving some human groups behind. Humans themselves probably overhunted large game, self-destructively changing their living conditions. Another line of argument that has recently gained credibility postulates that the food-collecting mode of life persisted as long as the population of hunters and gatherers remained small enough to exploit the resources of their habitats with reasonable ease. Since population increased slowly and since suitable habitats were numerous on a global scale, 2 million years passed before hunter-gatherers reached the "carrying capacities" of accessible environments through the increase of their own numbers and a resulting broadening of foraging activity. This account also explains the low rate of technological innovation prior to the late Paleolithic era: small populations blessed with ample resources were served well by their techniques and refined skills. Although Paleolithic peoples would have known that seeds grow and that gardening is possible (and it is likely they occasionally practiced it), they had no compelling incentive to revolutionize their way of life. Only when increasing population density could no longer be readily relieved by migration was the balance between needs and re-sources finally upset and plant and animal husbandry taken up as a new way of life.

Our ancestors did not give up their Paleolithic existence willingly. By abandoning, under pressure of ecological degradation, a nomadic lifestyle of food-collecting and adopting a mode of food-producing—by "progressing" from hunting and gathering to gardening and stock-raising—only then did humankind reluctantly fall out of a Garden of Eden and into the Neolithic era.

The Reign of the Farmer

At the end of the last Ice Age, around 12,000 years ago, the Neolithic Revolution began to unfold. This revolution, first and foremost a socioeconomic and technological transformation, involved a shift from food-gathering to food-producing. It originated in a few regions before eventually spreading around the globe. In habitats suitable only as pasture it led to pastoral nomadism or herding animal flocks; in others it led to farming and settled village life. Thus arose the Neolithic or New Stone Age.

Growing Your Own

A surprising but grand fact of prehistory: Neolithic communities based on domesticated plants and animals arose independently several times in different parts of the world after 10,000 BCE—the Near East, India, Africa, North Asia, Southeast Asia, and Central and South America. The physical separation of the world's hemispheres—the Old World and the New World—decisively argues against simple diffusion of Neolithic techniques, as do the separate domestications of wheat, rice, corn, and potatoes in different regions. On the time scale of prehistory the transformation appears to have been relatively abrupt, but in fact the process occurred gradually. Nonetheless, the Neolithic revolution radically altered the lives of the peoples affected and, indirectly, the conditions of their habitats. Although different interpretations exist concerning the origin of the Neolithic, no one disputes its world-transforming effects.

The Neolithic was the outcome of a cascading series of events and processes. In the case of gardening—low-intensity farming—we now know that in various locales around the world human groups settled down in permanent villages, yet continued to practice hunting, gathering, and a Paleolithic economy before the full transition to a Neolithic mode of production. These settled groups lived by complex foraging in limited territories, intensified plant collection, and exploitation of a broad spectrum of secondary or tertiary food sources, such as nuts and seafood. They also lived in houses, and in this sense early sedentary humans were themselves a domesticated species. (The English word "domestic" derives from the Latin word *domus*, meaning "house." Humans thus domesticated themselves as they domesticated plants or animals!) But the inexorable pressure of population against dwindling collectible resources, along with the greater nutritional value of wild and domesticated cereal grains, ultimately led to increasing dependence on food-production and a more complete food-producing way of life.

In most places in the world people continued a Paleolithic existence after the appearance of Neolithic settlements 12,000 years ago. They were blissfully unpressured to take up a new Neolithic mode of food-producing, and as a cultural and economic mode of existence even today a few surviving groups follow a Paleolithic lifestyle. As a period in prehistory, the Neolithic has an arc of its own that covers developments from the first simple horticulturists and pastoralists to complex late Neolithic groups living in "towns." In retrospect, especially compared to the extreme length of the Paleolithic period, the Neolithic of prehistory lasted just a moment before civilization in Mesopotamia and Egypt began to usher in further transformations around 5,000 years ago. But even in its diminished time frame the Neolithic spread geographically and persisted in particular locales over thousands of years from roughly 12,000 to 5,000 years ago, when the Neolithic first gave way to civilization in the Near East. To those experiencing it, Neolithic life must have proceeded over generations at a leisurely seasonal pace.

Two alternative paths toward food production led out of the Paleolithic: one from gathering to cereal horticulture (gardening), and then to plow agriculture; the other from hunting to herding and pastoral nomadism. A distinct geography governed these Neolithic alternatives: In climates with sufficient atmospheric or surface water, horticulture and settled villages arose; in grasslands too arid for farming, nomadic people and herds of animals retained a nomadic way of life. Of these very different paths, one led historically to nomadic societies such as the Mongols and the Bedouins. The other, especially in the form that combined farming and domestication of animals, led to the great agrarian civilizations and eventually to industrialization.

Opportunistic and even systematic hunting and gathering persisted alongside food-producing, but where Neolithic settlements arose the basic economy shifted to raising crops on small cleared plots. Gardening contrasts with intensified agriculture using irrigation, plows, and draft animals, which later developed in the first civilizations in the Near East. Early Neolithic peoples did not use the plow but, where necessary, cleared land using large stone axes and adzes; they cultivated their plots using hoes or digging sticks. In many areas of the

Fig. 2.1. Neolithic tools. Neolithic horticulture required larger tools for clearing and cultivating plots and for harvesting and processing grains.

world, especially tropical and subtropical ones, swidden, or "slash and burn," agriculture developed where plots were cultivated for a few years and then abandoned to replenish themselves before being cultivated again. The Neolithic toolkit continued to contain small chipped stones, used in sickles, for example, but was augmented by larger, often polished implements such as axes, grinding stones, and mortars and pestles found at all Neolithic sites. Animal antlers also proved useful as picks and digging sticks. And grain had to be collected, threshed, winnowed, stored, and ground, all of which required an elaborate set of technologies and social practices.

Human populations around the world independently domesticated and began cultivating a variety of plants: several wheats, barleys, rye, peas, lentils, and flax in Southwest Asia; millet and sorghum in Africa; millet and soybeans in North China; rice and beans in Southeast Asia; maize (corn) in Mesoamerica; potatoes, quinoa, manioc, and beans in South America. Domestication constitutes a process (not an act) that involves taming, breeding, genetic selection, and occasionally introducing plants into new ecological settings. In the case of wheat, for example, wild wheat is brittle, with seeds easily scattered by the wind and animals, a trait that enables the plant to survive under natural conditions. Domesticated wheat retains its seeds, which simplifies harvesting but leaves the plant dependent on the farmer for its propagation. Humans changed the plant's genes; the plant changed humanity. And, with humans raising the grain, the rat, the mouse, and the house sparrow "self-domesticated" and joined the Neolithic ark.

The domestication of animals developed out of intimate and longstanding human contact with wild species. Logically, at least, there is a clear succession from hunting and following herds to corralling, herding, taming, and breeding. The living example of the Sami (Lapp) people who follow and exploit semiwild reindeer herds illustrates how the shift from hunting to husbandry and pastoral nomadism may have occurred. As with plant culture, the domestication of animals involved human selection from wild types, selective slaughtering, selective breeding, and what Darwin later called "unconscious selection" from among flocks and herds. Humans in the Old World domesticated cattle, goats, sheep, pigs, chickens, and, later, horses. In the New World Andean communities domesticated only llamas and the guinea pig; peoples in the Americas thus experienced a comparative deficiency of animal protein in the diet.

Animals are valuable to humans in diverse ways. Some of them convert inedible plants to meat, and meat contains more complex proteins than plants. Animals provide food on the hoof, food that keeps from spoiling until needed. Animals produce valuable secondary products that were increasingly exploited as the Neolithic unfolded in the Old World. Cattle, sheep, pigs, and the rest are "animal factories" that produce more cattle, sheep, and pigs. Chickens lay eggs, and cows, sheep, goats, and horses produce milk. Treated and storable milk products in yogurts, cheeses, and brewed beverages sustained the great herding societies of Asia and pastoralists everywhere. Manure became another valuable animal product as fertilizer and fuel. Animal hides provided raw material for leather and a variety of products, and sheep, of course, produced fleece. (Wool was first woven into fabric on Neolithic looms.) Animals provided traction and transportation. The Neolithic maintained the close dependence on plants and animals that humankind had developed over the previous 2 million years. But the technologies of exploiting them and the social system sustained by those technologies had changed radically.

After a few thousand years of the Neolithic in the Near East, mixed economies that combined the technologies of horticulture and animal husbandry made their appearance. Late Neolithic groups in the Old World apparently kept animals for traction and used wheeled carts on roads and pathways that have been favorably compared to those of medieval Europe. The historical route to intensified agriculture and to civilization was through this mixed Neolithic farming. If biology and evolution were partly responsible for the character of our first mode of existence in the Paleolithic, then the Neolithic Revolution represents a change of historical direction initiated by humans themselves in response to their changing environment.

Complementing the many techniques and skills involved in farming and husbandry, several ancillary technologies arose as part of the shift to the Neolithic. First among these novelties was textiles, an innovation independently arrived at in various parts of the Old and New Worlds. Recent findings show that some Paleolithic groups occasionally practiced techniques of weaving, perhaps in basketry, but only in the Neolithic did the need for cloth and storage vessels expand to the point where textile technologies flourished. The production of textiles involves several interconnected steps and technological practices:

shearing sheep or growing and harvesting flax or cotton, processing the raw material, spinning thread (an ever-present part of women's lives until the Industrial Revolution 10,000 years later), constructing looms, dyeing, and weaving the cloth. In considering the advent of textile production in the Neolithic, one cannot overlook design considerations and the symbolic and informational role of dress in all societies. Then, as now, how people dress conveys a lot of information about who they are and where they come from.

Pottery, which also originated independently in multiple centers around the world, is another new technology that formed a key part of the Neolithic Revolution. If only inadvertently, Paleolithic peoples had produced fired-clay ceramics, but nothing in the Paleolithic economy called for a further development of the technique. Pottery almost certainly arose in response to the need for a storage technology: jars or vessels to store and carry the surplus products of the first agrarian societies. Neolithic communities used plasters and mortars in building construction, and pottery may have arisen out of plastering techniques applied to baskets. Eventually, "manufacturing centers" and small-scale transport of ceramics developed. Pottery is a "pyrotechnology," for the secret of pottery is that chemically combined water is driven from the clay when it is fired, turning it into an artificial stone. Neolithic kilns produced temperatures upwards of 900°C. Later, in the Bronze and Iron Ages, the Neolithic pyrotechnology of pottery made metallurgy possible.

In Neolithic settings, hundreds if not thousands of techniques and technologies large and small melded to produce the new mode of life. Neolithic peoples built permanent structures in wood, mud brick, and stone, all of which testify to expert craft skills. They twisted rope and practiced lapidary crafts, and Neolithic peoples even developed metallurgy of a sort, using naturally occurring raw copper. The technology of cold metalworking produced useful tools. The now-famous "Ice man," the extraordinary frozen mummy exposed in 1991 by a retreating glacier in the Alps, was first thought to belong to a Bronze Age culture because of the fine copper axe he was carrying when he perished. As it turns out, he lived in Europe around 3300 BCE, evidently a prosperous Neolithic farmer with a superior cold-forged metal tool.

The Neolithic was also a social revolution and produced a radical change in lifeways. Decentralized and self-sufficient settled villages, consisting of a dozen to two dozen houses, with several hundred inhabitants became the norm among Neolithic groups. Compared to the smaller bands of the Paleolithic, village life supported collections of families united into tribes. The Neolithic house doubtless became the center of social organization; production took place on a household basis. The imaginative suggestion has been made that living inside houses forced Neolithic peoples to deal in new ways with issues concerning public space, privacy, and hospitality. Neolithic peoples may have used hallucinatory drugs, and they began to experiment with fermented beverages. Although a sexual division of labor probably persisted in the Neolithic, horticultural societies, by deemphasizing hunting, may have embodied greater gender equality. A comparatively sedentary lifestyle, a diet higher in carbohydrates, and earlier weaning increased fertility, while freedom from the burden of carrying infants from camp to camp enabled women to bear and care for more children. And one suspects that the economic value of children—in tending animals or helping in the garden, for example—was greater in Neolithic times than in the Paleolithic. At least with regard to Europe, archaeologists have made compelling claims for the existence of cults devoted to Neolithic goddesses and goddess worship. There were doubtless shamans, or medicine "men," some of whom may also have been women. Neolithic societies remained patriarchal, but males were not as dominant as they would become with the advent of civilization.

In the early Neolithic, little or no occupational specialization differentiated individuals who earned their bread solely through craft expertise. This circumstance changed by the later Neolithic, as greater food surpluses and increased exchange led to more complex and wealthier settlements with full-time potters, weavers, masons, toolmakers, priests, and chiefs. Social stratification kept pace with the growth of surplus production. By the late Neolithic, low-level hierarchal societies, tribal chiefdoms, or what anthropologists call "big men" societies appeared. These societies were based on kinship, ranking, and the power to accumulate and redistribute goods sometimes in great redistributive feasts. Leaders now controlled the resources of 5,000 to 20,000 people. They were not yet kings, however, because they retained relatively little for themselves and because Neolithic societies were incapable of producing truly huge surpluses.

Compared to the Paleolithic economy and lifestyle, one could argue that the standard of living actually became depressed in the transition to the Neolithic in that low-intensity horticulture required more labor, produced a less varied and nutritious diet, and allowed less leisure than Paleolithic hunting and gathering in its heyday. But—and this was the primary advantage—Neolithic economies produced more food and could therefore support more people and larger population densities (estimated at a hundredfold more per square mile) than Paleolithic foraging.

Populations expanded and the Neolithic economy spread rapidly to fill suitable niches. By 3000 BCE thousands of agrarian villages dotted the Near East, usually within a day's walk of one another. Wealthier and more complex social structures developed, regional crossroads and trading centers arose, and by the late Neolithic real towns had emerged. The Neolithic town Çatal Hüyük in modern Turkey dates from 6000 BCE, but the classic example is the earlier and especially rich Neolithic town of Jericho. Neolithic settlements appeared along the Jordon River in the Middle East by 9000 BCE, and by 7350 BCE Jericho itself had become a well-watered, brick-walled city of 2,000 or more people tending flocks and plots in the surrounding hinterland. Jericho had a tower nine meters high and ten meters in diameter, and its celebrated walls were three meters thick, four meters high, and 700 meters in circumference. The walls were necessary because the surplus stored behind them attracted raiders. The later walled enclosure at Great Zimbabwe (c. 1300 CE) in Africa evidences similar forces at play. War-like clashes between Paleolithic peoples had undoubtedly occurred repeatedly over the millennia in disputes over territory, to capture females, or for cannibalistic or ritual purposes. But with the Neolithic, for the first time, humans produced surplus food and wealth worth stealing and hence worth protecting. Paleolithic groups were forced to adapt to the Neolithic economies burgeoning around them. Thieving was one alternative; joining in a settled way of life was another. In the long run, Neolithic peoples marginalized hunter-gatherers and drove them virtually to extinction. Idealized memories of the foraging lifestyle left their mark in "Garden of Eden" or "happy hunting grounds" legends in many societies.

Blessed or cursed with a new economic mode of living, humans gained greater control over nature and began to make more of an impact on their environments. The ecological consequences of the Neolithic dictated that the domestic replace the wild, and where it occurred the Neolithic Revolution proved irreversible—a return to the Paleolithic was impossible because Paleolithic habitats had been transformed and the Paleolithic lifestyle was no longer sustainable.

Moonshine

The Neolithic Revolution was a techno-economic process that occurred without the aid or input of any independent "science." In assessing the connection between technology and science in the Neolithic, pottery provides an example exactly analogous to making fire in the Paleolithic. Potters made pots simply because pots were needed and because they acquired the necessary craft knowledge and skills. Neolithic potters possessed practical knowledge of the behavior of clay and of fire, and, although they may have had explanations for the phenomena of their crafts, they toiled without any systematic science of materials or the self-conscious application of theory to practice or any higher learning to tap for practical purposes. It would denigrate Neolithic crafts to suppose that they could have developed only with the aid of higher learning.

Can anything, then, be said of science in the Neolithic? In one area, with regard to what can be called Neolithic astronomy, we stand on strong ground in speaking about knowledge in a field of science. Indeed, considerable evidence makes plain that many, and probably most,

Fig. 2.2. Jericho. Neolithic farming produced a surplus that needed to be stored and defended. Even in its early phases, the Neolithic settlement of Jericho surrounded itself with massive walls and towers, as shown in this archaeological dig.

Neolithic peoples systematically observed the heavens, particularly the patterns of motion of the sun and moon and that they regularly created astronomically aligned monuments that served as seasonal calendars. In the case of Neolithic astronomy, we are dealing not with the prehistory of science, but with science in prehistory.

The famous monument of Stonehenge on the Salisbury Plain in southwest England provides the most dramatic and best-understood case in point. Stonehenge, it has now been determined by radiocarbon dating, was built intermittently in three major phases by different groups over a 1,600-year period from 3100 BCE to 1500 BCE, by which time the Bronze Age finally washed across the Salisbury Plain. The word "Stonehenge" means "hanging stone," and transporting, working, and erecting the huge stones represents a formidable technological achievement on the part of the Neolithic peoples of prehistoric Britain.

A huge amount of labor went into building Stonehenge—estimates range to 30 million man-hours, equivalent to an annual productive labor of 10,000 people. In order to create a circular ditch and an embankment 350 feet in diameter 3,500 cubic yards of earth were excavated. Outside the sanctuary the first builders of Stonehenge erected the so-called Heel Stone, estimated to weigh 35 tons. Eighty-two "bluestones" weighing approximately 5 tons apiece were brought to the site (mostly over water) from Wales, an incredible 240 kilometers (150 miles) away. Each of the 30 uprights of the outer stone circle of Stonehenge weighed in the neighborhood of 25 tons, and the 30 lintels running around the top of the ring weighed 7 tons apiece. More impressive still, inside the stone circle stood the five great trilithons or three-stone behemoths. The average trilithon upright weighs 30 tons and the largest probably weighs over 50 tons. (By contrast, the stones that went into building the pyramids in Egypt weighed on the order of 5 tons.) The great monoliths were transported 40 kilometers (25 miles) overland from Marlborough Downs, although the suggestion has been made that ancient glaciers may have been responsible for moving them at least part way to Stonehenge. The architects of Stonehenge appear to have laid out the monument on a true circle, and in so doing they may have used some practical geometry and a standard measure, the so-called megalithic yard.

The labor was probably seasonal, taking place over generations. A stored food surplus was required to feed workers, and some relatively centralized authority was needed to collect and distribute food and to supervise construction. Neolithic farming and ranching communities appeared on the Salisbury Plain by the fourth millennium BCE and evidently reached the required level of productivity. Although Neolithic farming never attained the levels of intensification later achieved by civilized societies, Stonehenge and the other megalithic ("large stone") structures show that even comparatively low-intensity agriculture can produce sufficient surpluses to account for monumental building.

Recognition that Stonehenge is an astronomical device has been confirmed only in our day. As literate peoples encountered Stonehenge

Fig. 2.3. Stonehenge. Neolithic and early Bronze Age tribes in Britain built and rebuilt the famous monument at Stonehenge as a regional ceremonial center and as an "observatory" to track the seasons of the year.

Map 2.1. The Salisbury plain. Stonehenge was set among a cluster of Neolithic sites, indicating the relative wealth and resources of the region. Some of the smaller stones that went into making Stonehenge were transported 150 miles (240 km) by rollers and raft from Western Wales; some of the largest stones came from 25 miles (40 km) north of the site.

over the centuries, any number of wild interpretations emerged as to who built it and why. Geoffrey of Monmouth in his twelfth-century *History of the Kings of Britain* has Merlin from King Arthur's court magically transporting the stones from Wales. Other authors have postulated that the Romans or the Danes built Stonehenge. A still-current fantasy holds that the Druids built and used Stonehenge as a ceremonial center. (In fact, the Celtic Iron Age Druids and their culture only appeared a thousand years after Stonehenge was completed.) Even in the 1950s, when the possibility became clear that Neolithic peoples from the Salisbury Plain themselves were responsible for Stonehenge, there was considerable resistance to the idea that "howling barbarians" might have been capable of building such an impressive monument, and some supposed that itinerant contractors from the Near

Fig. 2.4. Midsummer sunrise at Stonehenge. On the morning of the summer solstice (June 21) the sun rises along the main axis of Stonehenge and sits atop the Heel Stone.

East built it. All scholars now agree that Stonehenge was a major ceremonial center and cult site built by the people of the Salisbury Plain. Its astronomical uses indicate that it functioned as a ceremonial center aligned around the motions of the sun and the moon and it provided the basis for a regional calendar.

The English antiquarian William Stukeley (1687–1765) was the first modern to write about the solar alignment of Stonehenge in 1740. The sun rises every day at a different point on the horizon; that point moves back and forth along the horizon over the course of a year, and each year at midsummer the sun, viewed from the center of the sanctuary at Stonehenge, rises at its most northern point, which is precisely where the builders placed the Heel Stone. The monument's primary astronomical orientation toward the midsummer sunrise is confirmed annually and has not been disputed since Stuckeley.

In the 1960s, however, controversy erupted over claims for Stonehenge as a sophisticated Neolithic astronomical "observatory" and "computer." The matter remains disputed today, but wide agreement exists on at least some larger astronomical significance for Stonehenge, especially with regard to tracking cyclical movements of the sun and the moon. The monument seems to have been built to mark the extreme and mean points of seasonal movement of both heavenly bodies along the horizon as they rise and set. Thus, the monument at Stonehenge marks not only the sun's rise at the summer solstice, but the rise of the sun at winter solstice and at the fall and spring equinoxes. It also indicates the sun's settings at these times, and it tracks the more complicated movements of the moon back and forth along the horizon, marking four different extremes for lunar motion.

The construction of Stonehenge required sustained observations of the sun and the moon over a period of decades and mastery of horizon astronomy. The monument embodied such observations, even in its earliest phases. The ruins testify to detailed knowledge of heavenly movements and to a widespread practice of "ritual astronomy." We have no access to what megalithic Europeans thought they were doing; their "theories" of the sun and the moon, if any, may have been utterly fantastic, and we would probably label their explanations more religious than naturalistic or scientific. Still, megalithic monuments embody a scientific approach in that they reflect understanding of regularities of celestial motions and they bespeak long-term systematic observations of nature. Paleolithic peoples knew of the periodic motion of the sun and the moon, of course, but to create a Neolithic monument like Stonehenge that records these longer-term motions along the horizon required careful observation and (presumably oral) recordkeeping over many years, probably over generations. In this way the knowledge accumulated and embodied in Stonehenge required a degree of organization and systematization not seen in the historical record to that point. Although religious elders, hereditary experts, or priestly keepers of knowledge doubtless built and tended Stonehenge, it probably goes too far to suggest that megalithic monuments provide evidence for a class of professional astronomers or for astronomical research of the sort that later appeared in the first civilizations. Stonehenge may better be thought of as a celestial orrery or clock that kept track of the major motions of the major celestial bodies and possibly some stars. In addition, Stonehenge certainly functioned as a seasonal calendar, accurate and reliable down to a day. As a calendar, Stonehenge kept track of the solar year and, even more, harmonized the annual motion of the sun with the more complicated periodic motion of the moon. It may even have been used to predict eclipses. In these telling ways—systematically observing the heavens, mastering the clock-like movement of the sun and the moon, gaining intellectual control over the calendar—it is possible and even necessary to speak of Neolithic "astronomy" at Stonehenge. The further development of astronomy awaited the advent of writing and cohorts of full-time experts with the patronage of centralized bureaucratic governments. But long before those developments, Neolithic farmers systematically investigated the panorama of the heavens.

On the other side of the globe the remarkable giant statues of Easter Island (also known as Rapa Nui) provide mute testimony to the same forces at play. Easter Island is small and very isolated: a 46-square-mile speck of land 1,400 miles west of South America and 900 miles from the nearest inhabited Pacific island. Polynesian peoples reached Easter Island by sea sometime after 300 CE and prospered through cultivating sweet potatoes, harvesting in a subtropical palm forest, and fishing in an abundant sea. The economy was that of settled Paleolithic or simple Neolithic societies, but local resources were rich, and even at slow growth rates over a millennium the founding population inevitably expanded, reaching 7,000 to 9,000 at the peak of the culture around 1200 to 1500 CE. (Some experts put the figure at over 20,000.)

Islanders carved and erected more than 250 of their monumental moai statues on giant ceremonial platforms facing the sea. Notably, the platforms possessed built-in astronomical orientations. Reminiscent of the works of the peoples of Stonehenge or the Olmecs of Central America, the average moai stood over 12 feet in height, weighed nearly 14 tons, and was transported up to six miles overland by gangs of 55 to 70 men; a few mammoth idols rose nearly 30 feet tall and weighed up to 90 tons. Hundreds more statues—some significantly larger still—remain unfinished in the quarry, where all activity seems to have stopped abruptly. Remote Easter Island became completely deforested because of the demand for firewood and construction material for sea-going canoes, without which islanders could not fish for their staple of porpoise and tuna. By 1500, with the elimination of the palm tree and the extinction of native bird populations, demographic pressures became devastatingly acute, and islanders intensified chicken-raising and resorted to cannibalism and eating rats. The population quickly crashed to perhaps one-tenth its former size, the sad remnant "discovered" by Europeans in 1722. Only 100 souls lived there in 1887. The wealth of the pristine island had provided rich resources where a human society evolved in a typically Neolithic (or settled Paleolithic) pattern. But human appetites and the island's narrow ecological limits doomed the

Map 2.2. Easter Island. This isolated speck of land in the South Pacific lies 1,400 miles off the coast of South America and 900 miles from the nearest inhabited island to the west. Polynesian seafarers, probably navigating by star charts and taking advantage of their knowledge of wind and current changes, arrived at Easter Island around 300 CE. Europeans "discovered" the island in 1722.

Fig. 2.5. Neolithic society on Easter Island. A society based on low-intensity agriculture flourished here for hundreds of years before it was extinguished by ecological ruin. During its heyday it produced megalithic sculptures called *moai* comparable in scale to Stonehenge and other monumental public works that are typical of Neolithic societies.

continuation of the stone-working, heaven-gazing, and wood-burning culture that evolved there.

In general, through observation of the sun and the moon Neolithic peoples around the world established markers, usually horizon markers, that monitored the periodic motion of these bodies across the horizon and sky, tracked the year and the seasons, and provided information of great value to communities of early farmers. In some cases the devices they created to reckon the year and predict the seasons became quite elaborate and costly and were possible only because of the surplus wealth produced in favored places.

Before Stonehenge and long before the settlement and ruination of Easter Island, in certain constricted environments growing populations pressed against even enlarged Neolithic resources, setting the stage in Egypt, Mesopotamia, and elsewhere for a great technological transformation of the human way of life—the advent of urban civilization.

Pharaohs and Engineers

Neolithic societies never reached the complexity of kingdoms. They never built large cities, or large enclosed structures like palaces or temples; they had no need for writing to keep records; and they never established a tradition of higher learning or institutionalized science. These features arose only when Neolithic societies coalesced into civilizations—a second great transformation in human social evolution.

This revolution is often referred to as the Urban Revolution or Urban Bronze-Age Revolution. Whatever its name, changes that began around 6,000 years ago in the Near East ushered in the first civilizations, replete with all the social and historical consequences accompanying cities, high population densities, centralized political and economic authority, the origin and organization of regional states, the development of complex and stratified societies, monumental architecture, and the beginnings of writing and higher learning. The transition was another techno-economic revolution, this time arising out of the need for intensified agricultural production to sustain increasingly large populations that pressed against the carrying capacities of their habitats. As an episode in human history and the history of technology, the Urban Revolution proved to be unrivaled in its consequences until the Industrial Revolution that took root in eighteenth-century Europe.

A new mode of intensified agriculture, distinct from Neolithic horticulture or pasturage, provided the underpinnings of the first civilizations. In that mode simple gardening was superseded by field agriculture based on large-scale water-management networks constructed and maintained as public works by conscripted labor gangs (the corvée) under the supervision of state-employed engineers. In the Old World the ox-drawn scratch plow replaced the hoe and digging stick. And subsistence-level farming gave way to the production of large surpluses of cereals (estimated at a minimum of 50 percent above Neolithic levels) that could be taxed, stored, and redistributed. Centralized political authorities dominated by a pharaoh or king came into being to manage these complex systems of agricultural production. Along with hydraulically intensified agriculture (generally artificial irrigation) and a centralized state authority, the Urban Revolution sustained much larger populations, urban centers, coercive institutions in the form of armies, tax collectors, and police, expanded trade, palaces, and temples, a priestly class, religious institutions, and higher learning. In such bureaucratically organized societies, cadres of learned scribes developed mathematics, medicine, and astronomy.

Taming the Rivers

The Urban Revolution unfolded independently in multiple centers across the Old and New Worlds. The same remarkable pattern of Neolithic settlements coalescing into centralized kingdoms based on intensified agriculture occurs at least six times in six different sites around the globe: in Mesopotamia after 3500 BCE, in Egypt after 3400 BCE, in the Indus River Valley after 2500 BCE, in China after 1800 BCE, in Mesoamerica at about 500 BCE, and in South America after 300 BCE The origin and development of these civilizations were essentially independent and not the result of diffusion from a single center, and hence they are known as the pristine civilizations.

Why did civilization arise independently and repeatedly on a worldwide scale after the fourth millennium BCE in those particular sites? Several explanations have been proposed. The precise processes involved in the leap to civilization are research questions actively debated by archaeologists and anthropologists, but many scholars emphasize the importance of hydrology and ecology, and they recognize that intensified agriculture, abetted by large-scale hydraulic engineering projects, was a key element in the formation of large, highly centralized bureaucratic states. The fact alone that pristine civilizations arose in hydrologically distressed regions—that is, where too little or too much water required hydraulic engineering for the successful practice of intensified agriculture—gives credence to what is called the hydraulic hypothesis, linking the rise of civilization with the technology of large-scale hydraulic systems. Under a hot, tropical or semitropical sun, irrigation agriculture is extraordinarily productive and yields that can sustain large populations become possible. Silt-laden rivers provide water for irrigation and, especially when controlled artificially, they enrich the soils around them. Irrigation agriculture and flood control required hydraulic engineering works and some level of communal action to build and maintain them and to distribute water when and where needed: marshes had to be drained; dams, dikes, canals, sluices, conduits, terraces, catchments, and embankments had to be built; and ditches had to be kept free of debris. Water disputes had to be settled by some authority, and grain surpluses had to be stored, guarded, and redistributed. The interacting effects of the geographical setting and

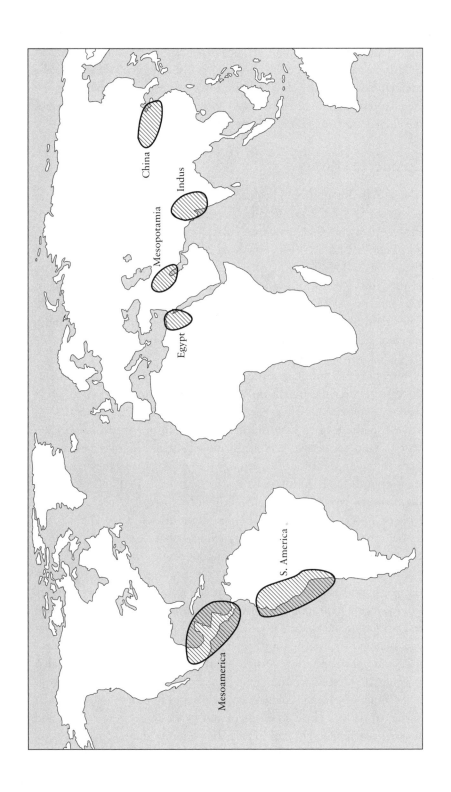

the techniques of hydraulic agriculture reinforced trends toward an authoritarian state.

Along these lines, the notion of "environmental circumscription" provides the key explanatory concept: civilizations arose in prehistoric river valleys and flood plains that were environmentally restricted agricultural zones beyond which intensive farming was impossible or impractical. In these constricted habitats, like the Nile River Valley, expanding Neolithic populations soon pressed against the limits imposed by desert, cataracts, and sea, leading to pressures to intensify food production. Warfare became chronic and developed beyond raiding to involve conquest and subjugation since, in a habitat already filled, the vanquished could no longer bud off and form a new agricultural community. Whereas previously in both the Paleolithic and Neolithic, defeated groups could generally move on to a new locale, in environmentally restricted areas around the Tigris or Euphrates Rivers, in the Nile River Valley and elsewhere agriculturalists had nowhere to go. Victors not only took over land and smaller irrigation works but subjugated and dominated defeated groups, sparing their lives in return for their labor as slaves and peasants in maintaining systems of intensified farming. Once this process started, the historical momentum favoring confederating and centralizing forces was irreversible. Neolithic communities thus became increasingly stratified, culminating in a dominant elite in command of an agricultural underclass as regional powers subsumed local ones. Time and again civilization and the state emerged wherever these ecological and demographic conditions occurred.

Further research will doubtless amplify this picture, but for now a common pattern with common characteristics seems apparent. History is too easily thought of as a sequence of unique events—what has been lampooned as "one damned thing after another." But the recurrent rise of civilizations in the Near East, in the Far East, and in the New World testifies to significant regularities in the historical record.

The model described above admirably fits the first human civilization arising on the flood plain between the Tigris and the Euphrates Rivers in present-day Iraq. This was ancient Mesopotamia, the land "between the rivers." By 4000 BCE Neolithic villages filled the Mesopotamian plain. Local authorities drained marshes in the lower delta and, later, installed extensive irrigation works on the flood plain upriver. Great walled cities such as Uruk, Ur, and Sumer, with populations between 50,000 and 200,000, arose after 3500 BCE, and the dynastic civilization of the Sumerians developed fully by 2500 BCE. Possibly because of the shifting and unpredictable courses and flood patterns of the Tigris and Euphrates, no single kingdom or polity dominated Mesopotamia as in Egypt, but rather a series of city-states along with empires based on them rose and fell over the succeeding millennia.

Mesopotamian civilization shows a great deal of continuity over

Map 3.1. (*opposite*) The earliest civilizations. The transition from Neolithic horticulture to intensified agriculture occurred independently in several regions of the Old and New Worlds. Increasing population in ecologically confined habitats apparently led to new technologies to increase food production.

Map 3.2. Hydraulic civilizations. The first civilizations formed in ancient Mesopotamia (modern Iraq) on the food plain of the Euphrates and Tigris Rivers, astride the Nile River in Egypt, and along the Indus River. The agricultural benefits of annual flooding were intensified by hydraulic management.

thousands of years, even though different groups, from different portions of Mesopotamia, took their turns at cultural, political, and military ascendance. When the Babylonians of central Mesopotamia became the dominant power, they absorbed a good deal of Sumerian culture and adapted Sumerian script for writing their own language. When Assyria (a kingdom in northern Mesopotamia) began to control the region, it similarly absorbed much of Babylonian culture.

All of these civilizations were based on irrigation agriculture. Main canals were upward of 75 feet wide and ran for several miles, with hundreds of connecting channels. All Mesopotamian civilizations developed centralized political authority and complex bureaucracies to collect, store, and redistribute agricultural surpluses. All are characterized by monumental building, including most notably great brick temple complexes and pyramids known as ziggurats. For example, Ur-Nammu's ziggurat of Third Dynasty Ur (dating to approximately 2000 BCE) formed part of a larger complex measuring 400 by 200 yards. Nebuchadnezzar's tower (600 BCE) rose over 90 meters (270 feet) and was, according to tradition, the basis of the biblical story of the Tower of Babel. Mesopotamian civilization also developed writing, mathematics, and a very sophisticated and mature astronomy.

Ancient Egypt illustrates a similar route to civilization. The Nile River Valley is a circumscribed strip of green hemmed in by a sea of desert to the east and west, mountains to the south, and the Mediterranean to the north; it forms a narrow ribbon 12–25 miles wide and hundreds of miles long. Neolithic settlements proliferated along the Nile, and already in the sixth millennium BCE kingdoms emerged; seven predynastic kingdoms have been identified down to roughly 3400–3200 BCE. (Egyptologists agree about the order of events, but they differ by centuries on dating, especially in the early dynasties and Old Kingdom Egypt.) Sometime in that period King Menes united the two kingdoms of Upper and Lower Egypt, thus becoming the first Egyptian pharaoh of what we know as the first dynasty. And, according to tradition, Menes also organized hydraulic works, having embanked the Nile at Thebes. The explosive growth of Egyptian civilization followed. Based on managing the annual flooding of the Nile, Egypt manifested all the earmarks of high civilization, including large-scale building in the great pyramids at Giza, which were early creations of Egyptian civilization. Centralized authority grew correspondingly at an early date; 20,000 soldiers came to compose the Egyptian army; the pharaohs became legal heirs to all property in Egypt and controlled absolutely their 2.5 million subject-tenants; bureaucracy, writing, mathematics, elementary astronomy, expanded crafts, and all the other complexities of civilization displayed themselves in turn.

Less is known of civilization in the Indus River Valley, but the outlines of its historical development are plain. Neolithic settlements appeared along the Indus by 7000 BCE. Civilization may have arisen indigenously or some of its incipient features may possibly have arrived with settlers or traders from Mesopotamia. One way or another, the alluvial flood plain of the Indus River Valley provided the indispensable setting for Indus civilization, and irrigation agriculture the requisite means. The cities of Mohenjo-daro and Harappa in modern-day Pakistan date to 2300 BCE. Harappan civilization, as it is known, thereafter spread inland and along the coast of the Arabian Sea. Peoples of the Indus River Valley farmed the arid plains, and they built embankments to protect cities against erratic, silt-laden floods. Indicative of strong central government, Harappan towns were rigidly planned walled communities with laid-out streets and blocks, towers, granaries, and sewers, and all the trappings of civilization. At the center of Mohenjo-daro, for example, stood an enclosed citadel (200 × 400 yards) with its 40-foot-high brick mound. Within, the Great Bath held a manmade pool 12 meters long, 7 meters wide, and almost 3 meters deep, and archaeologists have identified what may be priestly residences and an assembly hall. The population of Mohenjo-daro has been estimated at 40,000; Harappa at 20,000. Harappan metallurgists used copper, bronze, gold, silver, tin, and other metals; potters produced glazed pots; and writing and higher learning developed. Limited evidence suggests that even at an early period authoritarian regimes with a strong priestly-bureaucratic-military class already held command. But after 1750 BCE the original urban culture of the Indus declined, probably because of climate and ecological factors, including the changing course of the Indus River.

In China a similar pattern repeated itself along the Yellow River (the Hwang-Ho). By 2500 BCE thousands of late Neolithic villages spread out along the river, and as irrigation agriculture began to be practiced, kingdoms arose. Yü the Great, the putative founder of the semimythical first dynasty (Xia), is legendary in China as the ruler who "controlled the waters." The Shang (Yin) dynasty (1520–1030 BCE), which marks the documented beginning of Chinese civilization, made itself master of the Yellow River plain by dint of extensive irrigation works. Later, engineers brought irrigation techniques to the more southern Yangtze River. Rice cultivation spread northward from the far south of China, and rice production requires nontrivial hydraulic control. One of the roles of government throughout Chinese history was to build and maintain waterworks; as a result, dikes, dams, canals, and artificial lakes (such as the 165-acre Lake Quebei) proliferated across China. Deliberate government policies of water conservancy and agricultural improvement also involved drainage. To effect these installations, massive corvée labor was extracted from the peasantry.

The early Chinese built cities with protective walls, palaces, and ceremonial centers. Their society became highly stratified; Chinese emperors functioned as high priests, and great emphasis was placed on royal burials that included the emperor's entourage, sacrificed by the hundreds to accompany him. China was first unified in 221 BCE, and unprecedented authority became centralized in the emperor, backed administratively by an elaborate and formidable bureaucracy associated with royal courts. The population of China under the control of the emperor has been estimated at 60 million at the beginning of the common era. The early Chinese state built granaries and maintained standing armies. Sophisticated bronze metallurgy was also practiced, with the bronze tripod the symbol of administrative power invested in officials. As for monumental building, in addition to hydraulic works, the Great Wall of China has been hailed as the largest building project in history. Construction of the first 1,250 miles of the Great Wall (on the divide between steppe and arable land) began in the fourth and third centuries BCE and was finished in 221–207 BCE, coincident with the first unification of China. (In later historical times the total length of Chinese defensive walls extended to over 3,000 miles.) The Grand Canal (originally built in 581–618 CE), the interior waterway stretching 1,100 miles from Hangchow to Beijing, deserves mention as another example of monumental building associated with Chinese civilization. On the order of 5.5 million people labored on the project in which 2 million workers may have perished. No less characteristically, writing, mathematics, and astronomy came to be part of Chinese civilization.

Swamps and Deserts

The separate and independent rise of civilizations in the Old and New Worlds represents a great experiment in human social and cultural development. Despite departures in the New World, notably the absence of cattle, the wheel, and the plow, the independent appearance of civilization in the Western Hemisphere and the deep parallels among pristine civilizations in regions where water management was necessary lend support to the hydraulic hypothesis and the view that regularities in history derive from the material and technical bases of human existence.

Recent findings have confirmed that humans entered the Americas and hunted and gathered their way to southern Chile by at least 12,500 years ago. In Central (or Meso-) America, Paleolithic hunter-gatherers gave way to fully settled Neolithic villages by 1500 BCE. Increasingly complex Neolithic settlements filled the humid lowlands and coastal regions of Central America by 1000 BCE. Olmec culture flourished from 1150 to 600 BCE inland along rivers flowing into the Gulf of Mexico and is sometimes said to be the first American "civilization." But in fact the Olmecs seem to have been at a high Neolithic stage comparable to the megalithic culture at Stonehenge. Olmec "towns" held populations of fewer than 1,000. Nonetheless, they built ceremonial centers with burial mounds, and they are known for colossal Olmec stone heads, some over 20 tons in weight and transported 100 miles, according to one report. They developed a calendar and, suggestive of the origins of true civilization, hieroglyphic writing. The Olmecs declined after 600 BCE, but they provided cultural models that later, more fully formed American civilizations built upon.

Founded around 500 BCE, the first true city in the New World was at Monte Albán looking down on the semiarid Oaxaca Valley in Central Mexico. Small-scale irrigation agriculture was practiced in the valley, and Monte Albán was a planned city that possibly represented the confederation or consolidation of three regional powers into what became Zapotec civilization. Engineers leveled the top of the mountain for a large astronomically oriented acropolis, stone temples, pyramids, and a ball court. Two miles of stone walls encircled the city; 15,000 people lived there by 200 BCE, 25,000 by the eighth century CE. Before its subsequent decline, Zapotec scribes wrote with hieroglyphs and possessed a complex calendar.

Coexisting with Monte Albán but an order of magnitude larger, the huge city of Teotihuacán arose in the dry Teotihuacán Valley near modern Mexico City after 200 BCE. Estimates for the population of the city at its peak in the period 300–700 CE range from 125,000 to 200,000, making it the largest and most powerful urban center in Mesoamerica; it was the fifth largest city in the world in 500 CE, and it remained one of the world's largest urban centers for several hundred years. Oriented astronomically, the planned town of Teotihuacán covered eight square miles, and the main avenue ran for over three miles. The largest structure was the gigantic Temple of the Sun, a huge stepped pyramid nearly 200 feet high, 35 million cubic feet in volume, with a temple on top. There were 600 other pyramids and temples in Teotihuacán and several thousand apartment complexes. As in other early civilizations, hydraulic works and irrigation agriculture made

Fig. 3.1. Teotihuacán. Cities and monumental building are defining features of all civilizations. Here, the huge Temple of the Sun dominates the ancient Mesoamerican city of Teotihuacán.

Teotihuacán possible. In addition to farming land in the seasonally flooded upper valley, Teotihuacános built canals and installed extensive, permanent irrigation works along the San Juan River in the lower valley. Teotihuacán itself was well supplied with water by the river, canals, and reservoirs. Control over a highly developed obsidian trade also increased the prosperity of the city. What archaeologists have identified as a gigantic royal palace and a major bureaucratic/administrative center testify both to extreme social and economic stratification and to centralization of power into royal/priestly authority. At its height the civilization of Teotihuacán dominated the whole Valley of Mexico.

Contemporaneous with civilization in the dry valleys of central Mexico, Mayan civilization rose in the wet lowlands of Mesoamerica and flourished for a thousand years between 100 BCE and the ninth century CE. Until the 1970s the archaeology of Mayan civilization seemed to discredit any link between civilization and the taming of waters. But an interpretative revolution in Mayan studies followed from the discoveries of extensive Mayan engineering installations covering 741 acres at Pulltrouser Swamp in modern Belize. The problem for lowland Mayan agriculture was not too little water, as in Egypt, but too much water. This was a problem the Maya overcame by farming raised fields (3 feet high, 15–30 feet wide, and 325 feet long at Pulltrouser) with canals and drainage channels in between. The works drained water from fields, the muck in canals served as fertilizer, and the system overall proved capable of producing surpluses sufficient to support large populations. And it required collective effort to build and maintain. The distinctive Mayan form of intensified wetland agriculture now reveals the hydraulic underpinnings of Mayan civilization.

The largest Mayan city was Tikal, which had a population of 77,000 before its collapse about 800 CE. Population densities during the Maya Classic Period are estimated to have been 10 to 15 times greater than that supported in the remaining jungles of Central America today. Monumental building dominated Mayan cities, especially temple platforms and large stepped pyramids, similar to ziggurats, with a stairway leading to a temple on top. Political authority was centralized in noble classes and Mayan kings. And the Maya developed the most sophisticated mathematical, calendrical, and astronomical systems of any civilization in the Americas.

In the rise of civilization in South America, the pattern repeats itself yet again. Collectively covering millions of acres, Peruvian irrigation systems represent the largest archaeological artifact in the Western Hemisphere. The many short rivers flowing from the Andes Mountains to the Pacific across an arid coastal plain are now seen to form the ecological equivalent of the Nile River. Early village settlement arose in more than sixty of these extremely dry coastal valleys, and increasingly elaborate and well-engineered irrigation systems became essential to support the civilizations that developed there. One of the irrigation canals of the Chimu people, for example, ran 44 miles; their capital at Chan-Chan covered nearly seven square miles. In joining existing irrigation systems, Moche civilization expanded out of the Moche River Valley after 100 BCE, ultimately occupying 250 miles of desert coastline and up to 50 miles inland. The Moche urban center at Pampa Grande had a population of 10,000, and the Huaca del Sol pyramid, made of 147 million adobe bricks, stood 135 feet high. Moche civilization endured for nine centuries.

In southern Peru another center of civilization arose in the highlands around Lake Titicaca. There, based on the cultivation of potatoes, a fecund agricultural system of raised and ridged fields similar to Mayan wet farming fueled a line of civilizations. One report puts the population of the mountain city of Tiwanaku at 40,000–120,000 at the city's zenith between 375 and 675 CE. The succeeding Incas installed irrigation works and practiced water management on a larger scale than their predecessors, and militarily the Incas were the first to unite the productive resources of the coastal plains and the mountain highlands. At its peak in the fifteenth century CE, the Inca empire extended 2,700 miles and included 6 to 8 million people (some say 10 million). Monumental building is well represented in the Inca capital of Cuzco with its exquisite mortarless masonry and water supply and drainage systems, in remote Machu Picchu with its steeply terraced fields, and no less in the incredible system of roads that united the Inca empire. Two road systems—one coastal, one in the mountains—ran for 2,200 miles each, and altogether the Incas built 19,000 miles of path and road, a huge engineering achievement accomplished without metal tools. The state maintained an elaborate system of grain-storage facilities and redistribution mechanisms. The Inca emperor was the sacred focus of an absolutist state rivaling ancient Egypt in despotism, and like the Egyptian pharaohs, dead Inca emperors in Peru were mummified and worshiped.

Thus, time and again the Urban Revolution produced civilizations that depended on large-scale hydraulic engineering, and it repeatedly transformed human existence from Neolithic roots. The similarities of ancient American civilizations and those of the Old World have often been noticed and sometimes attributed to diffusion from the Old World to the New. But rather than invoking exotic contact across space and time to explain these parallels, would it not be less remarkable simply to say that similar material, historical, and cultural conditions produced similar civilizations?

Bronze and the Bronze Age

Based on the new technologies of irrigation and field agriculture, the worldwide rise of urban civilization marks a fundamental and irreversible turning point in the history of technology and in human affairs generally. A cascade of ancillary technologies accompanied the rise of civilization, including, at least in the Old World, bronze metallurgy. The Chalcolithic or Copper Age of smelting copper came first in the period 5000–3500 BCE. The mastery of bronze (copper alloyed with tin) thereafter lends its name to the new civilization as the Bronze Age. Metals offer several advantages over stone as tools and weapons, and in the long run metals replaced stone. Metalworking as it was first practiced in the pristine civilizations embodies a complicated set of technologies, including mining ore, smelting, and hammering or casting the product into useful tools and objects; and bronze metallurgy requires furnaces with bellows to raise temperatures to 1100°C. In the New World, bronze did not replace the digging stick, stone hammers, chisels, or the obsidian blade for tools, but highly expert gold and silver metallurgy developed nonetheless for decorative and ornamental purposes. The sophisticated gold craftsmanship of pre-Columbian Indians in Peru is justly renowned, and Chimu metallurgists apparently used techniques that amount to chemically electroplating gold. Sophisticated metalworking appears in sub-Saharan Africa early in the first millenium BCE; in addition to smelting and forging iron for weapons and trade, metallurgists of the Benin Empire used lost wax techniques to cast exquisite bronze commemorative plaques and other objects. The superiority of metal tools—first bronze and then iron after about

1200 BCE—especially as weapons transformed the landscape of the early civilizations.

Control over mineral resources thus became significant in the early civilizations. Sinai copper mines proved of great importance to Egyptian pharaohs; tin for making bronze had to be transported over long distances throughout the Near East; and, as mentioned, an extensive obsidian trade developed in Mesoamerica. Increased trade and expanded economic activity stand out among the earmarks of early civilizations. Occupational specialization and a sharpened division of labor likewise characterized civilized life from the outset. Craft production was no longer exclusively part-time or carried on as a household system of production, but rather became the business of specialized crafts whose practitioners earned their daily bread primarily in exchange for the practice of their craft skills. Certain "industrial" quarters of early cities were apparently given over to certain crafts and craft specialists. Quarrying, mining, gold- and metalworking, textiles and dying, making papyrus paper, and a thousand other new specialties arose and helped to maintain the complicated societies then forming. Among the new technologies of the Bronze Age, one might also mention brewing beer from bread, which became a noteworthy activity in Mesopotamia, where the famous Hammurabi Code regulated beer parlors in detail. Likewise in Inca Peru, ceremonial consumption of intoxicating beverages amounted to a redistribution of state-owned vegetable protein.

As a feature of the rise of state-level civilizations, humans began to exploit new sources of energy and power to do work. The muscle power of the ox (a castrated bull) was applied to pull the plow, and the horse was domesticated and entered humanity's service. The Hittites of second millennium BCE Anatolia first harnessed the horse and the ass to a wheeled cart, thus creating the chariot and transforming warfare throughout the Near East. The camel may have been domesticated in Africa as early as 3000 BCE and began to provide essential transport. So, too, did the llama in South America and the elephant in India and South Asia. Wind power became a new energy source tapped for the first time with the rise of civilization. The Nile River especially, with the current flowing north and the prevailing winds blowing south, became a highway for sailboats and a factor contributing to the unity of ancient Egypt. Boats also came to ply the waters between Mesopotamia and the Indus River Valley. Slavery arose coincident with civilization, and the corvée, while less coercive than slavery, fits into this same category of the human use of human beings.

Pyramids

Monumental architecture in the form of pyramids, temples, and palaces is diagnostic of high civilization and is remarkable in the history of technology, not only as a set of extraordinary technical accomplishments, but also as indicative of the institution and practice of architecture and the developed crafts and trades associated with engineering. The Egyptian pyramids provide the classic example of monumental building by an early civilization. The case is well documented, and it encapsulates the themes raised thus far regarding agriculture, civilization, and the Urban Revolution.

Consider first the sheer immensity of the Great Pyramid at Giza. Built on the west bank of the Nile during the zenith of the pyramid building era between 2789 and 2767 BCE (or possibly 2589–2566 BCE) by Khufu (Cheops), the first pharaoh of the Fourth Dynasty, the Great Pyramid is the largest solid-stone structure ever built: it consists of an unbelievable 94 million cubic feet of masonry, made up of 2.3 million blocks averaging 2.5 tons apiece, with a total weight of 6 million tons; it covers 13.5 acres, in 210 courses of stone, and stands 485 feet high and 763 feet on a side; chambers, buttresses, and passageways lie within. Sheathed with polished stone, the scale of the construction—not to mention the beauty of the finished structure—has not been surpassed in the nearly five millennia of human history since the Great Pyramid was built.

The Egyptian Imhotep (2635–2595 BCE) is famous for having served as engineer and chief architect to the third dynasty king Djoser. Imhotep build the Step Pyramid at Saqqara that served as Djoser's tomb. That we know Imhotep's name or that there was such a position under Djoser in the third millennium BCE is revealing of deep changes underway having to do with the solidification of civilization, the development of central authority, and the need for expertise, not least in monumental building. Like Imhotep the architects and engineers who built the Great Pyramid and the others like it commanded

Fig. 3.2. The Great Pyramid at Giza. An engineering marvel of the third millennium BCE, the Great Pyramid of Cheops (Khufu) at Giza culminated the tradition of pyramid building in Egyptian civilization. Some modern interpreters see it as a monumental exercise in political "state building."

some elementary and some not-so-elementary practical mathematics. Design and material requirements demanded such expertise, as did the very exact north-south and east-west alignment. Ancient Egyptian engineers and architects, expert agents of the state, understood the mathematics and appreciated the elegance of perfect pyramids, but the Egyptian pyramids (and monumental building generally) need to be seen primarily as stupendous engineering achievements.

According to a report by the fifth-century BCE Greek historian Herodotus, 100,000 people toiled for twenty years to build the Great Pyramid; perhaps 4,000–5,000 craftsmen worked at the site year round. The techniques of pyramid construction are now well understood, and excepting the possible use of a cantilevered machine to lift stones, no categorically new building methods developed compared to what one finds in Neolithic building techniques. Simple tools and practical procedures carried the day but, characteristic of the new powers of civilization, more people, by orders of magnitude, were deployed and construction completed that much faster than at Neolithic sites.

Such an extraordinary monument did not suddenly appear in the Egyptian desert. Rather, the Great Pyramid culminates a clear progression of pyramid building coincident with the growth and expansion of the Egyptian agrarian state.

Several fanciful theories have been put forward to explain why the Great Pyramid and preceding and succeeding pyramids were built, but the function of these structures as tombs for pharaohs seems irrefutable, even if it may not have been their only purpose. A problem exists, however: at some periods at least, the number of new pyramids exceeded the number of pharaohs; and several pyramids were built simultaneously by a single pharaoh. Moreover, most of the truly monumental pyramids came into being in just over a century in the late Third and early Fourth Dynasties. According to one account, in four generations over 112 years between 2834 and 2722 BCE, six pharaohs built thirteen pyramids. Clearly, something more than burying the dead is needed to explain the extraordinary sociocultural and technological phenomenon of the Egyptian pyramids.

One explanation of pyramid building from an engineering point of view attempts to explain the more or less continuous construction that took place on the west bank of the Nile during the heyday of pyramid building. In this interpretation, pyramid building was an activity pursued in its own right as an exercise in statecraft. The sequence of the early pyramids comprised giant public works projects designed to mobilize the population during the agricultural off-season and to reinforce the idea and reality of the state in ancient Egypt. More than one pyramid arose simultaneously because a labor pool—and surely an increasingly large labor pool—was available and because the geometry of pyramids dictates that fewer laborers are required near the top of a pyramid than at the bottom, thus permitting the transfer of labor to newly started projects. Monumental building was therefore a kind of institutional muscle-flexing by the early Egyptian state, a public work and welfare project somewhat akin to the arms industry today or building the Interstate Highway System in the 1950s and 1960s in the United States.

The engineering key to this argument comes from two particular pyramids. The first, the pyramid at Meidum, begun by the pharaoh Huni (Uni), who reigned for 24 years between 2837 and 2814 BCE, and continued by his son Sneferu. The Meidum pyramid stood 80 feet high and ran 130 feet on its side. It was to have been the first true pyramid with sheer, sloping sides and no visible steps. However, the pyramid at Meidum turned out to be an engineering disaster and a monumental structural failure, as the outer stone casing collapsed in rubble around the inner core of the pyramid. Designed with the evidently excessive slope of 54 degrees, the collapsed ruin may still be seen by the traveler.

The second pyramid at issue is the succeeding "Bent" pyramid at Dashur, also built by King Sneferu. It is a huge pyramid 335 feet high, 620 feet on a side, with a volume of 50 million cubic feet. Extraordinarily, the Bent pyramid is truly bent, angled, like Meidum, at 54 degrees on the lower half and 43 degrees on the top. One supposes that when the pyramid at Meidum failed, engineers reduced the slope of the Bent pyramid, still under construction, as a precaution. The next pyramid built by Sneferu, the Red pyramid, retained the safer slope of 43 degrees. (The Great Pyramid and later pyramids returned to increased elevations over 50 degrees, but used improved internal buttressing techniques.)

One does not have to follow every detail in order to accept the general point. The Egyptian pyramids were large state-run construction projects. A surplus of idle agricultural workers available seasonally for

Fig. 3.3. The pyramid at Meidum. Built at a steep angle, the outer casing of the pyramid at Meidum collapsed around its central core during construction.

Fig. 3.4. The Bent pyramid. The lower portion of this pyramid rises at the same angle as the pyramid at Meidum, but ancient Egyptian engineers reduced the slope for the upper portion to ensure its stability. The Bent and Meidum pyramids were apparently constructed concurrently with engineers decreasing the angle of the Bent pyramid once they learned of the failure at Meidum.

three months a year during the Nile floods provided the labor pool. (Agricultural productivity was thus not affected by the demand for labor for pyramid building.) Contrary to a once-common belief, forced slave labor did not build the pyramids, but labor was conscripted (like military conscription today) and organized in work gangs. Workers received food supplied by state granaries, and the completed pyramids served as tombs for departed pharaohs. Inevitably, elaborate theologies, priestly ceremonies, and ancillary technologies (such as mummifying) grew up around burying pharaohs. But in their construction the pyramids functioned primarily as gigantic public works projects, the effect of which helped maintain the economy of irrigation agriculture in the Nile River Valley and bolstered centralizing political and social forces, notably the state. Indeed, the heyday of pyramid building was the heyday of political centralization in Old Kingdom Egypt. The pyramids were symbolic as well as literal exercises in state building.

Writing

One earmark of the earliest civilizations, already alluded to, was the elaboration and institutionalization of higher learning—writing, recordkeeping, literature, and science. The fact that aspects of arithmetic, geometry, and astronomy originated in all of the earliest civilizations merits close attention, and it specifically suggests that such societies imposed a distinctive mark on the scientific traditions they fostered.

Knowledge in the first civilizations was subordinated to utilitarian ends and provided useful services in recordkeeping, political administration, economic transactions, calendrical exactitude, architectural and engineering projects, agricultural management, medicine and healing, religion, and astrological prediction. Since higher learning was heavily skewed toward useful knowledge and its applications, in this sociological sense practically oriented or applied science, in fact, preceded pure science or abstract theoretical research later fostered by the Greeks.

In what we are labeling the Babylonian model, the pristine state institutionalized and patronized expert knowledge in the service of governance. State and temple authorities patronized the acquisition and application of knowledge by cadres of learned scribes. The early states all created and maintained bureaucracies and a bureaucratic civil service which, in some measure, dealt with knowledge of mathematics and the natural world. A number of bureaucratic institutions prevailed in Mesopotamian city-states which employed learned civil servants, court astrologers, and specialized calendar keepers. Similarly in ancient Egypt, expert knowledge was institutionalized in the "House of Life," a scriptorium and center of learning that primarily maintained ritual knowledge and customs, but that harbored magical, medical, astronomical, mathematical, and possibly other lore and expertise. Archival halls and temple libraries also existed, and the record speaks of Egyptian savants, hierarchies of court doctors, magicians, and learned priests.

Again and again, higher learning with practical applications was supported by state and temple authorities and deployed to maintain the state and its agricultural economy. Knowledge became the concern of cadres of professional experts employed in state institutions whose efforts were bent to the service of sustaining society rather than to any individualistic craving for discovery. An additional characteristic of this bureaucratic pattern of science is the fact that scribal experts were anonymous; not a single biography of the individuals who over hundreds of years contributed to science in the first civilizations has come down to us.

Another odd characteristic of the first scientific traditions seems to be a penchant to record knowledge in the form of lists rather than in any analytical system of theorems or generalizations. Science in the first civilizations was characteristically pursued with a notable lack of abstraction or generality and without any of the naturalistic theory or the goal of knowledge as an end in its own right that the Greeks later emphasized.

Writing and reckoning were first and foremost practical technologies with practical origins meeting the practical needs of early civilizations. Centralized authority and bureaucracies responsible for redistributing large surpluses required the recording of verbal and quantitative information. All early civilizations developed arithmetical systems and systems of permanent recordkeeping. The archaeological discovery of what amount to ancient Mesopotamian invoices—insignia sealed in clay—underscores the economic and utilitarian roots of writing and reckoning. Eighty-five percent of cuneiform tablets uncovered at Uruk

(3000 BCE), for example, represent economic records, and Egyptian temple and palace records are similar. (Would a future archeologist find much different in our records today?) Ultimately writing came to supplant oral traditions and the skills and techniques of human memory. While the vast majority of early written records concern economic, legal, commercial, votive/religious, and administrative affairs, a significant literary component also came into being.

The scribal art was highly valued everywhere, and its practitioners enjoyed high social status. Educated scribes made up a privileged caste patronized by palace or temple, and literacy offered a pathway to power. It led to employment in huge and varied bureaucracies and often to high status in government. The large bureaucracies of the hydraulic civilizations, many of which left continuous records over thousands of years, provided civil service careers for junior and senior administrators, as well as specialized posts in specialized institutions as accountants, astrologer/astronomers, mathematicians, doctors, engineers, and teachers. No wonder that novice scribes were the sons (and occasionally the daughters) of the elite.

Civilization brought with it the first schools, institutions where writing was formally taught. In Mesopotamia scribal schools known as the é-dubba or "tablet house" taught writing, mathematics, and later a literature of myths and sayings. Many Mesopotamian tablets record the countless writing and calculating exercises performed by generations of students in schools that operated in the same location teaching the same curriculum for a thousand years and longer. In Egypt, writing was institutionalized in scribal schools and other institutions that

Fig. 3.5a–b. Babylonian and Egyptian writing systems. Different civilizations developed different techniques for recording information in writing. Most began with representations called pictographs. Many later used signs to represent sounds of a spoken language.

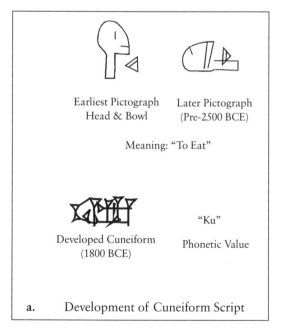

Earliest Pictograph
Head & Bowl

Later Pictograph
(Pre-2500 BCE)

Meaning: "To Eat"

Developed Cuneiform
(1800 BCE)

"Ku"

Phonetic Value

a. Development of Cuneiform Script

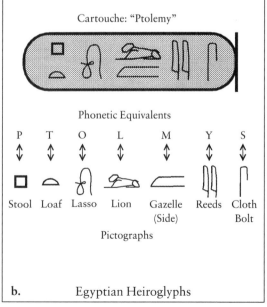

Cartouche: "Ptolemy"

Phonetic Equivalents

P	T	O	L	M	Y	S

Stool Loaf Lasso Lion Gazelle (Side) Reeds Cloth Bolt

Pictographs

b. Egyptian Heiroglyphs

contained scriptoria and libraries, and student exercises form a large part of the written records that have survived.

Although writing and recordkeeping are characteristic features of all civilizations, writing systems need to be thought of as technologies, and they have varied considerably. The earliest, the cuneiform system of writing on clay tablets, arose with Sumerian civilization in ancient Mesopotamia. Over the millennia of Mesopotamian civilization innumerable cuneiform clay tablets were dried or baked, stored, and catalogued in great libraries and archives, with tens of thousands ultimately preserved. Cuneiform—or wedge writing—is so called because Sumerian scribes used a wedge-shaped reed stylus to inscribe clay tablets. Sumerian scribes in the third millennium BCE self-consciously developed a sophisticated system of 600–1,000 signs (called ideograms) to represent the idea of a word or an action, as in "I ♥ my dog." Later, the number of Sumerian characters was reduced, but the scribal art remained very difficult to master and literacy remained restricted to a scribal profession. Cuneiform signs assumed sound (or phonographic) values at an early period and were written as syllables voicing the Sumerian language. Indeed, Old Babylonian (Akkadian), a different language from the original Sumerian, came to be written using Sumerian phonetic values. In other words, pictographs originally pictured things, whereas the signs later came to represent sounds of spoken languages. Sumerian continued to be taught in the é-dubba as a dead language after the eighteenth century BCE, similar to the way Latin was taught in European universities until the nineteenth and twentieth centuries. Sumerian and Babylonian languages had written grammars, and many tablets record word lists, bilingual lexicons, and bilingual texts.

Pictographic writing is known in Egypt from predynastic times, and the hieroglyphs ("sacred carvings") of ancient Egypt were used by the first dynasty, around 3000 BCE The idea of writing may have passed from Mesopotamia, but specific Egyptian writing developed independently. Hieroglyphs are ideographic, but from an early period Egyptian writing incorporated phonographic elements voicing the Egyptian language. Six thousand formal Egyptian hieroglyphs have been identified, but pharaonic engravers and scribes commonly used only 700–800 across the millennia. Formal hieroglyphs were obviously not easy to write, so scribes developed simpler scripts (called hieratic and demotic) for the day-to-day maintenance of Egyptian civilization. Among the technologies that made this possible was papyrus paper. The last hieroglyphic inscription dates from 394 CE, after which knowledge of ancient Egyptian writing was lost. Only the acclaimed Rosetta stone— an inscription dated to 196 BCE with its text written in hieroglyphics, demotic, and Greek—discovered by Napoleon's soldiers in 1799 and deciphered by J.-F. Champollion in 1824—allows us to read again the records of the ancient Egyptian scribes. It should also be noted that purely phonetic alphabets where the sign stands only for a vowel or consonant sound—such as the Greek or Roman alphabets—are a late historical development of secondary civilizations, first appearing after 1100 BCE with the Phoenicians.

Reckoning

Mathematical methods developed along with writing and out of the same practical needs. The ancient Greek historian Herodotus, for example, thought geometry (or "earth measure") originated in Egypt as a result of the need to resurvey fields after the Nile floods. Along these lines, with the agricultural surpluses generated by irrigation agriculture came the first money (in ancient Babylonia and in Shang dynasty China) and the first standardized weights and measures (in ancient Egypt, the Indus River Valley, and in early China). Although pure mathematics later became an abstract game played by mathematicians, the practical, economic, and craft roots of early mathematics remain visible in these applications.

Each of the early civilizations developed its own system of mathematics. The ancient Sumerians and Babylonians evolved a sexigesimal or base-60 system (in contrast with our own decimal or base-10 system). Although not entirely consistent and initially lacking a zero, it was the first place-value system, where the "digits" represented powers of 60. Sexigesimal remnants can be found today in the 60-minute hour, the 60-second minute, and the 360 degrees of the circle. In contrast, Egyptian numbers resembled later Roman numerals with separate signs for the decimal numbers and no place value. Such a number system was more cumbersome and less efficient in handling the calculating requirements of Egyptian civilization.

As for mathematical operations, Babylonian mathematicians, using tables of numbers—multiples, reciprocals, squares, cubes, Pythagorean triplets, and the like—could perform many complex calculations, including recipe-like procedures that calculated compound interest and solved quadratic and cubic equations. In ancient Egypt, the "method of duplication," that is, the process of multiplication by doubling and redoubling numbers, was especially handy with a Roman-style number system. Egyptian mathematicians arrived at a superior estimate of the value of π (256/81 or 3.16 compared to the rough value of 3 of Babylonian mathematics and the Bible), and they developed tables that facilitated working with fractions.

In every early civilization the problems tackled by mathematicians reflect the practical and utilitarian direction of their interests. Engineering and supply problems predominated, generally solved by mathematical recipes involving little or no abstract understanding of numbers. The solution was usually arrived at recipe-style ("add 2 cups of sugar,

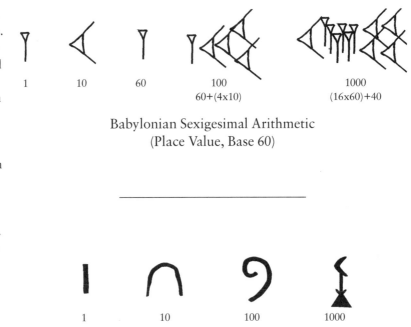

Fig. 3.6. Babylonian and Egyptian number systems. Different civilizations developed different numeral systems and reckoning methods. The Babylonian system was a base-60, place-value system with number signs for the values 1 and 10. Egyptian hieroglyphic numerals represented values of 10 in a manner reminiscent of later Roman numerals. No civilization was without a system to record numerical information.

1   10   60   100   1000
           60+(4x10)   (16x60)+40

Babylonian Sexigesimal Arithmetic
(Place Value, Base 60)

1   10   100   1000

Egyptian Hieroglyphic Numerals

1 cup of milk, etc."), much like a computer program would handle the underlying equation ("square *a*, multiply $a \times b$, add a^2 and ab"). Although we do not know how the recipes were concocted, they were computationally sound and gave correct answers.

The Greeks had yet to invent abstract mathematics, but in a few restricted instances some very esoteric nonutilitarian "playfulness" becomes apparent in the achievements of the early scribes. In Babylonia, for example, mathematicians calculated the square root of 2 to the equivalent of six decimal places, beyond any conceivable need in engineering or reckoning. Similarly in China expert mathematicians computed π to the very high and, practically speaking, useless accuracy of seven decimal places. However, as interesting as they are, even these steps toward abstract mathematics developed in the context of broad programs of study directed at practical ends and patronized by one authority or another. In ancient Mesopotamia tables of exponential functions that would appear to be as abstract as an excessively accurate approximation of the square root of 2 were, in fact, used to calculate compound interest, and quadratic equations were solved in connection with other problems. Linear equations were solved to determine shares of inheritance and the division of fields. Lists of coefficients for building materials may have been used for the quick calculation of carrying loads. Coefficients for precious metals and for economic goods presumably had equally practical applications. And calculation of volumes reflected no idle interest in geometry but was applied in the construction of canals and other components of the infrastructure.

Time, the Gods, and the Heavens

All agricultural civilizations developed calendrical systems based on astronomical observations, and in several of the first civilizations we can identify what can only be called sophisticated astronomical research. The utility and necessity of accurate calendars in agrarian societies seems self-evident, not only for agricultural purposes, but also for regulating ritual activities. The commercial and economic role of the calendar in, for example, dating contracts and future transactions likewise seems clear.

In Mesopotamia a highly accurate calendar was in place by 1000 BCE, and by 300 BCE Mesopotamian calendrical experts had created a mathematically abstract calendar valid for centuries ahead. Since they had adopted lunar calendars of 12 lunar months or 354 days, which is obviously out of sync with the solar year of 365¼ days, an extra lunar month occasionally had to be inserted (or intercalated) to keep lunar months and (seasonal) solar years in harmony; Babylonian astronomers inserted seven intercalary months over periods of 19 years. Ancient Egyptian priest/astronomers maintained two different lunar calendars, but a third solar/civil calendar governed official Egyptian life. That calendar consisted of 12 months of 30 days and five festival days. Each year the 365-day civil calendar thus deviated from the solar year by one-quarter day; and so over the long course of Egyptian history the civil year drifted backward and every 1460 years (4 times 365) completely circled the solar/agricultural year. The civil and solar calendars thus coincided in 2770 BCE and again in 1310. This unwieldy calendric confusion is resolved when one remembers that the central event in Egypt—the annual, highly regular Nile flood—could be predicted independently from the seasonal first appearance of the star Sirius above the horizon.

Calendars, astronomy, astrology, meteorology, and magic formed part of a general pattern, repeated in Mesopotamia, Egypt, India, China, and the Americas. Despite our modern biases it is not possible or justifiable to separate astronomy from astrology or astronomers from astrologers and magicians in these early civilizations, for the enterprises formed an inseparable unity. In predicting the fate of crops, the outcome of military action, or the future affairs of the king, astrology and occult learning were universally seen as useful knowledge. Indeed, along with calendrical astronomy (which, after all, predicted the seasons), they exemplify the pattern of knowledge of nature turned to state-supported practical ends.

Of all the ancient scientific traditions, Babylonian astronomy was the best developed, and it merits detailed attention. In ancient Babylonia a shift in divination from reading the entrails of animals to an astral religion may have encouraged the study of the heavens. Astronomical observations were recorded as early as 2000 BCE, and continuous observations date from 747 BCE. By the fifth century BCE Babylonian astronomers could track the principal heavenly bodies indefinitely into the future. Mesopotamian astronomers fully mastered solstices, equinoxes, and the cycles of the sun and moon. In particular, later Babylonian astronomy understood and could predict solar and lunar eclipses and eclipse magnitudes. Astronomers computed and extrapolated the risings, settings, and visibility of planets, especially Venus as a morning and evening star. The legacy of Babylonian astronomy and the sexigesimal system was great, not only for our measure of the circle in degrees, but also for the seven-day week and the identification of the planets. Indeed, many technical procedures of Babylonian astronomy were handed down and adopted by later Greek and Hellenistic astronomers. What needs emphasis here is the research conducted by Babylonian astronomers. Obviously, they observed the heavens, no doubt with sighting instruments, and kept accurate records. We now know that they did much more than observe and keep records; they also conducted systematic research to solve very specific scientific problems in astronomy.

It is instructive to examine the "new moon problem" as a case in point. For calendrical and religious reasons Babylonian astronomers needed to know the length of the lunar month in days. The interval between full moons or new moons varies between 29 and 30 days (the average is 29.53 days). Which was it going to be in any given month? Several independent variables affect the outcome: the relative distance between the sun and moon in the heavens as seen from the earth (AB on figure 3.7), the season of the year (the ancle α), and longer-term lunar cycles (CD). With these independent variables at play the reappearance of the new moon obviously becomes tricky to predict. Babylonian astronomers conducted research and mastered the "new moon problem" to the point of being able to create exact astronomical tables that reliably predicted when a new moon would be visible. They achieved this capability first by observing and recording planetary motion in the heavens and then by studying the tables of data they had compiled. They had formalized and systematized knowledge as never before. The "new moon problem" indicates active scientific research by ancient Babylonian astronomers on a very specific problem (29 or 30 days?). This research was based on observation, mathematical analysis, and modeling of the phenomena, and was theoretical insofar as more attention was paid to the abstract models of mathematical cycles than to what was visibly going on in the heavens.

Medicine and the social organization of medicine also formed a distinct feature of the bureaucratic model of state support for useful knowledge. Cadres of official medical practitioners appeared in every early state, and their practical and empirical knowledge of anatomy, surgery, and herbal medicines grew as a result of state support for medical learning. The Edwin Smith medical papyrus from the Egyptian

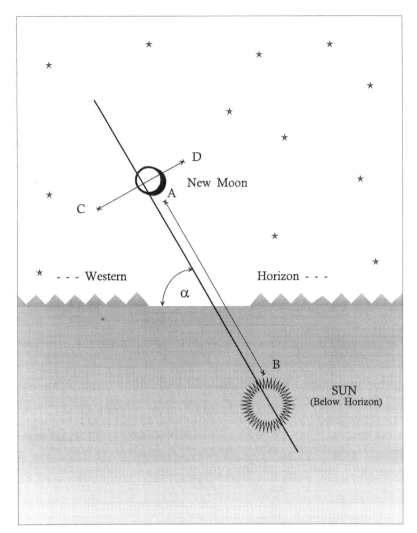

Fig. 3.7. The earliest scientific research. Ancient Babylonian astronomers systematically investigated variables determining the first appearance of the new moon each lunar month. More than simply observing the phenomena, Babylonian astronomers investigated patterns in the variations of these factors, reflecting a maturing science of astronomy.

New Kingdom (ca. 1200 BCE) is often cited for its "rational," nontheistic approaches to medical cases.

Similarly, alchemy and alchemical expertise began to be patronized at an early date in the first civilizations; the roots of alchemy doubtless lay in the practice of ancient metallurgy, a case, if ever there was one, of technology giving rise to science. Alchemy, like astrology, offered the promise of utility, and the theme of state support for alchemy winds its way through all cultures until the modern era. The distinction that we might draw between the rational and the pseudoscientific was not recognized. All of these investigations seemed to be fields of useful knowledge.

A cautious word needs to be added about the cosmologies and worldviews of the earliest civilizations. It seems safe to assume that these were all societies in which religion played a prominent role. For the most part their heavens were divine, magical, and inhabited by gods; heavenly bodies were often associated with sacred deities, and the heavens embodied myths and stories of gods. Thus in Egypt, the goddess Nut held up the sky, and deceased pharaohs became stars. In Babylonia the movement of the planets represented the movement of celestial gods. In ancient Mesoamerica, according to the Maya, the earth was a giant reptile floating in a pond. The Chinese held more organic and less pantheistic views of the cosmos. But none of the first civilizations developed any theoretical models of the cosmos as a whole, certainly no abstract, mechanical, or naturalistic ones. Little is recognizable in these cultures as independent naturalistic inquiries into the natural world or as a conception of "nature" to be studied abstractly.

The first civilizations tended to treat knowledge extensively, by drawing up encyclopedic tables and lists of words, numbers, gods, plants, animals, stones, cities, rulers, occupations, or scribes, sometimes indiscriminately. This manner of coping with and recording knowledge—what has been called the "science of lists"—may have been favored generally in societies that had not yet invented formal logic and analytical thought. The laborious drudgery that went into them, intellectually unrewarding to the individuals who compiled the data, may have been possible only where the state patronized battalions of scribes as civil servants.

In sum, deriving from practical necessity, science repeatedly emerged part and parcel with civilization. Writing and arithmetic were new technologies applicable to the solution of many practical problems. Institutions and the institutionalized status of specialized experts underwritten by the state served the same utilitarian purposes. The evidence of advanced calendars, sophisticated astronomical puzzle-solving, and occasional mathematical "playfulness" make plain the high level of scientific accomplishment in the first civilizations. Lacking was the abstract dimension of theory that we recognize as a further hallmark of science. What has to be explained, therefore, is the origin of scientific theory and the pursuit of natural knowledge for its own sake, what came to be called natural philosophy—the philosophy of nature. If science in the form of mathematics and astronomy arose independently and many times over with the first civilizations, natural philosophy originated uniquely with the Greeks.

Greeks Bearing Gifts

Ancient history displays a remarkable singularity in what has sometimes been termed the "Greek miracle." Just to the west of the Near
Eastern civilizations, around the shores of the Aegean Sea, Greek-
speaking peoples originated a unique civilization.

Given its proximity to Egypt and Mesopotamia, Greek civilization
derived some of its traits from its older neighbors. But those traits
took root in a habitat sharply different from the semiarid flood plains
of Egypt and Mesopotamia. Instead of a centralized kingdom, Greek
civilization arose as a set of decentralized city-states, and it retained
its loose structure until Alexander the Great (356–323 BCE) unified
Greece and the wider world in the fourth century BCE. The appearance
of Alexander the Great marks a major turning point in the history and
periodization of antiquity. The earlier pre-imperial period of ancient
Greece, from 600 to 323 BCE, is known as the Hellenic era, while the
period following Alexander's conquests has been designated as the Hellenistic. The Hellenic is the era of localized Greek city-states, the Hellenistic, or Hellenic-like, is that of large cosmopolitan cultures that
followed in the Near East.

During the Hellenic period Greek science took an unprecedented
turn as natural philosophers, unsupported by the state and uncom-
mitted to any program of useful knowledge, developed a series of abstract speculations about the natural world. Then, with Alexander's
conquest of the wealthy districts of the East, Greek science entered its
Golden Age through a merger of its theoretical spirit with the bureaucratic pattern of institutional patronage.

Several features characterize Hellenic science. The most remarkable
was the Greek invention of scientific theory—"natural philosophy"
or the philosophy of nature. Early Greek speculations on the cosmos
and the disinterested Hellenic quest for abstract knowledge were unprecedented endeavors. They added a fundamental new element to the
definition of science and shifted the direction of its history. In launch-

ing their novel intellectual enterprise, early Greek natural philosophers raised fundamental questions that proved highly influential and continue to be asked today.

A second notable feature of Hellenic science concerns its institutional status. At least in the period down to Alexander the Great, state patronage for Greek science did not exist and, unlike in the Near East, there were no scientific institutions. Some informal "schools"—intellectually very important ones—did appear in classical Greek culture, but these operated more in the vein of private associations or clubs rather than educational institutions. No public support or funding existed for schools of higher learning, libraries, or observatories, nor did scientists or natural philosophers receive public employment. Quite unlike his state-sponsored counterpart, the Greek natural philosopher was an independent operator. Although we know little of their private lives, it appears that early natural philosophers either possessed independent wealth or earned a living as private teachers, doctors, or engineers since there was no social role for natural philosophers or scientists as such. Hellenic science thus floated in a sociological vacuum to the point where the utterly impractical and apparently meaningless private investigations of its practitioners sometimes excited animosity and ridicule.

In the East, knowledge had been turned to practical ends and purposes. But in Hellenic Greece a distinctive model, the Hellenic model, developed for science and for science in society. Its ideology stressed the philosophical dimension of knowledge and a detachment from any social or economic objectives. In an influential passage in his Republic (c. 390 BCE), for example, Plato mocks the idea that one should study geometry or astronomy in pursuit of practical benefits for agriculture, military affairs, navigation, or the calendar. Plato insisted on separating the pursuit of natural knowledge from the lesser activities of the crafts and technology. In this regard it might be said that the Greeks undertook natural philosophy as play or recreation or to fulfill higher goals concerning the life of reason and philosophic contemplation. By contrast, no comparable disinterested intellectual endeavor had been evident in the scientific cultures of the ancient hydraulic civilizations. Finally in this connection, whereas a utilitarian pattern appeared in each of the pristine civilizations, Hellenic natural philosophy appeared once, in Hellas, the result of a singular set of historical circumstances. In sum, Hellenic natural knowledge represents a new sort of science and scientific activity—self-consciously theoretical inquiries into nature.

Recent research, while not taking away from the glories of early Greek natural philosophy, has tended to set the Greek scientific enterprise in a larger, more pluralistic cultural context. It used to be thought, for example, that science and rationality arose almost miraculously from the dark world of religion and myth prevailing before the Hellenic. Today, historians emphasize that ancient Greece was not culturally insulated from the East or from the "barbarian" world beyond Greece itself. In particular, recent interpretations stress the influence of Egyptian civilization on the development of Hellenic culture around the Aegean Sea. Within the Hellenic world the continuation of popular beliefs in magic, folklore, alchemy, astrology, and religious mysticism of one variety or another represented intellectual competition to relatively secularized scientific knowledge.

Roots

The appearance of Greek science and natural philosophy may seem less surprising than it once did, but the question remains of how to account for the rise of natural philosophy in ancient Greece. Greece was a so-called secondary civilization, arising on the periphery of Egypt and Mesopotamia and millennia after their establishment, but it was ecologically and economically very different from the principal centers of civilization in the Near East and elsewhere. (See map 4.1.) Whereas these pristine civilizations arose on the basis of hydraulic agriculture, food production and farming in the Greek city-states depended almost wholly on seasonal rainfall and runoff from mountain snow. The Greeks did not disdain waterworks, as research has shown, but these remained small scale since Greece lacked a great river and a large, productive flood plain. Furthermore, Neolithic deforestation and erosion had already degraded the ecology and productive capabilities of Greece to the extent that only comparatively low population densities could be supported. The spawning of scores of Greek colonies by a constant flow of emigrants around the Mediterranean in the eighth through sixth centuries BCE testifies to these ecological and cultural pressures. Classical Greece could not feed itself and depended on grain imports from abroad. The relatively poor agrarian economy of ancient Greece sustained itself on goat and sheep husbandry and on cultivating olive trees and grapevines, which flourish on marginal soils by tapping subsurface water. The secondary products of wine and olive oil gave the Greeks something to trade and, as a result, Hellenic civilization acquired a maritime, mercantile, and outward-looking cast.

Just as the mountains of Greece compartmentalized the land in separate valleys, Hellenic civilization was politically decentralized and fragmented into small, independent city-states. The government of a city-state in a region with a limited and eroded agricultural base could never concentrate enormous wealth like that of an Egyptian pharaoh to patronize a pervasive bureaucracy that bent every social and cultural activity toward the interests of the state.

The Greeks are famous for the level of their political debate about

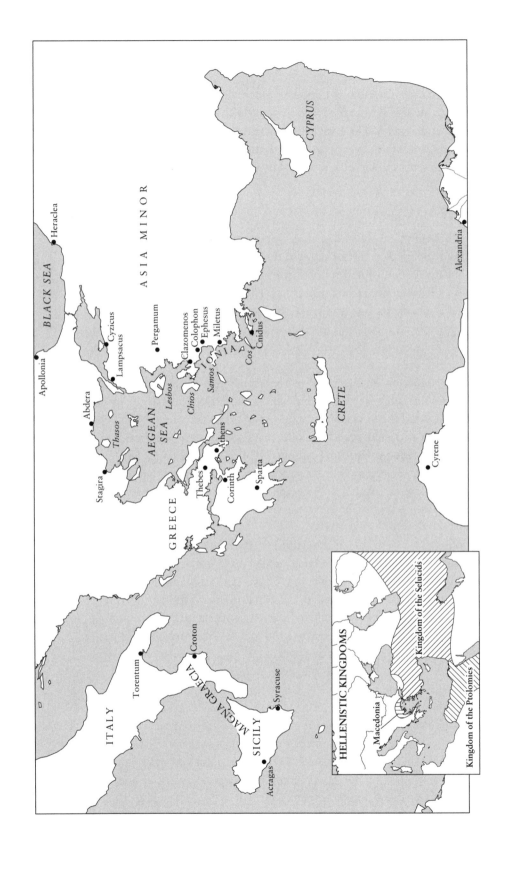

law and justice and for their analysis of kingdoms, aristocracies, democracies, tyrannies, and the like. A small step separates rational debate about political constitutions from inquiring into the constitution of nature—and vice versa, as the later history of science was to show. These political debates may indeed have provided one route to the origins of Greek science. It may be impossible to reach an understanding of exactly *why* a new scientific culture came into being in the unique habitat of Hellas. (If Ionia and Athens had remained as bereft of science as, say, Corinth and Sparta, would there be any grounds for surprise?) But once a scientific culture arose in ancient Greece it was shaped by a society that attached no social value to scientific research or instruction and that provided no public support for schools of higher learning.

Greek science did not originate in Greece, but in Asia Minor on the (then-) fertile Mediterranean coast of present-day Turkey, at first in the city of Miletus and later in several other cities of the region known as Ionia. In the seventh century BCE Ionia was the center of Greek civilization while the Greek mainland was decidedly the province. Lying on the eastern shores of the Aegean, it had more fertile land and received more rainfall than mainland Greece. Ionia remained more urbanized and economically superior to Greece proper for two centuries. Not surprisingly, the majority of the first natural philosophers hailed from Ionia.

The Ionians and the entire collection of early Greek natural philosophers are known as the pre-Socratics, that is, thinkers active in the formative period of Greek philosophical and scientific thought before Socrates (470?–399 BCE). (See table 4.1.) Greek natural philosophy is usually said to begin with Thales of Miletus, who lived from about 625 to about 545 BCE. Thales is a test case for historical interpretation, for we have nothing from Thales himself and are wholly dependent on secondary reports. Our view of Thales is thus refracted through both the biases of ancient commentators and our own interpretative frames. We do know that he came from Miletus, a vibrant trading city on the Ionian coast of Asia Minor, and that he was later crowned as one of the seven "wise men" of archaic Greece, along with his contemporary, the lawgiver Solon. Thales was probably rich, and he probably traveled to Egypt, from where he is said to have brought geometry to the Greek-speaking world. As Plato reports, maliciously perhaps, Thales and his philosophy earned a reputation for unworldliness: "A servant-girl is said to have mocked Thales for falling into a well while he was observing the stars and gazing upwards, declaring that he was eager to know things in the sky, but that what was behind him and just by his feet escaped his notice." By the same token, according to Aristotle, Thales exploited his knowledge of nature through an astute scientific observation of a forthcoming harvest in order to corner the market on olive presses and thus to demonstrate that philosophers could be rich and useful, if those were their concerns. Thales allegedly also applied

Map 4.1. (*opposite*) The world of ancient Greece. Greek civilization originated as a cluster of small city-states around the Aegean Sea. Greek science first arose in towns along the Ionian coast of Asia Minor. After the conquests of Alexander the Great in the fourth century BCE the Greek world stretched from Egypt to the borders of China, forming the largest empire in the ancient world. After Alexander's death in 323, his empire (*inset*) collapsed into three states: Macedonian Greece, Ptolemaic Egypt, and the Seleucid Kingdom in Mesopotamia.

Table 4.1
The Pre-Socratic Natural Philosophers

The Milesians	
Thales	fl. 585 BCE
Anaximander	fl. 555 BCE
Anaximenes	fl. 535 BCE
Empedocles of Acragas	fl. 445 BCE
The Pythagoreans	
Pythagoras of Samos	fl. 525 BCE
Philosophers of Change	
Heraclitus of Ephesus	fl. 500 BCE
Parmenides of Elea	fl. 480 BCE
The Atomists	
Leucippus of Miletus	fl. 435 BCE
Democritus of Abdera	fl. 410 BCE
Socrates of Athens	470?–399 BCE
Plato of Athens	428–347 BCE

his acute scientific knowledge in wartime to help King Croesus ford a river in 547 BCE. In the end, the social role of wise man or magus probably befits Thales better than that of the "first scientist," which he is often called, if by "scientist" one has more modern social models in mind.

That we know Thales's name and these details about his life unexpectedly reveals something significant about his natural philosophy and about the subsequent development of science. Thales's claims about nature were just that, *his* claims, made on his own authority as an individual (with or without other support). Put another way, in the tradition stemming from Greek science, ideas are the intellectual property of individuals (or, less often, close-knit groups) who take responsibility and are assigned credit (sometimes by naming laws after them) for their contributions. This circumstance is in sharp contrast with the anonymity of scientists in the ancient bureaucratic kingdoms and, in fact, in all pre-Greek civilizations.

Thales made claims about nature, including his idea that the south-blowing Etesian winds cause the Nile flood. Another theory of his held that the earth floats on water like a log or a ship and that the earth quakes when rocked by some movement of the water. Only a hundred years after Thales, Herodotus savagely attacked these ideas, and to the modern scientific mind they may seem oddly primitive notions. But they are nonetheless extraordinary in several important regards. For one, the explanations offered by Thales are entirely general; they seek to account for *all* earthquakes and *all* Nile floods, and not only a single case. In a related way, Thales invokes no gods or supernatural entities in his explanations; to use the stock phrase, he "leaves the gods out." Thus, "hail ruined my olive crop" not as punishment because I offended Zeus or Hera in a particular instance, but accidentally because in all instances—mine unfortunately included—hail results from natural processes that involve the freezing of water in the atmosphere. Note that a feature of Greek natural philosophy—the "discovery of nature"—required objectifying and demystifying nature, in order that theories might be proposed regarding nature in the first place. That is, "nature" had to be defined as an entity to be investigated; the concept may appear self-evident to us, but it was not necessarily so to our scientific forebears. "Naturalistic" explanations first posit the phenomenon in question to be a regular part of some external nature and hence a natural phenomenon, then account for the phenomenon also in terms of nature. Thus for the Nile, naturally occurring winds are invoked to explain the natural phenomenon of flooding. Interestingly in the case of earthquakes, Thales employs analogies to what we see in the world (ships, floating logs) in his explanation. It is far from the case, however, that Thales or (most of) his successors were atheists or irreligious; in fact, Thales also taught that the world is divine and "full of gods" and that the magnet possesses a "soul." There is no contradiction, however, for despite whatever homage is due the gods, Thales sets the natural world off both as somehow separate from the divine and as something comprehensible by the powers of human intellect.

Thales is known for his view that the world is composed of a primordial watery substrate. This deceptively simple pronouncement represents the first attempt to say something about the material "stuff" making up the world around us. It marks the beginning of matter theory—that line of scientific theorizing concerned with the makeup of the physical world below the level of ordinary perception. In asking about the material basis of things early in the sixth century BCE, Thales became the founding father of the first of the "schools" of Greek natural philosophy mentioned above, the Milesians. This Milesian school and its tradition of matter theory are an important element of pre-Socratic thought, but consideration of the intellectual dynamic driving the Milesians reveals another feature of the enterprise of early Greek science: the rise of science as rational debate. In a word, the Milesian philosophers disagreed, and they used reason, logic, and observation to attack the ideas of others and to bolster their own propositions.

Thales's notion that water is the primary substance had its problems, notably to explain how water could give rise to fire, its opposite, water and fire being mutually destructive, as in fire boiling away water or water quenching fire. Anaximander of Miletus (fl. 555 BCE) dealt with this problem in the generation following Thales by rejecting water as the underlying agent and by putting forth the much vaguer notion of some "boundless" or formless initial state (the *Apeiron*) out of which duality and the world grew. By allowing duality to emerge from unity, as it were, Anaximander's "boundless" explained hot and cold, which

Thales could not, but the concept of the "boundless" remained forbiddingly abstract and metaphysical. The next Milesian, Anaximenes, responded to this difficulty and to the same general question around 535 BCE. His answer was to suggest air (or the "pneuma") as the primeval element. More down to earth, this suggestion would also seem to suffer from the problem of opposites that troubled Thales's water theory, except that Anaximenes posited two conflicting *forces* in the universe, rarefaction and condensation, which variously condensed air into liquids and solids and rarefied it into fire. The tradition of the Milesian school culminated a century later with the thought of Empedocles (fl. 445 BCE), who as an adult lived in Greek Italy. In a theory that remained influential for 2,000 years Empedocles postulated four primary elements—earth, air, fire, and water—and the attracting and repelling forces of (what else?) Love and Strife.

The pluralistic and abstract character of natural knowledge among the early Greeks is no better illustrated than by another pre-Socratic "school," the cult of the Pythagoreans. The Pythagoreans, centered in Italy, formed an organized religious brotherhood and sect, and individual innovator-adepts submerged their contributions to the collectivity by giving credit to their founding guru, Pythagoras (fl. 525 BCE), originally from the island of Samos off the Ionian coast. The Pythagoreans embodied a certain "orientalism" reminiscent of the master's sixth-century Persian contemporary, Zoroaster.

The Pythagoreans are famed for introducing mathematics into natural philosophy. Their mathematics was not the crude arithmetic of the marketplace or the practical geometrical procedures of the surveyor or architect, or even the exact mathematical tools of Babylonian astronomers. Rather, the Pythagoreans elevated mathematics to the level of the abstract and the theoretical, and they made the concept of number central to their view of nature. In its way, number was the Pythagorean response to the Milesian question about the material stuff of the world. In focusing on number, the Pythagoreans introduced potent notions of idealism into natural philosophy and science—the idea that some more-perfect reality accessible through intellectual understanding underlies the observed world of appearances. Put crudely, the real world contains no perfect triangles, no absolutely straight lines or numerical abstractions; such entities exist only in the realm of pure mathematics. That the Pythagoreans and their intellectual successors thought that such mathematical perfection somehow constitutes the world (or even that it is useful to think so) inaugurated a whole new way of thinking about nature, and it launched the great tradition of mathematical idealism that has been so powerful a current in scientific thought since then.

Pythagoras is supposed to have achieved the profound insight of mathematical order in the universe in considering musical strings and the tones they sound; half the length producing the octave above,

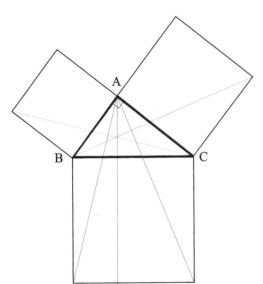

Fig. 4.1. The Pythagorean theorem. Although Pythagorean triplets (like 3-4-5) were recorded by Babylonian scribes, the Pythagorean theorem ($AB^2 + AC^2 = BC^2$) was first proved in Euclid's *Elements*. When the nineteenth-century philosopher Schopenhauer saw the diagram, he remarked "That's not a proof, it's a mousetrap."

one-third producing the higher fifth tone, and so on. Based on this unexpected correlation between small integers and the real world, Pythagoras and his followers extended their mathematical investigations. Some of their results, such as their classification of odd and even numbers, seem unexceptional to us; others, such as a sacred triangle (the *Tetratkys*) representing the sum of the numbers 1, 2, 3, and 4 (= 10), or their association of the institution of marriage with the number 5 in joining the 2 of femaleness with the 3 of maleness, reflect what we would all too easily consider a bizarre numerology.

Of course, Pythagoras is credited with the discovery of the theorem in geometry that bears his name. It says that for any right triangle (to use the algebraic formulation), $a^2 + b^2 = c^2$, where c is the hypotenuse of the triangle and a and b are the legs. (See figure 4.1.) Lurking in the Pythagorean theorem is a corollary that says that not all line lengths can be expressed as ratios or fractions of other unit lengths. Some pairs of lines (like a leg and the diagonal of a square) are incommensurable—that is, their ratio cannot be expressed by any pair of integers. To the Pythagoreans the square root of 2 was "alogon," the unutterable. The discovery of irrationality was subversive of the Pythagorean commitment to integers and the program of investigating mathematical harmonies in the world, and, supposedly, knowledge of the irrational was therefore held as the innermost secret of the Pythagorean cult.

The more fundamental point to be made about these discoveries is the role of mathematical proof in demonstrating their certainty. The invention of deductive reasoning and proof, wherein even the most skeptical auditor is forced step by step to the inevitable Q.E.D. ("thus proven") at the end, was a remarkable innovation in the histories of mathematics, logic, and science. The Egyptians knew of Pythagorean triplets (whole numbers obeying the Pythagorean theorem, as in 3-4-5 right triangles), and the Babylonians prepared tables listing them. But no one until the Pythagoreans saw in them a theorem to be proved. Rigorous mathematical demonstrations did not appear full-blown with the Pythagoreans, and the process of developing an axiomatic and deductive plane geometry continued until Euclid compiled his *Elements* around 300 BCE. Nevertheless, to the early Pythagoreans goes the credit for studying mathematics as natural philosophy, for turning Greek mathematics away from practical arithmetic to pure arithmetic and geometry, and for developing the proof as a means and model for justifying claims to knowledge.

The different traditions represented by the Milesians, the Pythagoreans, and their successors make plain that Greek natural philosophy in the pre-Socratic period lacked an agreed-upon unity and was fragmented into different schools of thought. In this connection two other major groups of pre-Socratic natural philosophers need to be mentioned at least briefly: the atomists and the so-called philosophers of change. The atomists, notably Leucippus of Miletus (fl. 435 BCE) and Democritus of Abdera (fl. 410 BCE), responded in their way to the Milesian challenge of a century earlier by imagining that the world is composed of atoms, the least reducible, indivisible particles of matter. These theorists supposed that differences in the shape, position, motion, and arrangement of atoms in the void are the root cause of the differences we see in objects around us. Ancient atomism faced a grave difficulty in explaining how random atoms could assume any coherent or lasting pattern in nature other than by cosmic accident, and atomist philosophy thereby earned a reputation for atheism. Some atomist demonstrations designed to illustrate the corporeality of air (a bottle of air held underwater) may be viewed as early scientific experiments, but ones whose purposes were to illustrate and not to test. Atomism attracted a small following, notably in the person of the Roman poet Lucretius, but the movement was decidedly a minor branch of thought until its revival in seventeenth-century Europe and the emergence of modern atomic theories in the nineteenth century. Indeed the attention usually given to ancient atomism reflects our interests more than the ancients'.

The pre-Socratics did not limit their inquiries to the inanimate world around them, but also initiated natural philosophical investigations of the living world. Alcmaeon of Croton (fl. 500 BCE), for example, reportedly undertook anatomical research and dissected merely for the sake of learning.

Heraclitus of Ephesus (fl. 500 BCE) and Parmenides of Elea (fl. 480 BCE) are labeled the "philosophers of change" because they initiated a great debate over the nature of what we experience as change in the world. Heraclitus held that change is perpetual, that everything flows, that the same river is never crossed twice. Parmenides countered with the radical notion that nothing changes, that change is nothing but an illusion, despite the apparent evidence of our senses. The debate was important because it made explaining change central to natural philosophy. While the Milesians and the Pythagoreans do not seem to have considered the question, after Parmenides it was unavoidable: not simply the world, but apparent *flux* in the world is what natural philosophy needs to explain. The Heraclitean-Parmenidean debate also raised fundamental questions about the senses and about how we can know things. In part these questions involved the psychology of perception (e.g., the stick that seems to bend in water, the red of a red apple) and the general reliability of the senses. On another level they dealt with whether and, if so, how knowledge can be based on the senses or indeed on anything at all. The consequence for natural science was that thenceforth not only did every claim to knowledge about nature formally have to be buttressed by its own internal evidence and reasoning, but it had also to be accompanied (either implicitly or explicitly) by a separate rationale as to why *any* evidence or reasoning might support any such claims.

The Hippocratic medical tradition arose in the Hellenic period—the collective body of medical literature credited to the great fifth-century physician Hippocrates of Cos (fl. 425 BCE). In the Hippocratic tradition, with its emphasis on reason, cautious observation, medical prognostication, and natural healing, one finds a good deal of natural knowledge and scientific thinking akin to that pursued by the natural philosophers. For example, articulating a view that remained influential into the nineteenth century of our era, Hippocratic theorists correlated the four elements (earth, air, fire, and water) with four bodily humors (blood, phlegm, yellow bile, and black bile), and then argued that health represents a balance between and among the humors. In this way Hippocratic medicine tied into the Aristotelian elements and his physics and cosmology, but also, as indicated in the previous figure, the Hippocratic tradition extended those notions into anatomy, physiology, and somewhat later human temperaments. Unlike medicine today where diagnosis or identifying the cause of disease is paramount or at least prior to treatment, for Hippocrates and his followers prognosis or predicting the course and outcome of an illness certified a physician's skill and provided the basis for restoring balance to patients. The skepticism of Hippocratic medicine—the doubt that certain knowledge is even possible—set it apart from most of the speculations of natural philosophy. Ancient medicine remained more tied to practice and craft than natural philosophy, and "scientific physicians," such as they were, competed alongside many "schools" and diverse forms of healing arts that included magic, incantations, and dream cures. The famous Hippocratic Oath, where prospective physicians pledge "to do no harm," was one way Hippocratic physicians stood out and one way physicians today do likewise.

Around the Greek world clearly identifiable medical institutions could be found, notably in the temples and cult centers devoted to Asclepius, the deified physician and supposed offspring of Apollo. Asclepieions, or healing centers, appeared in Cos, Epidauros, Athens, and elsewhere. Medical practice was not regulated in antiquity, and doctors were often itinerant. Medicine was a highly specialized trade, and practitioners could become wealthy. City-states contracted with doctors in wartime, but by and large Hippocratic and other doctors operated independently of the political state or any government bureaucracy.

Worlds of Pure Thought

Although early Greek natural philosophers initiated abstract inquiries into nature, there was no unity to their endeavors, and nothing like sustained scientific research is evident in their traditions. That changed in the fourth century BCE with the two great intellectual syntheses of Plato and Aristotle.

Before Plato there was no consensus in Greek cosmology or astronomical theory. Instead, the pre-Socratic tradition was notorious for the diversity of the models proposed. In the sixth century BCE Anaximander of Miletus had hypothesized that the earth is a disk, floating naturally in space with humans living on its flat surface. There are wheels of fire in the heavens and the luminous heavenly bodies we see are really holes in the fire wheels; the wheel of the stars is closest to the earth, the wheel of the sun farthest; eclipses result from holes becoming blocked; and certain mathematical proportions govern the location of the heavenly wheels. This cosmological model is remarkable for just that, being a model—some simplified simulation of the real thing, a likeness we might construct. Anaximander's view is more sophisticated than Egyptian and Mesopotamian cosmologies as well as the succeeding model of Anaximines (who held that the earth is a table held up by air), in that Anaximander can account for what supports the earth, that is, the earth positioned in the middle of nowhere. The model of the Pythagoreans displaced the earth from the center of the cosmos and held that it (and probably the sun) went around some vague central fire and an even more mysterious counter-Earth. The mechanical and vaguely mathematical character of these models made them distinctly Greek inventions, but their advocates did not pursue or elaborate any of the details.

The case of Plato of Athens (428–347 BCE) and his geometrical astronomy carries us past the founding era of the pre-Socratics and lands us solidly in classical fourth-century Greece. Plato was a pupil of Socrates, the fifth-century master who "called philosophy down from the skies." In his youth, Socrates is said to have been interested in natural philosophy, but he concluded that nothing certain was to be learned in the study of nature, and he focused his attentions instead on examining the human experience and the good life. But he offended the political authorities and was sentenced to death. After his execution in 399 BCE, the mantle of philosophy passed to Plato, who seemingly felt better prepared to make direct statements about the natural world. Plato formalized the enterprises of philosophy and natural philosophy by establishing a private school, his Academy at Athens (which survived for 800 years). Significantly, inscribed over the portals of the Academy was the motto, "Let no one enter who is ignorant of geometry."

Geometry was important to Plato and his philosophy as a form of intellectual discipline and as a model for all that was metaphysically abstract and perfect. Geometry was also key to Plato's matter theory, as he identified the fundamental elements of earth, air, fire, water, and an extra aether with the five so-called perfect solids, three-dimensional polyhedra each with identical regular polygonal faces, which geometers had proven could be only five in number. (See figure 4.2.) But Plato himself was a philosopher, not a serious geometer or mathematician. Nor was he an astronomer. He did not observe the heavens, and he disdained those who did. Nevertheless, in his *Timaeus* Plato presents a fairly complex model of the heavens, involving a central earth linked mechanically along a common axis to a series of spinning shells or spheres that carry around the various heavenly bodies. A mystical part of Plato's cosmology and a common philosophical opinion for centuries held that the heavens were alive and divine. Although the cosmology was influential, in most respects it was no advance over the previous models of the pre-Socratics. In one crucial particular, however, Plato exerted a profound and lasting effect on astronomy and the history of science: he set Greek astronomers to solving research problems.

Plato believed that the heavenly bodies revolve in circles around a stationary earth. He held this opinion not because he observed that the sun, moon, planets, fixed stars, and everything in the heavens move in circular arcs across the sky once every 24 hours, which sensory evidence confirms. Nor did his belief that the heavens were essentially unchanging apart from their motion rest only on the reported experience of previous generations. Rather, Plato held his views concerning celestial motion on first principles. Because of their majesty and virtually divine status Plato believed that the heavens represent an embodiment of the eternal, transcendent, and perfect world of pure Form. Plato's world of the Forms constitutes an unchanging ideal reality, of which our changing world is only a pale and imperfect reflection. Therefore, circular motion was the only motion suitable to the heavens because the circle is a figure of constant curvature with no beginning or end. Because they faithfully mirrored the perfection of the world of the Forms, Plato likewise concluded that the heavens must necessarily move uniformly; uniform motion does not speed up or slow down, betraying the imperfection of change, but remains constant and undevi-

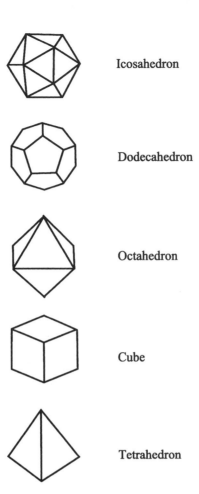

Icosahedron

Dodecahedron

Octahedron

Cube

Tetrahedron

Fig. 4.2. The Platonic solids. Plato knew there cannot be more than these five regular shapes (each with congruent equilateral polygonal faces), and he correlated these geometrical forms with the elements.

ating. Uniform circular motion of heavenly spheres was not questioned thereafter in antiquity.

While most motions in the heavens do seem to be circular, some motions are plainly not circular and equally plainly not uniform. The daily movement of the stars, the annual trip of the sun around the heavens, and the monthly revolution of the moon are apparently circular, but other movements in the heavens are not, notably the movement of the planets or "wandering stars" as observed over a period of months. Relative to the background of the fixed stars, the planets slow in their courses, stop, move backwards, stop again, and move forward again, sweeping out great, noncircular loops in the sky. This was the great problem of the "stations and retrogradations" of the planets that Plato had uppermost in mind when, to use the famous phrase, he enjoined astronomers to "save the phenomena" with circles and uniform perfect circular motion. Explaining the stations and retrogradations of the planets was the central problem in astronomy for nearly 2,000 years from Plato's era until after Copernicus in the sixteenth century CE.

Clearly something new presented itself with Plato's call and with theoretical and observational information systematized to this degree, and what we see is a new kind of scientific research. Planetary motions presented difficulties, Plato believing the planets move in one way (circularly), and observation showing they move in another way (in a loop); there was an obvious conflict to be worked out, an area for research. But note the crucial converse: there is nothing at all problematical about the observed stations and retrogradations unless, like Plato and his followers, one thought the planets ought to move otherwise than they appear, in this case with uniform circular motion. The astronomical paradigm initiated by Plato represents more than the onset of some straightforward "research" into self-evident phenomena; Plato's prior philosophical (theoretical) commitments to Forms and circles made manifest the phenomena to be investigated. Thus Plato defined a problem in natural philosophy where none existed before. But the import of Plato's paradigm in astronomy goes even further: he also defined for theorists and astronomers what constituted appropriate or acceptable solutions to the problem of the planets, that is, models that use uniform circular motion to produce apparently nonuniform appearances. Nothing else qualified as a solution to the puzzle.

Fourth-century astronomers took up the problem and formed a small but distinct tradition of research in astronomy and cosmology. Plato's student Eudoxus of Cnidus (fl. 365 BCE) was the first to respond. He proposed a model for the heavens which consisted of twenty-seven nested (homocentric) celestial spheres variously revolving around a central earth. The Eudoxean model made the universe into something akin to a grand cosmic onion. Some of the spheres were deployed to explain the apparent motion of the stars, sun, and moon, and each retrograding planet was assigned a system of four rotating spheres: one to account for daily motion, one for periodic motion through the heavens, and two, moving oppositely and tracing out a figure-8–like path of stations and retrogradations, known as the "hippopede." The model "worked," but there were problems with it. The observed inequality of the four seasons (they are not all the same length in days) was one, and to account for it a younger contemporary of Eudoxus,

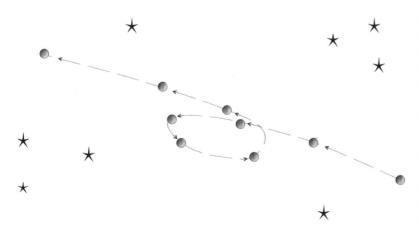

Fig. 4.3. Retrograde motion of Mars. To an observer on Earth, over a period of months Mars appears to reverse its direction as seen against the background of the fixed stars and then reverse it again to resume its forward trajectory. Accounting for these loops in terms of uniform circular motion defined a key problem that occupied astronomers for 2,000 years.

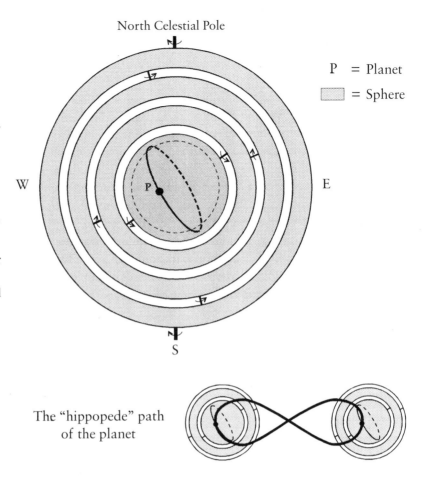

Fig. 4.4. Eudoxus's system of homocentric spheres. In Eudoxus's "onion" system, Earth is at rest in the center of the universe, and each planet is nestled in a separate set of spheres that account for its daily and other periodic motions across the heavens. From the point of view of an observer on Earth, two of the spheres produce the apparent "hippopede" (or figure-eight) motion that resembles the stations and retrogradations of the planets.

North Celestial Pole

P = Planet

= Sphere

W

E

S

The "hippopede" path of the planet

Callipus of Cyzicus (fl. 330 BCE), improved on the model by adding an extra sphere for the sun and raising the number of spheres to thirty-five in all. But the model was still imperfect, notably in not being able to explain how the universe could function mechanically with all those spheres spinning just below and above each other at different rates and inclinations. In the next generation Aristotle (384–322 BCE) tried his hand at this issue in technical astronomy, and, by inserting a number of counteracting spheres, he increased their number to fifty-five or fifty-six.

The Eudoxean model of homocentric spheres and the small research tradition associated with it hardly survived the Hellenic era, much less antiquity. In the final analysis the intellectual and conceptual problems afflicting Eudoxus's approach proved fatal. Those problems included difficulties explaining why the seasons are not the same number of days, why Venus varies in brightness, and why Venus, Mercury, and the sun should always stay close to one another. By the second century BCE astronomers were actively considering alternatives to homocentrism, and the culmination of ancient astronomy in the work of Claudius Ptolemy (fl. 150 CE) 500 years later, as we will see, shows only the vaguest relation to what Plato, Eudoxus, and their colleagues had in mind.

This research tradition was nonetheless notable in several key respects. For one, the case makes evident how much scientific research at this level depends on consensus among scientific practitioners. It makes no sense, in other words, for Eudoxus, Callipus, and Aristotle to have taken up the detailed investigations just described unless they agreed that Plato's approach was basically correct. The instance makes plain once again the community-based nature of the scientific enterprise in its Greek as well as its bureaucratic guise. In some larger sense groups, not individuals, practice science. Finally, like their counterparts among anonymous Babylonian astronomers and astrologers pursuing their particular researches, Eudoxus, Callipus, and Aristotle did not simply know things about nature, they were not simply manipulating nature, and they were not simply theorizing about nature. They were checking nature out in detail, along lines established by their general philosophical, metaphysical, and theoretical commitments. The arsenal of techniques pertinent to the human inquiry into nature had expanded considerably from the first Paleolithic lunar tallies.

Enter Aristotle

Aristotle marked a further watershed in the history of science. His work, which encompassed logic, physics, cosmology, psychology, natural history, anatomy, metaphysics, ethics, and aesthetics, represents both the culmination of the Hellenic Enlightenment and the fountainhead of science and higher learning for the following 2,000 years. Aristotle dominated scientific traditions in late antiquity, in medieval Islam, and in early modern Europe where his science and his worldview defined scientific methodology and the research agenda up to just a few centuries ago.

Born in the town of Stagira in Thrace in northern Greece in 384 BCE, Aristotle came from a privileged family, his father being royal physician to the king of Macedonia. In his teens Aristotle went to Athens to study with Plato, and he remained in Athens as a member of the Academy for 20 years until Plato's death in 347. He then traveled around the Aegean until 343, when King Philip II of Macedonia called him to his court to tutor his son, Alexander, who became Alexander the Great. After Alexander assumed the throne and began his world conquest in 336, Aristotle returned to Athens, where he founded his own school, the Lyceum. After Alexander's early death in 323, Aristotle found it politically prudent to leave Athens; he died the following year at the age of 62. The extensive writings that we commonly regard as Aristotle's were compiled to some extent during his lifetime and to some extent by disciples during the first two centuries after his death. In any event, several entire books have survived, unlike the fragments of the natural philosophers who preceded him. Indeed, Aristotle's commentaries on their work tell us much of what we know of his predecessors.

From a sociological point of view, as with all Hellenic scientists, Aristotle's research was undirected by any state authority, and he had no institutional affiliation. The Lyceum, his own place of teaching—a grove on the outskirts of Athens—was not formally established as a school during his lifetime. He was thus in some measure a footloose professor, one of the leisured intellectuals to whom, in fact, he attributed the achievements of theoretical science. The substance of his studies reflected his sociological standing—utterly abstract and of no possible use in engineering, medicine, or statecraft. Although he recognized the distinction between theoretical and practical knowledge, "speculative philosophers" and "[medical] practitioners," Aristotle confined his research to his private interests in the philosophy of nature. Even when he wrote on anatomy and biology, fields that might have lent themselves to useful studies applicable to the treatment of illness, he focused his objectives on the place of living beings in a rational cosmology. Similarly, his studies of the theory of motion, which remained influential until the seventeenth century, formed part of a program of purely theoretical research and were of no practical use in technical or economic applications.

Aristotle expressed himself in unambiguous terms concerning the relationship between science and technology. After humanity acquired the needed practical arts, leisured intellectuals cultivated pure science: "When everything [practical] had been already provided, those sciences were discovered which deal neither with the necessities nor with the enjoyment of life, and this took place earliest in regions where men had leisure." Curiosity provided the motivation for the development of pure science: "For men were first led to study [natural] philosophy, as indeed they are today, by wonder. . . . Thus, if they took to philosophy to escape ignorance, it is patent that they were pursuing science for the sake of knowledge itself, and not for utilitarian applications." Aristotle's opinions are thus consistent with our studies that show the ratio of theoretical to practical orientations among known Hellenic scientists to have been roughly 4 to 1.

For the generations of natural philosophers who followed Aristotle the beauty and power of his achievement stemmed in large measure from the unity and universality of his worldview. He offered a comprehensive, coherent, and intellectually satisfying vision of the natural world and humanity's place in it, a vision that remains unequaled in scope and explanatory ambition.

Aristotle's physics, and indeed all of Aristotle's natural philosophy, is rightly said to represent the science of common sense. Unlike Plato's transcendentalism, Aristotle held that sensation and observation are valid—indeed, they represent the only route to knowledge. Time

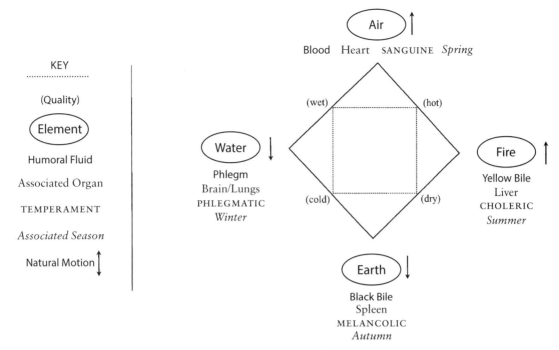

Air

Blood Heart SANGUINE *Spring*

(wet) (hot)

Water

Phlegm
Brain/Lungs
PHLEGMATIC
Winter

(cold) (dry)

Fire

Yellow Bile
Liver
CHOLERIC
Summer

Earth

Black Bile
Spleen
MELANCOLIC
Autumn

KEY

(Quality)

Element

Humoral Fluid

Associated Organ

TEMPERAMENT

Associated Season

Natural Motion

and again Aristotle's views conform with everyday observation and the commonplace experiences of the world we know (unlike modern science, which often contradicts plain observation and requires a re-education of the senses before it can be accepted). Aristotle emphasized the sensible *qualities* of things, in opposition to the quantitative and transcendental approaches of the Pythagoreans or Plato's followers. Aristotle's natural philosophy was therefore more commonsensical and scientifically promising.

Aristotle's theory of matter provides an easy entrée to his overall vision of the cosmos. He followed Empedocles and Plato in adhering to the four elements of earth, air, fire, and water. But unlike Plato, who believed the elements to be fashioned of abstract polyhedrons, Aristotle took them to be composed of pairs of even more fundamental qualities: hot, cold, wet, and dry, projected onto a theoretically quality-less "first matter" or *prima materia*. Thus, as figure 4.5 illustrates, the qualities wet and cold make up the element water, hot and dry = fire, wet and hot = air, cold and dry = earth. Ordinary earth and all other composite bodies are mixtures of the pure elements, which are never found in an isolated state. And, again unlike Plato who found reality only in the transcendent world of the Forms, Aristotle held that the world we experience is materially real because objects in the world (such as tables and trees) are inseparable amalgamations of elemental matter and Form. Aristotle's matter theory is eminently rational and conformable to experience in, for example, explaining the boiling of water as a transformation of water into "air" by the substitution of the quality of hot for the quality of cold. In this case the application

Fig. 4.5. The Aristotelian elements. In Aristotle's matter theory, pairs of qualities (hot-cold, wet-dry) define each of the four elements: earth, air, fire, and water. Substitute one quality for another, and the element changes. Each element has a natural tendency to move either toward or away from the center of the cosmos. Associated with each element is an elaborate set of humors, temperaments, and seasonal characteristics that connected matter theory not only to what we would call chemistry and physics but also to physiology and medicine and medical practice in the Western tradition for 2,000 years.

of fire replaces the cold and wet of water for the hot and wet of air. It should be noted that such a qualitative theory of the elements provides a theoretical basis for alchemy, in that qualities are projected onto a quality-less *prima materia* or "first matter" and it thus becomes theoretically possible to strip, say, lead of its qualities and substitute the qualities of gold. The theory as much as the authority of Aristotle thus legitimated the alchemical enterprise.

For Aristotle, the physics of motion—change of place—is only a special case of change or alteration in general, such as growth, fermentation, and decay. He associated a motion with each element according to its nature: earth and water, being heavy, naturally move to the center of the universe (that is, the earth); air and fire, being light, naturally move away from the center. Nothing else is required to explain this intrinsic motion, just as nothing else is required by modern physics to explain inertial motion. Accordingly, each element seeks a place in the universe, its so-called natural place: earth at the center layered with concentric shells of water, air, and fire. Thus, his theoretical analysis accords well with what we observe in nature, with lakes and oceans atop the earth, with bubbles of air rising in water, with the atmosphere atop the waters and the earth, and with fire seeming to rise in the air and meteors to spark in the sky. Indeed, theoretically, the reason concentric shells of earth, water, air, and fire that surround the cosmic center are not perfectly spherical is that the terrestrial region represents the realm of change, violence, imperfection, and corruption. On Earth things get jumbled up, unlike the perfect, unchanging, and incorruptible celestial regions. In support of these conjectures Aristotle alluded to experimental confirmation. If one attempts to submerge a bag or bladder of air in water one will sense resistance against displacing air from its natural place into the realm of water, and if the balloon is forcibly submerged and then released it will spontaneously return to the realm of air.

In Aristotle's scheme of the world, the earth we live on is essentially spherical and totally motionless at the center of the universe. If, in some extraordinary thought experiment, we could displace the earth from the center, it would naturally return to the center and reassemble there, just as stones fall back to the center through air and through water in order to return to their natural place. Thus, Aristotle's geocentric cosmology—the idea that the spherical earth remains unmoving at the center of the cosmos—is backed up by the authority of physics and confirms our sensory experience of the earth at rest and the heavens in motion. Aristotle confirmed the sphericity of the earth, for example, from the shadow it casts on the moon during lunar eclipses; and he offered common-sense arguments against a moving earth, such as the observation that a ball thrown straight up falls back to its point of origin and is not left behind as the earth turns underneath.

Since different natural motions occur in the region of the earth (ei-

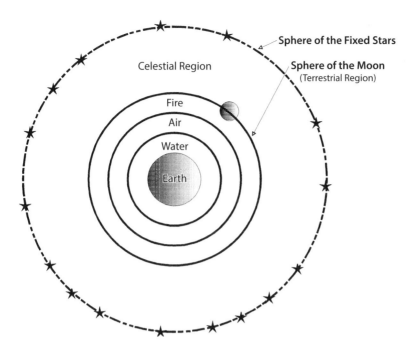

Fig. 4.6. The Aristotelian cosmos. According to Aristotle, each of the four elements has a "natural place" in the universe. In the terrestrial region (to the height of the moon), earth and water move "naturally" in straight lines downward toward the center of the cosmos (Earth), while air and fire "naturally" move in straight lines upward and away from the center. The sphere of the moon separates the terrestrial region, with its four elements (including fiery meteors and comets), from the celestial region—the realm of the fifth, "aetherial" element whose natural motion is circular. The stars and planets reside in the celestial region and take their circular motions from the aetherial spheres in which they are embedded.

ther up or down) and the region of the heavens (always circular), Aristotle's cosmology makes a sharp distinction between the physics of the two regions. When terrestrial objects move naturally, that is, without the motion being started or sustained by a living or external mover, they move either up or down, toward or away from the center of the earth, depending on whether they are heavy or light. The terrestrial or sublunary realm is defined as the world below the orbit of the moon, wherein the four elements seek their natural place. The heavens above the moon are the celestial realm of a fifth element—the quintessence, Aristotle's aether. This element never combines with the other elements and, unlike them, is incorruptible and exists only in the pure state, separately in its own realm in the heavens. Aristotle associated a natural motion with the aether, too, not straight-line motion toward or away from the center, but perfect circles around the center. This seemingly metaphysical doctrine of the perfection of the celestial region is also based on naturalistic observations—heavenly objects appear in fact to be spherical and seem (at least in daily motion) to move in perfect circles around the earth. The enduring and unchanging face of the heavens that we observe from our world of flux and change is related to the unchangeable character of the aether. This dual physics, with separate laws of motion for terrestrial and celestial realms, was consistent with everyday experience and observation and remained intact until the seventeenth century when it was superseded by Newton's laws of motion and universal attraction which postulated a single physics for the whole cosmos.

In addition to the naturally occurring up or down motion of bodies composed of earth, water, fire, and air, nonspontaneous motion observed in the world around us, such as the flight of an arrow, requires explanation. Aristotle envisioned all such motion as forced or violent (as against natural) motion. He proclaimed that such motion always requires an external mover, someone or something to apply an outside force of some sort to cause the motion in question. Moreover, the mover must be in constant contact with the object. In the vast majority of instances Aristotelian movers can be easily identified and the principle apparently confirmed: the horse pulls the cart, the wind blows the sail, and the hand guides the pen. But paradoxical counterexamples exist: the arrow or the javelin in flight after it has lost contact with its mover. Where is the mover in those cases? (Aristotle himself said the medium somehow does the pushing.) In addition, for Aristotle the apparently unmoved motion of animals or plants derives from the faculties of their souls—the animal or vegetable (or, in the case of human beings, the rational) souls they possess.

Except for the puzzling case of projectile motion Aristotle's theory appears to be consistent with at least casual observations of the physical world. Aristotle went beyond these general principles and postulated quantitative relationships between and among force, velocity, and resistance. His results were not self-evidently implausible. He gave the example of a boat dragged across a beach. Clearly, the boat will not move by itself; an external motive force is required. That force has to be sufficient to overcome the resistance of the friction between boat and sand before any motion can occur; and the speed with which the boat moves thereafter depends on how much force is applied beyond that minimum. The harder the haulers haul, the faster the boat will go; the greater the friction, the slower it will go. In the case of a falling body, the motive force is proportional to the *weight* of the body, so it follows that heavy bodies will fall downwards faster than light bodies (the more earthy matter a body has, the heavier it is, and the more easily it "divides the air" to descend to its natural place). This notion follows from Aristotle's principles and harmonizes with what we observe. For example, a heavy book falls faster than a light sheet of paper. Similarly, the same object falls more slowly in water than in air, and still slower in honey or in molten lead, where it may even float. In these and many other ways, Aristotle's notions hold for what we observe and experience. Aristotle's science is largely compatible with everyday common sense and what people then and now actually observe and for which empirical evidence through experiment or otherwise is easily obtainable. One can easily understand why his philosophy of nature prevailed for so long.

Another historically significant principle follows from Aristotle's law of motion, the idea that motion must take place in a medium of some appreciable density. In other words, motion in a vacuum is impossible. Motion in a vacuum implies motion without resistance; but if resistance tends toward zero, the velocity of a moving body becomes infinitely large, which implies that a body can move with infinite speed and can thus be at multiple places at the same time, an apparent absurdity completely inconsistent with all experience. A corollary of Aristotle's rejection of the vacuum was repudiation of atomism, denying the doctrine of empty space through which atoms supposedly moved. For Aristotle space must be completely filled. The power and comprehensiveness of Aristotle's conception of motion overcame the objections that were intermittently leveled against it. It would ultimately take a profound scientific revolution to overthrow Aristotelian views on motion in a medium and to replace them with an alternative doctrine. For two millennia Aristotle's views concerning the stuff of the world, the concept of place, cosmology, and the principles and observed effects of motion made great sense and were accordingly held and shared by those who studied natural philosophy in the Greek tradition.

It would be a mistake to overemphasize the physical sciences in analyzing Aristotle's thought, even though they were fundamental to his worldview. Aristotle was tremendously influential and highly skilled as an observational—one can almost say experimental—biologist and taxonomist. (We must remember, however, that the word *biology* did not come into existence until the nineteenth century of our era.) He conducted empirical investigations, carefully observing the development of the chick embryo, for example. Something like a third of his writings concern matters biological. Crucially, the model that Aristotle used to explain the all-important issue of change derives not from physics but from biology. The growth and development of living things provided a model of change for Aristotle with change embodying a process of becoming, of coming-into-being, the "actualization of that which is potential" in things, as in the classic example of the potential oak in the actual acorn. Growth or change merely brings out features that already exist potentially, thus avoiding the Parmenidian paradox of creating something from nothing. Furthermore, for Aristotle, the passing away of one form involves the coming-to-be of another, and therefore the cosmos must be eternal, with cycles of time repeating themselves *ad infinitum*.

In the details of his dealings with living things Aristotle became the pioneer of systematic taxonomy. He ranked life into a grand hierarchy, classifying animals into "bloodless" invertebrates and vertebrates with blood. His identification of three types of "soul" (nutritive, sensitive, and rational), corresponding to vegetable, animal, and the higher cognitive functions of humans, provided a link to anatomy and physiology, or considerations of how the body operates. Aristotle endorsed the concept of spontaneous generation, and he conceived reproduction as males contributing "form" and females only "matter" to the offspring. Over the ages, Aristotle proved as influential in the life sciences as he did in the physical sciences; in particular the later Greco-Roman physician and likewise influential theorist Galen (130–200 CE) began his work within the basic frame of reference that Aristotle had set down. Theophrastus of Eresus (371–286 BCE), Aristotle's successor as head of the Lyceum in Athens, extended the range of the master's investigations into botany in work that remained a standard source until the eighteenth century.

Aristotle was not a dogmatic philosopher and his word was not taken as gospel. Rather, while his basic tenets were retained, his work provided a springboard for scientific research and for traditions of inquiry that unfolded over the succeeding centuries. Theophrastus directed a trenchant criticism at Aristotle's doctrine of fire as one of the elements. With regard to local motion, by highlighting the phenomenon of acceleration, Strato of Lampsacus, successor to Theophrastus at the Lyceum from 286 to 268 BCE, criticized Aristotle's lack of attention to the speeding up and slowing down of bodies as they begin or end their motion. The Byzantine natural philosopher John Philoponus later added to this ongoing debate over Aristotle's theories of motion, and thinkers in the European Middle Ages intensified the controversies and eventually produced radical revisions of Aristotle's doctrines. This critical tradition evolved over a period of 2,000 years.

Aristotle's writings provided the basis of higher learning in the cultures of late antiquity, Islam, and the European Middle Ages. His cosmos remained at root theological, and, like Plato, he held the heavens to be animate and divine, kept in motion by the Unmoved, or Prime Mover. To this extent Aristotle's philosophy could be harmonized with the theologies of Judaism, Christianity, and Islam, and ultimately theologians of all three faiths bent their efforts to squaring their religious doctrines with Aristotle's teachings. By the same token, many Byzantine, Muslim, and Christian scientists found their inspiration in attempts to understand nature—what they believed to be God's handiwork. With its notions of hierarchy and chains of being, much else in Aristotle resonated with later Christianity and the political interests of ruling political authorities, a circumstance that doubtless also helped insure the long-term success of his natural philosophy.

The intellectual legacy of Aristotle's studies shaped the history of scientific thought in the civilizations that inherited Greek learning. The clarity of his analyses and the cosmological comprehensiveness of his views set the standard for the scientific cultures following the Hellenic Enlightenment. The coincidence that Aristotle and his pupil Alexander the Great died within a year of one another (322 and 323 BCE, respectively) seems emblematic, for in their ways they both transformed the contemporary world. The world that immediately followed their deaths was far different—scientifically and politically—than the one they had inhabited.

Alexandria and After

Ancient Greek civilization developed in two phases. In the first, the *Hellenic*, city-states arose in Ionia and on the Greek peninsula. They were semi-prosperous, each supported by an agricultural hinterland (and, generally, by imported food), and they remained independent—there was no overarching Greek king. Then, in the fourth century BCE, a second phase, the *Hellenistic*, took shape, marked successively by confederation, imperialism, and conquest. The result was a vast expansion of Greek culture and learning.

In Macedonia, the northern district of Greece, a local king, Philip II, gathered his forces, which included horse-mounted infantry and rock-throwing artillery, and began the unification of the Greek peninsula. When Philip was assassinated in 336 his son, Alexander, who became known to his contemporaries as "the Triumphant" and to us as "the Great," continued Philip's expansionist course and forged the most extensive empire in the ancient world. At its height it reached out from Old Greece and encompassed the great river valley civilization in Egypt, through the Mesopotamian heartland of the first civilizations on the flood plain of the Tigris and Euphrates Rivers, and on to the Indus River Valley in the east. The empire of Alexander the Great lasted only 11 years, from 334 when he defeated the Persians until his early death at the age of 33 in 323 BCE. After Alexander's death India returned to Indian control, and the empire collapsed into three kingdoms: Macedonia (including the Greek peninsula), Egypt, and the Seleucid Empire in Mesopotamia (see inset, map 4.1). They were eleven years that changed the world.

The onset of the Hellenistic marks a break in the historical chronology of ancient science. Hellenic natural philosophy, with its unpatronized individualists, gave way to the more cosmopolitan world of the Hellenistic—the Golden Age of Greek science—and a new mode of organization and social support of research. Hellenistic science represents the historical melding or hybridization of the tradition of

Hellenic natural philosophy with patterns of state-supported science (our Babylonian model) that had originated in the eastern kingdoms. Kings and emperors had patronized a bureaucratic science that leaned toward useful applications; Hellenic science was the work of solitary thinkers who immersed themselves in abstract thought. Hellenistic science in the lands of the ancient Near East combined those disparate traditions. State support and patronage for scientific theory and abstract learning were novelties of Hellenistic culture and remained part of the historical pattern of science in all subsequent societies that inherited the Greek tradition.

The roots of the new scientific culture were planted in Egypt, now governed by a Greek ruling class that promptly established itself as an Egyptian dynasty—Ptolemaic Egypt. The first Greek king of Egypt, Ptolemaios Soter, began the tradition of royal patronage of science and learning, and it fell to his successor, Ptolemaios Philadelphus, to found the famous Museum at Alexandria, a new town built as a port on the Mediterranean shore of the Nile delta during Alexander's lifetime. With varying degrees of official support and patronage the Museum existed for 700 years—into the fifth century CE, an existence roughly as long as the world's oldest universities today. The distinctive character of Hellenistic and Greco-Roman science derives at least in part from this institutionalization of pure science and natural philosophy.

In essence, the Museum at Alexandria was a research institution—an ancient Institute for Advanced Study. Unlike its present namesake, the Museum did not display collections of objects (a function museums acquired only in the European Renaissance). It was, instead, a temple dedicated to the nine mythical Muses of culture—including Clio, the muse of history, and Urania, the muse of astronomy. There, subsidized members combining Hellenic and Babylonian traditions conducted their own research fully supported by state resources. In the royal precinct, the Ptolemies and their successors maintained splendid quarters for the Museum and its staff which included rooms, lecture halls, dissection studios, gardens, a zoo, an observatory, and possibly other facilities for research. The Ptolemies added a glorious library that soon contained 500,000 or more papyrus scrolls. Patronage always proved fickle in antiquity, depending on the largesse of individual kings and emperors, but at any one time the Museum harbored upwards of 100 scientists and literary scholars who received stipends from the state and meals from the Museum's kitchen while being allowed a Hellenic-style freedom of inquiry, excused even from the obligation to teach. No wonder the record indicates that stipendiaries were the targets of envious attacks as "rare birds" fed in gilded cages—such is the cultural ambiguity provoked by state support for pure learning. The later Roman emperors of Egypt kept up this extraordinary tradition of state support no less than their Hellenistic predecessors, making Alexandria the most significant center of science in the Hellenistic and Greco-Roman eras.

The motives of the Ptolemies and other Hellenistic and Greco-Roman patrons of science and learning are not clear. Doubtless they sought practical returns, and institutional pressure was at least indirectly brought to bear on the scientists at the Museum to bend their research toward useful applications. The fact that the Museum supported anatomical-medical research lends support to that conjecture. Similarly the zoo sheltered the war elephants of the king, the Library collected books on government and contemporary "political science," and academicians pursued geography and cartography. Applied military research may also have taken place at the Museum. Data suggest that Hellenistic scientists were somewhat more practically oriented than their earlier Hellenic counterparts. Beyond any immediate utility, however, it would seem that fame, glory, and prestige accruing to patrons were major motives for supporting the rare birds who roosted in the Museum. Whether the Ptolemies or their Roman successors got their money's worth depends on one's assessment of the relative values of the abstract and practical products of research.

The Hellenistic pattern of support for learning was not limited to Alexandria, and several cities in late antiquity came to boast of museums and libraries, including the library at Pergamum, a city that rivaled Alexandria as a center of state-supported science and scholarship. At Athens the institutional status of Plato's Academy and Aristotle's Lyceum is revealing in this regard. These schools, too, acquired a Hellenistic dimension. We saw that they began in the Hellenic era as informal, entirely private associations of masters and students devoted to the study of their founders' thought. They received legal status, notably as religious associations, but got no public support at the outset, remaining self-supporting as schools and communities of scholars. The formal institutional character of the Academy and the Lyceum became strengthened when, in the Alexandrian mode, the Roman emperors Antoninus Pius and Marcus Aurelius endowed imperial chairs in Athens and elsewhere in the second century CE. The Lyceum in Athens and the Museum at Alexandria also shared contacts and personnel. The Lyceum continued to be active at least until the end of the second century CE, and the Academy survived into the sixth century, nearly a thousand years after its founding. Still, the Lyceum and the Academy in Athens were primarily schools with teaching the key activity; research itself remained incidental, unlike the extraordinary case of the Alexandrian Museum where scholars received support for unfettered research.

Although literary and language studies predominated at Alexandria, a historically unparalleled flourish of scientific activity also occurred there, especially during the first century of the Museum's existence, the third century BCE. A tradition of abstract, formal mathematics is the greatest and most enduring Alexandrian accomplishment. As exemplified by Euclid's geometry, Hellenistic mathematics was exceedingly formal and nonarithmetical, qualities that placed it far from the needs of artisans but squarely at the fountainhead of subsequent mathematical research. Euclid had probably studied at the Academy in Athens before he moved to Alexandria under the patronage of the Ptolemies. Apollonius of Perga (fl. 220–190 BCE) did most of his work there, too; he was known for his mastery of the conic sections (which found its first application in Johannes Kepler's astronomical theories 1,800 years later). To this tradition belongs Archimedes of Syracuse (287–212 BCE), probably the greatest mathematical genius of antiquity. Archimedes lived and died in Syracuse in Italy, but he traveled to Alexandria at one point and corresponded with the head of the Library, Eratosthenes of Cyrene (fl. 225 BCE). Eratosthenes, himself a multifaceted man of science, performed a famous observation and calculation to determine the circumference of the earth, and persons educated in the Greek tradition did not believe the earth to be flat; Eratosthenes also inaugurated notable work in geography and cartography. The latter fields of research continued in Alexandria down through the astronomer Ptolemy 400 years later. Innovative anatomical research also took place at the Museum, seen notably in the work of Herophilus of Chalcedon (fl. 270 BCE) and Erasistratus of Chios (fl. 260 BCE). Alexandrian anatomists evidently conducted human dissections and possibly vivisections as well. Other Alexandrian scientists undertook substantial research in astronomy, optics, harmonics, acoustics, and mechanics.

In astronomy, the Eudoxean model of geocentric spheres was challenged early in the Hellenistic period. The reader will recall the research tradition that stemmed from Plato's legendary injunction to "save the phenomena"—particularly the problem of the stations and retrogradations of the planets—and Eudoxus's geocentric solution in terms of his onion-skin universe and its rotating and counter-rotating spheres. But the model of nested homocentric spheres, even as refined by Aristotle, faced serious difficulties, notably in accurately reproducing the retrograde motions of planets. And the unequal lengths of the seasons, difficult to explain if the sun moves uniformly at a constant distance from a central earth, was another technical problem undermining the Eudoxean approach. Already in the fourth century—the century of Plato and Aristotle—Heraclides of Pontus (fl. 330 BCE) suggested that the apparent daily circling of the heavens could be accounted for by assuming that the heavens remained stationary as the earth spun on its axis once a day. The suggestion was generally considered implausible since it seemingly contradicted the direct sensory evidence that the earth is stationary.

Astronomical theory and cosmology posed questions that continued to excite the curiosity of many natural philosophers over subsequent centuries. One of those was Aristarchus of Samos (310–230 BCE), an expert astronomer and mathematician and, it seems, an associate of the Museum. According to Archimedes, Aristarchus espoused a heliocentric, or sun-centered, cosmology, not unlike the system proposed by Copernicus nearly 2,000 years later. He placed the sun at the center and attributed two motions to the earth: a daily rotation on its axis (to account for the apparent daily circuit of the heavens) and an annual revolution around the sun (to account for the apparent path of the sun around the zodiac).

Aristarchus's heliocentrism was known but overwhelmingly rejected in antiquity, not for some anti-intellectual bias but rather for its essential implausibility. The heliocentric theory, which in its essentials we hold today, faced so many scientific objections at the time that only a zealot would subscribe to it. If the earth whirled on its axis and raced around the sun, surely everything not nailed down would go flying off the earth or be left behind in a wake of debris, a conclusion contradicted by the sensible evidence of birds flying with equal ease in all directions and bodies projected directly upwards returning to where they began. In addition, the motion of the earth postulated by Aristarchus's heliocentrism plainly violated Aristotle's physics of natural motion. Earthy and watery things that make up the earth naturally tend to the center of the cosmos—to require the earth to spin like celestial matter or move otherwise through space is to ask it to undertake motions that Aristotle and all of science declared impossible. If the earth was displaced from the center, its parts would simply return and reorder themselves there. Rational scientists would never accept a theory that flew in the face of everyday observations and that violated long-held doctrines that formed the basis of ongoing, productive research. Today, we also become suspicious of people who propose ideas that violate the laws of physics.

A highly technical but scientifically more telling point also counted strongly against Aristarchus and his sun-centered theory, the problem of stellar parallax. To state the problem simply, if the earth orbits the sun, then the relative position of the stars ought to change over the course of six months as the earthbound observer viewed the heavens from widely different positions. But no such change was observed, at least not until the nineteenth century. (To observe parallax the reader might hold a finger in front of his or her nose and watch it "move" as the left and right eyes are alternately opened and closed.) Archimedes gives us Aristarchus's response to the difficulty: Aristarchus compared the size of the earth's orbit to a grain of sand, meaning that the diameter of the earth's orbit around the sun is so small in relation to the distance to the fixed stars that the change of stellar position would be too small to be observed. This was an ingenious answer for why stellar parallax cannot be observed (the same answer, incidentally, that

Copernicus later gave), but Aristarchus then faced the further objection that the size of the universe had to be expanded to extraordinary, (then) unbelievable proportions in order for heliocentrism to hold. The scientific problems facing the heliocentric hypothesis were formidable, and ancient astronomers stood on strong ground in repudiating it. Religious objections also arose against setting the corrupt and changeable earth in the divine and incorruptible heavens. Not surprisingly, Aristarchus was threatened with charges of impiety.

The difficult problem of accounting for planetary motions resulted in alternatives to the astronomies of Eudoxus and Aristarchus. Apollonius of Perga, the Alexandrian scientist mentioned above in connection with the conic sections, helped build an alternative means of "saving the phenomena" and preserving geocentrism. He developed two powerful mathematical tools that astronomers used to model observed motion in the heavens: *epicycles* and *eccentrics*. (See figure 5.1.) The epicycle model had planets orbiting on small circles that in turn moved on larger circles; the eccentric is simply an off-centered circle. Both the retrograde motion of the planets and the variable lengths of the seasons could be easily and accurately modeled using epicycles. By assigning different sizes, speeds, and directions to these circles, Hellenistic astronomers developed increasingly accurate models for heavenly motion.

Ancient astronomy culminated in the work of Claudius Ptolemy in the second century CE. Ptolemy lived and worked under Roman governance in Alexandria. Building on his predecessors' use of epicycles and eccentrics Ptolemy composed a massive and highly technical manual of astronomy, the *Mathematical Syntaxis*, the celebrated *Almagest* (so named by later Muslim scholars). The Almagest is premised upon geocentrism and circular motion in the heavens and is extremely mathematical and geometrical in its approach. To his arsenal of epicycles and eccentrics, Ptolemy added a third tool, the so-called *equant* point, necessitated by the still-elusive harmony between planetary theory and observation. An observer at the equant point would see the planet move with uniform circular motion, while it was in fact moving at a changing rate with respect to the earth. Ptolemy's equant violated the spirit, if not the letter, of Plato's injunction to "save the phenomena" using uniform circular motion, but the objection was abstruse even to astronomers and in no way undermined commitments to geocentrism. The equant proved a handy tool, and Ptolemy deployed it and other improvisations to create elaborate, if wholly abstract, mathematical constructions, celestial "Ferris Wheels" whose stately turnings charted the eternal and unchanging heavens. In theory a "Ptolemaic" system with appropriate epicycles, eccentrics, and equants can be designed today to match the accuracy of *any* observed orbit. Ptolemy's *Almagest* was a major scientific achievement. For 1,500 years it remained the bible of every astronomer whose work derived from Hellenistic sources.

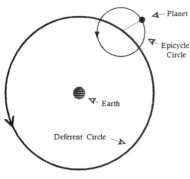

The Epicycle Model

Fig. 5.1. Ptolemy's astronomical devices. To reconcile observed planetary positions with the doctrine of uniform circular motion Ptolemy employed epicycles, eccentric circles, and equant points. The epicycle model involves placing circles on circles; eccentrics are off-centered circles; the equant point is an imaginary point in space from which uniform circular motion is measured.

The Eccentric Model

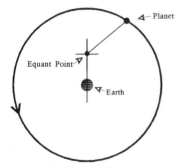

The Equant Model

Ptolemy also contributed to a Greek tradition of geometrical optics, notably incorporating experimental data into his study of refraction—the bending of light in different media. And his work in geography and cartography similarly proved influential. But one should not put too modern a spin on Ptolemy. For him, mathematical science was a form of philosophy and essentially an ethical and spiritual enterprise. He believed the heavens to be divine and, indeed, animate. For Ptolemy, the movement of the heavens self-evidently affected the sublunary world (through the tides or the seasons, for example). Thus, although Ptolemy distinguished between astrology and astronomy, he recognized the legitimacy of astrology and the effort to predict the future. In fact, he wrote a large and influential book on astrology, the *Tetrabiblos*,

Fig. 5.2. Ptolemy's model for Mercury. Ptolemy deployed epicycles, eccentrics, and equants in elaborate and often confusing combinations to solve problems of planetary motion. In the case of Mercury (pictured here) the planet is situated on an epicycle circle; the center of that circle revolves on a larger eccentric circle, the center of which moves in the opposite direction on an epicycle circle of its own. The required uniformity of the planet's motion is measured by the angle (α) swept out unvaryingly by a line joining the equant point and the center of the planet's epicycle circle. These techniques can be made to account for any observed trajectories. The intricate sets of solutions Ptolemy and his successors produced constituted gigantic "Ferris wheel" mechanisms moving the heavens.

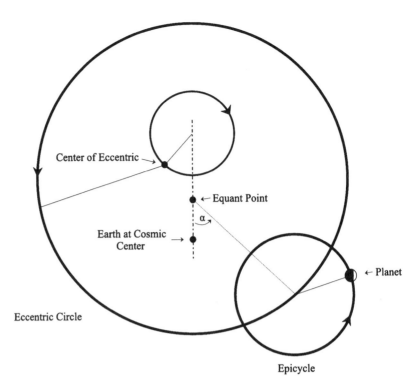

Center of Eccentric →

← Equant Point

α

Earth at Cosmic Center →

← Planet

Eccentric Circle

Epicycle

and, not least of his many accomplishments, he may fairly be said to have been the greatest astrologer of antiquity.

An upsurge of alchemy matched the strength of contemporary astrology. What came to be the foundational texts of a semisecret tradition were compiled in Hellenistic Alexandria and elsewhere. The tradition is labeled "Hermetic" because these compilations were attributed to its mythical founder, Hermes Trismegistus, a legendary Egyptian priest thought to be living around the time of Moses. This body of mystical work contained esoteric and supposedly divinely inspired doctrines pertaining to the secret workings of the universe. Although the idea and practice that base metals can be transmuted into gold and silver doubtless involved some amount of fraud in antiquity, the roots of alchemy lay in demonstrated metallurgical practice, and alchemical science, so to speak, evolved out of Bronze and Iron Age technologies involving metals. Alchemy offered the promise of utility, and in that sense it represents another early practical science, especially to the extent that rulers patronized it. But for serious practitioners, the alchemical quest for elixirs of immortality or the philosopher's stone that would transmute metals always entailed a spiritual dimension, wherein the alchemist sought to purify himself as much as he hoped to purify base metals. Ancient and medieval alchemy should not be thought of as pseudochemistry. Rather, alchemy needs to be understood on its own as a technically—and theoretically—based practical science that combined substantial mystical and spiritual elements.

The impact of alchemy was small, and on the whole Hellenistic science at Alexandria and elsewhere in the ancient world was not applied to technology or, in general, pursued for utilitarian ends. Natural philosophy remained largely insular, as it had been previously in the Hellenic. It remained isolated, not in any direct way connected or applied to the productive economy or the predominant practical problems of the age. In addition, the ideology stemming from Plato and the pre-Socratics that held manual labor in contempt and that repudiated any practical or economic utility to science continued in the Hellenistic. Ideology thus reinforced the existing separation of *theoria* and *praxis*.

Mechanics itself, however, was subject to scientific analysis by Hellenistic scientists in theoretical treatises on mechanics and the mechanical arts. Archimedes, above all, mastered the mechanical principles of the simple machines: the lever, wedge, screw, pulley, and windlass; and in analyzing the balance (including the hydrostatic balance), the ancients articulated a theoretical and mathematical science of weight. The practical possibilities of this mechanical tradition are evident in the work of Ctesibius of Alexandria (fl. 270 BCE), Philo of Byzantium (fl. 200 BCE), and Hero of Alexandria (fl. 60 BCE). Based on their knowledge of weight and pneumatics, these men designed ingenious mechanical devices—in the category of "wondrous machines" that could automatically open temple doors or pour libations, but whose purpose was to provoke awe and wonder, and not to contribute to economic progress. Hero, for example, contrived to spin a ball using

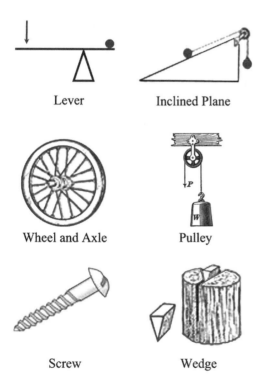

Lever

Inclined Plane

Wheel and Axle

Pulley

Screw

Wedge

Fig. 5.3. The simple machines. The ancient world depended on these devices to lift, move, and handle materials, and ancient scientists, notably Archimedes (287–212 BCE) and Hero of Alexandria (10–75 CE), understood the scientific principles behind the mechanical advantage the simple machines conferred in leveraging a force.

fire and steam, but no one in antiquity conceived or followed up with a practical steam engine. In a word, the study of mechanics in Alexandria was, like its kindred sciences, almost completely detached from the wider world of technology in antiquity.

But not completely. The Archimedean screw, for example, was a machine that lifted water; it was invented in the third century BCE purportedly by Archimedes, and it derived from this tradition of scientific mechanics. Archimedes, who died in 212 BCE during the defense of his native Syracuse against the Romans, became legendary for his technological wizardry with siege engines and war machinery. His published work remained abstract and philosophical, but even discounting as legend much that is credited to him, Archimedes probably did apply himself to engineering technology and practical achievement. He supposedly used his knowledge of mechanics in wartime, and in this capacity he acted as an ancient engineer (*architecton*), one of whose domains was military engineering.

The case of the ancient torsion-spring catapult is revealing. Weapons development was nothing new in antiquity, and, indeed, something like a technological arms race took place among craftsmen and sponsoring patrons to build the largest rowed warship. Philip II of Macedon and Greek kings in Syracuse, Rhodes, and elsewhere supported programs to develop and improve the catapult and a variety of ballistic machines. Sophisticated engineering research in the form of systematic tests took place at Alexandria to identify variables affecting the functioning of catapults and to create the most effective and efficient machines. The government sponsored this research, and scientists at Alexandria carried out some of it. Although the mechanical tradition at Alexandria was less otherworldly than once thought, one must qualify the ways in which catapult research represents applied science in antiquity. Overall, the tests seem to have been entirely empirical, that is, executed by scientist-engineers perhaps, but without the application of any scientific theory or the exploitation of theoretical knowledge. After decades of patient effort and recordkeeping the scientist-engineers at Alexandria created a practical and mathematically exact "catapult formula" that involved extracting cube roots and that specified the optimal dimensions for any ballistic machine and its projectile. With this formula Archimedes himself purportedly built the largest stone-throwing catapult on record. But the formula is simply a rule of thumb expressed in mathematical terms. Development of the catapult is better thought of as applied engineering research.

Some scientific instruments existed in antiquity—notably finely crafted astronomical clockwork and other observing devices, where science and technology combined in the service, not of warfare or the larger economy, but of the scientific enterprise itself. As interesting and historically revealing as all these examples are, they do not belie the general point that ancient science on the whole had very little practical

Torsion
Spring

Rigid Bow Arm

Universal Joint

Pedestal

Slider

Claw and Trigger

Circular
Ratchet

Winch

Stock

Fig. 5.4. The torsion-spring catapult. The Greeks under Philip of Macedon had begun to use ballistic artillery in the form of machines that could hurl large projectiles at an enemy. In some designs the action was produced by twisting and suddenly releasing bundles of elastic material. This large Roman model fired stones weighing 70 pounds. Hellenistic scientist-engineers experimented to improve the devices.

import, was not as a rule directed to practical ends, and had no significant impact on ancient engineering as a whole.

Technology in antiquity needs to be seen as a domain separate from ancient science, a robust world of farming, weaving, potting, building, transporting, healing, governing, and like myriads of crafts and techniques great and small that made up and maintained Hellenistic and Greco-Roman civilization. Hundreds of small new technologies and technological refinements occurred in the 800 years of the Hellenistic and Greco-Roman periods (such as a kickwheel added to the potter's wheel), but overall the technological bases of production did not change fundamentally during the period. Some industrial-style production occurred in a few fields like mining; and long-distance commercial movement of people and goods took place regularly. But most production remained craft-based and local, and artisans, traditionally secretive about knowledge of their skills, tended to monopolize practice without the benefit of writing, science, or natural philosophy.

While ancient science formed part of civilized life in towns, technology and engineering practice could be found everywhere in the ancient world, vigorously and expertly developed in great cities and towns, to be sure, but also in the countryside, where the practice of science and natural philosophy was notably absent. The engineer (*architecton*) was a recognized and employable social type in antiquity. A handful of individuals stood at the top rank of ancient engineering. The Roman Vitruvius, for example, worked as architect/engineer to the first Roman emperor, Augustus, at the turn of the first century CE, and he contributed to an engineering literature. However, most engineers and, indeed, most artisans were anonymous practitioners plying trades at great remove socially, intellectually, and practically from the scientific world of Alexandria.

The Romans were the greatest technologists and engineers of the ancient world, and one can argue that Roman civilization itself represents one grand technological achievement. In the first centuries BCE and CE Roman military and political power came to dominate the whole of the Mediterranean basin and much of the Hellenistic world that had arisen in the east. (Mesopotamia remained beyond the reach of Rome.) The Roman empire grew up around several technologies. Military and naval technologies created the disciplined Roman legion and the Roman navy. The extensive systems of Roman roads and aqueducts provided an essential infrastructure. The expertise and sophistication of the Romans in matters of formal law may also be thought of as a social technology of no small import in running the empire. Less lofty perhaps, but no less important as a building block of Roman civilization, the invention of cement was a key new technology introduced by the Romans, one that made stone construction much cheaper and easier, and it literally cemented the expansion of the Roman empire. The Roman arch likewise revolutionized architecture and construction. The fact that Rome produced known engineers, a few of whom wrote books (an uncommon practice among engineers), such as Vitruvius and Frontinus (35–103 CE), likewise testifies to the significance of engineering and technology to Roman civilization and vice versa.

While Roman engineering flourished, there was little Roman science. Very little Greek science was ever translated into Latin. For the sake of tradition, Roman emperors patronized the Museum in faraway Alexandria, but the Romans did not value, indeed they spurned, science, mathematics, and Greek learning in general. Some privileged young Romans learned Greek and toured and studied in Greece. But Rome itself produced no Roman scientist or natural philosopher of the first or even the second rank. This circumstance has proved a puzzlement for those who see science and technology as always and necessarily linked. The temptation has been to overemphasize those exceptional Romans who did write on matters scientific. The notable Roman poet Lucretius (d. 55 BCE), whose long poem *On the Nature of Things* advanced

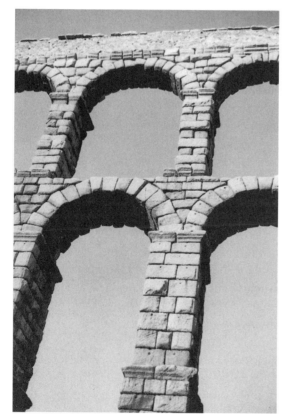

Fig. 5.5. Roman building technology. Roman engineers were highly competent in the use of the wedge arch in the construction of buildings, bridges, and elevated aqueducts. This Roman aqueduct in Segovia, Spain, is an outstanding example. The invention of cement greatly facilitated Roman building techniques.

atomist notions, is one example. The great Roman compiler Pliny the Elder (24–79 CE), whose multivolume *Natural History* summarized as much of the natural world as he could document (the fabulous along with the commonplace), is another. For better or worse, Pliny's work remained the starting point for the study of natural history until the sixteenth century; that he devoted considerable space to the practical uses of animals and that he dedicated his *Natural History* to the emperor Titus suggests that, insofar as Roman science existed at all, familiar social forces were at play.

Greeks scholars lectured in Rome. Greek doctors, particularly, found employment in Rome, more for their clinical skills than for their theoretical knowledge. The illustrious scientist-physician Galen of Pergamum (ca. 130–200 CE) was Greek—born, raised, and trained in Asia Minor and Alexandria. But Galen climbed the ladder of medical success in Rome through gladiatorial and court circles, becoming court physician to the Roman emperor Marcus Aurelius. The bloody combat of gladiators at the Coliseum in Rome and elsewhere provided an opportunity for Galen and others to consider the interior of the human body and its functioning, and Galenic medicine differed from the Hippocratic tradition in allowing surgery. Galen produced a large and influential body of work in anatomy, physiology, and what we would call biology. He built on Aristotle and the Hippocratic corpus and he articulated rational and comprehensive accounts of the workings of the human body based on detailed anatomical analysis, and Galen more than anyone else was responsible for introducing the notion of human temperaments into medical and therapeutic consideration.

Galen's anatomy and physiology differ markedly from succeeding views in the European Renaissance and those held today, but that fact should not detract from the power and persuasiveness of his understanding of the human fabric. For Galen and his many successors, three different vital systems and as many *pneuma* operated in humans. He held that the liver and venous system absorbed nutrients and distributed a nourishing blood throughout the body. The brain and nerves distributed a psychic essence that permitted thought. The heart was the seat of innate heat which distributed a third, vital fluid through the arteries, thus enabling movement; the lungs regulated respiration and the cooling of the heart's innate heat. The circulation of the blood was a conceptual impossibility for Galenists because they believed that, except for a minor passageway in the heart where the nutrifying blood provided the raw material for the arterial spirit, veins and arteries were two entirely separate systems.

Galen was a prolific author, supposedly writing some 500 treatises, of which 83 survived antiquity. He remained the undisputed authority in anatomy and physiology into the early modern era. Galen exemplifies the continuing interaction between medicine and philosophy in the Hellenistic and Greco-Roman eras, but Galen was Greek, and the tradition out of which he emerged and to which he contributed was Hellenistic and not Roman. The phenomenal lack of any Roman tradition in mathematics or the natural sciences contrasts strongly not only with Roman engineering but also with the substantial record of Roman literary and artistic accomplishment in poetry, the theater, literature, history, and the fine arts. The names of Cicero, Virgil, Horace,

Brain/Nerves
distribute
"Animal Spirits"

Air

(Lungs Cool
Heart)

Heart/Arteries
distribute
"Vital Spirits"

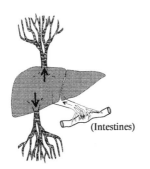

(Intestines)

Liver/Veins
distribute
"Nutrifying Spirits"

Fig. 5.6. Galenic physiology. Ancient physicians and students of anatomy separated the internal organs into three distinct subsystems governed by three different "spirits" functioning in the human body: a psychic essence permeating the brain and the nerves, a vivifying arterial spirit arising in the heart, and a nutrifying venous spirit originating in the liver.

and Suetonius alone suffice to indicate the extent to which literary and learned culture held a valued place in Roman civilization generally. The Roman case shows that a civilization of great social and technological complexity could thrive for centuries essentially without theoretical science or natural philosophy.

Decline and Fall

The causes of the marked decline of science and natural philosophy at the end of the Greco-Roman era have long been debated among historians of science. Not all agree even about the facts. Some claim the decline can be dated from 200 BCE in the Hellenistic era; others say it only began after 200 CE in the Greco-Roman period. Certainly, not all scientific and natural philosophical activity came to a halt after the second century CE. Still, ancient science seems to have run out of steam in late antiquity. Generally speaking, less overall activity took place, and the level of scientific originality declined as time went on. Intellectual labor was increasingly directed less toward discovering new knowledge than toward preserving old knowledge. This characteristic state of affairs gave rise to generations of compilers and commentators. Oribasius at Constantinople, for example, in the middle of the fourth century CE, wrote a formidable medical compendium of seventy volumes. (It is notable, but hardly surprising in this regard, that medicine displayed greater historical continuity in antiquity than did ancient science or natural philosophy.) Whatever animated the pursuit of science seems to have disappeared. Eventually, the desire merely to preserve past knowledge fell off. Increasing skepticism arose about even the possibility of secure knowledge, and magic and popular superstitious beliefs gained ground. The substance and spirit of Greek scientific accomplishment in its Hellenic and Hellenistic modes gradually faded away in late antiquity.

Several explanations have been proposed to explain why. One possible explanation points to the lack of a clear social role for science and scientific careers. Science was weakly socialized and institutionalized in the ancient world, and it largely lacked an ideological or material basis of support in society. No employment was available for individuals in their capacities as scientists or as natural philosophers, and the separation of science and natural philosophy from *philosophy* itself that developed in the Hellenistic period further undercut any social role for the scientific enterprise.

A related explanation has to do with the ancient economy and the separation of science and technology in antiquity. That is, given slavery, the relative cheapness of human labor, and the ideology that natural knowledge should not be applied to practical ends, little incentive existed to employ scientists or to search for practical outcomes of abstract understandings of nature. In other words, by excluding the possible utility of natural knowledge, the social role and social support for science were undermined.

Historians have also made a strong case that the flourishing of various religious cults and sects in late antiquity did much to weaken the authority and vitality of ancient scientific traditions. To varying degrees the many cults of late antiquity were anti-intellectual in their approach, and they represented intellectual and spiritual competition with traditional knowledge of nature. The cult of the Greek fertility goddess Demeter and the cult of the Egyptian goddess Isis attracted wide followings. Popular among officials of the Roman empire, Mithraism, a late oriental mystery cult worshiping the Persian god of light, Mithras, embodied arcane and secret astrological and astronomical knowledge having to do with the precession of the equinoxes. And growing out of messianic Judaism, the most successful new cult was Christianity.

Official toleration of Christians in 313 CE, the emperor Constantine's conversion to Christianity in 337, and the declaration in 391 of Christianity as the official state religion of the Roman Empire marked the extraordinary social and institutional success of the Christian church. Experts debate the effect of Christianity on ancient science, but with its heavy theological orientation, its devotional stress on the religious life, and its emphasis on revelation, the afterlife, and the second coming of Christ, the early Christian church and the church leaders who fashioned it displayed greater or lesser degrees of hostility, skepticism, ambivalence, and/or indifference toward pagan culture in general and toward science and inquiries into nature in particular. To cite only one example, Saint Augustine (354–430 CE) railed against natural philosophy and "those whom the Greeks call *physici*." On a more mundane level the church became firmly institutionalized in ancient civilization and a formidable institutional presence at every level of society. Church bureaucracy and administration offered employment and careers, which had the effect of draining talent, intellectual and otherwise, which previously might have been recruited for the Museum at Alexandria or for science in general.

Historians of technology have asked why no industrial revolution developed in antiquity. The question is a poor one, as we will see further, but the simple answer seems to be that there was no need, that contemporary modes of production and the slave-based economy of the day satisfactorily maintained the status quo. The capitalist idea of profit as a desirable end to pursue was completely foreign to the contemporary mentality. So, too, was the idea that technology on a large scale could or should be harnessed to those ends. An industrial revolution was literally unthinkable in antiquity.

The ultimate destruction of the Museum and Library at Alexandria proved an incalculable loss. Although the details remain obscure, Alexandria and its intellectual infrastructure suffered many blows from the late third century onward. Much of the town was destroyed in 270–75 in Roman efforts to reconquer the city after its momentary capture by Syrian and Arab invaders. Christian vigilantes may well have burned books in the fourth century, and in 415 Christian fanatics murdered the pagan Hypatia, the first known female mathematician and the last known stipendiary of the Museum. The centuries-old Museum thus came to an end, although a subsidiary library may have continued. Alexandria fell to Muslim forces in 642, conquerors who may have effected further depredations on what remained of the ancient Library at Alexandria. Elsewhere, in 529 the Christian Byzantine emperor Justinian ordered the closing of the Platonic Academy in Athens.

The Roman Empire split into its western and eastern divisions in the fourth century CE (see map 5.1 of the Byzantine Empire). In 330 CE, Constantine the Great transferred the capital of the empire from Rome to Constantinople, modern-day Istanbul. Waves of barbarian tribes pressed in on the western empire from Europe. Visigoth invaders sacked Rome for the first time in 410 CE. Other Germans deposed the last Roman emperor in Italy in 476 CE, a date that marks the traditional end of the Roman Empire. While the latinized West Roman

Empire fell, the Hellenized East Roman Empire—the Greek-speaking Byzantine Empire—centered in Constantinople, endured, indeed flourished. But, the fortunes of Byzantium declined in the seventh century as its glory and granaries contracted before the armed might of ascendant Islamic Arabs. Pouring out of Arabia after 632 CE, the followers of the Prophet Mohammed conquered Syria and Mesopotamia. They captured Egypt and Alexandria in 642 CE and pressed in on Constantinople itself by the end of the century. Science and civilization would continue to develop in Muslim Spain, in eastern regions, and throughout the Islamic world, but by the seventh century CE the era of Greek science and antiquity itself had clearly come to an end.

The Roman West, which included much of Europe, had always been underdeveloped compared to the East. Decline, intellectual and otherwise, at the end of antiquity affected the West much more than the East, where greater continuity prevailed. Indeed, the words *disruption* and *discontinuity* aptly describe "western civilization" at the end of Greco-Roman antiquity. The population of Italy, for example, dropped by 50 percent between 200 and 600 CE. An era had ended, and surely to contemporaries no promise of renewal seemed to be forthcoming. The late Roman author and senator Boethius (480–524 CE) knew that he stood at a historical crossroads, and his case is a poignant one on that account. Boethius was exceptionally well educated and fully the inheritor of the classical tradition of Greek and Latin antiquity that stretched back a millennium to Plato, Aristotle, and the pre-Socratics. Yet he held office and attended not a Roman emperor, but the Ostrogoth king in Rome, Theodoric. Imprisoned for many years by Theodoric, Boethius made every effort to pass on to the succeeding age as much of antiquity's accumulated knowledge as he could. He wrote short elementary handbooks on arithmetic, geometry, astronomy, mechanics, physics, and music. In addition, he translated some of Aristotle's logical treatises, some Euclid, and perhaps Archimedes and Ptolemy. In prison, he also wrote his immortal meditations, *On the Consolation of Philosophy*, which proved small consolation indeed. Theodoric had Boethius executed in 524.

Historians interested in the European Middle Ages and in the history of science in the Middle Ages often point to Boethius and his compeers to indicate the extent to which knowledge from classical antiquity passed directly into the stream of European history and culture. Cassiodorus (488–575), a Roman like Boethius, who influenced the early monastic movement, is regularly cited in this connection, as are the later learned churchmen Isidore of Seville (560–636) and the Venerable Bede (d. 735). There is much of interest about these men and their circumstances, but the Latin West inherited the merest crumbs of Greek science. From a world perspective, what needs to be emphasized is the utterly sorry state of learning among the Christian barbarians of Europe and the Latin West in the early Middle Ages. After the fall of

Rome literacy itself virtually disappeared, and knowledge of Greek for all intents and purposes vanished in Western Europe. Isidore of Seville apparently thought the sun illuminated the stars. Two eleventh-century European scholars, Regimbold of Cologne and Radolf of Liège, could not fathom the sense of the elementary proposition from geometry that "the interior angles of a triangle equal two right angles." The terms "feet," "square feet," and "cubic feet" had no meaning for them.

How and why the scientific traditions of Greek antiquity took hold in western Europe centuries later require separate explanations and a return to the world stage.

Part II

Thinking and Doing among the World's Peoples

After the Hellenistic merger of the institutional patterns of the ancient Near East with the abstract intellectual approach of the Hellenic Greeks, Hellenistic-like scientific traditions took root in a series of Near Eastern empires in late antiquity and the Middle Ages: in Byzantium, in Sassanid Persia, and in the vast region that formed the Islamic conquest. At the same time, indigenous traditions of scientific research developed independently in China, India, and Central and South America. In a Babylonian mode monarchs in control of rich agricultural lands patronized experts in calendrical astronomy, astrology, mathematics, and medicine in the hope and expectation that these fields of research would produce useful knowledge. In the meantime, science and the traditional crafts continued everywhere as essentially separate enterprises. Part II investigates these developments across the thousand years of 500 to 1500 of the common era.

The Enduring East

Byzantine Orthodoxy

After Rome fell in 476 CE, the empire's eastern districts with their capital at Constantinople gradually metamorphosed into the Greek-speaking Byzantine Empire (see map 6.1). A Christian state, headed by an emperor and governed by elaborate and scheming bureaucracies (hence "byzantine"), the Byzantine Empire endured for a thousand years before being overrun by the Ottoman Turks in 1453. In possession of the Egyptian breadbasket the empire flourished and wealthy emperors continued to patronize many old institutions of higher learning.

Science in the Byzantine Empire remains to be studied by historians in greater detail. Byzantine civilization is often criticized as anti-intellectual and stifled by a mystical Christianity that was imposed as a state religion. That the emperor Justinian (r. 527–565) closed Plato's still-functioning Academy at Athens along with other schools in 529 CE is commonly seen as evidence of the state's repressive posture toward science. Yet, to dismiss Byzantium from the history of science would be to overlook continuations of Hellenistic traditions and the ways, quite typical of eastern bureaucratic civilizations, in which science and useful knowledge became institutionalized in society.

Even after the Justinian closures, state schools and church schools provided instruction in the mathematical sciences (the quadrivium: arithmetic, geometry, astronomy, and music), the physical sciences, and medicine; libraries existed as centers of learning. The true hospital, as an institution of in-patient medical treatment (and Christian mercy), was a notable Byzantine innovation. It was, like the hospital today, primarily a center of medical technology, not science. As hospitals arose throughout the Byzantine Empire through the largesse of government, church, and aristocratic patrons, in some measure they also became centers of medical research. Byzantine medicine fully assimilated the medical and physiological teachings of Galen and Hippocrates,

Sassanid Empire

Byzantine Empire to 527

Justinian's Acquisitions to 554

Caspian Sea

Jundishapur

Persian Gulf

Tigris R.

Euphrates R.

Red Sea

Black Sea

Nile R.

Alexandria

Constantinople

Mediterranean Sea

Rome

while some hospitals maintained libraries and teaching programs and even fostered some original investigations and innovative techniques. Learned Byzantine doctors turned out influential medical and pharmacological tracts, albeit with much repetition of Greek knowledge. Veterinary medicine was a notable aspect of scientifico-medical activity in Byzantine civilization, one heavily supported by monarchs who had an interest in the well-being of their war horses since cavalry and the cavalry charge formed the basis of the Byzantine army and military tactics. As a result, Byzantine veterinarians produced many veterinary manuals, occasionally on a high level of originality.

In the fields of the exact sciences the Greek-speaking Byzantine scholars inherited much Greek learning from antiquity. They knew their Aristotle, Euclid, and Ptolemy, and Byzantine astronomers and mathematicians themselves sometimes produced sophisticated tracts based on earlier Greek and contemporary Persian sources. In addition to calendar work, a strong element of astrology, reflecting that venerable and inextinguishable desire to know the future, characterized Byzantine astronomy. Experts likewise studied music and mathematical music theory, perhaps for liturgical purposes. And finally, Byzantine alchemy and alchemical mineralogy cannot be overlooked as areas of both considerable research activity and of perceived practical utility.

The most notable natural philosopher of the early Byzantine era was John Philoponus. A Christian, Philoponus lived and worked in Alexandria under Byzantine governance during the middle years of the sixth century CE, and he launched the most sweeping attack on Aristotelian physics prior to the Scientific Revolution in Europe. In various commentaries he developed trenchant critiques of Aristotle and several aspects of Aristotelian physical theory. In his ingenious analysis of projectile motion, for example—motion Aristotle had lamely explained by invoking the ambient air as the required mover—Philoponus suggested that the thrower endowed the projectile with a certain power to move itself. Philoponus's views in turn sparked critical responses from other commentators, and because he so focused the debate on specific problems in Aristotle's writings on natural philosophy, Philoponus was later influential among Islamic and European natural philosophers when they came to review Aristotle's work. Having studied the science of the Greeks and writing in Greek, his career and his accomplishments are landmarks in the tradition of Byzantine science.

A full social history of Byzantine science would display the subject in a more favorable light than does intellectual history alone, which stresses originality and pure theory. Such a social history would pay close attention to intellectually unambitious medical tracts, to treatises published by veterinary surgeons retained by Byzantine monarchs, to the many farmers' manuals and herbals produced under Byzantine governance, as well as to astrology and alchemy. In a society where

Map 6.1. (opposite) Byzantium and Sassanid Persia. In late antiquity two derivative civilizations took root in the Middle East—the Byzantine Empire centered in Constantinople and Sassanid Persia in the heartland of ancient Mesopotamia. Both assimilated ancient Greek science and became centers of learning.

bureaucratic centralization was extreme, support came precisely for encyclopedists, translators, and writers of manuals on subjects useful and mundane. And it is precisely the kind of work that historians intent on detecting theoretical novelty tend to neglect.

The loss of Egypt and the productive resources of the Nile Valley to invading Arabs in the seventh century was a severe setback to the economy and society of Byzantium. Yet a reduced Byzantine civilization maintained itself, its cities, its institutions, and its science for hundreds of years. Inevitably, however, decline set in after the year 1000, as Byzantium faced challenges from the Turks, from the Venetians, and from friendly and not-so-friendly European Christians on crusade. In 1204 Crusaders pillaged Constantinople and occupied it until 1261. Finally, in 1453 the city and the empire fell to the Turks. The story of science in the empire of the Ottoman Turks that followed is a chapter being written by up-and-coming scholars.

Although Byzantium never became a center of significant original science, it did not repudiate the tradition of secular Greek learning. Indeed, it tolerated and even preserved that tradition alongside the official state religion of Christianity.

Mesopotamia Revisited

In the heartland of ancient Mesopotamia the Sassanid dynasty created a typical Near Eastern system of scientific institutions along with a typical Near Eastern economy based on hydraulic agriculture and the restoration and maintenance of the old irrigation systems. The Sassanid dynasty was founded in 224 CE, and was characterized by a strong central government and a bureaucratic caste that included scribes, astrologers, doctors, poets, and musicians. By the sixth century the royal residence at Jundishapur, northeast of present-day Basra, had become a cultural crossroads where many different learned traditions mingled: Persian, Christian, Greek, Hindu, Jewish, and Syrian. Persian cultural life became enriched when Nestorian Christians—bringing Greek learning with them—fled Byzantium after their center at Edessa in Turkey was shut down in 489. A significant translation effort, centered in Jundishapur, rendered Greek texts into Syriac, the local language. Texts deemed to contain useful knowledge were generally chosen for translation—mainly the medical arts, but also scientific subjects, including Aristotle's logical tracts, mathematics, and astronomy. Jundishapur also became the site of a hospital and medical school, enriched by the presence of Indian medical masters; later taken over by Arab-Islamic caliphs, the medical school at Jundishapur continued to flourish until the eleventh century. Persian government authorities also sponsored astronomical and astrological investigations. While recent reappraisals have tempered its importance, Jundishapur was nonetheless a cosmopolitan intellectual center and a center of scientific patronage for several centuries before Persia fell to the forces of Islam in 642 CE.

Sassanid civilization illustrates once again that centralization of authority to manage a hydraulic agricultural economy fostered scientific institutions. Its culture in some measure hybridized the institutional and intellectual traditions of the ancient Oriental kingdoms and those of classical, Hellenic Greece, and it produced state-dominated institutions, in some of which a Greek tradition of pure science found a niche. Once again, enormous wealth in the form of large agricultural surpluses generated by intensified irrigation agriculture made such institutional patronage possible. The case also confirms that prior to the development of modern science in western Europe Greek scientific influence flourished predominantly in the East.

Under the Banner of Islam

The Middle East produced still another scientific civilization, this time under the aegis of Islam. The flight of the Prophet Mohammed from Mecca in 622 CE marks the traditional beginning of the Muslim era. The word *Islam* means submission to the will of God, and Muslims (or Moslems) are those who submit. Arabs are the peoples of Arabia, and out of the Arabian desert and a nomadic pastoral society of the seventh century the faith of Islam spread to many different peoples east and west. Within three decades Islamic armies conquered Arabia, Egypt, and Mesopotamia—replacing Persian power and severely reducing the Byzantine Empire. In slightly more than a century they established an Islamic commonwealth stretching from Portugal to Central Asia. A unified sociocultural domain, Islam prospered as a great world civilization, and its scientific culture flourished for at least five centuries.

The success of Islam depended as much on its faithful farmers as on its soldiers. The former took over established flood plains in Mesopotamia and Egypt, and in what amounted to an agricultural revolution they adapted new and more diversified food crops to the Mediterranean ecosystem: rice, sugar cane, cotton, melons, citrus fruits, and other products. With rebuilt and enlarged systems of irrigation, Islamic farming extended the growing season and increased productivity. That Islamic scientists turned out an uninterrupted series of treatises on agriculture and irrigation is one indication of the importance of these endeavors. So, too, are the specialized treatises on camels, horses, bees, and falcons, all animals of note for Islamic farmers and Islamic rulers.

The effects of such improved agricultural productivity were typical: unprecedented population increases, urbanization, social stratification, political centralization, and state patronage of higher learning. Baghdad, founded in 762 on the Tigris, became the largest city in the world in the 930s with a population of 1.1 million. Córdoba in southwestern Spain reached a population close to 1,000,000 under Islamic rule,

SPAIN
Toledo
Cordoba
Seville

BERBERS

Tunis

Rome

Mediterranean
Sea

Constantinople

Black Sea

Caspian
Sea

Edessa
Antioch
SYRIA
Damascus
PALESTINE
Jerusalem

Alexandria
Cairo
EGYPT

Nile R.

PERSIA

Tigris R. Baghdad
Jundishapur
Euphrates R.

Persian Gulf

Medina

Mecca

ARABIA

Extent of Islam, 632

Conquests to 750

Byzantine Empire, c. 1000

and several other Islamic cities had populations between 100,000 and 500,000 during a period when the largest European cities had populations numbering fewer than 50,000.

Islam was and is based on literacy and the holy book of the Quran (Koran); and, although policy vacillated, Islam showed itself tolerant toward Christians and Jews, also "people of the book." Thus, in contrast to the barbarian farmers of Europe who pillaged and destroyed the high civilizations they encountered, the nomadic and pastoral Arabs established conquest empires by maintaining and assimilating the high cultures they encountered. Early Islamic rulers encouraged the mastery of foreign cultural traditions, including notably Greek philosophy and science, perhaps in order to bolster the logical and rhetorical position of their new religion in the face of more highly developed religions and critical intellectual traditions. The result was another hybrid society, the cultural "hellenization" of Islam and its typically bureaucratized institutions, funded by wealthy monarchs and patrons who encouraged useful knowledge along with a dash of natural philosophy.

Medieval Islam became the principal heir to ancient Greek science, and Islamic civilization remained the world leader in virtually every field of science from at least 800 to 1300 CE. The sheer level of scientific activity makes the point, as the number of Islamic scientists during the four centuries after the Prophet matched the number of Greek scientists during the four centuries following Thales. Islamic scientists established the first truly international scientific community, stretching from Iberia to Central Asia. Yet, despite considerable scholarly attention, medieval Islamic science is sometimes still dismissed as a conduit passively "transmitting" ancient Greek science to the European Middle Ages. A moment's thought, however, shows how ahistorical it is to evaluate the history of Islamic science only or even largely as a link to European science, or even to subsume Islamic science into the "Western tradition." Medieval Islam and its science must be judged on their own terms, and those terms are as much Eastern as Western.

Only a small fraction of Islamic scientific texts have been published. Most remain unstudied and in manuscript. Scholarly emphasis to date has been on classic texts, on the "internal" history of scientific ideas, on biographies, and on "precursor-itis," or identifying Arabic scientists who were precursors of ideas that were of later importance to European science. The institutional aspects of Islamic science are only beginning to be studied with scholarly rigor, and nothing like a full historical survey exists for the Islamic case.

Furthermore, the field divides into two divergent interpretative schools. One school argues for a "marginality" thesis, holding that the secular, rational sciences inherited from Greek civilization—known in Islam as the "foreign" (*aw'il*) sciences—never became integrated into Islamic culture, remaining only on the cultural margins, tolerated at best, but never a fundamental part of Islamic society. The "assimila-

Map 6.2. (opposite) Islam. Following the birth of Islam in the seventh century, the Islamic conquest stretched from the Atlantic Ocean almost to the borders of China. In capturing Egypt and the resources of the Nile, the forces of Islam dealt a severe blow to Byzantine civilization.

tionist" school, on the other hand, contends that the foreign sciences became woven into the fabric of Islamic life. Neither interpretation quite fits the facts, but the presentation favored here leans toward the assimilationists, especially in tracing the institutional basis of Islamic science and in recognizing a similarity between the social function of science in Islam and in other Eastern civilizations.

Islamic scientific culture originated through the effort to master the learning of more established civilizations, and that first required the translation of documents into Arabic. Given the early conquest of Jundishapur, Persian and Indian influences, rather than Greek, were more influential in the early stages of Islamic civilization. Already in the 760s, for example, an Indian mission reached Baghdad to teach Indian science and philosophy and to aid in translations of Indian astronomical and mathematical texts from Sanskrit into Arabic. Later, Muslim men of science traveled to India to study with Indian masters.

In the following century, however, the translation movement came to focus on Greek scientific works. The governing caliph in Baghdad, Al-Ma'mun, founded the House of Wisdom (the *Bayt al-Hikma*) in 832 CE specifically as a center of translation and mastery of the secular foreign sciences. Al-Ma'mun sent emissaries to collect Greek scientific manuscripts from Byzantine sources for the House of Wisdom where families of scholar-translators, notably Ishāq ibn Hunayn and his relatives, undertook the Herculean task of rendering into Arabic the Greek philosophical and scientific tradition. As a result, virtually the entire corpus of Greek natural science, mathematics, and medicine was brought over into Arabic, and Arabic became the international language of civilization and science. Ptolemy's *Almagest*, for example—the very title, *al-Mageste*, is Arabic for "the greatest"— appeared in several translations in Baghdad early in the ninth century, as well as Euclid's *Elements*, several works of Archimedes, and many of Aristotle, beginning with his logical treatises. Aristotle became the intellectual godfather of Islamic theoretical science, spawning a succession of commentators and critical thinkers. A measure of the effort expended on translating Greek texts is that, even now, more Aristotelian writings—the works of Aristotle and his Greek commentators— supposedly are available in Arabic than in any European language.

Al Ma'mun supported his translators and the House of Wisdom, not merely out of the love of learning, but for practical utility deemed directly useful to the monarch, notably in such fields as medicine, applied mathematics, astronomy, astrology, alchemy, and logic. (Aristotle became assimilated initially for the practical value of his logic for law and government, and only later did the entire body of his scientific and philosophical works find its way into Arabic.) Medicine was the primary field naturalized by Islamic translators; Ishāq ibn Hunayn alone supposedly translated 150 works of Galen and Hippocrates. Thus, by

900 CE, while Europe possessed perhaps three works of Galen, transcribed by solitary scholars, Islam had 129 produced under government patronage. The basis had been established for a great scientific civilization.

In the Islamic world the secular sciences were generally not valued for their own sakes, but rather for their utility; secular knowledge was normally not pursued by individualistic natural philosophers as an end in itself as in Hellenic Greece or later in Christian Europe. To this extent, the "marginality" thesis provides a degree of insight into the place of pure science in Islamic society. Nevertheless, such a view slights the ways in which science became patronized and institutionalized in a variety of social niches in Islamic culture. As social history, the "assimilationist" thesis more properly portrays the role and institutionalized character of science and natural knowledge in Islam.

Each local mosque, for example, was a center of literacy and learning, albeit largely religious. But mosques also had official timekeepers (the *muwaqqit*) who set times for prayer. This recondite and exact procedure could only be effected by competent astronomers or at least trained experts. Thus, for example, afternoon prayers occur when the shadow of an object equals the length of its shadow at noon plus the length of the object. Several esoteric geographical and seasonal factors determine these times, and the *muwaqqit* used elaborate timekeeping tables, some with upwards of 30,000 entries, supplemented by instruments such as astrolabes and elaborate sundials to ascertain when prayer should take place. (The astrolabe became a highly developed instrument capable of solving 300 types of problems in astronomy, geography, and trigonometry.) Similarly, the faithful prayed in the direction of Mecca, and therefore geographical knowledge also had to be applied locally to discover that direction. Astronomers determined the beginning of Ramadan, the month-long period of daily fasts, and the hour of dawn each day. Along these lines, each local Islamic community possessed mathematically and legally trained specialists, the *faradi*, who superintended the division of inheritances.

The Islamic legal college, or *madrasa*, was an institution of higher learning wherein some "foreign sciences" were taught. Widespread throughout the Islamic world, the madrasa was primarily an advanced school for legal instruction in the "Islamic sciences"—law, not theology, being the preeminent science in Islam. The madrasa should not be equated with the later European university, in that the madrasa was not a self-governing corporation (prohibited in Islam). It did not maintain a standard curriculum, and it did not confer degrees. Technically a charitable endowment rigidly bound by its founding charter and prohibited from teaching anything contrary to the fundamental tenets of Islam, the madrasa operated more as an assemblage of independent scholars with whom students studied on an individual basis and where instruction emphasized memorization, recitation, and mastery of authoritative texts. Endowments supported instructors and paid the tuition, room, and board of students.

The secular sciences found a niche in these institutions of higher learning. Logic, for example, was taken over from Greek traditions, and arithmetic was studied for the purposes of training the *faradi* or experts for handling inheritances. Similarly, geometry, trigonometry, and astronomy, although tightly controlled, likewise came within the fold of Islamic studies because of the religious needs of determining proper times for prayer and the direction of Mecca. While not publicly professed, specialists also offered private instruction in the "foreign sciences" outside the formal setting of the madrasa. And secular scientific and philosophical books could be found in public libraries associated with madrasas and mosques. In a word, then, the student who wished to learn the natural sciences could do so at a high level of sophistication in and around the institution of the madrasa.

The library formed another major institution of Islamic civilization wherein the natural sciences were nurtured. Often attached to madrasas or mosques, usually staffed by librarians and open to the public, hundreds if not thousands of libraries arose throughout the Islamic world. Córdoba alone had seventy libraries, one containing between 400,000 and 500,000 volumes. Thirty madrasas existed in Baghdad in the thirteenth century, each with its own library, and 150 madrasas operated in Damascus in 1500 with as many libraries. The library deep in Islamicized Africa at Timbuktu in Mali held 30,000 tomes. The library attached to the observatory in Maraghah reportedly contained 400,000 volumes. Another House of Wisdom (the *Dār al-'ilm*) in tenth-century Cairo contained perhaps 2 million books, including some 18,000 scientific titles. One collector boasted that it would take 400 camels to transport his library; the estate of another included 600 boxes of books, manhandled by two men each. The tenth-century Persian physician Ibn Sīnā (980–1037), known in the West as Avicenna, writing in Arabic left an account of the impressive quality of the royal library in Muslim Bukhara on the Asian outskirts of Islam:

> I found there many rooms filled with books which were arranged in cases, row upon row. One room was allotted to works on Arabic philology and poetry, another to jurisprudence and so forth, the books on each particular science having a room to themselves. I inspected the catalogue of ancient Greek authors and looked for the books which I required; I saw in this collection books of which few people have heard even the names, and which I myself have never seen either before or since.

In sharp contrast, libraries in medieval Europe numbered only hundreds of items, and as late as the fourteenth century the library collection at the University of Paris contained only 2,000 titles, while a century later the Vatican library numbered only a few hundred more.

But the love of learning alone could not have accounted for Islamic libraries. The formation of huge collections was clearly dependent on the willingness of caliphs and wealthy patrons to underwrite the costs. It was also dependent on paper-making, a new technology acquired from the Chinese in the eighth century which allowed the mass production of paper and much cheaper books. Paper factories appeared in Samarkand after 751, in Baghdad in 793, in Cairo around 900, in Morocco in 1100, and in Spain in 1150. In Baghdad alone 100 shops turned out paper books. Ironically, when the printing press appeared in the fifteenth century Islamic authorities banned it for fear of defiling the name of God and to prevent the proliferation of undesirable materials.

Although astronomers had previously observed the heavens, Islamic civilization created a new and distinctive scientific institution: the formal astronomical observatory. Underwritten by ruling caliphs and sultans, observatories, their equipment, and staffs of astronomers discharged several practical functions. Astronomers prepared increasingly accurate astronomical handbooks (*zij*) for calendrical and religious ends—to fix the times of prayer and of religious observances such as Ramadan. The Islamic calendar was a lunar calendar, like that of ancient Babylonia, of 12 months of 29 or 30 days unfolding over a 30-year cycle, with trained observers determining when the new moon commenced. Geography was also closely connected to astronomy and, beginning with Ptolemy's *Geography*, Muslim astronomers developed navigational and geographical techniques serviceable to both sailors and desert travelers.

Islamic authorities formally distinguished between astronomy as the study of the heavens and astrology as investigating heavenly influence on human affairs. The distinction may have facilitated the social integration of astronomy, but the strongest single motive behind royal patronage of astronomy remained the putative divinatory power of astrology. Despite its occasional condemnation by religious authorities on the grounds that it misdirected piety toward the stars rather than God, astrology remained the most popular of the secular sciences, and it flourished especially in court settings, where regulations and exams fixed the qualifications, duties, and salaries of astrologers. Elsewhere, the local chief of police regulated astrology as a marketplace activity. Along with Ptolemy's *Almagest*, Muslim astronomers/astrologers had available his astrological treatise, the *Tetrabiblos*, and many used it and like volumes to cast horoscopes and gain patronage as court astrologers.

Observatories arose throughout the Muslim world. Al-Ma'mun founded the first around 828 in Baghdad. The best known, established in 1259, was the observatory at Maraghah in a fertile region near the Caspian Sea. It was formed in part to improve astrological prediction. A substantial library was attached to the observatory and actual in-

Fig. 6.1. An astrolabe. This multifaceted device was invented in the Islamic world to facilitate astronomical observation and to solve problems relating to timekeeping, geography, and astronomy.

struction in the sciences was offered there with government support. Expert astronomers made up what can only be called the Maraghah school, and such men as al-Tūsī (d. 1274), al-Shīrāzī (d. 1311), and their successor, Ibn al-Shātir (d. 1375), far surpassed ancient astronomy and astronomical theory in perfecting non-Ptolemaic (although still geocentric) models of planetary motion and in testing these against highly accurate observation. But, the observatory at Maraghah, like many others, proved short-lived, lasting at most 60 years. Even though protected by non-Islamic Mongol rulers, the Maraghah observatory and several other Islamic observatories were closed by religious reaction against impious study of astrology.

Farther north and east, in fifteenth-century Samarkand, sustained by irrigated orchards, gardens, and cropland, the celebrated Muslim scholar-prince Ulugh Beg (1393–1449) founded a madrasa and a major observatory. The importance that Islamic astronomers attached to the precision of their observations necessitated the use of exceptionally large instruments, such as the three-story sextant at Samarkand with a radius of 40 meters (132 feet). These large instruments, along with the observatory structures, the staffs of astronomers and support personnel, and their affiliated libraries entailed costs so high that they could only be met through government support. Through its observatories medieval Islam established a tradition of observational and theoretical astronomy unequaled until the achievements of European science in the sixteenth and seventeenth centuries.

Islamic mathematics, while justly renowned, consistently displayed a practical trend in its emphasis on arithmetic and algebra rather than on the formal theoretical geometry of the Greeks. Medieval Islamic mathematicians also developed trigonometry, which greatly facilitated working with arcs and angles in astronomy. The adoption of easily manipulated "Arabic numerals" from Indian sources further reflects this practical orientation. While Islamic mathematicians solved what were, in effect, higher-order equations, many problems had roots in the practical world dealing with taxes, charity, and the division of inheritances. The ninth-century mathematician al-Khwarizmi, for example, who originally introduced "Arabic numerals" from India, wrote a manual of practical mathematics, the *al-Jabr* or what came to be known in the West as the *Algebra*. Not coincidentally, al-Khwarizmi worked at the court of al-Ma'mun.

Islamic medicine and its institutionalized character deserve special attention. The Arabs had their own medical customs, and the Quran (Koran) contains many sayings of the Prophet regarding diet, hygiene, and various diseases and their treatment. The Arabic translation movement made available to physicians all of the Hippocratic canon and the works of Galen, notably through the texts of ancient Greek medicine preserved at Alexandria. Islamic medicine also assimilated Persian and Indian traditions, in part from having taken over the medical school at Jundishapur and in part from direct contact with India through the drug and perfume trades. The resulting medical amalgam became thoroughly naturalized and integrated into the social fabric of Islam.

A handful of madrasas specialized in medical training, but the hospital became the primary institutional locus of Islamic medicine. Government-supported hospitals existed throughout the Islamic world, with especially notable medical centers in Baghdad, which eclipsed Jundishapur, Damascus, which saw the foundation of six hospitals between the thirteenth and fifteenth centuries, and Cairo. Many hospitals came to possess elaborate medical staffs, specialized medical wards, attached medical libraries, and lecture halls (*majlis*). Islamic hospitals thus evolved as centers of teaching and research, as well as dispensaries of medical treatment, including medical astrology. And, whereas guilds and corporate structures were never recognized in Islamic societies, governments licensed physicians through local police officials. Islamic doctors, such as al-Rāzī (Rhazes, 854–925), al-Majūsī (Haly Abbas, d. 995), Ibn Sīnā (Avicenna), and others engaged in medical experimentation and developed unprecedentedly sophisticated and expert understanding of diseases and medical treatments.

The medical dimension may help explain a particular strength of Islamic science in optics. Especially in Egypt, where desert conditions contributed to eye ailments, a strong medical literature developed in ophthalmology, and Islamic physicians became expert in the treatment of the eye and the anatomy and physiology of vision. Although not a physician, the great Islamic physicist Ibn al-Haytham (Alhazen, 965–1040) worked in Egypt and wrote on eye diseases. His *Optics* is only the best known and most influential of a series of Islamic scientific works—many with an experimental approach—concerned with vision, refraction, the camera obscura, burning mirrors, lenses, the rainbow, and other optical phenomena.

Physicians enjoyed high public regard, and many Muslims who made scientific and philosophic contributions earned their living as court physicians or court-appointed administrators and legal officials. For example, Averroës (Ibn Rushd, 1126–98), known as "The Commentator" on Aristotle, worked as a court physician and religious jurist in Spain. The Islamic polymath Avicenna (Ibn Sīnā), renowned as the "Galen of Islam," accepted patronage as a physician in various courts in order to pursue philosophy and science. The noted Jewish philosopher and savant Moses Maimonides (Musa ibn Maymun, 1135–1204) acted as physician to the sultan at Cairo. In a word, court patronage provided institutionalized positions where physician-scientists could master and extend the secular sciences, and court positions afforded a degree of insulation from the dominant religious institutions and the supremacy of religious law in Islamic society at large.

Closely associated with courts and the patronage of rulers, a highly developed tradition of Islamic alchemy involved many scientists. Alchemy ranked among the sciences, being derived from Aristotle's matter theory. In the search for elixirs of immortality, Islamic alchemy also seems to have been influenced by Chinese alchemy, and it likewise subsumed work on mineralogy, which showed Indian and Iranian influences. Alchemy was a secret art, and adepts attributed some 3,000 alchemical texts to the founder of Islamic alchemy, the ninth-century figure Jābir ibn Hayyān, known as Geber in the Latin West. On one level, no doubt the one most appreciated by patrons, the transformation of base metals into gold and the creation of life-giving elixirs represented the goals of alchemy. To many practitioners, however, Islamic alchemy became a highly intellectual endeavor that primarily involved the spiritual refinement of the individual alchemist. In pursuing their science, Islamic alchemists invented new equipment and perfected new techniques, including distillation. Residues of Islamic alchemy remain in Arabic-derived terms, such as the word *alchemy* itself, *alcohol*, *alkali*, and *alembic*. Indeed, in such terms as *algebra*, *azimuth*, *algorithm*, and a host of others, the language of science to this day maintains the linguistic imprint of Arabic and the history of Islamic science.

The sheer institutional density of Islamic science accounts for some of its achievements and characteristics. Scholars and scientists staffed schools, libraries, mosques, hospitals, and especially observatories with their teams of astronomers and mathematicians. The opportunities and support that these institutions offered men of science produced a remarkable upsurge of scientific activity, as measured by the number of Islamic scientists which surpassed by an order of magnitude the handful of Europeans pursuing science before 1100 CE. Another result was a characteristic research profile, like that of the ancient bureaucratic kingdoms, which exaggerated utility, public service, and the interests of the state.

Technology and industry in medieval Islam gave as little to and received as little from the realm of science as they had in the Greco-Roman world. Islamic science embraced much of ancient Greek learning, as we have seen, but Islamic technology remained more akin to that of Rome and the eastern kingdoms. In architecture the Muslims employed the Roman arch rather than the Greek post and lintel system of building. And agriculture depended heavily on hydraulic engineering as it had in the Roman provinces and in all civilizations in the Near East. Indeed, the Islamic conquest maps closely onto regions that lent themselves to hydraulic intensification; Greece and Italy, where artificial irrigation was less important, did not become Islamicized, while Spain saw a dramatic development of hydraulic technology under Muslim rule. The construction of large dams, waterwheels, and qanats (underground channels with earthenware pipes designed to tap ground water) all formed part of the Islamic engineering repertoire. In Africa, Islamic centers at the walled city of Khartoum and at Timbuktu were situated near water sources. In Iran qanats supplied fully half of the water used for irrigation and urban needs. Such were the feats of craftsmen and artisans divorced from the bookish worlds of theology and science.

Scholars disagree on when the vitality of scientific activity started to decline in the Islamic world. Some say that the decline began after the twelfth century, especially in the Western regions; others say that important new science continued to be done in the East until the fifteenth and sixteenth centuries. However, no one denies that Islamic science and medicine reached their historical golden age in the centuries surrounding the year 1000 and that decline in the creative level of original work eventually set in. It should be noted that such a consensus

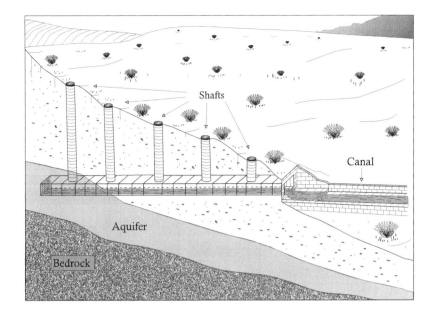

Fig 6.2. Qanat technology. Artificial irrigation sustained Islamic agriculture and civilization. Islamic engineers developed sophisticated hydraulic techniques, including qanats, which tapped underground water sources.

has tended to obscure the ways in which knowledge in mosques and madrasas continued to function in Islamic society for centuries, irrespective of any "decline" in the quality of original science. That point notwithstanding, several suggestions have been offered to account for the eventual decline of the Islamic scientific traditions, all of them external and sociological, for nothing in the internal logic of scientific ideas can account for the loss of vigor of Islamic science.

The main thesis has centered on the ultimate triumph of religious conservatives within Islam. As a religion, Islam emphasizes submission before the divine and unknowable nature of God/Allah. Thus, according to the "marginality" thesis, the cultural values and legal tenets of Islam proved such that secular philosophy and learning were always suspect to varying degrees and remained peripheral to the mainstream of Islamic society. Individual jurists and religious leaders, for example, could and did sometimes issue religious rulings (*fatwas*) for impiety against those who became too expert in the secular sciences. Different factions within Islam contended over the value of human reason and rationality in pursuing understanding, but ultimately, so the argument goes, religious conservatives prevailed, and with increasing intolerance the creative spirit of Islamic science evaporated. Why it flourished and why it declined when it did lie beyond the reach of marginalist explanations.

A related suggestion notes that Islamic civilization was more pluralistic at its outset and that science declined as the Islamic world became culturally more homogeneous. In many conquered areas religious believers were initially in the minority. Islam began as a colonial power, and especially at the edges of the Islamic imperium multicultural societies flourished at the outset, mingling diverse cultures and religions—Persian, Indian, Arab, African, Greek, Chinese, Jewish, and Christian. As time went on, conversions increased, and Islam became religiously more rigid and culturally less heterogeneous. Not until the fourteenth century was Islamicization fully complete in many areas. Consequently, the cultural "space" for creative scientific thinkers narrowed and so, again, the scientific vitality of Islam weakened commensurately. However, this account flies in the face of the fact that in its heyday Islamic science often flourished in the most Islamicized centers, such as Baghdad.

War and sociocultural disruptions occasioned by war have likewise been invoked as factors in the decline of Islamic science. In Spain the Islamic world began to be pressured by Christian Europe in the eleventh century, with Toledo falling in 1085, Seville in 1248, and the *reconquista* completed in 1492. In the East, Mongol armies from the steppes of Asia attacked the Islamic caliphate, invading and capturing Baghdad in 1258. Mongol invaders led by Timur (Tamerlane) returned to the Middle East at the turn of the fifteenth century, destroying Damascus in 1402. Although Islamic culture and institutions in the East quickly rebounded from these invasions, the overall effect, or so it is claimed, reinforced religious conservatism and disrupted the conditions necessary for the pursuit of science.

Other experts have focused on the economic decline of Islamic civilization after 1492 as a contributing factor in the cultural decline of its science. That is, once European seafaring traders penetrated the Indian Ocean in 1497, the Islamic world lost its monopoly on the valuable East Asian spice and commodity markets. In such shrinking economic circumstances, the argument suggests that science could hardly have been expected to flourish, especially since it leaned heavily on government support.

Each of these interpretations doubtless possesses some truth, and further historical research will shed more light on understanding the decline of Islamic science. But commentators have also wanted to explain not the decline of Islamic science but the very different question of why modern science did not emerge within the context of Islamic civilization. The question often posed is why, given the advanced state of Islamic science, no Scientific Revolution developed within Islam— why did Islamic scientists not repudiate the earth-centered cosmology of antiquity, expound modern heliocentrism, and develop inertial, Newtonian physics to account for motion in the heavens and on Earth? Much intellectual energy has been expended in dealing with the Islamic "failure" to make the leap to modern science. But to undertake to explain in retrospect the absolute myriad of things that *did not* happen in history confounds the enterprise of historians, who have a difficult enough time rendering plausible accounts for what *did* happen. As evident in this chapter, Islamic science flourished for several centuries, securely assimilated in observatories, libraries, madrasas, mosques, hospitals, and ruling courts. That was its positive achievement. Islamic scientists all labored within the pale of Islam, and they continued to do so for several centuries following the peak of Islamic scientific achievement. To suggest that science somehow "ought" to have developed as it did in the West misreads history and imposes chronologically and culturally alien standards on a vibrant medieval civilization.

The Middle Kingdom

Although borders and political units fluctuated, Chinese emperors controlled a huge, densely populated territory about the size of Europe. Even China proper (excluding Manchuria, Mongolia, Tibet, and western regions) encompassed about half the area of Europe and was still seven times the size of France (see map 7.1). From its first unification in 221 BCE China was the world's most populous country, and except for being briefly eclipsed by the patchwork Roman Empire, the successive empires of China stood as the largest political entities in the world. The population of China proper reached 115 million in 1200 CE, twice that of contemporary Europe and with nearly five times Europe's population density.

Geography isolated China from outside influences more than any other Old World civilization. Nomadic and pastoral peoples to the north and west had a large effect on Chinese history, to be sure, but mountains, deserts, and inhospitable steppe ringed China on the southwest, west, and north, and impeded contact with cultural and historical developments in West Asia and Europe. The earliest Chinese civilization arose in the valley of the Hwang-Ho (the Yellow River), and only later in historical periods did civilization spread to the lower valley and flood plain of the Yangtze River. China represents an archetypal hydraulic civilization whose cultural orientation faced eastward along these and related river and lake systems.

The technology of writing developed independently in China. A complex ideographic type of writing can be seen in the "oracle bone script" in the Shang dynasty (~1600-1046 BCE). It became highly developed with upwards of 5,000 characters by the ninth century BCE, and characters became standardized by the time of China's unification. Hundreds of basic signs could be combined into thousands (indeed, tens of thousands) of different characters. Because of this complexity and because each Chinese written word embodies phonetic and pictographic elements, Chinese writing was (and is) difficult to master. In

Map 7.1. China. Chinese civilization originated along the Hwang-Ho (Yellow) River in the second millennium BCE. Mountains, deserts, and steppe regions in the north and west effectively cut China off from the rest of Asia. The first unification of China occurred in the third century BCE, producing the largest and most populated political entity in the world. The map shows two of the great engineering works of Chinese civilization: the Great Wall and the Grand Canal.

adhering to the ideographic mode, Chinese writing did not simplify phonetically or syllabically as did ancient Egyptian, Sumerian, and Old Babylonian, but that "obstacle" did not impede the long, unbroken tradition of Chinese literacy and the impressive Chinese literary and scientific record from the second millennium BCE.

China embodies thousands of years of cultural continuity, and one cannot adequately trace here the intricate social and political changes observable in China's history as various empires rose and fell (see table 7.1) Nevertheless, the Song dynasties (960–1279 CE) and the "renaissance" accompanying the Song command attention. In many ways the Song period represents the zenith of traditional China. The several centuries of Song rule formed the golden age of Chinese science and technology, and they provide an effective point of contrast with contemporary developments elsewhere in the world.

The flowering of China under the Song resulted from agricultural changes, notably the upsurge of rice cultivation in South China and in the Yangtze basin beginning in the eighth century. Rice paddies produce a higher yield per acre than any other cultivated crop, so the mere introduction of rice inevitably produced significant social and cultural consequences. After 1012 the government introduced and systematically distributed new varieties of early-ripening and winter-ripening rice from Indochina. Some varieties ripened in 60 days, allowing two and even three harvests a year in favored locales. Other varieties required less water, which meant that new lands could be brought under cultivation. The Song made major efforts to extend rice production by reclaiming marshlands and lakesides, by terracing, and by improving irrigation, all under government direction. The new technique of planting out seedling rice plants eliminated the need for fallow, and the introduction of new tools for cultivating rice, such as the rice-field plow and paddle-chain water-lifting devices, likewise improved efficiency and productivity enough to provide increasingly large surpluses.

The consequences proved dramatic. The population of China more than doubled from 50 million in 800 CE to 115 million (one census reports 123 million) in 1200. The center of gravity of Chinese civilization shifted south, with more than twice as many Chinese living in the south than the north by 1080. Urbanization likewise skyrocketed. According to one report, Song dynasty China contained five cities with populations of more than a million, and another estimate puts the urban population at 20 percent of the total, a remarkably high figure for an agrarian society, one not reached in Europe until the nineteenth century. A system of imperial roads and way stations connected and

Fig. 7.1. Chinese pictographic writing. Written Chinese languages derive from pictographic antecedents. Various characters can be combined to form new word signs. In some cases parts of characters indicate the sound of the word and/or distinguish the general class of things to which the word in question belongs. Unlike other languages with pictographic origins, written Chinese never simplified into a wholly phonetic or sound-based script, and literacy in China still entails mastery of hundreds of separate word signs. Such difficulties did not prevent the Chinese language from developing sophisticated technical and scientific vocabularies.

Table 7.1
A Chronology of Chinese Dynasties

Early China	
Xia Dynasty	21st–17th centuries BCE
Shang Dynasty	~1600–1046 BCE
Zhou Dynasty	
Western Zhou	1046–771 BCE
Eastern Zhou	770–256 BCE
Warring States Period	475–221 BCE
Early Imperial China	
Qin Dynasty	221–206 BCE
Han Dynasty	206 BCE–220 CE
Three Kingdoms	220–280 CE
Jin Dynasties	265–420
Northern and Southern Dynasties	
Southern Dynasties	420–589
Northern Dynasties	386–581
Classical Imperial China	
Sui Dynasty	581–618
Tang Dynasty	618–907
Five Dynasties	907–960
Liao Dynasty	907–1125
Jin Dynasty	1115–1234
Song Dynasty	
Northern Song	960–1127
Southern Song	1127–1279
Later Imperial China	
Yüan (Mongol) Dynasty	1279–1368
Ming Dynasty	1368–1644
Qing (Manchu) Dynasty	1644–1911
Post-Imperial China	
Republic of China	1912–
People's Republic of China	1949–

organized the vast network. A leisured middle class arose along with the commercialization of agricultural commodities, increased trade, and expanded manufacturing.

Centralization of power in the hands of the emperor and rule by a governing bureaucracy—the mandarinate—reached new heights under the Song. The "mandate of heaven" dictated that the Chinese emperor rule all of China, and an improved civil service to support that mandate proved pervasive in Chinese life. The bureaucracy was huge and monolithic; a later report from Ming times puts the number of state civil servants at 100,000, exclusive of military officers. Such organization allowed direct control by the emperor down to the village level. No intermediary or independent bodies existed in China to challenge the authority of the emperor and the mandarinate. Different traditional provinces and linguistic regions acted as something of brakes to centralizing forces, but no other formal centers of power existed. Towns and cities were neither autonomous nor separate administrative units. Such an exclusive and centralized administration prevented the rise of independent institutional entities, notably colleges or guilds. The omnipresence of the Chinese mandarinate seems also to have restricted any neutral social or intellectual space for science or technology outside of official channels.

The teachings of the Chinese sage Confucius (551–479 BCE) dramatically shaped Chinese high culture, particularly through the official state ideology of Neo-Confucianism elaborated by commentators in Song times. The Confucian outlook focused on the family, humanity, and society, not nature or the world outside of human affairs. Confucianism was a practical philosophy that emphasized the ethical and moral dimensions of behavior and statecraft and the maintenance of a just and harmonious society. Thus, custom, etiquette, virtuous behavior, filial piety, respect for one's elders, submission to authority, the moral example of the sage, and justice (but not law) became the watchwords of Confucianism in Song times. In these ways Confucianism sustained the status quo and the paternalistic and patriarchal society of the day.

The power and appeal of the imperial bureaucracy drained talent that might have flowed into science. The bureaucracy skewed scholarship toward the humanities and the Confucian classics, and it helped enforce a deep divide between learned culture and the crafts. Under the Song, the imperial bureaucracy operated as a true meritocracy open to talent. The state recruited functionaries not through political or hereditary connections, but rather based on ability and performance on exacting state civil-service exams, which provided virtually exclusive access to political power. Already in Han times (206 BCE–220 CE) Chinese officials instituted the system of state examinations, one effect of which was to restrict the political power of the nobility. The Song dynasties reformed the system, and it reached its high point under their rule and continued in effect in China until 1904.

An official board of examiners offered three levels of examination (local, regional, and national) every two to three years. Some unfortunate students devoted their whole lives to taking and retaking the rigorous exams. Passage of even the lowest level exam brought special privileges, such as exemption from the forced labor of the corvée. Passing at the high levels incurred obligations since laureates could not refuse employment in the bureaucracy. Based on a standardized subject matter, the exams focused on Confucian classics, on esoteric literary and humanistic studies, and, under the Song, on administrative problems. Memorization took pride of place, along with recitation, poetry, and calligraphy. With the emphasis on moral learning and the goal of producing a scholar-gentry to rule the country, the civil service exams shaped the values and efforts of the best and brightest Chinese minds for nearly 2,000 years. Certain exceptions aside, science and technology did not figure in the exam system.

Outside the bureaucracy, other elements of society lacked the power and autonomy to be independent sources of any nascent scientific tradition. If the exam system effectively precluded rule by nobles, civilian authority also managed to subordinate military and merchant classes to its power. From the third century BCE China possessed large armies—on the order of a million soldiers. (Song armies numbered 1,259,000 men in 1045.) Yet, the military remained subject to civilian control. Military power was divided, units split up, and overlapping commands established. Merchant activity was likewise tightly controlled so that, unlike in Europe, merchants never rose to social or institutional positions of consequence. From the Confucian point of view, merchant activity, profit, and the accumulation of private wealth were disdained as antisocial vices. Merchants occasionally flourished and achieved great wealth, but periodic prosecutions and confiscations ensured the marginality and low status of merchants as a class in Chinese society. Likewise, the suppression of religious institutions in 842–845 CE, following a period of Buddhist prominence, meant that no clergy could challenge the predominance of the bureaucracy.

The Flowering of Chinese Technology

Learned culture in traditional China was largely separate from technology and the crafts. Calendrical astronomy benefited the state and society, and mathematics played a role in the solution of practical problems, but economic, military, and medical activities were, on the whole, carried out on the strength of traditional techniques that owed nothing to theoretical knowledge or research. Craftsmen were generally illiterate and possessed low social status; they learned practical skills through apprenticeship and experience, and they plied their trades without benefit of scientific theory. Scholars and "scientists," on the other hand, were literate, underwent years of schooling, enjoyed high social status, and remained socially apart from the world of artisans and engineers. The exam system and the bureaucracy, by institutionally segregating scholar-bureaucrats from artisans, craftsmen, and engineers, strengthened the separation of science and technology. The value system of traditional China, like that of Hellenic Greece, looked down upon crass technology. Scholars and literati repudiated working with their hands and preferred more refined concerns such as poetry, calligraphy, music, and belles-lettres.

In considering Chinese technology one must be wary of a tendency to record the priority of the Chinese over other civilizations for this or that invention: the wheelbarrow, the south-pointing chariot, lacquer, gunpowder, porcelain china, the umbrella, the fishing reel, the seed drill, the rotary winnowing fan, the crossbow, suspension bridges, and so on. While such "firsts" are interesting, they are of limited analytical value in historical inquiry. Rather, the starting point for any investigation of Chinese technology must be the realization that the totality of its advanced technologies, regardless of their originality or priority, made China a world leader in technology through the Song era and beyond.

Government control of industries was a characteristic feature of Chinese technology. The government nominally owned all resources in the country, and it monopolized production in key sectors by creating government workshops and state factories for such industries as mining, iron production, salt supply, silk, ceramics, paper-making, and alcoholic beverages. Through these monopolies run by bureaucrats, the Chinese state itself became a merchant producer, in large part to provide for its enormous military needs. The government commanded a vast array of specialized craftsmen, and anyone with technical skills was ostensibly subject to government service. Porcelain manufacturies turned out pieces by the millions; bolts of silk issued in like numbers. The Yüan emperors, for example, enlisted 260,000 skilled artisans for their own service; the Ming commanded 27,000 master craftsmen, each with several assistants; and in 1342, 17,000 state-controlled salt workers toiled along the lower Yangtze.

State management of technology and the economy reached a high point in the Song period when more government income came from mercantile activity and commodity taxes than from agricultural levies. One result was the spread of a monied economy. Coinage issuing from government mints jumped from 270,000 strings (of a thousand coins) in 997 to 6 million strings in 1073. As a result of that increase, the Song began issuing paper money in 1024, and paper money became the dominant currency in twelfth- and thirteenth-century China. The technology of paper money is significant not as a world historical "first," but because it facilitated the growth and functioning of Chinese civilization.

Hydraulic engineering represents another basic technology underpinning Chinese civilization. We earlier encountered the essential role of irrigation agriculture in discussing the initial rise of civilization along the Hwang-Ho river in the second millennium BCE. While many canals and embankments existed in China from an early date, the first elements of an empire-wide inland canal system appeared about 70 CE. Engineers completed the nearly 400 miles of the Loyang to Beijing canal in 608 CE and by the twelfth century China possessed some 50,000 kilometers (31,250 miles) of navigable waterways and canals. Completed in 1327, the Grand Canal alone stretched 1100 miles and linked Hangchow in the south with Beijing in the north, the equivalent of a canal from New York to Florida. After the Ming took power they repaired 40,987 reservoirs and launched an incredible reforestation effort in planting a billion trees to prevent soil erosion and to supply naval timber. Of course, such impressive engineering was impossible without the central state to organize construction, to levy taxes, and to redistribute the agricultural surplus. Canals allowed rice to be shipped from agricultural heartlands in the south to the political center in the north. One report has 400,000 tons of grain transported annually in the eleventh century. In Ming times 11,770 ships manned by 120,000 sailors handled inland shipping. Considerable maintenance and dredging were obviously required, all of it carried out by corvée labor, and the neglect of hydraulic systems inevitably led to famine and political unrest.

Pottery was an ancient craft that reached unprecedented artistic heights after the eleventh century. The imperial government owned its own industrial-scale kilns and workshops which came to employ thousands of craftsmen mass-producing both commonplace and luxury items. The Chinese originated porcelain—a mixture of fine clays and minerals fired at a high temperature—at the end of Han times and perfected porcelain wares in the twelfth century. The enduring art and technology of Chinese porcelain represent one of the great cultural achievements of the Song and Ming eras. They bespeak a wealthy and cultivated society, and, indeed, ceramics became a major item of internal and international commerce and of tax income for the state. Chinese pottery made its way through the Islamic world and to Africa. From the Middle Ages onward Europeans came to covet Chinese porcelains, and efforts to duplicate Chinese ceramic technology proved a spur to the pottery industry in Europe at the time of the Industrial Revolution in the eighteenth century.

Textiles constitute another major industry in traditional China. One twelfth-century Song emperor, for example, purchased and received in taxes a total of 1.17 million bolts of silk cloth. The Chinese textile industry is especially notable because of its mechanized character. Sources document the presence of the spinning wheel in China from 1035 CE, and Chinese technologists also created elaborate, water-powered reeling machines to unwind silkworm cocoons and wind silk thread onto bobbins for weaving into cloth. And paper manufacturing, possibly evolving out of the textile industry, provided a product that facilitated the administration of imperial China. Solid evidence exists for paper from late Han times early in the second century CE, although the technology may have originated several centuries earlier.

Chinese bureaucracies depended on writing, literary traditions, and libraries, which already existed in the Shang dynasty in the second millennium BCE. Although paper entered Chinese society at an early date, the technology of taking rubbings from carved inscriptions may have delayed the advent of printing until the first decade of the seventh century. Printing—block printing—at first simply reproduced seals for religious charms. The first book printed by means of whole pages of carved woodblock appeared in 868 CE, and the technology of printing soon recommended itself to government authorities who used it to print money, official decrees, and handbooks, particularly useful ones in medicine and pharmacy. An official printing office supplied printed copies of the classics to be studied for the civil-service exams, and overall the Chinese government produced an impressive output of printed material in the service of the bureaucracy. The first emperor of the Song dynasty, for example, ordered a compilation of Buddhist scripture, and the work, consisting of 130,000 two-page woodblocks in 5,048 volumes, duly appeared. In 1403 an official Chinese encyclopedia numbered 937 volumes and another of 1609 comprised 22,000 volumes written by 2,000 authors. Printed popular literature also circulated among the elites.

The Chinese invented movable type around 1040, first using ceramic characters. The technology developed further in Korea where the Korean government had 100,000 Chinese characters cast in 1403. But movable type proved impractical compared to woodblock printing, given the style of Chinese writing with pictograms and the consequent need for thousands of different characters. Block printing thus proved not only cheaper and more efficient, but it allowed illustrations, often in many colors. The ability to reproduce pictures put China well ahead of the West in printing technology even after Gutenberg developed movable type in Europe.

Chinese superiority in iron production likewise helps account for the vibrancy of its civilization. Possibly because of limited resources of copper and tin for bronze, Chinese metallurgists early turned to iron. By 117 BCE iron production had become a state enterprise with 48 foundries, each employing thousands of industrial laborers. Production zoomed from 13,500 tons in 806 CE to 125,000 tons in 1078 in the Song period, doubtless because of increased military demands. (By contrast England produced only 68,000 tons of iron in 1788 as the Industrial Revolution got under way in Europe.) Technologically innovative and advanced, the Chinese iron industry used water-powered bellows to provide a blast and smelted the ore with coke (partially combusted coal) by the eleventh century, some 700 years before like processes arose in Europe. By dint of such superior technology Song military arsenals turned out 32,000 suits of armor and 16 million iron arrowheads a year, as well as iron implements for agricultural use.

The invention of gunpowder in mid–ninth-century or early tenth-century China, and, more significantly, the application of gunpowder to military ends beginning in the twelfth century redirected the course of Chinese and world history. Gunpowder seems to have emerged from traditions of Chinese alchemical research and so is a notable historical instance of applied science or a new technology stemming from the world of learning. Its initial use in fireworks was intended not as a military device but as a means to ward off demons. Only as they became threatened by foreign invasion did Song military engineers improve the

Fig. 7.2. Chinese geomancer. Prior to laying out a new city, an expert geomancer or *fêng shui* master consults a compass-like device to ascertain the flow of energy (*chi*) in the locale. He will then use his readings to situate artificial structures in harmony with their natural surroundings.

formula for gunpowder and develop military applications in rockets, explosive grenades, bombs, mortars, and guns.

Unlike paper, the magnetic compass was a technology that Chinese civilization could manage without, but along with gunpowder the case illuminates the few ties between science and technology in traditional China. The mysterious properties of the loadstone—the natural magnetism of the mineral magnetite—were known by 300 BCE and first exploited for use as a fortuneteller's device. Knowledge attained by 100 BCE that a magnetic needle orients itself along a north-south axis was then applied in geomancy or *fêng-shui*, the proper siting of houses, temples, tombs, roads, and other installations. An elaborate naturalistic theory later arose to explain the movement of the compass needle in response to energy currents putatively flowing in and around the earth, an example of how, contrary to conventional wisdom today, technology sometimes fosters speculations about nature rather than the reverse.

Sources fail to attest to the use of the compass as a navigational tool at sea until Song times early in the twelfth century. China entered late as a major maritime power, but from the period of the Southern Song through the early Ming dynasties, that is, from the twelfth through the early fifteenth centuries, China developed the largest navy and became the greatest maritime power in the world. Hundreds of ships and thousands of sailors composed the Song navy. Kublai Khan, Mongol founder of the Yüan dynasty, attempted an invasion of Japan in 1281 with a navy of 4,400 ships. The Ming navy in 1420 counted 3,800 ships, of which 1,300 sailed as combat vessels. The Ming launched an official shipbuilding program and constructed 2,100 vessels in government shipyards between 1403 and 1419. With compasses, watertight compartments, up to four decks, four to six masts, and the recent invention of a sternpost rudder, these were the grandest, most seaworthy, technologically sophisticated vessels in the world. The largest approached 300 feet in length and 1,500 tons, or five times the displacement of contemporary European ships. Armed with cannon and carrying up to 1,000 sailors, they were also the most formidable.

The Ming used their powerful navy to assert a Chinese presence in the waters of South Asia and the Indian Ocean. From 1405 to 1433 they launched a series of seven great maritime expeditions led by the Chinese admiral Cheng Ho (also known as Zheng He). With several

Fig. 7.3. European and Chinese ships. Chinese civilization abandoned maritime exploration and left the Indian Ocean early in the fifteenth century, just decades before European sailors entered the region. The ships of the Chinese admiral Cheng Ho were much larger than European vessels, as shown in this imaginary comparison.

dozen ships and more than 20,000 men on each voyage, Cheng Ho sailed to Vietnam, Thailand, Java, and Sumatra in southeast Asia, to Sri Lanka and India, into the Persian Gulf and the Red Sea (reaching Jedda and Mecca), and down the coast of East Africa, possibly as far as Mozambique. The purpose of these impressive official expeditions seems to have been political, that is, to establish the authority and power of the Ming dynasty, and on at least one occasion Cheng Ho used force to assert his authority. With these initiatives, the Ming acquired a number of vassal states, and at least two Egyptian diplomatic missions wound their way to China.

Then, abruptly, the extraordinary maritime thrust of the Ming came to an end. Official shipbuilding ceased in 1419, and a decree of 1433 put an end to further Chinese overseas expeditions. No one can say whether the course of world history would have been radically different had the Chinese maintained a presence in the Indian Ocean and rebuffed the Portuguese when they arrived with their puny ships at the end of the same century. Several explanations have been offered to account for the stunning reversal of Chinese policy. One notion suggests that the Chinese repudiated overseas adventures because Cheng Ho was a Muslim and a eunuch, qualities reminiscent of the oppressive Mongol/Yüan years and greatly in disfavor among the nationalistic Ming. Another envisions the expeditions as merely the somewhat idiosyncratic initiative of two Ming emperors, and not as growing organically out of contemporary Chinese society and economy. A strong technical argument has also been advanced. Restoration of the Grand Canal began in 1411–15, and in 1417 the construction of deepwater ("pound") locks on the Grand Canal allowed a year-round link between the Yangtze and Yellow Rivers. Accordingly, the Ming transferred their capital from Nanking in the south to Beijing in the north, and as a result, the need for a strong navy or foreign adventures supposedly disappeared.

One way or another, Ming China turned inward, and a degree of technological stagnation set in. China remained a great and powerful civilization, but the dynamism and innovative qualities of the Song era no longer obtained. Only with its encounter with the West beginning in the seventeenth century would technological innovation once again move China.

The World as Organism

In approaching the subject of the natural sciences in traditional China, one must avoid the tendency, similar to that already observed with regard to Chinese technology, to place undue emphasis on a search for "first" honors in scientific discovery: first recognizing the nature of fossils, first using Mercator projections in maps and star charts, discovering Pascal's triangle and the mathematics of binomials, foreshadowing the even-tempered musical scale, or, particularly far-fetched, crediting alternations of yin and yang as anticipations of the "wave theory" of today's quantum physics. Such claims reflect a perverse judgmentalism and a desire, in the name of multicultural relativism, to inflate the accomplishments of Chinese science while devaluing those of the West. Instead, the present section emphasizes the social history of Chinese science rather than a chronology of discovery, and it strives to show that the relationship between science and society in traditional China parallels the other primary civilizations of the Old World: useful knowledge patronized by the state and developed in support of statecraft and civilization generally.

Any historical evaluation of Chinese science must overcome several further obstacles. The Western concept of science or natural philosophy remained foreign to intellectual thought in traditional China. As one author put it, "China had sciences, but no science." That is, learned experts pursued various scientific activities—in astronomy, astrology, mathematics, meteorology, cartography, seismology, alchemy, medicine, and related studies—but nothing united these separate endeavors into a distinct enterprise of critical inquiry into nature. Indeed, the Chinese language possessed no single word for "science." China, like Egypt and the other bureaucratic civilizations, lacked natural philosophy in the Hellenic sense, and one gathers that Chinese thinkers would have been perplexed by the notion of pure science pursued for its own sake. Chinese society provided no social role for the research scientist, and no separate occupation or distinct profession of science existed. Instead, elite amateurs and polymaths pursued scientific interests, often, perhaps furtively, when employed to gather and apply useful knowledge in a bureaucratic setting.

The traditional Chinese outlook conceived of nature in more holistic and organismic terms than did the West. Already in Han times, a conception emerged that envisaged the universe as a vast, single organism in which the world of nature and the social world of humankind merge in a complete unity. Heaven and Earth along with man and nature harmoniously coexisted, the celestial and the human linked through the person of the emperor. From the Chinese philosophical perspective, the two great complementary forces of yin and yang governed change in nature and in human affairs. In addition, the constituent five "phases" or elements of wood, fire, earth, metal, and water played dynamic roles in making up the world. The outlook was qualitative, and it emphasized recurring cycles, as yin, yang, and one or another of the elemental "phases" assumed predominance over the others. Notably, this five-element (Wuxing) theory further correlates associations with the cardinal directions, climate and the seasons, the major planets, parts of the body, and much more. A medical dimension is thus part of this picture, and traditional Chinese medicine derived diagnoses and treatments from this view of the world. This vision of the five elements

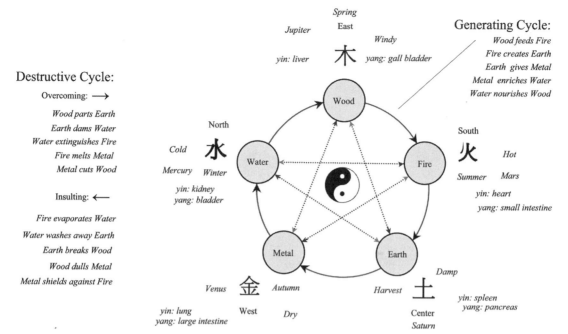

Destructive Cycle:

Overcoming: ⟶

Wood parts Earth
Earth dams Water
Water extinguishes Fire
Fire melts Metal
Metal cuts Wood

Insulting: ⟵

Fire evaporates Water
Water washes away Earth
Earth breaks Wood
Wood dulls Metal
Metal shields against Fire

Generating Cycle:

Wood feeds Fire
Fire creates Earth
Earth gives Metal
Metal enriches Water
Water nourishes Wood

Spring
Jupiter East
Windy
yin: liver 木 *yang: gall bladder*

Wood

North
Cold 水
Mercury *Winter* Water
yin: kidney
yang: bladder

South
火 *Hot*
Fire
Summer *Mars*
yin: heart
yang: small intestine

Metal

Earth

Damp
Venus 金 *Autumn*
Harvest 土
yin: spleen
yang: pancreas
yin: lung West *Dry*
yang: large intestine
Center
Saturn

Fig. 7.4. The Five Phases. The traditional Chinese view of the material world and the dynamics of change. Elaborate cycles of yin and yang produce a continually waxing and waning of the elements of wood, fire, earth, metal, and water. These changes further connected to the seasons, to astrology, and to the theory and practice of traditional Chinese medicine.

resembles that of the Aristotelian elements and associated Greek humoral medicine we saw earlier, but it would be a mistake to think of these two highly complex conceptualizations of the nature of nature as in any way the same or causally connected. In considering Chinese scientific thought, then, one must acknowledge that Chinese intellectuals lived in a world separated from the West by more than geography.

On a more mundane level, although schools abounded in China, Chinese educational institutions did not incorporate or provide instruction in the sciences. Founded in the eighth century CE, an Imperial Academy in the capital topped a complex educational structure, with a central Educational Directorate superintending a standardized Confucian curriculum for the empire. A host of private academies following the standard curriculum also blossomed. Unlike European universities, none of these schools possessed a legal charter granting them a permanent, independent existence. All existed by tradition and the will of the emperor. They could be and were closed simply by decree. Furthermore, the entire focus of these schools—public and private alike—was careerist and directed to preparing students to take the state civil-service exams. None granted degrees. Even the Imperial Academy was merely another bureau at which scholarly functionaries taught for limited periods of time, and only one such academy existed in the whole of China, compared to Europe in the following centuries with its scores of autonomous colleges and universities. Although authorities established separate schools of law, medicine, and mathematics around the year 1100 CE, none survived very long. The sciences simply did not figure in Chinese education or educational institutions.

These cultural and institutional impediments notwithstanding, the necessities of imperial administration dictated that from its inception the Chinese state had to develop bureaucratically useful knowledge and recruit technical experts for its service. In a typical fashion, like writing, applied mathematics became a part of the workings of Chinese civilization. By the fourth century BCE the Chinese developed a decimal place-value number system. Early Chinese mathematics used counting rods and, from the second century BCE, the abacus to facilitate arithmetical calculations. By the third century BCE Chinese mathematicians knew the Pythagorean theorem; they dealt with large numbers using powers of 10; they had mastered arithmetic operations, squares, and cubes, and, like the Babylonians, they handled problems we solve today with quadratic equations. By the thirteenth century CE the Chinese had become the greatest algebraists in the world.

While the record occasionally indicates Chinese mathematicians engaged in the seemingly playful exploration of numbers, as in the case of the calculation of π to 7 decimal places by Zu Chougzhi (429–500 CE), the overwhelming thrust of Chinese mathematics went toward the practical and utilitarian. The first-century CE text, *Nine Chapters on the Mathematical Art* (*Jiu Zhang Suan Shu*), for example, took up 246 problem-solutions dealing with measurements of agricultural fields, cereal exchange rates, and construction and distribution problems. To solve them, Chinese mathematicians used arithmetic and algebraic techniques, including simultaneous "equations" and square and cube roots. Indian influences made themselves felt in Chinese mathematics in the eighth century, as did Islamic mathematics later. Characteristically, Chinese mathematicians never developed a formal geometry, logical proofs, or deductive mathematical systems such as

Chinese Numerals
(Base 10)

Fig. 7.5. Chinese numerals (base 10). Chinese civilization developed a decimal, place-value number system early in its history. Supplemented with calculating devices such as the abacus, the Chinese number system proved a flexible tool for reckoning the complex accounts of Chinese civilization.

those found in Euclid. The social history of Chinese mathematics reveals no reward system for mathematicians within the context of the bureaucracy. Mathematicians worked mostly as scattered minor officials, their individual expertise squirreled away in separate bureaus. Alternatively, experts wandered about without any institutional affiliation. The three greatest contemporary Song mathematicians (Ts'in Kieou-Chao, Li Ye, and Yang Housi), for example, had their works published, but did not know each other, had different teachers, and used different methods. In considering the character and social role of Chinese mathematics, one must also factor in a strong element of numerology and traditions of mathematical secrets, all of which tended to fragment communities and disrupt intellectual continuity.

A pattern of state support for useful knowledge, characteristic of centralized societies, is nowhere more evident than in Chinese astronomy. Issuing the official calendar was the emperor's exclusive prerogative, a power apparently exercised already in the Xia dynasty (21st to 17th centuries BCE). Like their Mesopotamian counterparts, Chinese calendar-keepers maintained lunar and solar calendars—both highly accurate—and they solved the problem of intercalating an extra lunar month to keep the two in sync like the Babylonians by using the so-called Metonic cycle of 19 years and 235 lunations, that is, twelve years of twelve lunar months and seven years of thirteen lunar months.

Because disharmony in the heavens supposedly indicated disharmony in the emperor's rule, astronomy became a matter of state at an early period and the recipient of official patronage. Professional personnel superintended astronomical observations and the calendar even before the unification of China in 221 BCE, and soon an Imperial Board or Bureau of Astronomy assumed jurisdiction. Astronomical reports to the emperor were state secrets, and because they dealt with omens, portents, and related politico-religious matters, official astronomers occupied a special place in the overall bureaucracy with offices close to the emperor's quarters. Chinese astronomers played so delicate a role that they sometimes altered astronomical observations for political reasons. In an attempt to prevent political tampering, astronomical procedures became so inflexible that no new instruments or innovations in technique were permitted without the express consent of the emperor, and edicts forbade private persons from possessing astronomical instruments or consulting astronomical or divinatory texts.

Marco Polo (1254–1324), the Italian adventurer who served for 17 years as an administrator for the Mongol/Yüan dynasty, reported that the state patronized 5,000 astrologers and soothsayers. Special state exams—given irregularly outside the standard exam system—recruited mathematicians and astronomers for technical positions within the bureaucracy. Unlike the rest of the bureaucracy, families tended to monopolize positions requiring mathematical and astronomical expertise, with jobs handed down from one generation to another. The rules prohibited children of astronomers from pursuing other careers and, once appointed to the Astronomical Bureau, individuals could not transfer to other agencies of government.

The Chinese developed several theories of the cosmos, including one wherein the celestial bodies float in infinite empty space blown by a "hard wind." From the sixth century CE the official cosmology consisted of a stationary earth at the center of a great celestial sphere. Divided into twenty-eight "lunar mansions" corresponding to the daily progress of the moon in its monthly passage through the heavens, this sphere turned on a grand axis through the poles and linked heaven and earth. The emperor, the "son of heaven," stood as the linchpin of this cosmology, while China itself rested as the "middle kingdom" among the four cardinal points of the compass.

Although weak in astronomical theory, given the charge to search for heavenly omens, Chinese astronomers became acute observers. With reliable reports dating from the eighth century BCE and possibly from the Shang dynasty several centuries earlier, the range of Chinese observational accomplishments is impressive. The richness of documentary material reveals that, already in the fourth century BCE, Chinese astronomers measured the length of the solar year as 365¼ days. The north star and the circumpolar stars that were always visible in the night sky received special attention from Chinese astronomers who produced systematic star charts and catalogues. Chinese astronomers recorded 1,600 observations of solar and lunar eclipses from 720 BCE, and developed a limited ability to predict eclipses. They registered seventy-five novas and supernovas (or "guest" stars) between 352 BCE and 1604 CE, including the exploding star of 1054 (now the Crab Nebula), visible even in the daytime but apparently not noticed by Islamic or European astronomers. With comets a portent of disaster, Chinese astronomers carefully logged twenty-two centuries of cometary observations from 613 BCE to 1621 CE, including the viewing of Halley's comet every 76 years from 240 BCE. Observations of sunspots (observed through dust storms) date from 28 BCE. Chinese astronomers knew the 26,000-year cycle of the precession of the equinoxes. Like the astronomers of the other Eastern civilizations, but unlike the Greeks, they did not develop explanatory models for planetary motion. They mastered planetary periods without speculating about orbits.

Government officials also systematically collected weather data, the earliest records dating from 1216 BCE; to anticipate repairs on hydraulic installations, they gathered meteorological data on rain, wind, snowfalls, the aurora borealis ("northern lights"), and meteor showers. They also studied the composition of meteorites and compiled tide tables beginning in the ninth century CE. The social utility of this research is self-evident.

Three waves of foreign influences impacted on Chinese science. The first wave broke in the years 600–750 CE, coincident with Buddhist and Indian influences in T'ang times. Chinese Buddhists undertook pilgrimages to India from the early fifth century CE in search of holy texts. A significant translation movement developed, wherein over time nearly 200 teams of translators rendered some 1,700 Sanskrit texts into Chinese. As part of this movement, the secular sciences of India, including works in mathematics, astrology, astronomy, and medicine, made their way to China.

A second wave of foreign influence (this time Islamic) had a strong impact beginning with the Mongol conquest of China by Kublai Khan in the thirteenth century. Although not Muslims themselves, Mongol rulers employed Islamic astronomers in the Astronomical Bureau in Beijing and even created a parallel Muslim Bureau of Astronomy alongside the one for traditional Chinese astronomy; and later Ming emperors continued the tradition of a parallel Muslim Astronomical Bureau. Muslim astronomers deployed improved astronomical instruments, including a 40-foot-high gnomon, sighting tubes, and armillary spheres and rings adjusted for the Chinese (and not Western) orientation to the north celestial pole. Across the greater Mongol imperium, reciprocal Chinese-Persian contact developed in Yüan (Mongol) times (1264–1368) that included Chinese contact with Islamic astronomers at the Maraghah observatory. This tie put Chinese astronomers in touch with the works of Euclid and Ptolemy, but, consistent with their indifference to abstract science, the Chinese did not translate or assimilate these pillars of Western science before the third wave and the arrival of Europeans in the seventeenth century.

Before and after the Mongols, the Chinese used complex astronomical clocks and planetaria known as orreries. About 725 CE a Chinese artisan-engineer, Liang Ling-Tsan, invented the mechanical escapement, the key regulating device in all mechanical clocks. Using the escapement, a small tradition of clock and planetarium construction thereafter unfolded in China. This tradition reached its height at the end of the eleventh century when Su Sung (1020–1101), a Song dynasty diplomat and civil servant, received a government commission to build a machine that would replicate celestial movements and correct embarrassing shortcomings in the official calendar then in use. The Jurchen Tartars moved Su Sung's tower in 1129 after they captured Kaifeng from the Song. Finally, lightning struck it in 1195, and some years later, for want of skilled mechanics, Su Sung's great machine fell into complete disrepair. With it, Chinese expertise in mechanical horology declined, to the point where officials expressed amazement at Western clocks when they came to China in the seventeenth century. Su Sung's clock and like instruments did not seriously affect traditional practices within the Chinese Astronomical Bureau, but the case represents another historical example, not of technology derived from abstract knowledge of nature, but, quite the converse, of an independent technology applied in the service of science and scientific research.

Fig. 7.6. Su Sung's astronomical clock. Built in 1090 Su Sung's clock was an impressive feat of mechanical engineering and the most complex piece of clockwork to that point in history. Housed within a 40-foot-high tower, powered by a waterwheel, and controlled by complicated gearing, Su Sung's machine counted out the hours and turned a bronze armillary sphere and a celestial globe in synchrony with the heavens.

Earthquakes seriously affected China—800,000 people are reported to have died in a catastrophic earthquake in 1303, for example. Because it fell to the government to provide relief to outlying areas, the study of earthquakes became a practical matter of state. Earthquake records date from 780 BCE, and from Han times state astronomers of the Astronomical Bureau had the duty of recording them. Pursuant to that charge, in the second century CE Chang Heng created the remarkable "earthquake weathercock," an ingenious seismograph or earthquake detector. Many such machines existed in traditional China, and later in Mongol times they passed to Islam and the Maraghah observatory.

Cartography or map-making became yet another notable dimension of Chinese scientific expertise developed and deployed in the service of state administration. Chinese map-makers created highly accurate maps of the Chinese empire using various grid systems including what became known in the West as Mercator projections with unequal spacing of latitudes. They also produced relief maps, and in 1027 under the Northern Song they designed a wagon for measuring distances overland, and Ming cartographers produced several atlases after Cheng Ho's maritime expeditions into the Indian Ocean.

As befitted a highly centralized society, medicine was strictly regulated by the state and the practice of medicine was considered a form

Fig. 7.7. Chinese seismograph. Earthquakes regularly affected China, and the centralized state was responsible for providing relief for earthquake damage. As early as the second century BCE, Chinese experts developed the device depicted here. An earthquake would jostle a suspended weight inside a large bronze jar, releasing one of a number of balls and indicating the direction of the quake.

of public service. An Imperial Medical College came into existence in T'ang times (seventh to the tenth century CE), and physicians had to pass strict examinations. Court physicians occupied well-paid positions, and medical, like astronomical, expertise ran in families. Hospitals, or at least hospice-like organizations, arose in China out of Buddhist and Taoist philanthropic initiative, but these became state institutions after the suppression of religious foundations in 845 CE. To guide physicians, the central government issued many official textbooks dealing with general medicine, pharmacy, pediatrics, legal medicine, gynecology, and like subjects. One Song pharmaceutical document dating from around 990 CE contained 16,835 different medical recipes. The numerous botanical and zoological encyclopedias also deserve note, in part for their medicinal advice; a government official, Li Shih-Chen, compiled the *Pen Tsao Kang Mu*, or *Classification of Roots and Herbs*, which listed 1,892 medicaments in fifty-two volumes. Illustrations graced many of these books. The fact that works of natural history seem to take a special interest in insects, notably the silkworm, or that artificial breeding programs for the silkworm began early in Chinese history make plain once more that the state exploited useful knowledge across a wide range of applications.

Finally along these lines, one must not overlook magic, alchemy, and the occult sciences in traditional China. An element of the magical and the divinatory ran through Chinese medicine, astronomy, geography, and mathematics, the latter especially concerned with propitious numbers. Chinese alchemy became the most developed branch of esoteric knowledge, closely associated with Taoist religious philosophy. Popular from Han times, alchemy in the East, as in the West, was a practical science concerned with making elixirs of immortality and transmuting base metals into silver and gold, but Chinese adepts engaged in these efforts less for crass monetary benefit than from contemplative, spiritual motivations and the goal of spiritual transcendence. In some instances at least, alchemy attracted official patronage, as in the case of the Northern Wei emperor who supported an alchemical laboratory from 389 to 404 CE. Alchemists sought to duplicate natural processes carried on within the earth. They built elaborate furnaces and followed intricate alchemical procedures, and, as we saw, gunpowder emerged as an inadvertent by-product of alchemical experimentation.

As in so much else in Chinese history, a certain rigidity and decline began to affect Chinese science, medicine, and technology during the Ming dynasty in the fourteenth and fifteenth centuries CE. The reasons may well have been political. Unlike the expansive and innovative Song or the internationally open Mongols, Ming China turned inward and developed isolationist, conservative policies. Two centuries after the apogee of Chinese algebra under the Song, for example, Chinese mathematicians could no longer fathom earlier texts. A century after the great clockmaker Su Sung died, to repeat an example, no one could repair, much less duplicate, his handiwork. By the time Europeans arrived in China at the turn of the seventeenth century, the stagnation from the glory days of the Song had taken its toll for several centuries.

The third wave of foreign influence impacting on Chinese science emanated from Western Europe. The Jesuit scientist and missionary Matteo Ricci (1552–1610) arrived in Macao on the Chinese coast in 1582 and finally gained admission to Beijing in 1601. The Ming emperor, the court, and Chinese society generally remained hostile to Ricci's religion and his efforts to win converts, but they took special interest in what he could communicate of Western mathematics, astronomy, the calendar, hydraulics, painting, maps, clocks, and artillery, and the ability he brought to translate Western technical treatises into

Chinese. Indeed, Ricci himself became a court astronomer and mathematician and the titular deity of Chinese clockmakers. With Ricci leading the way, the Jesuits succeeded in their mission in China primarily because of their greater calendrical and astronomical expertise. In fact, the emperor handed over operational control of the Astronomical Bureau to the Jesuits. Ironically, Ricci brought with him not the new heliocentric astronomy of Copernicus, Kepler, and Galileo but, instead, perfected forms of Ptolemaic astronomy that Europeans had derived from Islamic sources and antiquity. In other words, the European science Ricci brought to China cannot be retrospectively praised because it was more "correct" than contemporary Chinese science. Rather, his Chinese hosts and employers valued it by the only measure that counted, its superior accuracy and utility in a Chinese context.

With the arrival of Ricci in China the subsequent history of Chinese science largely becomes its integration into ecumenical, world science.

Illicit Questions

As the diversity and sophistication of Chinese scientific traditions have become more evident to scholars over the last decades, a fundamental explanatory question has emerged: why the Scientific Revolution did not occur in China. As detailed in part 3, the umbrella term "Scientific Revolution" refers to the historical elaboration of modern science and the modern scientific worldview in Europe in the sixteenth and seventeenth centuries: the shift to a sun-centered planetary system, the articulation of a universal principle to explain celestial and terrestrial motion, the development of new approaches to the creation of scientific knowledge, and the institutionalization of science in distinct institutions. Since medieval China was scientifically and technologically more developed than Europe in many fields, it does indeed seem surprising that the Scientific Revolution unfolded in Europe and not in China. Over and over again, therefore, the question has arisen of what "went wrong" in China, what "handicapped" Chinese science, or what "prevented" the Scientific Revolution from happening there.

Historians to date have introduced several different explanations of why the Scientific Revolution failed to occur in China, but they all miss the point that science and learning were seamlessly integrated into the fabric of contemporary Chinese society and culture. They miss the point, too, that the historian's job is to explain what happened and not what didn't happen.

Regardless, the complexities of written and spoken Chinese may have made it less than an ideal medium for expressing or communicating science. That is, because mandarin Chinese and related languages are monosyllabic and written in pictographs, they are ambiguous and ill-suited as precise technical languages for science. But other experts dispute this suggestion, pointing to exact technical vocabularies in Chinese.

Chinese "modes of thought" may have proved inimical to logical, objective scientific reasoning of the sort that developed in the West. Historians have identified a persistent cultural pattern in China variously labeled as analogical reasoning or as correlative or "associative" thinking. This style of thinking, it is said, strove to interpret the world in terms of analogies and metaphorical systems of paired correspondences between diverse things (such as virtues, colors, directions, musical tones, numbers, organs, and planets) based on the fundamental forces of yin and yang and the five "phases" of metal, wood, water, fire, and earth. Yin and yang thus parallel female and male, day and night, wet and dry, the emperor and the heavens; "wood" becomes associated with "spring" and the cardinal direction "east," and so on. In a related way, the famous divinatory work, the "Book of Changes," the *I Ching*, purportedly exercised a negative influence on Chinese thought in that it rigidly defined analytical categories and unduly dominated the attention of Chinese intellectuals by promoting analogical reasoning.

Commentators have also blamed the related lack of a scientific method in China for the stagnant quality of Chinese science. They point to the suppression of two early schools of thought in China, the Mohists and the Legalists, whose tenets resembled Western scientific approaches and whose methods conceivably could have engendered Western-style science and a Scientific Revolution in China. The Mohist school, derived from the thought of Mo-Ti (fifth century BCE), primarily dealt with political matters, but its followers, together with a related group known as the Logicians, emphasized logic, empiricism, and deduction and induction as means of knowing, and thus conceivably could have given rise to a scientific tradition akin to what developed in the West. Gaining prominence in the fourth and third centuries BCE, the other school of thought, the Legalists, sought to develop a universal law code. Their efforts at classification and quantification, had they succeeded politically, might also have established a basis for the rise of modern science in China. The harsh approach of the Legalists won them little favor, however, and with the advent of the Han dynasty in 206 BCE both they and the Mohist school found themselves repudiated and replaced by the more mainstream but less strictly scientific philosophies of Taoism and Confucianism.

Traditional Chinese thought also lacked a concept of "laws of nature." Unlike Islam or the Christian West, Chinese civilization did not entertain notions of a divine, omnipotent lawgiver who issued fixed commandments for humans and for nature. Especially after the failure of the Legalists, Chinese society by and large was not subject to strictly defined positive law and law codes; the more flexible concepts of justice and custom generally governed Chinese legal proceedings. As a result, it made no sense for Chinese intellectuals to inquire into laws of nature or to find motivation for scientific efforts to discover order in God's handiwork.

Another notion advanced to explain the so-called failure of Chinese science concerns the felt cultural superiority of the Chinese. That is, China was a great and ancient civilization, culturally homogeneous, inward-looking, with a long written tradition and with a strong emphasis on traditional wisdom. China thus had no reason to overturn its traditional view of the world or to investigate or assimilate scientific knowledge of "barbarians" outside of China.

The dominant philosophies of Confucianism and Taoism likewise have been censured for stultifying scientific inquiries in traditional China. Several features of the Confucian outlook did indeed prove antithetical to the pursuit of science in the Western manner: the focus on society and human relations (and not a separate "nature"), the disdain of the practical arts, and the repudiation of "artificial" acts (i.e., experiment). Based on the Tao—"the way"—and the idea of universal harmony through cooperation, the Taoist outlook dictated that followers take no action in conflict with or contrary to nature. The very idea of any special inquiry into an "objective" nature, much less a prying, experimental prodding of nature, was foreign to Taoism. From these points of view, the Western conception of nature and scientific inquiry remained alien to the Chinese experience.

A final proposal suggests that because the merchant classes remained largely peripheral to Chinese civilization, modern science could not emerge in traditional China. Had entrepreneurs and free-market capitalism been encouraged in China and not subordinated to monolithic bureaucratic control, then, this argument suggests, perhaps more of a free market of ideas might have evolved, independent institutions akin to the university might have developed, and modern science conceivably resulted.

Each of the preceding explanations of why the Scientific Revolution did not unfold in China doubtless reflects some insight into circumstances in China before the coming of Europeans. However, akin to the previously encountered case of Islamic science, it is crucial to repeat that the negative question of why the Scientific Revolution did not occur in China is foreign to the historical enterprise and not one subject to historical analysis. The number of such negative questions is, of course, infinite. This particular question retrospectively and fallaciously presupposes that somehow China *should* have produced the Scientific Revolution and was only *prevented* from doing so because of obstacles or because China lacked some elusive necessary condition. It is a gross mistake to judge Chinese science by European standards, and only a retrospective projection of later European history onto the history of Chinese science would demand that China necessarily could and should have taken the same path as Europe. Quite the contrary, despite its comparative limitations, science in traditional China functioned perfectly well within its own bureaucratic, state, and social contexts. Such is not a moral judgment of the high and ancient civilization of China, just good history.

The question thus remains why the Scientific Revolution unfolded in Europe rather than why it did not happen elsewhere. Perhaps it is not too early to suggest that in an ecological context where government support but also government control was less pervasive, individual thinkers had more space and freedom to apply critical faculties to abstract questions.

Indus, Ganges, and Beyond

South Asia as a Cultural Sphere

Urban civilization existed continuously on the Indian subcontinent for at least 1,500 years before the first university appeared in Europe. The great world religions of Hinduism, Buddhism, and Jainism arose there and spread across the landmass south of the Himalayas. Great kingdoms came and went. Great art and music flourished in royal and princely courts. Complex socties and a distinct high culture developed in medieval India. As we might expect, Indian experts pursued professional and highly exact work in mathematics, astronomy, medicine, and several other sciences.

In recent decades the scholarly study of science and civilization in China has influenced historians concerned with the history of science and technology in India. But, alas, no comprehensive synthesis has yet appeared to match the studies of China. Historians have examined the texts of Indian astronomers, mathematicians, and doctors, sometimes with a now-familiar attitude that attributes all sorts of "firsts" to early Indian scientists. Although circumstances are changing, much more research remains to be done to fathom the Indian case. Here we can only suggest that the earmarks of a typical bureaucratic civilization again present themselves in India: irrigation agriculture, political centralization, social stratification, urban civilization, monumental architecture, and higher learning skewed toward utility.

Compared to China or the Islamic world, traditions of research in the natural sciences were less vigorous in India. In part at least, the otherworldly, transcendental character of Indian religions militated against the direct study of nature. In various ways, the major religions of Hinduism, Buddhism, and Jainism envision the everyday world as a grand illusion with a transcendental theological reality underlying an ephemeral world of appearances. In these philosophies, unlike Platonic or later Christian traditions, no correspondence unites the world we see with the abstract realm of a greater reality. Truth, then, remains wholly metaphysical and otherworldly, and the goal of knowledge becomes not the understanding of the ordinary world around us, but rather to transcend this world, to escape its debilitating karma, and to ascend to a higher plane. Spiritually very rich, such views did not focus traditional Indian thinkers on the natural world itself or on any underlying regularities in nature or nature's laws.

Civilization arose along the Indus River Valley in the third millennium BCE (see chapter 3), but declined after 1800 BCE for reasons that remain unclear but that probably resulted from shifting ecological patterns and the erratic and destructive course of the Indus. The society that followed was not an urban civilization but rather consisted of decentralized agricultural communities, tribally organized and each headed by a king and chief priest. In time, settlements spread from the Indus to the Ganges basin in eastern India. Four orders or estates constituted early Indian society: priests, warrior-nobles, peasants or tradesmen, and servants, a social division out of which later emerged the full complexities of the Indian caste system. This fourfold division tended to break down local or regional identities in favor of "class" identities. The priestly class (the Brahmin or Brahman) guarded lore and ritual expertise, without which the universe would supposedly collapse. The Brahmin monopolized education, enacted ceremonies, advised kings, participated in statecraft, and drank of the hallucinogenic beverage *soma*.

The historical picture of India down to the sixth century BCE remains fuzzy, depending wholly on literary evidence from religious texts known as the Vedas of the period 1500–1000 BCE and auxiliary Brahmanic commentaries compiled in the succeeding 500 years. The *Rig Veda* is an example of a Vedic text. Originally oral works, they became codified only with the advent of writing in India in the sixth century BCE. Certain obscurities aside, these early texts reveal the existence of scientific knowledge directed at the maintenance of the social and cosmic orders.

Given the centrality of the sacred Sanskrit texts and the "magical" power of their oral recitation, linguistics and grammatical studies became the first "sciences" to develop in India. The Sanskrit language and the Vedas formed the basis of all study, and many grammatical and linguistic guides were produced to lead novices and experts through their intricacies. The fifth-century BCE Sanskrit grammar of Panini, for example, set out 3,873 aphoristic rules concerning grammar, phonetics, meter, and etymology. The importance of oral recitation of the Vedic scriptures likewise led to traditional studies of acoustics and analyses of musical tones.

A smaller, subsidiary group of Vedic and Brahmanic texts concerned astronomy and mathematics. They make plain that a high-status professional class of priests, astrologers, stargazers, and calculators functioned within Vedic society. Experts created and maintained a calendar in order to regulate ceremonies and sacrifices that had to take place on specific days and in specific months and years. They developed multiple solutions for dividing the solar year into months and for intercalating an extra month to keep the religious calendar in synchrony with the solar year. The moon possessed astrological significance, and, like the Chinese, early Indian astrologers divided its monthly course across the heavens into twenty-seven (sometimes twenty-eight) constellations or "lunar mansions" (*naksatras*). The Vedic-Brahmanic calendar thus united lunar and solar cycles. The construction and orientation of altars was a related affair of high importance, for which mathematical competence proved essential. At early stages of Indian history, Indian mathematicians also explored very large numbers in keeping with Hindu and Buddhist notions of great cosmic cycles, giving names for numbers up to 10^{140}.

The Indian subcontinent was more open to outside influences than was China, and so, too, were Indian scientific and technical traditions. The invasion of India by the Persians in the sixth century BCE and their subsequent 200-year occupation of the Indus Valley opened the door to Persian and Babylonian influences on Indian astronomy. Similarly, the invasion by Alexander the Great in 327–326 BCE allowed Greek science to begin to diffuse into India. Conversely, Indian scientific and technical accomplishments influenced developments in the Islamic world, China, and Europe.

At least one relatively powerful kingdom (the Magadha) arose in India by the fourth century BCE. Until then no single polity united India but, triggered by the invasion of Alexander the Great, the Indian adventurer Chandragupta Maurya forged the first unified empire on the subcontinent, reigning as the first Mauryan king from 321 to 297 BCE (map 8.1). His grandson Aśoka expanded the realm during his reign from 272 to 232 BCE. One study claims that the contemporary Mauryan empire, centered on the Ganges, was then the largest empire in the world.

With the advent of the Mauryan empire greater clarity emerges in the historical record. The Mauryan empire was first and foremost a great hydraulic civilization. A Greek traveler, Megasthenes, spent time at Chandragupta's court around 300 BCE and relates that more than half of the arable land was irrigated and that, as a result, Indian agriculture produced two harvests a year. A special department of state supervised the construction and maintenance of a well-developed irrigation system with extensive canals and sluices, and the same bureau planned and directed the settlement of uncultivated land. Land and water were regarded as the property of the king, and the Mauryans levied charges on water taken for irrigation. With no intervening agencies between peasants and state tax collectors, peasants held lands in a kind of tenancy, and henceforth in Indian history the state received

Map 8.1. India. One of the major civilizations, India spread from its initial origin in the Indus River Valley eastward to the Ganges River and to the south of the Indian subcontinent. In the third century BCE India became united under Chandragupta Maurya. The map shows the extent of the Mauryan empire under his grandson Aśoka.

its main revenues in the form of ground rents. Irrigation thus proved essential to both food production and state revenue, and it also fortified political centralization. Archaeological evidence of ancient irrigation systems remains elusive, largely because rivers have so changed their courses since the onset of historical times in India. Documentary evidence, however, reflects the importance of the hydraulic infrastructure—it tells that under the Mauryans breaching a dam or tank became a capital offense, punishable by drowning.

The Mauryan empire did not lack the other hallmarks associated with hydraulic civilizations. An elaborate bureaucratic structure administered the empire. In addition to the department concerned with rivers, "digging," and irrigation, a number of regional and urban superintendents—all salaried officials of the king—dealt with commerce, weights and measures, excise, the mint, registration of births and deaths, supervision of foreigners, and the overseeing of such state industries as weaving, salt provision, mining, and iron-making. State control of the economy was a characteristic feature of Mauryan society and, indeed, artisans owed some royal service. Mauryan political success originated with and depended upon its military strength, and a complex war office with six subsidiary departments administered and provisioned a paid standing army of nearly 700,000 men and thousands of elephants. The existence of an elaborate bureaucracy of royal spies bolstered the autocratic nature of Mauryan government.

The growth and increasing wealth of cities under the Mauryans are additional earmarks of a developing civilization. Sixty-four gates, 570 towers, and a 25-mile defensive wall guarded the capital city at Pataliputra (present-day Patna) at the confluence of the Ganges and Son Rivers. Within the city, amid two- and three-story houses, the Mauryans erected a monumental wooden palace, replete with gilded pillars and an ornamental park with lakes and an arboretum. The Mauryans undertook other public works, including a communication system linking the empire with tree-lined roads, public wells, rest houses, and a mail service.

Although the details remain sketchy, it seems evident that expert knowledge continued to be deployed under the Mauryans. The social position of the Brahmin with their priestly expertise was not seriously undermined during the period, despite Asóka's conversion to Buddhism. Mauryan cities became centers of arts, crafts, literature, and education; the administration of the empire clearly required literacy and numeracy. We know that the superintendent of agriculture, for example, compiled meteorological statistics and used a rain gauge. One of Asóka's rock-edicts—carved inscriptions erected across his empire—also refers to his having established infirmaries for people and animals. And Babylonian and Hellenistic influences came to be felt in India at this point, especially in astrology. For example, the Greco-Babylonian zodiac of twelve houses or signs of 30 degrees each entered Indian astronomy and helped establish its astrological nature. Doubtless, further research will reveal more of Mauryan astronomers and astrologers and their attachment to powerful patrons.

The Mauryan empire declined after Asóka's death, and India splintered into a host of smaller kingdoms and principalities. More than 500 years passed before India once again regained a unified status, this time under the reign of the Guptas in the fourth century CE. The founder of this dynasty, Chandragupta (not to be confused with Chandragupta Maurya), ruled from 320 to 330, and his better-known grandson Chandragupta II (Chandragupta Vikramditya) held power from 375 to 415. The period of the Guptas continued until roughly 650 with some discontinuities and represents the golden age of classical Indian civilization. The Gupta empire resembled that of the Mauryans in its strong central power, public works, regulation of trade, and revenues from ground rent. The Gupta period is noted for the flourishing of Hindu art and literature, for traditions of liberal royal patronage, and for systematic scholarship in astronomy, mathematics, medicine, and linguistics. It formed the high-water mark of classical Indian science.

No less than earlier, Indian astronomy under the Guptas remained a practical activity. Trained professionals created calendars, set times for religious exercises, cast horoscopes, and made astrological predictions, with "lucky days" for agriculture as well as personal fortune. Indian astronomy was not especially observational or theoretical, and it did not delve into the physics of celestial movements. The emphasis remained entirely on astrological prediction and computational expertise. Furthermore, because of its putative roots in the ancient Vedas, Indian astronomy remained a conservative, backward-looking enterprise that placed no premium on theoretical innovation. Isolated from the rest of Indian intellectual life, astronomers acted more like specialized priests, with technical expertise passing down in families from one generation to another. Unlike astronomy in China, the Islamic world, or Europe, where consensus generally united scientific traditions, some six regional schools of Indian astronomy-astrology competed for intellectual allegiance and material patronage.

Despite its limitations and divisions, Indian astronomy became highly technical and mathematical in the period of the Guptas. From the fourth through the seventh centuries various Indian astronomers produced a series of high-level textbooks (*siddhānta* or "solutions") covering the basics of astronomy: the solar year, equinoxes, solstices, lunar periods, the Metonic cycle, eclipses, planetary movements (using Greek planetary theory), seasonal star charts, and the precession of the equinoxes. Aryabhata I (b. 476 CE) lived in Pataliputra, composed a *siddhānta*, trained students, and held the unorthodox view that the earth rotates daily on its axis (despite his knowledge of Ptolemy's *Almagest*). In his *siddhānta* in the following century the astronomer Brahmagupta (b. 598 CE) repudiated Aryabhata's notion of a moving earth on the grounds that it violated common sense and that, were it true, birds would not be able to fly freely in every direction. Brahmagupta's estimate of the circumference of the earth was one of the most accurate of any ancient astronomer.

Indian astronomy depended on precise arithmetical calculations, and Aryabhata and Brahmagupta obtained renown as mathematicians no less than as astronomers. Algebraic and numerical in character, Indian mathematics by and large reflected practical concerns and eschewed general solutions in favor of "recipes." Aryabhata employed a place-value system and decimal notion using the nine "Arabic" numerals and zero in his work. (The appearance of zero within the context of Indian mathematics may possibly be due to specifically Indian religio-philosophical notions of "nothingness.") He calculated the value of π

to four decimal places, a value later Indian mathematicians extended to nine decimal places. In his *siddhānta* Brahmagupta extended earlier work on measurement, algebra, trigonometry, negative numbers, and irrational numbers such as π. Indian mathematical work became known to the West primarily through reports by the eleventh-century Islamic scientist al-Bīrūnī in his *History of India*.

As in civilizations elsewhere, the world of doctors and medicine became solidly institutionalized and developed. Wealthy and aristocratic families patronized physicians, and court physicians possessed especially high status, in part because of their expertise regarding poisons and snakebites. Top-level physicians seemingly differentiated themselves from empirics through training and licensing. A traditional medical text, the *Charaka Samhitā*, for example, speaks of a process of apprenticing with a master physician-teacher and getting royal permission before practicing medicine. The religious center at Nalanda flourished as a medical school from the fifth through the twelfth centuries CE. Thousands of students (reports vary from 4,000 to 10,000) and hundreds of teachers studied and taught at this vast complex, more than a mile square with 300 lecture rooms and a large library. Tuition was free, supported by the king and by rich families. Other teaching centers existed at Taxila and Benares. As mentioned, the Mauryan king Aśoka established medical infirmaries, and charitable dispensaries also existed in the Gupta period. Not surprisingly, veterinary medicine for war horses and elephants reached a high level of competence in India from the fourth century BCE.

Medical theory and practice became quite developed early in Indian history. The Vedic oral tradition reported anatomical information, particularly of the sacrificial horse, based on dissection, as well as botanical information and descriptions of diseases. The tradition known as the *Ayurveda*—or the "science of life"—began to be codified in the sixth century BCE, and it came to include sophisticated medical and physiological theories and treatments that involved maintaining equilibrium balances between and among various bodily humors. Ayurvedic medicine is famous for its rational approaches to diseases and their cures and, indeed, it possessed a self-conscious epistemological dimension in assessing the processes of medical reasoning and judgment. The standard medical compendium by the physician Charaka (the *Charaka Samhitā*) appeared around the first century CE. Reflecting the Hindu penchant for naming and listing, the *Charaka Samhitā* identifies 300 different bones, 500 muscles, 210 joints, and 70 "canals" or vessels; its associated nosology of diseases was no less elaborate. A related "collection" by the physician Susruta (the *Susruta Samhitā*) became a bible for Indian surgery. At their heights Indian medicine and surgery were probably the most developed and advanced of any contemporary civilization.

Alchemy, another science deemed to be useful, also flourished in

India, perhaps having arrived from China. Closely associated with medicine and the sect of Tantric Buddhism, Indian alchemical treatises focused on various forms of mercury, on preserving health, and on the creation of an undecayable body. Practitioners came to master a sophisticated corpus of chemical knowledge that found applications in medicine through elixirs, aphrodisiacs, and poisons.

Quite apart from these scientific developments, traditional India became a highly evolved technological civilization. Indeed, although not heavily mechanized, India has been labeled "an industrial society" for the period before European colonialism and the Industrial Revolution established themselves on the Indian subcontinent. The major industry in India was textiles, and India was then the world's leading textile producer. The caste of weavers, for example, stood second in numbers only to the agricultural caste, and textile production gave rise to subsidiary industries in chemicals and the dyeing and finishing of cloth. Shipbuilding, which supplied ocean-going vessels for the substantial Indian Ocean trade, was likewise a major industry of traditional India. Indian shipwrights developed construction techniques especially suited to monsoon conditions of the Indian Ocean, and the importance of the Indian shipbuilding trade actually increased after Europeans entered those waters. Although iron smelting in India dates from 1000 BCE, it was practiced on a comparatively small scale until the Islamic Mughal empire and the advent of gun manufacture in the sixteenth century. The competence of Indian foundrymen is no better illustrated than by the commemorative iron pillar 24 feet high made at Delhi under Chandragupta II in the fourth century CE. (It reportedly shows no sign of rust even to this day.) Indian artisans also engaged in pottery-making, glass-making, and a myriad of other practical crafts befitting a great civilization. Given its technological sophistication, India actually underwent an astonishing process of what amounted to its deindustrialization with the coming of formal British rule in the nineteenth century.

The caste system became more rigid in the Gupta period, with the definition of some 3,000 different hereditary castes. While the significance of the caste system for the history of technology in India was probably less than previously thought, the system remains noteworthy in that different technical crafts and craft traditions became socially separated into distinct castes and guild-like bodies, it being nominally forbidden to ply a trade outside of one's caste. Although caste barriers were sometimes breached, the separation of technology from scientific traditions is as evident in India as in China or ancient Greece.

A Hun invasion of Gupta territory in 455 CE proved disruptive, and a partial breakup of the empire ensued in the decade 480–90. Subsequent sixth-century Indian kings reestablished the empire, but the unity of classical Indian civilization collapsed completely in 647 after the death of the heirless king Harsha. A succession of minor Hindu

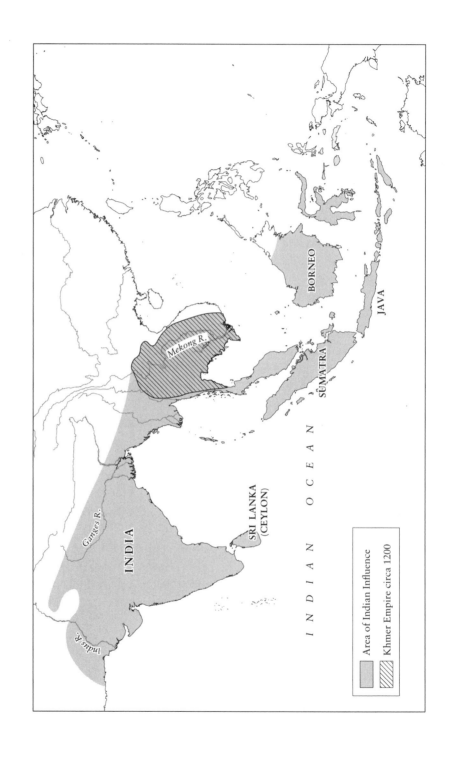

states followed, and Islamic influences and incursions began to be felt in India after 1000 CE. Islam exercised wide appeal, in part because it repudiated caste divisions. An independent Delhi sultanate ruled over the Indus and Ganges in northern India from 1206 to 1526, and, by dint of superior cannon technology, the Islamic Moghal empire governed North India from 1526 nominally to 1857. Muslim rule brought improved irrigation and hydraulic technology to North Indian agriculture, including the use of artificial lakes. The great Moghal emperor Akbar (1556–1605) established a special government canal department, and at the height of the Moghal empire, fully one-third of the water used for irrigation flowed in manmade canals. As part of the Islamic imperium, India also fully assimilated Islamic science, most visibly in the spread of Islamic astronomical observatories. The cultural and institutional success of Islam spelled the end of traditional Hindu science and learning in those areas touched by the teachings of the Prophet.

Traditional Indian culture continued in the non-Islamic south within the borders of states and wealthy cities dependent on intensified agriculture. The Chola kingdom, for example, flourished from 800 to 1300. Chola engineers built irrigation works on a huge scale, including the damming of rivers and the creation of an artificial lake 16 miles long. Bureaucratic supervision is evident in the special committees in charge of irrigation tanks. These tanks captured monsoon rain in the rainy season to use in the dry season. In Mysore in South India 38,000 tanks remained in the eighteenth century, and 50,000 tanks in Madras in the nineteenth. This tantalizing evidence notwithstanding, the major manifestation of centralized agriculture, science, and state patronage occurred not on the Indian subcontinent itself, but rather in the spread of Indian civilization offshore to Sri Lanka and to Southeast Asia.

Greater India

The correlations between science and hydraulic civilization are evident in the case of Buddhist Sri Lanka (ancient Ceylon). Founded by legendary "water kings," a quintessential hydraulic civilization arose on the island after invasions from the Indian mainland in the sixth century BCE, and a distinctive Sinhalese civilization maintained itself there for 1,500 years. Using thousands of tanks and catchments to collect monsoon rainfall, irrigation agriculture and grain production spread in the dry zone in the north of the island. The hallmarks of hydraulic civilization likewise appeared: centralized authority, a government irrigation department, corvée labor, agricultural surpluses, and monumental building, including shrines, temples, and palaces built with tens of millions of cubic feet of brickwork on a scale equaling that of the Egyptian pyramids. Large urban population concentrations inevitably accompanied these developments. Indeed, the main city of Polonna-

Map 8.2. (opposite) Greater India. An Indian-derived civilization arose on the island of Sri Lanka (Ceylon), and Indian-inspired cultures also developed in Southeast Asia, notably in the great Khmer Empire that appeared in the ninth century BCE along the Mekong River.

ruwa reportedly ranked as the most populous city in the world in the twelfth century CE.

The details remain sketchy, but records indicate royal patronage of expert knowledge in ancient Sri Lanka for work in astronomy, astrology, arithmetic, medicine, alchemy, geology, and acoustics. A bureaucratic caste, centered on temple scholars, also seems to have existed, with the chief royal physician a major government figure. Following the pattern established in India by Aśoka, the state diverted considerable resources to public health and medical institutions such as hospitals, lying-in homes, dispensaries, kitchens, and medicine halls. In all, Sri Lanka reveals a typical pattern of useful, patronized science.

From an early date in the first millennium CE Indian merchants voyaged eastward across the Indian Ocean. By dint of extensive trade contact and sea links to Sumatra, Java, and Bali in Indonesia and through cultural contact with Buddhist missionaries from Sri Lanka, a pan-Indian civilization arose in Southeast Asia and Malaysia. A third-century account by a Chinese traveler, for example, reported the existence of an Indian-based script, libraries, and archives in the Funan kingdom in what is modern Vietnam. China exerted its cultural influence from the north to be sure, but Indian influence predominated in the region and increased in the fourth and fifth centuries CE. Brahmin from India were welcomed as local rulers bringing with them Indian law and administrative procedures. The caste system came along too. Sanskrit became the language of government and learned religious commentaries, while Hinduism and Buddhism coexisted as the dominant faiths.

The great Cambodian or Khmer empire provides the most remarkable and revealing example of this extension of Indian cultural influence. A prosperous and independent kingdom for over six centuries from 802 to 1431, the Khmer empire at its height under King Jayavarman VII (who reigned 1181–1215) was the largest political entity ever in Southeast Asia, covering parts of modern Cambodia, Thailand, Laos, Burma, Vietnam, and the Malay Peninsula.

The Khmer empire arose along the alluvial plains of the lower Mekong River, and the great wealth of Khmer society derived from the most substantial irrigation infrastructure in Southeast Asian history. The annual monsoon flooded the Mekong and its tributaries, with the Cambodian Great Lake (Tônlé Sap) turning into a natural reservoir. Using impounded water techniques, Khmer engineers built an enormous system of artificial lakes, canals, channels, and shallow reservoirs with long embankments (called *barays*) to control the river system and to hold water for distribution in the dry season. By 1150 CE over 400,000 acres (167,000 hectares) were subject to artificial irrigation. The East Baray at Angkor Wat alone stretched 3.75 miles (6 km) long and 1.25 miles (2 km) wide. Hydrologic conditions along the Mekong were ideal for cultivating rice, that exceptionally bountiful crop that

Map 8.3. The Khmer Empire. Based on rice production and the hydrologic resources of the Mekong River and related tributaries, this Indian-inspired empire flourished magnificently in the twelfth and thirteenth centuries. Based on substantial irrigation and impounded water technologies, the Khmer Empire constituted the largest political entity in Southeast Asian history. It exemplified the typical trappings of high civilization, including monumental building, literacy, numeracy, astronomical knowledge, and state support for useful science. With the demise of its irrigation infrastructure Khmer civilization disappeared in the early fifteenth century.

produced dramatic effects whenever it was introduced, as we saw in the case of China. Such a productive capability supported dense population concentrations, an immense labor force, and a wealthy ruling class.

Yet again social and scientific patterns associated with hydraulic civilization repeated themselves in the Khmer empire. Khmer kings, living deities like Egyptian pharaohs, exercised a strong centralized authority. A complex theistic bureaucracy, headed by an oligarchy of learned Brahmins and military officers, ran the day-to-day affairs of the empire, including in a paramount way the building and maintenance of the hydraulic infrastructure. One source labels this bureaucracy a welfare state, perhaps because Jayavarman VII supposedly built 100 public hospitals as well as other public works. We see urbanization in the urban center of Vasodharapura. Various libraries and archives also testify to the bureaucratic nature of the state and of higher learning. In addition to irrigation projects and a highway system (with rest houses) linking various parts of the empire, Khmer royal power directed prodigious construction projects, notably in the capital district of Angkor, built up over a 300-year period. As an urban center Angkor covered

Fig. 8.1. Angkor Wat. Completed in 1150 CE, the huge and elaborately decorated temple complex at Angkor Wat stood at the ceremonial and administrative heart of the Khmer Empire in modern-day Cambodia. A moat and basins linked Angkor Wat to the nearby Siem Reap River and other regional waterworks and fields where rice production sustained a large population base. With the collapse of these irrigation systems in the fourteenth century the Khmer Empire itself collapsed. Angkor Wat was abandoned in 1444 and overrun by the surrounding forests, only to be revived as a world heritage site in the nineteenth and twentieth centuries.

60 square miles (15,000 hectares) and consisted of a whole series of towns along coordinate axes running 19 miles (30 km) east-west and 12 miles (19 km) north-south. Among the 200 temples in the region, each with its own system of reservoirs and canals of practical and symbolic significance, the complex at Angkor Wat is the largest temple in the world. Surrounded with a moat almost 660 feet wide (200 m), the temple is made of as much stone as the great pyramid at Cheops, and with virtually every square inch of surface area carved in bas-relief. The complex itself was completed in 1150 CE, after fewer than 40 years of construction. Nearly a mile square, Angkor Wat itself contained twelve major temples, and its central spire soared to nearly 200 feet (60 m). Even more formidable, the magnificent administrative and temple complex at Angkor Thom, finished in 1187, enclosed a walled city of almost four square miles (1000 hectares). Among their other uses, these great temples functioned as mausoleums for Khmer kings.

The Khmer court attracted Indian scholars, artists, and gurus, and with them Indian astronomy and alchemy came to Cambodia and Southeast Asia. Alongside ruling Brahmins and military leaders, a separate caste of teachers and priests plied their trades, teaching Sanskrit texts and training new generations of astrologers and court ceremonialists. The existence of Khmer "hospitals" suggests the organization of medical training and practice at a high level within the empire. The unity of astronomy, calendrical reckoning, astrology, numerology, and architecture is evident in the structure of Angkor Wat, meticulously laid out along lines dictated by Indian cosmology, with special moats and an architectural sacred mountain. Several of the thousands of bas-reliefs carved into the buildings indicate concern with elixirs of immortality. The complex also has built-in astronomical sight lines recording solar and lunar motion along the horizon. The spring equinox, which evidently marked the onset of the calendar year, receives special emphasis. Given these sight lines built into the monument, eclipse prediction is possible at Angkor Wat, but whether Khmer astronomers actually predicted eclipses is a matter of speculation.

Overbuilding may have exhausted the state and sapped the vitality of Khmer civilization. Beginning in the fourteenth century, the Khmer empire suffered repeated invasions from neighboring Thai and Cham (Vietnamese) peoples. These attacks destroyed the irrigation infrastructure on which Khmer civilization depended: maintenance activities ceased, war created damage, and demands for soldiers reduced the corvée. As a result, populations collapsed, and so did the Khmer empire itself. The Thais conquered; Sanskrit ceased to be the learned language of Southeast Asia; a new, less ornate style of Buddhism prevailed; and Angkor itself was abandoned in 1444, to be swallowed by the encroaching jungle. The French "discovered" and brought to the world's attention the ruins of Angkor and of Khmer civilization only in 1861. Although lost for four centuries, Khmer civilization testifies to the now-familiar pattern of agricultural intensification, bureaucratic centralization, and patronage of useful sciences.

The New World

Roughly coincident with Old World scientific civilizations in Islam, China, and India, a series of cultures arose in the New World: the Maya, the Aztec, and the Inca. The pattern of convergence with Old World civilizations is especially striking because developments in America unfolded along a separate technological trajectory, without plows, draft animals, or bronze or iron metallurgy. But like their ancient counterparts on the other side of the globe, these American civilizations came to deploy scientific experts and expertise in the running of their states. And they did so without the distinctive element of disinterested natural philosophy that was the fruit of the Hellenic tradition.

Lands of the Jaguar

Mayan civilization was based on antecedent American cultures. It depended on a form of intensified agricultural production and growing corn that exploited the productive capabilities of lowland swamps. Mayan civilization arose after 100 BCE and flourished for a thousand years in the area radiating outward from modern Belize in Central America (see map 9.1). In a characteristic fashion and more than in any other American civilization, the Maya came to deploy institutionalized knowledge for the maintenance of their societies.

As a result of the vandalism by Spanish conquerors in the New World, what we know of Mesoamerican "glyph" writing remains limited. The Spanish destroyed thousands of codices—fanfold bark- and deerskin-paper books, of which only four Mayan examples survive. The primary source of our knowledge consists of the 5,000 Mayan texts engraved on stone stelae and architectural elements, some with hundreds of glyphs or carved signs. As a result of significant recent advances in deciphering ancient Mayan, passages are being translated at an accelerated rate, and over 90 percent can now be read accurately. Based on Olmec roots, Mayan writing is now known to embody pho-

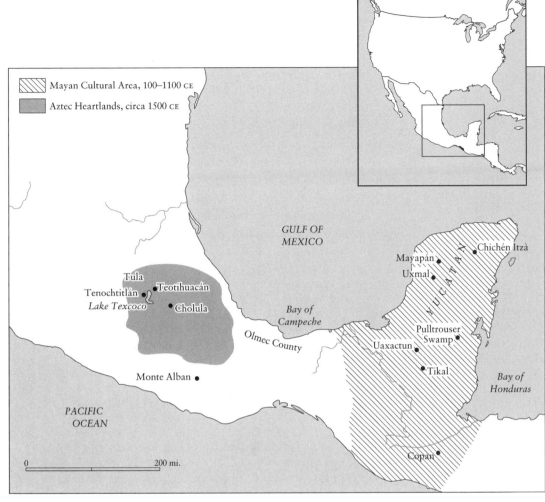

Map 9.1. Civilization in Mesoamerica. High civilization arose independently in Central America. Mayan civilization centered itself in the humid lowlands surrounding the Bay of Honduras, and, later, the Aztec empire arose in a more desert-like setting around a lake where modern Mexico City stands.

netic and pictographic components of a distinct Mayan language. The 287 hieroglyphic signs now deciphered, of which 140 carry phonetic meaning, presently give a clear picture that ritualistic sculpture-writing primarily recorded historical events—the reigns and deeds of kings, dynasties, and ruling families. Undoubtedly, records of dynastic legends are overrepresented in the medium of public sculpture. If the written records produced over hundreds of years had not been destroyed, a more mundane picture of Mayan society would emerge.

The inherent difficulty of mastering Mayan writing would suggest the existence of a specialized caste of Mayan scribes and the necessity of extensive training to join such a cadre. Other evidence shows that the scribal profession formed an exclusive, specialized occupation, one that enjoyed great status and prestige at the top of a highly stratified society. Scribes were drawn from the upper ranks of Mayan nobles and were often the princely second sons of kings. Their positions were probably hereditary; they functioned as high courtiers and royal confidants, closely associated with ruling kings, and apparently they sometimes vied for political power. At least in some instances, such as the late Mayan center of Mayapán, a separate "academy" existed for training priests and scribes. The scribal caste had its own patron deity, the high-god Itzamná, the inventor of writing and the patron of learning. Scribes wore a special headdress as a symbol of their profession and used special utensils, and they were sometimes buried with codices and with great pomp. Given what is known of Mayan scribes and their accomplishment, one may well speak of a class of Mayan intellectuals.

The Maya used a vigesimal, or base-20 system of arithmetic in which dots represent 1s and bars represent 5s. The suggestion has been made that the choice of the five-unit and the vigesimal system stemmed from a correspondence with the five digits of the hand and the total of twenty fingers and toes. In any event, the Maya developed a place-value sys-

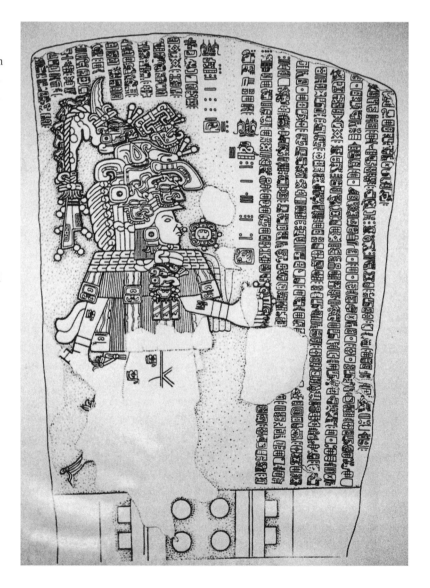

Fig. 9.1. Pre-Mayan stela. The complex writing system used in Mayan civilization has now been largely deciphered. Based on earlier antecedents like the second-century CE stone carving shown here, the Mayan form of writing used glyphs that possessed phonetic or sound values. Mayan inscriptions had a public character, commonly memorializing political ascendancy, warfare, and important anniversaries.

Fig. 9.2. Mayan word signs. The originally pictographic character of pre-Mayan and Mayan languages came to assume phonetic values. These signs could be combined to represent other words and concepts.

"ix"
A Day Name

"mol"
A Month Name

"cu - tz(u)"
Turkey

"tzu - f(u)"
Dog

"chu-ca-h(a)"
To Capture

"chu"   "ca"   "ha"

tem with a sign for zero, and they used it to reckon extraordinarily large numbers. They did not develop fractions, but like their Babylonian counterparts several millennia earlier, Mayan mathematicians created tables of multiples to aid in calculation. Mayan mathematical expertise functioned primarily in connection with numerology, ritual astronomy, and an elaborate calendrical system.

Based on Oaxacan antecedents, the Mayan calendar and dating systems became quite complex, the most complex in the Americas and probably ever in the world. The Maya juggled four or five different timekeeping systems simultaneously, and an earlier view that the Maya were obsessed with cycles of time has not been seriously undermined by the more recent understandings of the historical content of Mayan writing.

The most important of the Mayan calendars was the so-called *tzolkin*, a 260-day sacred cycle, itself generated by thirteen 20-day periods. The 260-day cycle is evident in Mayan history as early as 200 BCE and may be related to the period of human gestation; later Maya developed related, more elaborate cycles based on multiples of 260. Along with the *tzolkin* the Maya inherited a "Vague Year" calendar consisting of eighteen 20-day months and a period of five unlucky days for a separate 365-day year. The Maya did not rectify the quarter-day discrepancy with the actual solar year, and, as a result, their "Vague Year" calendar, like that in ancient Egypt, gradually lapped the seasons over a period of 1,460 years. Meshing the *tzolkin* and "Vague Year" cal-

1	5	19	100	1000
		(3x5)+4	(5x20)+0	(2x400)+(10x20)+0

Fig. 9.3. The Mayan vigesimal number system. The Mayan number system was a base-20 place-value system with separate signs for 0, 1, and 5. Note that the "places" were stacked on top of one another.

endars, the Maya produced the so-called Calendar Round, a fantastic combination of the 365-day and 260-day cycles that turned on each other like gears in a great clock repeating every 52 years. The Calendar Round operated as an elaborate fortune-telling machine. Each day in the long cycle possessed its own name and became associated with various omens, and specialized priests used the Calendar Round for astrological divination, research, and prophecy.

Mayan calendar keepers and astronomers also maintained a separate lunar calendar and a fourth timekeeping system, a sequential count of days, the so-called Long Count that enumerated a linear succession of days in six units of time ranging from one day to almost 400 years. The starting point for the Maya Long Count has been calculated to have been August 13, 1314 BCE (alternatively, sources give 1313 BCE), and the end of the world was foreseen for December 21, 2012. That date came and went with some fanfare and a movie in 2012, but such end-of-the-world scenarios missed the point that for the Maya the whole machinery would start over again, something in the manner of an odometer turning over. In other mythological constructs Mayan time-reckoners employed time in lengths that ran into millions of years.

Closely linked to the calendar, Mayan astronomy formed part of a single astrological enterprise devoted to sacred purposes. Anonymous cadres of Mayan court astronomers observed the heavens in state institutions such as the well-known "observatory," the Caracol at Chichén Itzá. The Caracol was aligned on the horizon points marking the extremes of Venus's rising and setting in the year 1000, and it incorporated built-in astronomical sight lines for the equinoxes, the summer solstice, lunar settings, and cardinal points of true south and true west. Other Mayan urban centers at Copán, Uaxactún, and Uxmal embodied related sight lines. Indeed astronomical orientations seem to be a primary element in all Mayan public architecture and city planning, as evident in the seemingly skewed axes of Mayan buildings and towns that mirror the risings and settings of the planet Venus as well as the solstices and equinoctial points. No less important are zenith passage markers, which track the sun as it seasonally passes highest in the sky at noon, the event most likely used to monitor the agricultural cycle.

A handful of surviving codices make plain that Mayan astronomy became a highly sophisticated research endeavor. Mayan astronomers computed the length of the solar year to an accuracy greater than 365 1/4 days, although they used a 365-day year for calendar purposes. Judging from a stele at Palenque, Mayan astronomers in the seventh century CE determined the length of the lunar month to the equivalent of three decimal places or 29.530 days. (At Copán in the eighth century, another virtually identical calculation of the lunar month exists.) Given their mastery of the periodic motion of the sun and the moon, the Maya possessed the knowledge to solve the Babylonian "new moon" problem discussed earlier. Like the Babylonians, the Maya could predict eclipses accurately, and they created eclipse tables that calculated probabilities of eclipses.

Fig. 9.4. Mayan Calendar Round. Like their counterparts in all other civilizations, Mayan astronomers developed a complex and reliable calendar. Theirs involved elaborate cycles of days, months, and years. It took 52 years for the Mayan Calendar Round to repeat itself.

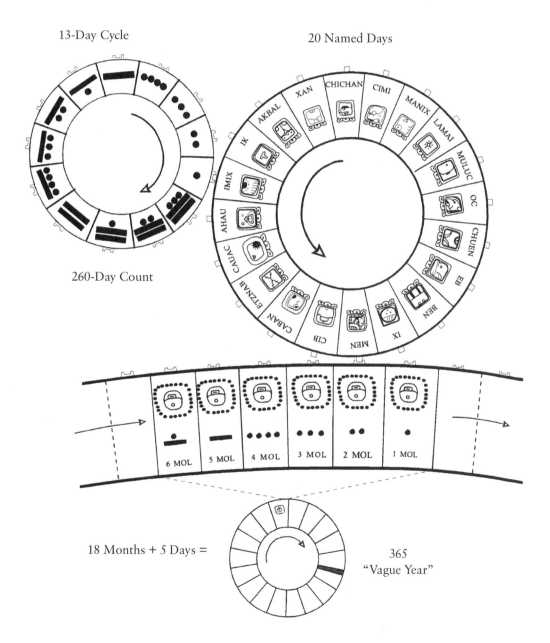

13-Day Cycle

20 Named Days

260-Day Count

18 Months + 5 Days =

365 "Vague Year"

Fig. 9.5. The Mayan "observatory" at Chichén Itzá. From the observing platform of the Caracol ancient Mayan astronomers had a clear line of sight above the treetops. The various windows in the structure align with the risings and the settings of Venus and other heavenly bodies along the horizon.

The planet Venus, the object of special veneration in Mayan culture, was carefully monitored. Mayan astronomers kept a separate Venus calendar and actively undertook research to improve the accuracy of their Venus tables to within two hours in 481 years. Like their earlier Babylonian counterparts, Mayan astronomers harmonized cycles of Venusian years and solar years, and they elaborated even more complex cycles, including one that integrated 104 solar years, 146 sacred *tzolkin* cycles, and 65 Venus years. Experts may well have produced tables for Mars and for Mercury. Other inscriptions indicate that Jupiter possessed astrological significance; certain stars had special meaning as well. Clearly, Mayan astronomers engaged in specialized research at a high level of precision and expertise.

A range of utilitarian motivations prompted this mass of esoteric astronomical work. Calendrical mastery at the most simple level gave Mayan rulers an understanding of the seasons and agricultural cycles. In its more complex formulations the Mayan calendar governed elaborate religious and ritual activities, and the full complexities of Mayan astrology produced magic numbers, prognosticated the fates of individuals, and predicted propitious and unfavorable times for a host of activities. That the Venus cycle was used to time military campaigns, for example, gives an indication of the political uses of learning and the ways in which Mayan astronomy and astrology were integrated into the power structure and ruling ideology of Mayan society.

The high culture of the Maya came under tremendous stress around 800 CE, and the major centers in the central regions of Mayan civilization collapsed entirely around 900. A resurgence of Mayan power developed on the Yucatan peninsula—notably in Chichén Itzá—in the following period to 1200, but from the eleventh century the Long Count fell into disuse, and the record indicates a decline in the rigor of training for scribes and priests. Thereafter the high civilization of the Maya passed into history. An array of causes have been put forward and debated to explain the protracted death of Mayan civilization. Endemic warfare among confederations of city-states may have played a role, and the inevitable pressures of population measured against a fragile system of food production may have produced radical demographic fluctuations. Compounding such possibilities, two centuries of drought—the worst in 8,000 years—affected the Mayan lowlands in 800–1000 and no doubt took a heavy toll. Researchers have recently highlighted the problems of deforestation stemming from Mayan techniques of using large amounts of wood to make lime for stucco with which Mayan monumental architecture was plastered. Deforestation probably disrupted rainfall patterns and, at least in places, led to soil erosion and thereby the ruination of agriculture. Mayan civilization gradually waned, and with it the exquisite system of understanding nature that the Maya had achieved.

Cactus and Eagle

Central America also saw the rise of Toltec and Aztec civilizations. Based on irrigation agriculture, between 900 and 1100 CE the Toltec city of Tula had 35,000–60,000 inhabitants, and the Toltecs built what is technically the largest pyramid in the world, a manmade mountain of 133 million cubic feet (3.8 million meters³), 1,000 feet (300 m) on a side, and 170 feet (52 m) high covering 45 acres (18 hectares) at Cholula.

The Aztecs began as a seminomadic tribe, and in the fourteenth and fifteenth centuries they established the most powerful empire in Central America. They built their city of Tenochtitlán in 1325 on a lake where Mexico City presently stands; according to legend, an omen—an eagle perched on a cactus—drew them to the lake. The Aztecs proved to be master hydraulic engineers. The lake, Lake Texcoco, was saltwater, and Aztec engineers built a huge dike across it to separate a fresh-water portion (fed by springs) from the brine; they also installed floodgates to regulate lake levels and aqueducts to bring additional fresh water to the lake. Each year millions of fish and ducks were taken from lakes, which also provided a nutritious algae paste. The Aztecs developed an intensive style of lake-marsh (or lacustrine) agriculture that involved dikes, dams, drainage canals, and land reclamation, all produced as public works under state management. Agricultural production was literally based on floating paddies known as *chinampas*. Measuring 100 meters by 5–10 meters (328 feet by 16–33 feet) and fertilized by human waste and bat guano, these plots proved extraordinarily fruitful and produced up to seven crops a year and the surplus which supported urbanized Aztec civilization. On the eve of contact with the Europeans, Aztec farmers cultivated over 30,000 acres (over 12,000 hectares) of *chinampas*.

The chief city of Tenochtitlán covered five square miles (1300 hectares) and became filled with monumental pyramids, palaces, ceremonial centers, ball courts, markets, and roads built with corvée labor. (Ballgames and ball courts were part of the sociology and architecture of many Mesoamerican cities.) Aqueducts brought fresh water into the city, and 1,000 public employees swept and watered the streets daily. On the eve of the European conquest Tenochtitlán had a population estimated at 200,000 to 300,000, the largest ever in the Americas.

Predictably, Aztec kings, a full-time priest caste, a bureaucracy, and the military ruled this most powerful state in pre-Columbian Mesoamerica. They controlled an empire of 5 million people, possibly twice the population of ancient Egypt. The bureaucracy ran local and imperial administrations, collected taxes and tribute, and performed judicial duties; positions in the bureaucracy were hereditary in aristocratic lineages, and separate schools existed for commoners and for the nobility and priests. The Aztecs formed a militaristic society heavily involved in ritual sacrifice and probably cannibalism—Aztec priests offered up tens of thousands of human sacrifices a year. But the Aztecs also developed extensive trading and mercantile communities, which required mathematics and recordkeeping like other civilizations. The chocolate bean served as a unit of currency.

The Aztecs shared common writing and number systems, astronomy, and theologies with earlier Mesoamerican societies. The Aztec system of writing was less developed than the Mayan, being more pictographic, but it possessed some phonetic elements. The subjects of surviving Aztec books and codices concern religion, history, genealogy, geography, and administrative records, including tribute lists, censuses, and land surveys; some Aztec books were specialized instruction manuals for priests. Aztec numeration was a simple dot system. They inherited the Mayan 52-year Calendar Round, with its 260-day and 365-day cycles. Architect-astronomers aligned the *Templo Major* in Tenochtitlán to the setting sun, and other buildings were sited along equinoctial lines. The calendar governed an elaborate train of seasonal festivals and feasts, wherein offerings to various manifestations of the Aztec sun god Tezcatlipoca played a central role. Indeed, theology required the ritual sacrifice of blood in order to keep the sun in its course and the earth bountiful.

The Aztecs also possessed sophisticated botanical and medical knowledge. Priests functioned as medical specialists, and medical knowledge passed from fathers to sons. Based on empirical researches, Aztec doctors developed an elaborate and apparently effective medical pharmacopeia that was at least the equal of that of their Spanish conquerors. (Aztec life expectancy exceeded that of Europeans by 10 years or more.) Aztec medicine and astronomy were related through the belief that the heavens and the human body were mutually linked. In a pattern of patronage that is now familiar, in 1467 the Aztec emperor Montezuma I established a botanical and zoological garden wherein expert knowledge was cultivated and transmitted. In this connection one might mention the domestication of the red cochineal insect used as a dye for textiles throughout Mexico, Central America, and in Europe after the Conquest.

The Spanish adventurer Hernán Cortés landed on the coasts of Mexico in 1519 with 500 fellow conquistadors. They arrived with horses and cannon brought on their great ships, the fruit of the military revolution that had unfolded in Europe over the previous century and a half. Cortés also brought with him cultural practices (such as more effective communications and organizational capabilities) that put him at an advantage in the struggle for conquest he and his compatriots launched as they landed. With the help of anti-Aztec native allies, Cortéz's ragged band of Spaniards, having sailed across the great ocean of the Atlantic, subdued the great civilization of the Aztecs in two years.

Andean Empires

A similar story of cultural and scientific development followed by doom repeated itself in South America. There a series of independent civilizations arose that mirrored the pattern of their sister societies around the globe. Cultural development in South America reached its peak in the great civilization of the Inca (or Inka) that spread for more than 2,000 miles up and down the west coast of South America between the Andes and the Pacific in the thirteenth and fourteenth centuries CE. In an earlier chapter we had occasion to mention the hydraulic basis of the Inca empire, with its highly stratified and bureaucratized society and monumental building.

Although less is known about science and civilization in pre-Columbian South America than most of the other pristine civilizations, all evidence conforms to the now-familiar emergence of science and expert learning in the service of the state. The ancient Inca did not develop writing or formal mathematical systems, but with their *quipu* or strands of knotted string the Inca devised an effective method of recording information, and recent research has suggested that quipu may have encoded a written language. To this extent the Inca were not an exception to the common pattern of mathematical and recordkeep-

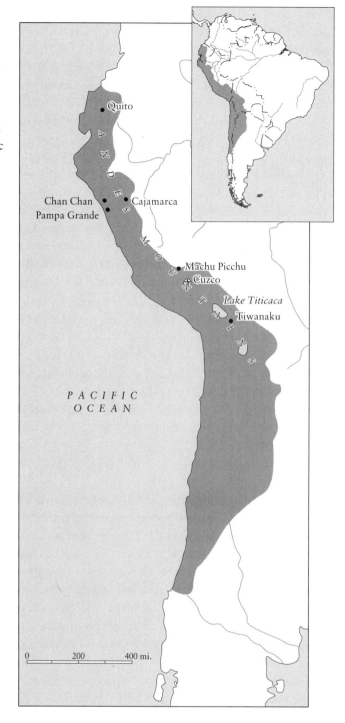

Map 9.2. The Inca empire circa 1530. On the west coast of South America the Inca built on cultural developments of the previous 1,000 years and created a high civilization between the Andes Mountains and the Pacific Ocean where engineers terraced mountains and built irrigation systems tapping numerous short rivers to intensify agriculture.

ing systems in the first civilizations. The Inca empire was organized socially on the basis of a decimal system, with units of 10 to 10,000 people, and the Inca possessed a set of standard weights and measures. Quipu recorded information that included tax and census records and imperial history, and as part of the enormous imperial bureaucracy a hereditary class of accountants memorized information contained in quipu.

Inca astronomer-priests divided the heavens into quarters according to seasonal changes in the inclination of the Milky Way across the southern sky. The mountains of the Andes created natural horizon markers against which the Inca could track the periodic motion of the sun, moon, planets, constellations, and even apparently void spaces referred to as "dark clouds." The Inca also built artificial stone pillars along the horizon to mark solstices. Forty-one sight lines (*ceques*) radiated from the grand ceremonial center of the Coricancha temple in Cuzco, along which other markers indicated lunar positions, water sources, and political subdivisions of the Inca empire. In other words, the Inca incorporated a calendar and map of their empire into the design of the city of Cuzco—it became an architectural quipu. Other Inca sites likewise had astronomical orientations, such as the summer solstice indicator built at Machu Picchu.

The Incas reportedly employed lunar and solar calendar systems, but because calendrical knowledge was not written down, local lords

Fig. 9.6. Inca recordkeeping. All high civilizations developed systems of recordkeeping, usually in the form of writing. The Inca used knotted strings called *quipu* for recording numerical and literary information.

instituted a variety of timekeeping systems. In the Inca capital of Cuzco authorities maintained a seasonal, 365-day solar calendar of twelve 30-day months (each of three 10-day weeks) and five festival days; they corrected the quarter-day discrepancy with the solar year by resetting the calendar at the summer solstice (in December). The Inca also kept track of the zenith motion of the sun, and sources speak of Inca observatories and an office of state astrologer. They also maintained a 12-month lunar calendar of forty-one eight-day weeks comprising a year of 328 days. (Other sources speak of an Inca lunar calendar of 12- and 13-month years.) Observations of the seasonal first rise of the constellation of the Pleiades in the night sky corrected the inherent discrepancies in a 328-day lunar year. The annual appearance of these stars regulated other ritual and seasonal events, including the deployment of corvée labor, probably because it coincided with the coming of rains in August.

Like that of the Aztecs, Incan medical and botanical knowledge was quite sophisticated. Specialized classes of "doctors" and surgeons existed along with state-appointed collectors of medicinal herbs. Inca medical personnel performed amputations and in urgent cases trepanned patients (i.e., cut holes in their skulls), presumably to forestall the fatal effects of brain swelling. Like the ancient Egyptians, the Incas mastered the arts of mummification.

Inca civilization fell in 1532 to an invading Spanish force led by the conquistador Francisco Pizarro. Along with the collapse of the Aztecs a decade earlier the history of the Americas and the history of science and technology in the Americas thereby became inextricably linked to developments in Europe.

In the meantime recent research has shown that the human impact on vast expanses of the Amazon basin was more substantial than previously realized. Humans following Paleolithic and then Neolithic economies slowly filled up and transformed the great rainforest into an artifact of human habitation. Cultural influences may well have seeped from Central America or across the Andes into this vast arena or even the other way (witness the cacao plant, for example). As interesting and indicative as these developments are of the great themes of population, ecology, and production that have driven human history, the change-over to intensified agricultural production and more complex societies did not occur in the Amazon basin until after the area was overtaken by other forces from the outside.

Sun Daggers

In contrast to Central and South America, the full panoply of an indigenous high civilization did not arise in pre-Columbian North America. The continent did indeed have great rivers, but in its eastern two-thirds these flowed through vast, unrestricted expanses of temperate forest and plain. Population densities never crossed critical thresholds, and bureaucratically centralized societies did not arise. At first, a Paleolithic economy universally governed the lives of North American hunters and gatherers. In certain areas an intensified form of the Paleolithic way of life came to involve systematic exploitation of deer, fowl, wild grain, and nuts. Then, from roughly 500 BCE, as groups began to cultivate beans and as the practice of growing corn and squash spread outward from Central America, characteristically Neolithic societies appeared in North America. As one might anticipate, the increased wealth of the Neolithic mode of production resulted in somewhat more stratified societies; permanent settlements, towns, and ceremonial centers; trading networks; and larger-scale building. Commonly known as Mound Builders, these cultures, such as the Hopewell and Adena, maintained themselves over centuries and are characterized by the large earthen works they left behind. The Great Serpent Mound in modern Ohio is a well-known example. It and many related constructions served as mortuary sites and may also have been redistribution centers. Such works and the cultures that gave rise to them are strongly reminiscent of Neolithic Stonehenge in terms of social complexity. The so-called Mississippian culture that flourished in the area of the American Midwest from 750 into the 1600s represents a high point of social evolution in pre-Columbian North America. Based on more complex agricultural systems for producing corn, Mississippian groups created a true city, Cahokia, in modern Illinois; it covered six square miles (1500 hectares) with hundreds of mounds and temples and a population of 30,000–40,000 in 1200 CE.

Patterns of culture in North America thus paralleled developments seen elsewhere in particular locales around the world with successive Paleolithic, intensified Paleolithic, Neolithic, and intensified Neolithic stages of production. Each stage evolved a set of appropriate technologies, knowledge systems, and cultural forms necessary to maintain the populations involved, often for extended periods. Not unexpectedly, Amerindian groups developed their own practical astronomies. The solstice orientation of many of the works of the Mound Builders, the extensive petroglyph astronomical record that includes a possible observation of the Crab Nebula explosion of 1054, and the stone circles known as "medicine wheels" of the Plains Indians with their solstice markers—all these evidence a diverse but familiar response of hunter-gatherers, pastoralists, and Neolithic farmers to problems posed by the natural world around them.

In the desert regions of the American Southwest a further social development occurred that displays a partial, but revealing interplay of ecology, technology, and science similar to other early civilizations. As Amerindian groups such as the Hohokam and the Anasazi utilized intermediate levels of irrigation in their agriculture, they consequently achieved intermediate levels of development between Neolithic and

Fig. 9.7. Great Serpent Mound. Like Neolithic societies elsewhere around the world, native American groups regularly built substantial structures, often with astronomical orientations. Here an early illustration depicts the archaeological site known as Great Serpent Mound (Adams County, Ohio), probably built and used by the Adena culture between 100 BCE and 700 CE. Perched on a bluff overlooking Brush Creek, the four-foot-high mound structure uncoils to a length of a quarter of a mile. The effigy represents a snake ingesting an egg. The site was probably a regional center that several separate Neolithic groups came to for trading and ceremonial purposes.

high civilization. What might be termed "incipient hydraulic society" produced intermediate political centralization, intermediate social stratification, intermediate population densities, intermediate monumental building, and intermediate levels of scientific development. This American case is revealing of the forces and factors at play in the initial appearance and character of pristine and other high civilizations around the world that we have surveyed.

The Hohokam peoples migrated from Mexico to Arizona around 300 BCE, bringing irrigation technology with them. Hohokam engineers tapped the Gila River, and by 800 CE they completed a substantial irrigation system that involved 50 kilometers (some sources say "several hundred miles") of canals, including main feeder canals several meters wide that flowed for 16 kilometers (9 miles). Coordinated effort was obviously required to build and maintain these irrigation works. Hohokam irrigation technology watered 30,000 acres (12,000 hectares) and produced two crops a year. Increases in agricultural productivity on this scale produced predictable social repercussions for the Hohokam: population growth, political centralization, public building, and territorial expansion.

After centuries of interaction with the Hohokam, the Anasazi cul-

Map 9.3. Civilization in the American Southwest. Around 1,500 years ago civilizations began to form north of Mexico. The Anasazi Indians flourished in the American Southwest for 200 years around 1050 CE. A regional intensification of agriculture led to dispersed population centers and an extensive network of roads centering on Chaco Canyon in modern New Mexico.

tural group (also known as the Ancestral Publeoan Peoples, or Ancient Ones) coalesced around 700 CE and flourished over two centuries from 950 to 1150 further north in the Four Corners area of the American Southwest. The area is barren desert that receives an average of only nine inches of rain a year and suffers extreme conditions of freezing winters and sweltering summers. Yet here the Anasazi settled and produced a vibrant and expansive society. At its peak Anasazi culture consisted of 75 relatively small communities more or less evenly spread out over an area of 25,000 square miles (6.5 million hectares). The total population numbered on the order of 10,000 people who resided in distinctive cliff "cities." Anasazi architects also fashioned an 800-unit stone and masonry structure of four and five stories at the main Anasazi center at Chaco Canyon. (They brought in timber for beams and construction from a distance of 80 kilometers [50 miles].) With several large ceremonial buildings, the Chaco Canyon complex could harbor 7,000 temporary or permanent residents. A system of regional roads hundreds of miles/kilometers in length linked Anasazi settlements. These roads were wide—up to nine meters (30 feet) across—and deliberately engineered with roadbeds. Agricultural production was possible only by means of irrigation, and with extensive works that tapped the San Juan River Basin and seasonal streams and that involved canals, check dams, and terraced hillsides, Anasazi agriculture yielded corn production at modern levels. Because of the persistent uncertainty of the weather and the harvest, it would seem that Anasazi opted to spread population and production over a wide area, so that if the crop failed in one place, success in another could maintain the whole group. The creation of such sizable residences and ceremonial complexes and irrigation and transportation-communication systems could not have been the work of individuals or even local segments of the Anasazi population.

What about Anasazi "science"? Would this small group of Amerindians dependent on irrigation agriculture in a marginal ecological zone display the same earmarks of scientific knowledge and expertise characteristic of more affluent civilizations elsewhere? A major archaeological find in 1977 proved such to be the case. Archaeologists discovered an exquisite Anasazi seasonal observatory that accurately indicated the summer and winter solstices and the spring and fall equinoxes. (See figure 9.9.) As the sun passes overhead at these seasonal junctures, beams of light created by the manmade rock formation move across a spiral design inscribed on the facing stone wall. Perched on Fajada Butte 450 feet (140 m) above the valley floor in Chaco Canyon, the Anasazi "Sun Dagger" is distinctive among archaeoastronomical artifacts in marking these seasonal passages at the sun's zenith and not along the horizon. The Fajada Butte construction also records the maxima and minima points of the moon's 18.6-year cycle. Other research elsewhere in Chaco Canyon has shown that the Anasazi ceremonial centers or

Fig. 9.8. The Great Kiva at Pueblo Bonito, Chaco Canyon. The Anasazi peoples settled in the American southwest in eighth century CE. Given the ecological uncertainties of their desert setting, their society flourished based on water management techniques. They built cliff dwellings to house much of the population and large ceremonial centers, known as kivas, to store surplus corn and to conduct ritual activities. Not surprisingly, these and other Anasazi constructions embody astronomical knowledge.

kivas likewise had an astronomical orientation and embodied astronomical knowledge. Built as a circle to mirror the heavens, the main door of the Great Kiva at Chaco aligned with the North Star, and at the summer solstice the sun's rays would pass through a window and strike a special niche. Clearly, the Anasazi, like so many groups before them faced with the necessity of gaining control of the calendar, developed the requisite expertise, and some among them became masters of observational astronomy.

Because of the limits imposed by the ecological setting, the levels of intensified agricultural production, population density, monumental building, political centralization, and institutionalized scientific expertise among the Anasazi never rose to match those of fully developed

Fig. 9.9. Anasazi astronomy. Like all agricultural civilizations the Anasazi developed reliable systems of calendrical reckoning. At the top of Fajada Butte in Chaco Canyon in New Mexico archaeologists have found a unique Anasazi astronomical marker. Because of the way the Anasazi angled giant slabs of rock against the side of the butte, at noon on the day of the solstices and equinoxes "daggers" of light are cast on the opposite wall and slice across the incised spiral pattern in distinctive ways, as shown in the figure, to indicate solstices and equinoxes.

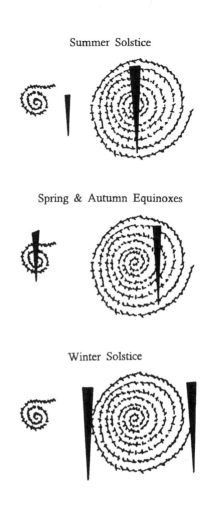

Summer Solstice

Spring & Autumn Equinoxes

Winter Solstice

civilizations elsewhere. Nevertheless, that the Anasazi displayed elements of a comparable pattern speaks volumes about the associations between cultural development and scientific expertise. Unfortunately for the Anasazi, a severe drought affected their region in the years 1276–99, and, given the marginal viability of their mode of existence in the first place, this drought spelled the end of their remarkable cultural achievements.

Intermission

Let us step back and briefly consider the state of science and systems of natural knowledge on a world scale at roughly the year 1000 CE. Plainly, no cultural group was without some understanding of the natural world. The point applies to the not-insignificant number of peoples who continued to live by Paleolithic rules and to forage for their food, as well as to tribes of nomadic pastoralists great and small, to villages of simple farmers who followed in the footsteps of their Neolithic ancestors, and to the centers of urban civilization in the Islamic world, classical India, Sung China, Mesoamerica, and Peru.

What distinguishes the science and scientific cultures of these latter civilizations is that they *institutionalized* knowledge and patronized the development of science and scientific expertise in order to administer the comparatively huge social, political, and economic entities that constituted their respective complex civilizations. Europe in 1000 offered little of scientific or technological interest to the wider world or to historians looking back on Europe in the early Middle Ages.

In the year 1000 CE none of the worldviews held by any of the world's peoples envisioned the earth other than at the center of their respective universes. Similarly, nowhere except in the Islamic world—and there its status proved increasingly precarious—did savants pursue the intellectual game of disinterested theoretical inquiry into nature uniquely initiated by the ancient Hellenic Greeks.

A related but separate pattern pertains to the nature and distribution of technology on a world level at the outset of the second millennium CE. No society lacked technology—the very notion is absurd. "Paleolithic" groups lived by means of appropriate "Paleolithic" technologies. "Neolithic" groups lived by means of appropriate "Neolithic" technologies. And the universally more diverse and complex character of urban civilization depended on a myriad of specialized crafts and trades that kept the machinery of cities and civilization going.

Only in those handful of subject areas where societies required and patronized specialized knowledge—astrology/astronomy, literacy, numeracy, aspects of engineering, medicine, law, and perhaps priest lore to name the major ones—is it at all meaningful to speak of a limited existence of science and expert knowledge turned to practical ends. Otherwise, the worlds of technology and learned science remained sociologically and institutionally poles apart. The vast bulk of technology did not derive from the world of science and had developed according to sociologically distinct craft traditions.

To measure the full historical force of the pattern of hydraulic civilization seen in the Islamic world, in China, in India, pan-India, and in contemporary American civilizations and to fully evaluate their attendant scientific cultures, one must compare these great civilizations with the rise of a secondary civilization in rain-watered Europe where the ecological conditions did not call for any government management or control of the basic agrarian economy. Yet, a train of novelties would transform Europe and bring it onto the world stage with world-transforming effects.

Part III

Europe and the Solar System

Compared with the East and with medieval Islam, Christian Europe at the turn of the first millennium CE was truly an "empty quarter." Latin Christendom numbered 22 million souls in 1000, versus 60 million in the heartland of China, 79 million on the Indian subcontinent, and perhaps 40 million under Islamic governance. By the year 1000 the population of Rome had declined to 35,000 from its peak of 450,000 in antiquity; only 20,000 people lived in Paris, and only 15,000 inhabited London. In contrast, the population of Córdoba in Islamic Spain topped 450,000 (some estimates range as high as 1 million), Constantinople held 300,000, Kaifeng in China reached 400,000, and the population of Baghdad — the largest city in the world — was nearly 1 million. Contemporary Europe was a cultural, intellectual, economic, technological, and demographic backwater that paled before the technological and scientific vitality of contemporary centers of civilization in Islam, Byzantium, India, China, Mesoamerica, and South America.

Through the first millennium of the Christian era Europe was dotted with rural settlements and displayed only a thin veneer of literate culture. A patchwork of tribal societies existing under an essentially Neolithic economy knitted together western Europe in the early Middle Ages, and repeated Viking incursions after the ninth century put at risk what fragile social and institutional bonds there were. Prior to the "Renaissance of the twelfth century," the state of learning in Europe north of the Alps also remained meager. Lacking large urban centers, significant ports, wealthy royal or aristocratic patrons, and highly developed cultural institutions, Europe stood in sharp contrast to both contemporary civilizations and those that had come before.

Civilization arrived in Europe by a unique route and developed under unique physical and social conditions. In the East the first civilizations formed in constricted semiarid river valleys where centralized governments managed the basic agricultural economies. In Europe, where rain fell in the spring and summer, no such governmental inter-

vention established itself, nor indeed did an urbanized civilization arise at all until around the tenth century, when a unique form of agricultural intensification began to change Europe into an urbanized civilization. Once it found a way to intensify its agriculture Europe achieved a very different character. Demographically, its population swelled to match India and China, and technologically, economically, and politically Europe became a major player on the world scene. Beginning in the fifteenth century, based on its mastery of the technologies of cannons and the oceangoing ship, European power began to spread beyond its own borders and to establish world-transforming overseas empires. Western Europe also became the world center of scientific learning and research. Indeed, the origin of modern science unfolded in the Scientific Revolution in Europe in the sixteenth and seventeenth centuries and nowhere else.

These remarkable historical developments raise several questions. First, how could the destitute "empty quarter" of Europe transform itself materially and intellectually in such profound and historically far-reaching ways? How and why as part of the Scientific Revolution did European scientists come to propose and accept heliocentrism and the earth as a planet spinning through space? How did the modern enterprise of science come to be defined, and how can we account for science becoming such a part of European society in early-modern history? Part III addresses itself to this set of questions.

Plows, Stirrups, Guns, and Plagues

A series of interlocked technical innovations—an agricultural revolution, new military technologies, and a dependence on wind and water for the generation of power—shaped the history of medieval Europe. This technical perspective enables us to address the questions of why and how Europe transformed itself from a cultural backwater based on an economy scarcely more advanced than that of traditional Neolithic societies to a vibrant and unique, albeit aggressive, civilization that came to lead the world in the development of science and industry.

"Oats, Peas, Beans, and Barley Grow"

The Agricultural Revolution of the Middle Ages represents a response to problems that resulted from a rising population in combination with a land shortage. The population of Europe as a whole rose 38 percent between 600 and 1000. The increase in France stood closer to 45 percent, and geographical pockets undoubtedly experienced larger increases. In medieval Europe land was put to many uses, not only to produce food and fiber on cropland but also to pasture dairy animals for food, cattle and horses for traction, and sheep for wool, while expanding cities reduced the acreage available for agricultural production. Moreover, tracts of forestland provided timber for construction, shipbuilding, and both heating and industrial fuel. In particular, wood was used as the fuel in making iron, an industry that consumed vast quantities of timber and placed a heavy strain on land use. By the ninth century the people of Europe began to face the kind of ecological crises that thousands of years earlier had induced the Neolithic settlers of the river valleys of the East to intensify their agriculture and make the transition to civilization.

In Europe, agricultural intensification did not and could not follow the same pattern that it had in the ancient East where artificial irrigation provided a technological solution. Europe was already irrigated naturally by sufficient rainfall which fell during the spring and summer. Instead, the European farmer could increase his production only by plowing the heavy soils of districts that had resisted the encroachments of the light Mediterranean scratch plow. The unique constellation of technological innovations adapted to the unique ecological conditions of Northern Europe produced the European Agricultural Revolution.

The first innovation involved the introduction of the heavy plow or moldboard plow. This behemoth of wood and iron, mounted on wheels and armed with an iron cutter, tore up the soil at the root line and turned it over, forming a furrow and eliminating the need for cross-plowing. The heavy plow was resisted by enormous friction and therefore had to be pulled by as many as eight oxen. By contrast, the Mediterranean scratch plow, adapted to light soils, was essentially a hoe dragged through the ground by one or two oxen, with fields plowed twice. The heavy plow, which the Romans had invented but rarely used, increased agricultural production by allowing the farmer to cultivate the wet lowlands of Europe.

A second innovation that contributed to an increase in agricultural production involved the substitution of the horse, with its greater speed and endurance, for the ox as a draft animal. The traditional neck harness, which the ox with its short neck could tolerate, was unsuitable for the horse. Europeans, it seems, adapted the horse collar from the Chinese who employed it several centuries before. The device transferred the pressure points from the windpipe to the shoulders and thereby increased the horse's traction four- or fivefold. In combination with the iron horseshoe, another European innovation, it resulted in a shift to the horse from the ox as the principal draft animal.

Still another component of the Agricultural Revolution of the Middle Ages was the development of the three-field rotation system. The classic two-field farming system of the Mediterranean regions of antiquity typically involved farming one field while leaving another fallow. In the new three-field pattern that arose on the European plain, arable land was divided into three fields with plantings rotated over a three-year cycle: two seasonal plantings employed two of the fields, a winter wheat crop and a spring crop of oats, peas, beans, barley, and lentils, with the third field left fallow.

These new technologies produced a variety of social consequences, both progressive and problematical. The deep plow made it possible to farm new lands, particularly the rich, alluvial soils on the European plain, and this ability helps account for the northward shift of European agriculture in the Middle Ages. Then, because the heavy plow and its team of oxen was an expensive tool, beyond the capacities of individual peasant farmers to own, it brought collective ownership and patterns of communal agriculture and communal animal husbandry, thus solidifying the medieval village and the manorial system as the bedrock of European society at least through the French Revolution.

Similarly, the shift to horses allowed for larger villages because of the larger "radius" horses could work, presumably leading to socially more diverse and satisfying village life. Horses also decreased the cost of transporting goods, so that a greater number of villages could participate in regional, national, and international economies.

The three-field rotation system produced important advantages. The spring crop of vegetables and oats significantly improved the diets of the common people of Europe, an improvement captured in the old English folksong, "Oat, Peas, Beans, and Barley Grow." The three-field system nominally increased the productive capabilities of European agriculture upward from 33 percent to 50 percent, an extraordinary surplus of food production that fed the rise of Europe and European cities in the High Middle Ages.

A richer, more productive, more urbanized Europe resulted from the Agricultural Revolution of the Middle Ages, a Europe destined to create modern science and lead the world in technological progress, but also a Europe that contained the seeds of many future problems—land shortage, timber famine, population pressure, imperial ferocity, devastating epidemics, world war, and finally, as a result of its technological success, ecological disruption on a global scale.

By 1300 the population of Europe to the Ural Mountains trebled to 79 million from a low of 26 million in 600. Paris increased in population by more than 10 times to 228,000 in 1300, and then to 280,000 in 1400. Coincident with urbanization and this growth in population came a wave of cathedral building (Canterbury begun in 1175, Chartres in 1194), university founding (Bologna in 1088, Paris in 1160) and the full complement of high medieval culture including the ethos of chivalry and poetic tales of damsels in distress and knights in search of the Holy Grail. The mechanical clock tellingly came to adorn cathedrals and city halls and regulate the populace in cities and towns from the 1300s onward.

Agriculture was not the only activity in which technology contributed to the rise of medieval Europe. In military affairs technological innovations produced some of the unique developments that characterize European feudalism and that begin to account for Europe's eventual global dominance. One of the definitive figures of European feudalism, the armored knight mounted on an armored charger, was created by a key piece of technology—the stirrup. In Europe prior to the eighth century the mounted warrior remained mounted only until he reached the field of battle, where he dismounted and fought on foot. Without stirrups to provide stability only the most skilled horsemen could fight as true cavalry and could swing a sword or stretch a bow without losing his mount. The Chinese invented the stirrup in the fifth century CE, and it thereafter diffused westward. With no moving parts, the stirrup is a deceptively simple piece of technology, but in stabilizing a warrior on his horse it allows for fighting on horseback without dismounting. With stirrups, a rider with a lance hooked to his armor became a formidable unit where momentum replaced muscle in a new style of warfare—"mounted shock combat." The European knight evolved into the medieval equivalent of the tank, with ever more heavily armored knights and horses forming the most powerful arm on the battlefield.

The new European technology of mounted shock combat meshed easily with the manorial system brought about by the Agricultural Revolution. The knight replaced the peasant-soldier common in the early Middle Ages, and being a knight became a full-time job. The cost of equipping the traditional knight in shining armor, while substantial, lay within the means of a local lord. The system resulted in truly feudal relations, wherein vassal knights pledged their loyalty and their arms to a higher feudal lord in exchange for part of the lord's domain to be governed and taxed in the lord's name. Such local relations were especially apt for the decentralized character of European societies in the Middle Ages. No strong central government, comparable to those of bureaucratic civilization, was required to manage an agricultural economy that needed no hydraulic infrastructure. The manorial system was well adapted to the European ecology. The advent of the knight and European feudalism further forged appropriately local relations between villages and the knights and lords who governed them. The knight-village relation became characteristic of European feudalism and the manorial system, wherein the village owed "dues" to the church and the knightly manor house. Transformed through the Agricultural Revolution, the village now produced the surplus needed to support cadres of knights, and those knights policed, taxed, and enforced justice on a local level.

Given primogeniture, the custom of feudal lands passing to the first-born son, the number of landless knights rose, and ultimately more knights populated Europe than could be easily accommodated. As a result, a first wave of European expansion erupted with the Crusades. Pope Urban II launched the first Crusade in 1096. These unruly expeditions continued to pour forth from Europe for nearly 200 years; the seventh and last Crusade began in 1270. Since the European invaders encountered civilizations that were technologically their equal and culturally their superiors—Byzantine and Islamic civilizations—there was little chance that they would prevail.

Coincident with these changes European engineers developed a fascination for new machines and new sources of power, and they adopted and developed novel methods of generating and harnessing it. Indeed, medieval Europe became the first great civilization not to be run primarily by human muscle power. The most outstanding example concerns the development of water-powered machines and their incorporation into the fabric of village life and European society generally. The waterwheel became widely used to wring energy from the profusion of streams that run through many parts of Europe, and it powered a variety of other machines including mills to grind grain, sawmills, and hammer mills. The mill was often owned by the local lord and operated by his much-disliked miller, at least as the miller is depicted in Chaucer's *The Cantebury Tales* written at the end of the fourteenth century. In some districts windmills increased cropland by reclaiming land from the sea. The need for water-driven mills may be attributed to a lack of surplus labor and to the increased production levels generated by the Agricultural Revolution. That is, with more grain to grind, a widespread shift to water- or wind-powered milling machines was only to be expected. The mill existed in antiquity, but saw limited use perhaps because of the availability of slaves and human muscle power to do the work required to grind grain. It is no coincidence, therefore, that slavery withered away in western Europe coincident with the advent of labor-saving machines.

Anonymous medieval engineers also used wind to turn windmills and tidal flow to drive tidal mills. In so doing they mastered older kinds of mechanical gearing and linkage and invented new ones. Europeans perfected water- and wind-driven mills, the spring catapult (or trebuchet), and a host of other devices, and in so doing they drew on new sources of nonhuman motive power. Their civilization was literally driven by comparatively more powerful "engines" of wind and water which tapped more energy of one sort or another than anywhere else

Fig. 10.1. Water power. Europeans began to use wind and water power on an unprecedented scale. Undershot waterwheels were constructed at many sites along the profuse streams with which Europe is favored.

in the world. Medieval Europeans have been described as "power-conscious to the point of fantasy," and by dint of the medieval fascination with machines, more than other cultures European civilization came to envision nature as a wellspring of power to be exploited technologically for the benefit of humankind. This distinctive attitude toward nature has had powerful and increasingly dire consequences.

The impressive array of technological innovations that led to the transformation of European society and culture owed nothing to theoretical science, in large measure because science had little to offer. The science ultimately inherited from the ancient world and from medieval Islam had even less applicability in medieval Europe than it had in those earlier civilizations. Some knowledge of geometry (but not proofs of theorems) and calendrical reckoning proved useful, but none of it had any application in the development of the machines and techniques for which medieval Europe became justly famous.

But the development of European civilization created new external conditions for science and natural philosophy and set the stage for a vital new culture of learning that emerged in Europe. In what is known as the "Renaissance of the twelfth century," European scholars came into contact with, and themselves began to build on, the philosophical and scientific traditions of antiquity and Islam. And just as Europe was singular in its method of intensifying agriculture and in its use of machinery, it was also singular in founding an institution to harbor higher learning—the university.

Books and Gowns

Europe north of the Alps had never been the scene of much higher learning prior to the twelfth century, and, hence, it is misleading to speak of Europe as having fallen into a "Dark Age." Since Roman times a veneer of literate culture, manifested in monastery schools and the occasional scholar, covered a core of essentially Neolithic village life in northern Europe. Monasteries spread across Europe after 500 CE, and with their scriptoria (or rooms for copying texts) and libraries they maintained themselves as minor centers of learning. Catholic priests had to be minimally literate, and in 789 the Frankish king and later Holy Roman emperor Charlemagne issued an ordinance establishing "cathedral schools" in each bishopric in order to guarantee a supply of literate priests for an otherwise illiterate society. Of necessity, the level of learning and instruction in the cathedral schools and monasteries of the early Middle Ages remained quite low, centered essentially on rudiments of the "Seven Liberal Arts" (grammar, rhetoric, logic, arithmetic, geometry, music, and astronomy) inherited from classical antiquity. Some knowledge of astronomy was needed for astrological and calendrical purposes, especially for setting the date for Easter. But beyond these elementary approaches to higher learning in the early Middle Ages, the intellectual emphasis remained on theology and religious affairs rather than on science. Almost no original scientific research took place.

Paradoxically, at the very edge of Europe, monasteries in Ireland achieved a reputation for their theological sophistication and general learning, including knowledge of Greek, which was otherwise essentially lost to Europeans. In this way, it is said, the Irish saved civilization. And every now and again a truly learned individual appeared on the scene, such as Gerbert of Aurillac (945–1003), who became Pope Sylvester II in 999. In addition to mastering the Bible, the writings of the Church Fathers, and what little of classical pagan knowledge had passed to Europe, Gerbert studied the mathematical sciences in monasteries in northern Spain, where Islamic learning had filtered through and from where he brought back to France knowledge of the abacus and the astrolabe, that convenient device for sighting on stars and performing simple astronomical calculations. Gerbert's idiosyncratic and still narrow mastery of the mathematical sciences notwithstanding, intellectual life in early medieval Europe remained comparatively elementary and socially insignificant.

Against the background of weakly organized learning in the early Middle Ages, the appearance of the European university in the twelfth century and its rapid spread across Europe mark an institutional watershed in the history of science and learning. Instruction in medicine arose in the independent principality of Salerno in Italy in the ninth century, but the union of students and faculty founded at Bologna in 1088 usually ranks as the first university in Europe. The University of Paris followed about 1170, Oxford by 1220, and perhaps eighty additional universities appeared by 1500. The rise of the European university coincided with burgeoning cities and growing wealth made possible by the Agricultural Revolution, for universities were decidedly urban institutions, not rural like the monasteries, and they depended (and depend) on an idle student body with the means to pay for and the job prospects to justify attending universities.

Be it noted that in the era before printing, books were rare and expensive items, all painstakingly copied by hand and sometimes handsomely illuminated. Books were chained down in medieval libraries, and the circle of those who had access to books was restricted. By the same token it was possible in a lifetime to read everything written.

Despite occasional claims to the contrary, the European university was a unique institution. Modeled after the craft guilds of medieval Europe, universities evolved as nominally secular communities of students and master teachers, either as guilds of students (as in Bologna) who employed professors or as guilds of master teachers (as in Paris) who took fees from students. Moreover, universities did not depend on state or individual patronage like the scribal schools of antiquity or the Islamic *madrasa*. They were not state organs, but rather re-

ALEXAND. IIMOL

mained independent, typically feudal institutions—chartered corporations with distinct legal privileges under only the loose authority of the church and state. Privileges included the institutional right to grant degrees and freedom from town control. As essentially autonomous, self-governing institutions, universities thus fell in a middle ground between the total state control typical of the bureaucracies of the great empires and the wholly individualistic character of Hellenic science.

The early universities were vibrant institutions serving the invigorated societies of late medieval Europe. They functioned mainly to train the clergy, doctors, lawyers, administrators, and teachers increasingly required to run the affairs of the state, the church, and the private sector as Europe in the Middle Ages continued to flourish. The graduate faculties of theology, law, and medicine provided instruction and training in those advanced subjects to select students, while the undergraduate liberal arts faculty taught all students at the first stages of their university careers. The natural sciences found a secure home within the arts faculty in the core curriculum concerned with logic and the last four of the seven liberal arts (the quadrivium), devoted to arithmetic, geometry, astronomy, and music. Baccalaureate arts graduates who went on to pursue advanced degrees in theology, law, or medicine typically took a master's degree in arts, where natural philosophy was a prime component of the course of study. Master's-level students often taught in the undergraduate arts faculty while undertaking graduate studies. In this way the natural sciences became subjects for more intense study by some scholars within the university context as a preliminary stage in a larger program of professional training. Unlike universities today, the medieval university was not primarily a research institution, nor was science pursued primarily as an end in itself.

Before Greek and Islamic science could provide the basis of a scientific curriculum for the university, that corpus of knowledge had to be made available through extensive programs of translation into Latin. The great Muslim city of Toledo fell to the Christians in 1085 (another indication of the new power of an expansive European civilization), and Toledo became the center of translation activity where teams of translators rendered classic scientific and philosophical texts from Arabic into Latin. Jewish intellectuals in Spain played an important role in this activity by translating from Arabic into Hebrew for themselves and into Spanish for their Christian collaborators and patrons, who would further render the work from Spanish into Latin. Translations also took place in southern Italy and in Sicily ("liberated" by Norman knights in the second half of the eleventh century), where scholars rendered Latin versions from Greek as well as Arabic originals. Note that the motivation for these remarkable efforts was not entirely an abstract love of knowledge, for the sought-after documents primarily concerned the putatively useful sciences of medicine, astronomy, astrology, and alchemy.

As a result, by 1200 Europeans recovered much of ancient science along with the several centuries of scientific and philosophical accomplishment produced within the Islamic world. Adelard of Bath (fl. 1116–1142) translated Euclid's *Elements* (from Arabic) and other Arabic mathematical texts in the 1120s. The most noted translator, Gerard of Cremona (1114–1187), traveled to Spain around 1140 to locate a copy of Ptolemy's *Almagest* and stayed for 40 years to translate not only the *Almagest* in 1175, but also—while heading a team of translators—a total of seventy to eighty other books from Arabic originals, including many major Islamic treatises along with works of Archimedes, Galen, Hippocrates, Aristotle, and Islamic commentaries on Aristotle. Previously, Europe knew Aristotle only from a few minor works in logic. After 1200 his paramount importance as "the Philosopher" emerged clearly. Later in the Renaissance better translations would be made from older, more authentic Greek originals, but by 1200 the "Western" tradition finally made it to western Europe.

If the twelfth century was a period of translation, the thirteenth represents a period of assimilation wherein the learned of Europe began to absorb the scientific and philosophical traditions of antiquity and medieval Islam. Much of the process of assimilation amounted to attempts to reconcile a traditional Christian worldview with Aristotle and other pagan Greek traditions. The great intellectual synthesis of Thomas Aquinas (1224–1274) in large measure completed this process of assimilation. Whether Aquinas Christianized Aristotle or Aristotelianized Christianity, or both, matters little, for one way or another Aristotle came to provide a complete intellectual system upon which

Fig. 10.2. (*opposite*) The medieval university. The university was a new type of institution that appeared in Europe in the eleventh and twelfth centuries and whose numbers exploded thereafter. Centers of teaching and higher learning, universities were guilds of scholars and students that operated largely independently of civil and religious authority. Master teachers gave instruction in the liberal arts to undergraduates, and the graduate faculties of law, medicine, and theology awarded professional degrees. The university performed an essential social function in developing and providing these cadres. The medieval jurist Alexandre de Omola (1424–77) is pictured in the original of this sixteenth-century engraving lecturing on law to students at the University of Bologna, the first university, founded in 1088. The classroom lecture does not seem to have changed much in a thousand years.

medieval scholastics raised the edifice of rational thought about God, man, and nature. Aristotle's logic and analytical categories became virtually the exclusive conceptual means of investigating any subject. The elaboration and defense of Aristotle's works became a mission of the universities, and the resulting intellectual amalgam of Christian theol-

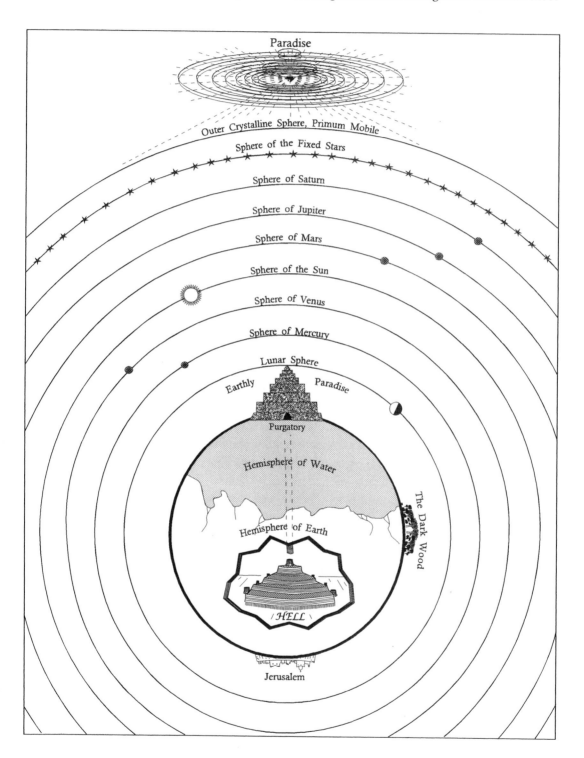

ogy and Aristotelian science produced a coherent and unified vision of the world and humanity's place in it.

Consider, for example, the vision of the cosmos revealed in *The Divine Comedy*, the famous poem by the medieval Italian poet Dante Alighieri (1265–1321). The earth remains motionless and fixed at the center of the world. The sublunary realm of flux—the earth with its four elements and their natural and violent motions—provides the setting for the human drama. In the heavens celestial spheres carry the planets and the stars in their courses. Hell sits at the center, purgatory in the middle, and paradise beyond. A great hierarchical chain of being unites all creatures in ranked order, from the lowest worm to the noblest king or pope and then, through a series of angels and archangels, to God. The physical laws are Aristotelian. The divine laws are God's. The whole arrangement is temporary, created at a particular moment in the past and awaiting the end of time. Such a powerful and unified vision must have provided considerable intellectual and spiritual satisfaction to common people and intellectuals alike.

Medieval scholars interpreted the world primarily from a theological point of view, but they believed that reason can play a role in human understanding of the divine and that we can learn of the existence and nature of God from his work as well as from his word—that is, from the study of nature as well as from the study of the Bible. But in the overall context of the medieval worldview secular natural science took second place whenever Aristotelian natural philosophy clashed with traditional Christian theology.

Since the Enlightenment of the eighteenth century it has become customary in Western democracies to emphasize the importance of religious freedom and the separation of church and school. In medieval Europe, unity—not separation—was required. As Aristotelian teachings began to infiltrate Europe and the new European universities in the thirteenth century, faith and reason had to be harmonized. Certain points in Aristotle clearly conflicted with traditional Catholic teachings—his views that the world was eternal, that there was no creation, that the human soul was not immortal, that limits existed on Divine potency.

Institutional conflicts compounded the intellectual problems of assimilating Aristotle into standard theology, as teachers in the arts faculty promoted philosophy and reason as independent and equally valid routes for arriving at truth, as opposed to the theology faculty, which naturally resisted the threat of secular philosophy and the natural sciences as competing systems of knowledge. A series of intellectual skirmishes between theologians and philosophers unfolded across the thirteenth century, culminating in the condemnation of 1277, wherein the bishop of Paris with the backing of the pope condemned the teaching of 219 "execrable errors" held by some Aristotelians and subjected anyone who held or taught them to excommunication.

Fig. 10.3. (opposite) Dante's scheme of the universe. In *The Divine Comedy* the early-fourteenth-century Italian poet Dante Alighieri developed the medieval portrait of the cosmos as a Christianized version of Aristotle's worldview.

On the surface, the condemnation seems to represent a decisive victory for conservative theology, the suppression of liberal and independent philosophy and science, and the institutional subordination of the arts faculty to the theology faculty within the university. Some commentators, however, see the condemnation of 1277 as only a minor outbreak of hostilities which in the end produced harmony rather than discord, especially given that the condemnation held only for a few decades at the University of Paris, while at Oxford less restrictive measures operated, and elsewhere none at all. Still other investigators go further and argue that by freeing medieval thinkers from the yoke of strict obedience to Aristotle, the condemnation of 1277, in effect, liberated them to conceive new alternatives in solving long-standing problems in Aristotelian science and natural philosophy. From this point of view, the Scientific Revolution did not begin with Copernicus in the sixteenth century, as is usually held, but 250 years earlier with Catholic scientific intellectuals and their response to the condemnation of 1277.

Questions concerning continuities versus discontinuities in late medieval and early modern history and the history of science have long been debated and remain far from settled among scholars even today. The thesis regarding the essentially modern character of late medieval science remains a notable focus of those debates. For the present we can refer to a middle-ground interpretation of the condemnation of 1277 which sees its effect neither as wholly squelching scientific inquiry nor as launching the Scientific Revolution outright. The condemnation produced a paradoxical effect, in that by subordinating philosophy to theology and by forcing the admission that God could have fashioned the world in any number of ways, given his omnipotence, a path opened for masters in the arts faculty to consider any and all scientific possibilities, as long as they stayed out of theology and did not claim that their intellectual games had any necessary relation to the world as God's artifact. An extraordinary flourish of what might be termed "hot house" science resulted, theologically inoffensive work wherein scholastic natural philosophers entertained all variety of scientific possibilities, but only hypothetically, on the basis of "suppositions" or thought experiments, or as products of their ingenious imaginations. Thus, for example, Jean Buridan (1297–1358) and Nicole Oresme (1320–82), among other medieval scientists, examined the possibility of the earth's daily rotation on its axis, and each offered what seem like compelling rational arguments for admitting that such motion occurs in nature. Each believed that science on its own might well lead to that conclusion. Yet both men ended up rejecting the possibility of a moving earth. Oresme reached his conclusion not on scientific grounds, but on the basis of apparent conflicts between that hypothesis and passages in the Bible and because of his view of the inherent superiority of theology as a means of arriving at truth.

By the fourteenth century the major question facing scientific intellectuals of the European Middle Ages no longer concerned simply uncovering new texts or assimilating Aristotle's natural philosophy to scripture or even of purging Aristotle of his anti-Christian elements, but rather building on the Aristotelian paradigm and breaking new ground. Under the general conceptual framework provided by Aristotle, scholastic natural philosophers actively and creatively pursued a wide range of scientific investigations. With two translations of Ptolemy's *Almagest* appearing by 1175, for example, an indigenous tradition of observational and mathematical astronomy arose in western Europe. The Alfonsine Tables (ca. 1275), calculated in an extra-university setting by the astronomers of the king of Castile, was one result, as were pathbreaking though ineffectual calls for calendar reform. Geoffrey Chaucer's *Treatise on the Astrolabe* figures among the astronomical works of the fourteenth century, Chaucer being better known, of course, for his poetical writings. Translation in the 1130s of Ptolemy's great work in astrology, the *Tetrabiblos*, actually preceded his purely astronomical text by half a century. Serious and sustained research in astrology arose alongside medieval astronomy, often associated with medicine and medical practice. Moreover, building on a strong Islamic tradition and propelled by the religious connotations associated with "light," medieval investigators carried on research in optics, including improved understanding of vision and the rainbow. In the realm of mathematics, Leonard of Pisa (ca. 1170–1240), known as Fibonacci, introduced Europeans to "Arabic" (actually Indian) numerals and sophisticated algebraic problems in his *Liber abaci* of 1228. A number of works attributed to Jordanus de Nemore (ca. 1220) took up mechanical questions concerning statics and a "science of weight" in the thirteenth century. The translation and assimilation of the medical treatises of Galen, the great Roman physician of late antiquity, revitalized medical theory and practice after 1200, as did the creation of medical faculties, which became separate centers of science within the medieval university alongside the arts faculty. Closely associated with developments in medieval medicine and Aristotelian traditions in biology and natural history, a number of more narrowly scientific texts appeared which touched on the life sciences, notably in works by Albert the Great (Albertus Magnus, 1200–1280), *On Vegetables* and *On Animals*. And since little distinction was drawn between rational and occult knowledge, one cannot dismiss significant efforts directed toward alchemy and natural magic by medieval alchemists, philosophical magicians, and doctors. Needless to say, women were not part of the medieval university scene, but some, like Hildegard of Bingen (1098–1179), served in positions of authority as abbesses and collected unto themselves bodies of useful knowledge of nature.

The men who undertook inquiries into nature in the Middle Ages were hardly monolithic in approach or slavishly Aristotelian in outlook. Rather, the record reveals a diversity of often conflicting points of view and approaches to studying nature. For example, the University of Paris, dominated by the religious order of the Dominicans, tended to be more purely Aristotelian and naturalistic in outlook, while at Oxford the Franciscan fathers shaded more toward the Platonic and abstract, with consequent differences in how they viewed the role of mathematics in interpreting nature. Scholastics also divided over the role of experiment and hands-on approaches to discovering new knowledge. In opposition to the traditional "scholastic" preoccupation with book learning, Robert Grosseteste (1168–1253), the first chancellor of the university at Oxford, argued for active investigations of nature and on that account is sometimes hailed as the father of the experimental method in science. Greatly influenced by Grosseteste, Roger Bacon (ca. 1215–92) (not to be confused with the later seventeenth-century propagandist for science Francis Bacon, 1561–1626) proposed that human ingenuity ought to be applied to the creation of useful mechanical devices such as self-propelled ships and carts. Similarly, in his *Letter on the Magnet* of 1269 Petrus Peregrinus emphasized the value of experiment in discovering new facts about nature. Retrospectively, the views of these men are suggestive to us today because they seem to anticipate a later experimental style in science. However, in their own context they represent minority approaches among savants of the Middle Ages. Franciscan authorities, for example, restricted the circulation of Roger Bacon's experimental researches, possibly because of their associations with magic. And none of the experimentalists of the Middle Ages questioned the general theological outlook of their era or suggested that the natural sciences were anything but the subordinate handmaiden to theology.

Two historically significant examples can illustrate the character and accomplishment of medieval scientific thought. The first concerns the explanation of projectile motion defended by Jean Buridan, the celebrated fourteenth-century master of arts at the University of Paris. The reader will recall that Aristotle's physics required a mover to be in contact with the object moved for each and every case of forced (i.e., nonnatural) motion. The question of identifying a mover for the apparently anomalous case of projectiles (such as arrows, javelins, chalk thrown at students) stimulated a small research tradition within the broad sweep of the Aristotelian paradigm. Building on earlier medieval commentators and the work of the sixth-century Byzantine natural philosopher John Philoponus, Buridan proposed that an internal motive agent he called "impetus," implanted in a projectile by its mover, provided a motive quality that drove projectiles after they lost contact with any evident mover. Buridan envisioned his impetus as applying not only to projectiles, but also to the free fall of bodies and the perpetual revolution of the heavenly spheres. As a kind of self-propelling quality, Buridan's impetus at first glance resembles Newton's principle of inertia, which postulates that bodies remain in motion (or at rest) unless otherwise acted upon by an outside force. However, such a superficial resemblance to the later idea of inertia belies a vast conceptual gulf separating Buridan's impetus and, indeed, medieval physics, from Newton and the Scientific Revolution. That is, for early modern physics, as we will see, projectiles move of their own accord and require no cause of any description. From Newton's point of view, what needs to be explained is not motion itself but rather *change* in motion, why a projectile starts or stops moving or changes speed or direction. But Buridan approached projectile motion from a diametrically opposite point of view in quite conventionally seeking to identify outside Aristotelian *movers* and an effective *cause* of motion for projectiles while in flight. In other words, rather than a radical break that equating impetus with inertia retrospectively and misleadingly suggests, Buridan placed his inventive explanation squarely within the tradition of Aristotelian science and Aristotelian research puzzles.

A second example of medieval scientific accomplishment concerns the great Parisian philosopher and church doctor Nicole Oresme—the outstanding man of science of the European Middle Ages. In his *On the Configuration of Qualities*, written around 1350, Oresme created visual constructs—graphs—to represent qualities and qualitative change geometrically. The somewhat modernized diagram shows how Oresme depicted uniformly accelerated motion, motion that he referred to as "uniformly difform" and that we see in falling bodies (see figure 10.4). The horizontal axis represents time; the vertical axis represents the degrees of speed possessed by a uniformly accelerating body; the area under *AB* (the triangle *ABC*) represents the total distance covered by the body. Inherent in the diagram lie several mathematical laws pertaining to motion that Oresme and his compeers understood perfectly: for example, that the distance covered by a uniformly accelerated body equals that of another body moving with constant speed equal to one-half of the final speed of the accelerated body. (The constant motion may be represented by the horizontal line *DE*; because the triangle *ADF* equals triangle *FBE*, the area under *DE* equals the area under *AB*, and hence the distances covered by uniform motion and uniformly accelerated motion are equal.) Also lurking in the graph is the notion that the distance a body covers in uniformly accelerated motion is proportional to the square of the time a body is accelerated ($s \propto t^2$).

The formula $s \propto t^2$ expresses Galileo's law of falling bodies, a law Galileo formulated 250 years after Oresme. The obvious question arises why Galileo and not Oresme gets credit for uncovering this fundamental law of nature. The answer is revealing, for after 1277 Oresme in no way associated his abstract inquiries into the character of accelerated motion with any motion in the real world. His interest lay in understanding accelerated and other motions and qualities on their own terms, on a wholly abstract basis, and purely as a theoretical intellec-

Fig. 10.4. Uniformly difform motion. In classifying different types of motion, the medieval scholastic Nicole Oresme (1320–82) depicted a motion that increases uniformly over time. He called it uniformly difform motion; we call it constant acceleration. Implicit in the diagram is the notion that the distance an accelerated body travels is proportional to the square of the time elapsed. Galileo later transformed Oresme's abstract rules into the laws of motion that apply to falling bodies.

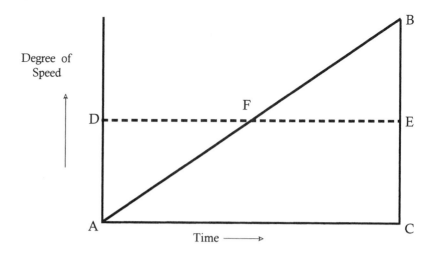

tual exercise. In other words, that Oresme's "uniformly difform" motion characterizes the fall of bodies in the real world apparently never occurred to him or his contemporaries. Oresme's was a phenomenal intellectual achievement of a prodigious scientific imagination, but it remained a precocious one that by itself did not lead to Galileo or the Scientific Revolution.

Considering its rise out of the impoverished institutional and intellectual circumstances of the early Middle Ages in Europe, late medieval science seems remarkably productive in the rational study of nature and in exploring the limits of Aristotelian natural philosophy. The European Middle Ages created a new institutional foundation in the European university and an intellectual foundation in bringing a critical review of Aristotelian science to Europe. But the historical significance of the medieval scientific accomplishment may be less in what it meant for the Middle Ages than for having laid the institutional and intellectual foundations for further developments during the Scientific Revolution of the sixteenth and seventeenth centuries.

A series of disruptions ravaged large parts of Europe in the fourteenth century and in ecological and demographic terms interrupted several centuries of European prosperity that was characteristic of the later Middle Ages. The climate in Europe became cooler and wetter, disastrously affecting harvests and agricultural productivity. Unprecedented famines erupted across Europe in the years 1315–17, and the resulting economic depression, seriously exacerbated by an international banking crisis in 1345, lasted well into the next century. The epidemic of bubonic and pneumonic plagues—the Black Death—sweeping Europe in 1347–48 wiped out a quarter to a third of the population. The plague hit urban centers and university towns especially hard with resulting high death rates for the literate classes. Tens of thousands of European villages simply disappeared, and devastating outbreaks of plague recurred until the eighteenth century. Some experts put the death toll as high as 40 percent, with demographic recovery not occurring until 1600. Less dramatic in its import, but wrenching nonetheless, the removal of the papacy from Rome to Avignon for most of the fourteenth century broke the unity of Christendom and divided the allegiances of Catholics along several papal lines. And the Hundred Years' War between England and France, breaking out in 1338, periodically devastated the heartland of France through the 1450s. Peasant revolts and threats of social unrest were no less characteristic of the decades on either side of 1400. These doleful developments fell hardest on the lower orders and affected science directly through the mortality among scientists and indirectly through institutional closures and educational disruptions. Taken together, they mark a watershed in the development of European material culture and contribute to a sense of discontinuity in reviewing the history of medieval and early modern science.

After the disruptions of the later Middle Ages the essentials of European agriculture and feudal society based upon that hardy system of production revived. Universities resumed their growth and expansion, and if the number of scholars pursuing scientific inquiries had dipped after 1350 their numbers not only recovered but, as our data in figure 10.4 indicates, ultimately rebounded as if the demographic horrors of the fourteenth century never occurred. Perhaps the alluring artistic accomplishments of the European Renaissance have seduced us into seeing more of a historical break at the end of the Middle Ages than is warranted.

Cannons and Sailing Ships

By the fourteenth century Europe had recapitulated some but not all of the earmarks of the earlier civilizations. Agriculture had been intensified, the population grew, urbanization took hold, building (in the form of the soaring cathedrals) became ever more monumental, and higher learning was institutionalized. But in a rainfall environment that forestalled the need for public works to build and maintain a system of hydraulic agriculture, neither a centralized authority nor a universal corvée came into being. Only later, beginning in the sixteenth century, did these components of civilization arrive on the European scene. The historical dynamic that produced those consequential innovations was a sweeping military revolution that was, like Europe's agricultural system, cathedrals, and universities, a unique development.

Gunpowder technologies originated in Asia. The Chinese invented gunpowder in the ninth century CE and developed fireworks and rockets before 1150. By the mid-1200s Chinese armies employed Roman candle–style "fire lances" and explosive bombs thrown by catapults,

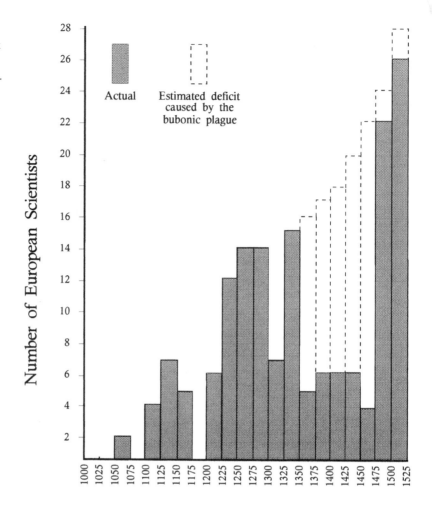

Fig. 10.5. The Plague. Also known as the Black Death, the plague struck Europe in 1347, depressing the amount of scientific activity. After more than 100 years Europe and European science began to recover.

and by 1288 the Chinese created metal-barreled guns. In an early instance of technology transfer the Mongols acquired gunpowder technology from the Chinese, whence it probably passed into Europe across the steppes of central Asia. Possibly through contact with Chinese engineers and technicians, the technology also passed into the Islamic world, as gunpowder was used against European Crusaders in 1249. Europeans may also have learned of the technology from travelers in the East like Marco Polo who worked for the Mongol emperor in China from 1275 to 1292.

While gunpowder and early firearms technology originated in China, large cannon seem to have originated in Europe in the decade 1310–20. The technology then spread quickly back to the Middle East and Asia, cannon finding their way to Islamic lands by the 1330s and to China by 1356. By 1500 the manufacture of guns had become a universal technology in the Old World with centers in China, the Moghal empire in India, the Ottoman empire, and Europe, and these powers further spread cannon technology to client-states throughout the Old World.

Early cannon and bombards were huge. Turkish guns deployed in the siege of Constantinople in 1453, for example, were so large they were cast on the site and could not be moved. "Mons Meg," a cannon cast for the duke of Burgundy in 1449, measured almost 10 feet long, weighted 17,000 pounds, and threw a stone ball of nearly two feet in diameter. Probably because they became engaged in a more intense cycle of competition, European military engineers and foundrymen actively developed the technologies of cannon-making, and with a superiority of gun design, Europeans soon surpassed their Asian counterparts from whom they initially learned the technology. The great size of guns—suitable for battering walls of established fortifications—gave way to smaller, relatively portable bronze and then to much less costly cast-iron cannon, especially after 1541 when the English mastered techniques of casting iron cannon under the initiative of King Henry VIII. Smaller cannon not only proved more mobile on land, they produced powerfully changed conditions when brought aboard ships.

Already in the fifteenth century, gunpowder and firearms began to play a decisive role on the battlefields of Europe, and by the end of the century they had transformed the politics, sociology, and economics of war. The "gunpowder revolution" undermined the military roles of the feudal knight and the feudal lord and replaced them with enormously expensive gunpowder armies and navies financed by central governments. The knight was not made useless by the new weaponry. In fact, knights continued to proffer their services and to maintain retinues of archers and pikemen. But neither knights nor noblemen could master the economic dimension of the revolution in warfare, for the new artillery lay beyond the means of any individual captain or lord and could be financed only by royal treasuries. At the outset of the Hundred Years' War (1337–1453), the primary means of combat remained the longbow, the crossbow, the pike, and the armored knight mounted on an armored charger. At the end, gunpowder artillery won out.

The military career of Joan of Arc (1412–31) nicely illustrates this transition in military and technological history. Joan of Arc, an illiterate peasant girl of 17, could defeat experienced English commanders partly because artillery was so new that previous military experience carried little advantage, as is the case with any new technology not derived from accumulated knowledge and tradition. Indeed, her fellow commanders praised Joan especially for her keen ability to place artillery in the field. (In what has been labeled the "Joan of Arc syndrome," it would seem that whenever a new technology appears on the scene, e.g., with computers, the young often surpass the old and may make significant contributions.)

The new weaponry appearing in Europe in the fifteenth century required large increases in the budgets of European governments. During the second half of the fifteenth century, for example, as the Military Revolution took hold, tax revenues in western Europe apparently doubled in real terms. From the 1440s to the 1550s, as a further example, the French artillery increased its annual consumption of gunpowder from 20,000 pounds to 500,000 pounds, and the number of French gunners rose from 40 to 275. Spanish military expenses went from less than 2 million ducats in 1556 to 13 million in the 1590s. To cope with these increasing expenditures Philip II of Spain tripled the tax burden on Castile and repeatedly repudiated the state debt, without ever meeting his military payroll on time.

The musket was introduced in the 1550s, and in their reforms Maurice and William Louis of Nassau instituted volley fire by extended rows of musketeers using standardized muskets and coordinated routines of loading and firing. Those reforms and standardized field artillery made for potent new armies from 1600. In the face of muskets and artillery, longbows, crossbows, broadswords, cavalry, and pikemen exercised diminished roles or vanished entirely from the scene of battle. Infantry, now bristling with handguns, once again became a dominant arm on the field of battle. As a result, over the next two centuries the size of standing armies of several European states jumped dramatically from the range of 10,000 to 100,000 soldiers. During the last 70 years of the seventeenth century alone the French army grew from 150,000 to perhaps 400,000 under the Sun King, Louis XIV.

Because cannon could reduce medieval castles and old-style city walls with relative ease, they mandated new and even more expensive defensive countermeasures that took the form of earthen ramparts studded with star-shaped masonry bastions known as the *trace italienne*. Fortified with guns of their own, these installations allowed a defender to rake the attacking formations. European governments poured money into the development of these new and expensive systems of fortifications, but they strained the resources of even the richest European states. Offense and defense alternated in an escalating pattern of challenge and response. Relentlessly, costs mounted and warfare became the province of centralized states.

Only larger political entities, notably centralized nation-states with taxing power or other mercantile wealth, could afford the new weaponry and its attendant fortifications. The Military Revolution, therefore, shifted power from local feudal authorities to centralized kingdoms and nation-states. The kingdom of France, for example—the most powerful of the early modern states in Europe—only emerged as an entity after the Hundred Years' War in the fifteenth century. The related development of the musket and the standing armies that resulted after 1550 reinforced this trend, insofar as central governments ended up as the only agencies that could afford standing armies and that possessed the bureaucratic capabilities to organize, equip, and maintain them.

The substantial government assistance and intervention required by this historically unique military technology led European societies to-

Fig. 10.6. Military drill. Although the musket had a lower rate of fire than the bow and arrow, it struck with much greater force. The complex firing procedure led to the development of the military drill. Organizing, supplying, and training large numbers of soldiers armed with guns reinforced political trends toward centralization.

ward centralized authority. The Military Revolution introduced competition between states and dynamic social mechanisms that relentlessly favored technical development. The centralizing effects—social, political, and economic—were akin to those called forth by irrigation agriculture in the great hydraulic civilizations we have previously examined. Thus, Europe acquired, like Egypt and China thousands of years before, the full panoply of civilized institutions. From the fifteenth century onward, the creation of national armies and royal navies resulted in political centralization as inevitably as had the hydrau-

Fig. 10.7. Fortification. As the castle fell to the cannon it was replaced by a complex system of fortification with projections known as the *trace italienne*. While effective as a defensive system it proved to be extraordinarily expensive.

lic projects of the ancient and medieval East. Arsenals, shipyards, and fortresses were maintained as state-owned and state-controlled public works, comparable to the dams and canals of hydraulic societies. And, beginning in the seventeenth century with the Swedish army of Gustavus Adolphus and later established as a national principle by the French Revolution, universal military conscription became the modern equivalent of the ancient corvée.

While Europe became increasingly centralized as a result of the Military Revolution, its ecology and geography precluded the rise of a cohesive European empire of the sort found in China, India, or the Islamic world. In comparison with the large-scale irrigation works of the East, which encompassed whole regions—the Nile Valley and the Euphrates-Tigris flood plain—a typical European military-political system, based on rain-watered agriculture, remained a more local affair in which a variety of ethnic, linguistic, and geographical factors combined to define a nation-state. As the primary outcome of the Military Revolution, then, a group of relatively centralized, competing nation-states emerged in Europe. They became locked in political, military, and economic competition, none sufficiently powerful to wholly dominate the others. More than anything else, then, the jostling that ensued between and among Spain, Portugal, France, England, the Low Countries, Prussia, Sweden, and later Russia created unique conditions that made Europe a hotbed of conflict and, at the same time, technologically poised for a world-historical role.

Other evidence supports the argument for limited centralization in Europe. The decentralized character of feudal Europe prior to the Military Revolution provides one point of contrast. For another, the new military technology was inherently innovative and, in a complete reversal of the traditions that had characterized hydraulic civilizations, the risks among European states were highest for those that failed to change. Lesser political units or nations (such as Poland) that did not or could not adapt to the Military Revolution simply disappeared as political entities, swept up by larger, more powerful neighbors. Most telling in this regard is the general absence of pan-European institutions that could conceivably have united a European empire. The papacy and the Holy Roman Empire—that nominal remnant of the West Roman Empire that persisted in Europe from its creation in 800 until its dissolution in 1806—represent the most potent of such institutions. Had the ecology of Europe demanded centralized authority, either the papacy or the Holy Roman Empire might have supplied a supranational structure of governance. In the event, however, neither brought about a unified Europe, and both remained weak institutions compared to the nation-states that emerged. Even when hegemony became artificially imposed on Europe, it was inevitably short-lived, as in the case of the Napoleonic Empire, which barely lasted a decade before collapsing in 1812.

A second major outcome of the Military Revolution, alongside political centralization within Europe, was a wave of European colonialism and the beginnings of European global conquest. A revolution in naval warfare which accompanied military changes on land formed the technological basis of the process. In part, this revolution in naval warfare entailed the creation of a new type of ship and new techniques of naval engagement. The comparatively lightly manned, wind-powered Portuguese caravel and its successor, the galleon, came to replace the human-powered oared galley, which required large crews and was typical of warships plying the Mediterranean. Similarly, whereas naval tactics for galleys such as that deployed by Turks and Christians in the battle of Lepanto in 1571 involved ramming and boarding, the emerging new style involved heavily armed gunned ships, artillery fire from broadsides and at a distance, and tactics to prevent boarding. Experts credit the defeat of the Spanish Armada sent against England in 1588 in part to the English having adopted broadside artillery fire "along the line" while the Spanish clung to the ramming and boarding techniques that had served them so well in the Mediterranean.

The development of the caravel/galleon shows how complicated technological change is in general. We must keep in mind that the players, the "actors," do not know the outcomes beforehand. Some shipbuilder did not explicitly set out to build an oceangoing gunned ship. Rather, technical, social/cultural, and geophysical factors interacted to produce the new ship. For example, sails and rigging had to be improved, gun ports cut, and gun carriages developed and installed. Captains had to master the compass (another Chinese invention) and determine their latitude at sea (i.e., how far north or south they were), techniques not wholly mastered for sailing south of the equator until the 1480s. It was not enough, of course, to sail to Africa, the Indian Ocean, or the Americas, the trick was to get back to Europe, and for Vasco da Gama, as for Columbus, that involved mastery of the technique known as the *volta*, whereby ships sailing northward along the African west coast headed *westward* back out into the Atlantic until they picked up winds to blow them eastward back to Iberia. Technological change embodies complex social processes, wherein strictly technical issues (in shipbuilding, say) interact with social factors of all sorts to produce technical and social outcomes that cannot be foreseen in advance. The case of the gunned ship likewise makes clear that we cannot isolate an autonomous "technology" in the world and then investigate its separate impact on society.

The global results of this new technological capability were stunning. The Portuguese made their first contacts along the sub-Saharan coast of Africa in 1443 and reached the Cape of Good Hope in 1488. Vasco da Gama's first voyage to the Indian Ocean by way of the Cape in 1497–98 involved four small ships, 170 men, and 20 cannon; he returned to Portugal with his holds full of spices extracted by force from Muslim and Hindu traders. Columbus sailed to the Indies in three small caravels. Hernán Cortés conquered Mexico in 1518 and 1519 with an expeditionary force of 600 men, seventeen horses, and ten cannon. Later European voyages deployed larger and more heavily armed flotillas, but Columbus's, da Gama's, and Cortés's small forces were sufficient for successful voyages and set the pattern for European mercantilism and colonialism for 300 years.

Portugal and Spain were the early entrants, and later France, the Netherlands, England, and others joined the game. Their colonies, colonial rivalries, and mercantilist activities set the tone for European struggles abroad and at home through the eighteenth century. In 1797 the French colonial historian Moreau de Saint-Méry wrote that the great tall ships of his day carrying African slaves to the colonies and colonial products back to Europe were "the most astonishing machines created by the genius of man." European navies provided the technical means for the West to force itself on the larger world. Experts posit that by 1800 European powers dominated 35 percent of lands, peoples, and resources of the earth.

What role did scientific thought play in this immensely consequential European development? The answer is essentially none. Some of the basic inventions (such as gunpowder and the compass), as we have seen, originated in China, where they were developed independently of any theoretical concerns. In Europe itself, with its established tradition of Aristotle, Euclid, and Ptolemy, none of contemporary natural philosophy was applicable to the development of the new ordnance or any of its ancillary techniques. In retrospect, theoretical ballistics might have been useful, but a science of ballistics had not yet been deduced; it awaited Galileo's law of falling bodies, and even then the applicability of theory to practice in the seventeenth century can be questioned. Metallurgical chemistry could have been useful to foundrymen, but prior to the nineteenth century theory was limited, and alchemy seemingly had nothing to offer. Hydrodynamics, which might have applied to ship design, also lay in the future. The mechanics of materials, which later became a pivotal engineering science, was reconnoitered by Galileo and only applied in the nineteenth century. Scientific cartography probably did play a supporting role in early European overseas expansion, but navigation remained a craft, not a science. The gunners, foundrymen, smiths, shipbuilders, engineers, and navigators all did their work and made their inventions and improvements with the aid of nothing more (and nothing less) than experience, skill, intuition, rules of thumb, and daring.

Indeed, the causal arrow flew from the world of technology to the world of science, as European governments—like their Eastern counterparts—became patrons of science and provided government support for scientific research in the hope of technical and economic benefits. It is no accident, therefore, that the institutionalization and bureaucratization of science by European governments appears first in early modern Portugal and Spain. While now thought of more as a late medieval crusader and less as the great humanist patron of scientific exploration, the Portuguese prince Henry the Navigator (1394–1460) was responsible for the historic series of fifteenth-century Portuguese voyages of exploration along the coasts of West Africa. He did much to promote navigation and to launch Portugal's maritime empire and, driven by the spice trade, Lisbon soon became the world's center of navigational and cartographical expertise. The ruling Portuguese court patronized various royal mathematicians, cosmographers, and professors of mathematics and astronomy, and it created two government offices charged to administer Portuguese trade and to prepare maps. Expert Portuguese cartographers obtained employment internationally.

In Spain, from 1516 to 1598 the Holy Roman emperor Charles V, and his son, Philip II, governed the largest contemporary European empire. Stemming from its colonial rivalry with neighboring Portugal and from technical issues over demarcating the borders of the empire halfway around the world, in the sixteenth century Spain replaced Portugal as the predominant center of scientific navigation and cartography. In Seville the government-sponsored Casa de la Contratación (House of Trade, 1503) kept and continually revised the master maps of Spanish overseas expansion. The position of Pilot-Major (1508) at the Casa superintended training pilots in navigation; the Cosmographer Royal (1523) was responsible for nautical instruments and charts; and in 1552 Philip II established a royal chair at the Casa in navigation and cosmography. Complementing the Casa de la Contratación, the Council of the Indies, the government body established in 1523 to administer colonial development, came to possess its own corps of royal cosmographers and administrators charged with various scientific and practical duties in the expansion of the Spanish empire. Spanish support for cartography and navigation culminated in the creation of an Academy of Mathematics founded in Madrid by Philip II in 1582 which taught cosmography, navigation, military engineering, and the occult sciences. A sign of the times: the engineering professor of fortifications received twice the salary of the leading university professor of philosophy.

The Spanish government also sponsored systematic and comprehensive geographical and natural history surveys of Spain and the Indies. Officers in the Council of the Indies circulated printed questionnaires and collected useful information on an unprecedented scale, and in the 1570s Philip II sent a historic scientific expedition under Francisco Hernández to the New World to collect geographical, botanical, and medical information. Spain and Portugal were the first European powers to deploy scientific expertise in the service of colonial development. Every succeeding European colonial power—Dutch, French, English, Russian—in turn followed the pattern of state support for science and colonial development established by Spain and Portugal. Thus, by dint of the Military Revolution, the institutionalization of science in early modern Europe began to resemble patterns established in other great civilizations.

Copernicus Incites a Revolution

On his deathbed in 1543 Nicholas Copernicus received the first published copy of his book, *De revolutionibus orbium coelestium* (*On the Revolutions of the Heavenly Spheres*). In this seminal work Copernicus proposed a sun-centered or heliocentric cosmology with a moving earth rotating once a day on its own axis and orbiting the sun once a year. In 1543 every culture in the world placed the earth instead at the center of its cosmology. In breaking so radically with geocentrism, received astronomical wisdom, and biblical tradition, Copernicus launched the Scientific Revolution and took the first steps toward the formation of the modern scientific worldview.

The Scientific Revolution represents a turning point in world history. By 1700 European scientists had overthrown the science and worldviews of Aristotle and Ptolemy. Europeans in 1700—and everyone else not long afterwards—lived in a vastly different intellectual world than that experienced by their predecessors in, say, 1500. The role and power of science, as a way of knowing about the world and as an agency with the potential of changing the world, likewise underwent profound restructurings as part of the Scientific Revolution.

The historical concept of *the* Scientific Revolution of the sixteenth and seventeenth centuries emerged only in the twentieth century. The Scientific Revolution was initially thought of as an intellectual transformation of our understanding of nature, a conceptual reordering of the cosmos that entailed, in the felicitous phrase, moving from a closed world to an infinite universe. As scholars have delved deeper into the subject, the unquestioned unity and reality of *the* Scientific Revolution or a Scientific Revolution began to break down. The Scientific Revolution as simply an episode in the history of scientific ideas is long a thing of the past. For example, any treatment of the Scientific Revolution must now address not just a triumphant astronomy or mechanics but the "occult" sciences of magic, alchemy, and astrology. Ideological arguments for the social utility of science prove to be a fundamental feature of the Scientific Revolution, and the emergence of new scientific methods—notably experimental science—likewise seems a key property of the "new science" of the sixteenth and seventeenth centuries. Changes in the social and institutional organization of contemporary science are now seen as additional defining elements of the Scientific Revolution. The current interpretative stance rejects any simple notion of the Scientific Revolution as a unitary event with clearly defined chronological or conceptual boundaries. Historians now tend to treat the Scientific Revolution as a useful conceptual tool, setting the episode in a broader historical context as a complex and multifaceted phenomenon to be studied through a variety of approaches.

The New World of the European Renaissance

The social context for science in Europe in the sixteenth and seventeenth centuries had changed in several dramatic ways from the Middle Ages. The Military Revolution, the European voyages of exploration, and the discovery of the New World altered the context in which the Scientific Revolution unfolded. The discovery of the Americas generally undermined the closed Eurocentric cosmos of the later Middle Ages, and the science of geography provided a stimulus of its own to the Scientific Revolution. With an emphasis on observational reports and practical experience, new geographical discoveries challenged received authority; cartography thus provided exemplary new ways of learning about the world in general, ways self-evidently superior to mastering inherited dogma from dusty manuscript books. Many of the scientists of the Scientific Revolution seem to have been involved in one fashion or another with geography or cartography.

But printing was a different matter. In the late 1430s Johannes Gutenberg, apparently independently of developments in Asia, invented printing with movable type, and the spread of this powerful new technology after 1450 likewise altered the cultural landscape of early modern Europe. The new medium created a "communications revolution" that increased the amount and accuracy of information available and made scribal copying of books obsolete. Producing some 13,000 works by 1500, printing presses spread rapidly throughout Europe and helped to break down the monopoly of learning in universities and to create a new lay intelligentsia. Indeed, the first print shops became something of intellectual centers themselves with authors, publishers, and workers rubbing shoulders in unprecedented ways in the production of new knowledge. Renaissance humanism, that renowned philosophical and literary movement emphasizing human values and the direct study of classical Greek and Latin texts, is hardly conceivable without the technology of printing that sustained the efforts of learned humanists. Regarding science, the advent of printing and humanist scholarship brought another wave in the recovery of ancient texts. Whereas Europeans first learned of ancient Greek science largely through translations from the Arabic in the twelfth century, in the later fifteenth century scholars brought forth new editions from Greek originals and uncovered influential new sources, notably Archimedes. Similarly, printing disseminated previously recondite handbooks of technical and magical "secrets" that proved influential in the developing Scientific Revolution. And, notably, the technology of printing produced a huge impact on contemporary science without any corresponding input from science on printing technology.

Particularly in Italy, the revival of cultural life and the arts in the fourteenth and fifteenth centuries commonly known as the Renaissance must also be considered as an element of changed conditions of the early modern period. The Italian Renaissance was an urban and comparatively secular phenomenon, aligned with courts and courtly patronage (including patronage by church prelates), but not the university. One associates the great flourish of artistic activity of the Renaissance with such painting talents as Donatello (1386–1466), Leonardo da Vinci (1452–1519), Raphael (1483–1520), and Michelangelo (1475–1564). In comparison with medieval art, the use of perspective—a projection system that realistically renders the three dimensions of space onto the two dimensions of a canvas—represents a new feature typical of Renaissance painting, and through the work of Leon Battista Alberti (1404–1472), Albrecht Dürer (1471–1528), and others, artists learned to practice mathematical rules governing perspective. The architect Filippo Brunelleschi (1377–1446) contributed to these developments, and he extended Renaissance sensibilities and energy into architecture, most famously in the great dome he erected on the cathedral in Florence. So noteworthy was this development that historians have been inclined to place Renaissance artists at the vanguard of those uncovering new knowledge about nature in the fifteenth and sixteenth centuries. Whatever one may make of that claim, early modern artists needed accurate knowledge of human muscular anatomy for lifelike renditions, and an explosion of anatomical research in the Renaissance may be attributed to this need in the artistic community.

Indicative of these changing times, the great Renaissance anatomist Andreas Vesalius (1514–1564) published his influential anatomical opus, *De humani corporis fabrica* (*On the Fabric of the Human Body*), in 1543, the same year that Copernicus published his tome on the heavenly spheres. Vesalius was a military surgeon and his anatomical expertise probably owed as much to the Military Revolution and to the new sorts of wounds inflicted by firearms as it did to any aesthetic requirements of Renaissance art. Vesalius's drawings were crafted to convey information about the body rather than to depict the body as it might actually appear. Other Italian anatomists continued to refine their skills and make anatomical discoveries. Bartolomeo Eustachi (d. 1574) and Gabriel Fallopius (1523–1562) gave their names to previously unknown tubes in the body, and in 1559 Realdo Colombo (1520–1560) postulated the lesser or pulmonary circulation of the blood from the heart through the lungs. Fabricius of Acquapendente (1537–1619) uncovered valves in the veins. These anatomical developments were capped by the discovery of the circulation of the blood by the English physician William Harvey (1578–1657). Harvey studied in Italy and was elected a fellow of the Royal College of Physicians in London where he lectured on anatomy. By careful observations of the slowly beating hearts of dying animals, along with estimates of the quantity of blood that leaves the heart, Harvey arrived at the conclusion that the arterial and venous blood vessels form a connected circulatory system. The publication of his discovery in 1628 was a revolutionary outcome of the fertile tradition of Renaissance anatomy. Indeed, these revisions of the anatomical doctrines inherited from Galen and Aristotle reflect the comprehensiveness of the Scientific Revolution in the Europe of the sixteenth and seventeenth centuries.

Magic and the occult sciences in the Renaissance constituted a defining element of contemporary science and natural philosophy. The importance of magic was overlooked in earlier histories of science, but more recent scholarship has made magic more central to the story of the Scientific Revolution. The occult sciences of the Renaissance included astrology, alchemy, demonology, divination, magic, Neoplatonism, Rosicrucianism (which involved secret societies and occult symbols), and the Cabala (concerning secret mysteries in the Bible). The range of magical activities varied considerably in the early modern period, from proscribed contact with the forces of evil through black magic to "natural" or "mathematical" magic, which had to do with remarkable machines or technical processes (such as burning mirrors or magnets) that produced astounding effects. Despite our prejudices against magic and the occult as irrational delusion and charlatanry, at its highest levels Renaissance magic and associated knowledge systems were serious spiritual and intellectual enterprises that embodied learned understanding of the natural world. The very notion of the occult involved a dual meaning, both as secrets shared among adepts and as secrets hidden in nature.

With the elaborate given name of Philippus Aureolus Theophrastus Bombastus Von Hohenheim the Swiss physician and alchemist Paracelsus (1493–1541) vehemently challenged the medical establishment and the reigning scholastic medical tradition. Paracelsus rejected most of ancient and Islamic medicine, and in so doing he struck at the heart of Aristotelian natural philosophy. His famous dictum that "the dose makes the poison" destroyed notions of innate properties, and his occult and practical knowledge of metals led him to their application in medical treatments, such as his use of mercury to treat syphilis, a disease new to Europe. Paracelsus's own notions proved influential, and in his way Paracelsus, like his contemporary Nicholas Copernicus, has

TERTIA
MUSCULO.
RVM TA
BVLA.

Fig. 11.1. The new anatomy. With the creation of gunpowder weapons, physicians and surgeons were confronted with treating more severe wounds and burns. A military surgeon, Andreas Vesalius, produced the first modern reference manual on human anatomy in 1543, the same year that Copernicus published his book on heliocentric astronomy.

to be seen as an originator of the Scientific Revolution. Widely traveled and a learned man, Paracelsus worked as a military surgeon and then taught for many years at the university in Basel.

The occult sciences gained legitimacy and momentum with the recovery and translation of the so-called Hermetic corpus in the mid-fifteenth century. A fundamental principle of the Hermetic philosophy linked the microcosm (or "small world") of the human body with the macrocosm (or "large world") of the universe as a whole through a system of occult (or "hidden") correspondences and relations of "sympathy" and "antipathy." The world, therefore, took on an emblematic quality, replete with hidden meanings, associations, and occult symbolism. In addition to astrological beliefs, Hermeticism affirmed the magical power of an enlightened magician or magus to change the course of nature. (The principle of Renaissance magic that postulated a universe pulsating with "forces" the magus could learn to command flowed into the modern notion, enunciated by Newton, of a universal force of gravity.) Hermeticism thus saw a transcendental, divine order in nature framed by underlying mathematical realities, and held the optimistic vision that humans could both understand nature and, through a technology of magic operate upon it in their own interests. These characteristics align Renaissance magic with many of the same individuals and historical forces that gave rise to the Scientific Revolution. The anti-Aristotelian and extra-university nature of these movements should also not be overlooked, nor should the opportunities they invited for patronage. The relative decline of magic later in the seventeenth century and the transition to more "open" systems of knowledge represent a major transition in the Scientific Revolution, but in the meantime Renaissance magic offered allegedly useful and practical powers.

A monumental historical dislocation, the Protestant Reformation broke the spiritual and political unity of the Catholic Church in the West in the sixteenth century. The Reformation called into question received religious authority, notably that of the Vatican. In retrospect it represents a major step in the secularization of modern society—that is, the historical shift from ecclesiastical to lay, civil authority governing society. The Reformation began when Martin Luther nailed his Ninety-Five Theses, which were controversial religious propositions, to the door of the church at Wittenberg in 1517, setting off a period of often bloody religious struggle that racked Europe through the Thirty Years' War that ended in 1648. The Scientific Revolution unfolded against the background of the Reformation, and many of its key figures—Johannes Kepler, Galileo Galilei, René Descartes, and Isaac Newton, to name just a few—became deeply affected by religious issues sparked by theological foment.

To this list of changed circumstances facing scientists in the period of the Scientific Revolution, a comparatively minor, yet increasingly irritating problem needs be added: calendar reform. Instituted by Julius Caesar in 45 BCE, the Julian calendar of 365¼ days (with its added full day every fourth February) is longer than the solar year by roughly 10 minutes. By the sixteenth century, the Julian calendar was out of sync with the solar year by some 10 days. Such an inconvenient disjuncture between civil and celestial time exacerbated the already tricky problem of setting the dates for Easter. Pope Sextus IV attempted calendar reform in 1475, but nothing came of it. Pope Leo X raised the issue again in 1512. Consulted on the matter, Nicholas Copernicus expressed his opinion that astronomical theory must be attended to before practical calendar reform was possible.

The Timid Revolutionary

Born in Poland, Copernicus (1473–1543) lived most of his life on the fringes of contemporary scientific civilization, working as a church administrator (canon) in a position gained through family connections. Apparently a timid person, submissive to authority, Copernicus seems an unlikely character to have launched any sort of revolution. In 1491 he matriculated at the University of Cracow, and he spent a decade around 1500 at various universities in Italy, where, in addition to formally studying law and medicine, he developed his interest in astronomy and in general absorbed the cultural ambience of the Italian Renaissance. Indeed, in a typical humanist exercise while still a student, Copernicus translated an otherwise obscure, noncontroversial Greek poet, Theophylactus.

The key to understanding Copernicus and his work comes with the recognition that he was the last of the ancient astronomers, not the first of the moderns. A conservative, he looked backward to ancient Greek astronomy, not forward to some new tradition. He worked as a successor to Ptolemy, not as a precursor of Kepler or Newton. He was at most an ambivalent revolutionary. His object was not to overthrow the old system of Greek astronomy, but rather to restore it to its original purity. In particular, he took seriously the injunction issued nearly 2,000 years earlier to "save the phenomena" and to explain the movements of the heavenly bodies strictly in terms of uniform circular motion. For Copernicus, Ptolemaic astronomy failed to provide a satisfactory account of the stations and retrogradations of the planets; with its elaborate geometrical constructions, it was an astronomical "monster." In particular, he repudiated Ptolemy's equant point—that arbitrary mathematical point in space whence astronomers measured the uniform circular motion of bodies. Uniformity of motion for orbits based on equants was merely a fiction; in fact, as long as astronomers deployed equants, they implied that the planets moved with nonuniform speeds. There had to be a better way, one more consistent with uniform circular motion and ancient tradition.

For Copernicus, that better way turned out to be heliocentrism or placing the sun at (or at least near) the center of the solar system and making the earth a planet. He first proposed heliocentrism in an anonymous manuscript tract, the "Commentariolus," which he circulated among professional astronomers after 1514. But he held off publication of his great work, *De revolutionibus*, possibly because he felt such secrets should not be revealed and certainly for fear, as he put it in his dedication to the pope, of being "hissed off the stage" for such an "absurd" theory. A younger German astronomer and protégé, Rheticus, saw Copernicus's manuscript and published a notice of it, the *Narratio prima* or "First Account," in 1540. With the way apparently cleared, Copernicus approved publication, and his *De revolutionibus orbium coelestium* duly appeared in 1543 just before his death.

Copernicus did not base his astronomy on any new observations. Nor did he *prove* heliocentrism in *De revolutionibus*. Rather, he simply hypothesized heliocentrism and worked out his astronomy from there. In the manner of Euclid's geometry Copernicus posited heliocentrism in a handful of axioms and developed propositions concerning planetary motion under the assumed conditions. He made these bold assumptions for essentially aesthetic and ideological reasons. For Copernicus the heliocentric system possessed greater simplicity and harmony in its proportions; it was intellectually more refined—more "pleasing to the mind"—and economical than what he regarded as the inelegant system of Ptolemy.

The greater simplicity of heliocentrism lay primarily in how it explained the stations and retrogradations of the planets which remained so awkward to explain in geocentric accounts. In the Copernican system such motion is an illusion resulting from the relative motion of the earth and the planet in question against the background of the fixed stars. That is, from a moving earth a moving planet may appear to stop, move backward, and then move forward again, while actually both the observed and the observer circle the sun without any backward motion. With heliocentrism the appearance of the stations and retrogradations of the planets remains, but the problem vanishes: "retrograde" motion automatically follows from the postulate of heliocentrism. The revolutionary character of Copernicus's achievement is nowhere more evident than in the fact that with the adoption of heliocentrism the central theoretical problem in astronomy for two millennia simply disappears.

The Copernican hypothesis was simpler and aesthetically more appealing on additional grounds. It explained why Mercury and Venus never stray farther from the sun than an angular distance of 28° and 48°, respectively. The Ptolemaic system adopted an ad hoc, unsatisfying solution to the problem, while for Copernicus, because the orbits of Mercury and Venus fall within the orbit of the earth, those planets must remain visually in the vicinity of the sun. Similarly the Copernican system dictated a definite order to the planets (Mercury, Venus, Earth, Mars, Jupiter, Saturn), while the matter remained uncertain in Ptolemaic astronomy. Using the Copernican planetary order, observed planetary positions, and simple geometry, astronomers could calculate the relative distance of planets from the sun and the relative size of the solar system.

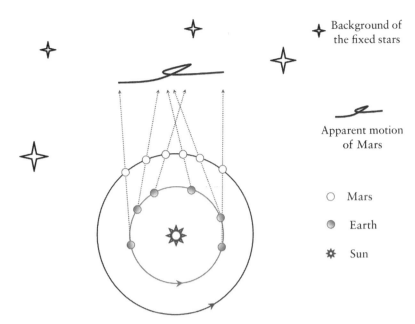

Background of
the fixed stars

Apparent motion
of Mars

○   Mars

◉   Earth

☀   Sun

Fig. 11.2. Retrogradation of the planets in the heliocentric system. Copernicus provided a simple explanation for the age-old problem of stations and retrogradations of the planets. In the heliocentric system a planet's looping motion as seen against the background of the fixed stars is only apparent, the result of the relative motion of the earth and the planet in question (the superior planet Mars in this case).

For Copernicus and like-minded astronomers, the sun occupied a position of paramount importance. In an oft-quoted passage in *De revolutionibus*, one redolent of Neoplatonism if not actual sun-worship, Copernicus wrote:

> In the middle of all sits the Sun enthroned. In this most beautiful temple could we place this luminary in any better position from which he can illuminate the whole at once? He is rightly called the Lamp, the Mind, the Ruler of the Universe; Hermes Trismegistus names him the Visible God, Sophocles' Electra calls him the All-seeing. So the Sun sits as upon a royal throne ruling his children the planets which circle round him . . . Meanwhile the Earth conceives by the Sun, and becomes pregnant with an annual rebirth.

For Copernicus the earth rotates once a day on its axis, thus accounting for the apparent daily motion of everything in the heavens, and the earth revolves around the sun once a year, accounting for the sun's apparent annual motion through the heavens. But Copernicus ascribed not two, but three motions to the earth, and to understand Copernicus's "third motion" reveals the essence of his worldview. In a word, Copernicus held that the planets orbit the sun not in empty or free space but embedded in the crystalline spheres of traditional astronomy. Thus, the spheres in the title of his magnum opus, *On the Revolution of the Heavenly Spheres*, refer not to the spheres of the planets—Earth, Mars, Venus, and so on—but to the crystalline spheres that carry the planets!

That being the case, a serious problem arose for Copernicus, for if the earth were carried around the sun in a solid crystalline sphere, the earth's north-south axis would not maintain its constant "tilt" of

23½ degrees toward the pole star (Polaris), and therefore no changes in seasons would occur. Introducing another "conical" motion of the earth's axis, Copernicus kept the earth pointed to the same spot in the heavens and thus accounted for seasonal changes while having the earth carried around by its celestial sphere. (See figure 11.3.) In addition, by making this annual third motion of the earth slightly longer than the annual period of the earth's orbit of the sun, Copernicus explained yet another tricky phenomenon, the precession of the equinoxes or the separate motion of the sphere of the fixed stars over a 26,000-year period.

Of course, like Aristarchus before him, Copernicus had to respond to the traditional objections to the idea of a moving earth, and he offered a modified version of standard Aristotelian physics to account for the phenomena. For Copernicus, circular motion is natural to spheres; therefore the earth rotates by its very nature and, like the other planets, is carried around the sun by the inherent natural motion of its crystalline sphere. Material particles naturally aggregate into spheres; hence, objects fall downwards on earth, not to the center of the universe, but only to the center of the earth. Bodies do not fly off the earth, given its diurnal and annual motions, because they share in the circular motions of their "mother." Qualitatively, it all works wonderfully well, and the aesthetic superiority shines through in the first twelve folios (twenty-four pages) of the first book of *De revolutionibus*, wherein Copernicus presents the general outline of his system.

The other five books and 195 folios of *De revolutionibus* are quite a different matter. There one finds a highly technical reform of mathematical astronomy, as rigorous and abstruse as Ptolemy's *Almagest*. In fact, superficial comparison can hardly distinguish the works. Copernicus did not intend his work for popular consumption, wanting only

Fig. 11.3. Copernicus's third ("conical") motion of the earth. To account for the fact that the earth's axis always points in the same direction, Copernicus added a third motion for the earth in addition to its daily and annual movements.

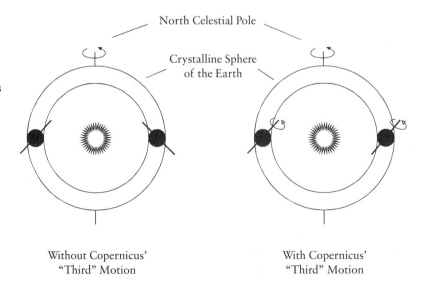

North Celestial Pole

Crystalline Sphere
of the Earth

Without Copernicus' "Third" Motion

With Copernicus' "Third" Motion

to be judged by other professional astronomers. Indeed, he said of his audience that "mathematics is written for mathematicians," and verso to his title page he had printed Plato's motto, "Let no one ignorant of geometry enter here."

Considered as a technical treatise for professional astronomers, therefore, *De revolutionibus* loses much of its aesthetic appeal. As it turns out, the sun stands only near the center of the solar system, not at the center. Copernicus avoided the dreaded equant, to be sure, but, committed to circular motion, he was forced to retain an elaborate apparatus of epicycles and eccentrics in order to explain the remaining irregularities in the apparent speed of the planets as they circled the sun. In the final analysis, as the technical details piled up, Copernicus's astronomy was not any more accurate or more simple than Ptolemy's. Although he eliminated large epicycles, depending on which circles one counts, Copernicus may have actually employed more epicycles than the corresponding contemporary version of Ptolemy.

Copernican astronomy also faced several nettlesome technical problems that undermined its appeal considerably. The most serious was the problem of stellar parallax, the same problem that subverted Aristarchus and heliocentrism in Greek antiquity. As noted in discussing Aristarchus's proposal, as the earth revolves around the sun, the stars ought to change their apparent relative positions. But astronomers observed no such stellar parallax.

The phenomenon of stellar parallax is in fact a very subtle one, never observable in naked-eye astronomy and not actually demonstrated until 1838. The discovery of stellar aberration by the English Astronomer Royal, James Bradley, in 1729 demonstrated the earth's annual motion but, amazingly, only in 1851 did the physicist J.-B.-L. Foucault definitively prove the daily rotation of the earth by using a giant pendulum. By the eighteenth and nineteenth centuries Ptolemaic astronomy had all but ceased to exist; by that time astronomers universally held to the earth's diurnal motion and to heliocentrism. Can it be that such definitive proofs are not what is needed to persuade converts to a new science?

Be that as it may, the Copernican explanation for the lack of observed stellar parallax resembled Aristarchus's: he assumed the stars were very far away and hence the parallax remains too small to be observed. But this hypothesis produced further problems, notably that the size of the universe ballooned to incredible proportions and the size of the stars (extrapolating from their apparent size) likewise became unbelievably immense. Ptolemaic astronomy had set the sphere of fixed stars at a distance of 20,000 Earth radii. For Copernicus the stars had to lie at least 400,000 Earth radii away, an apparently absurd distance in the context of sixteenth-century astronomy.

The fact that falling bodies do not appear to be left behind as the earth allegedly moves was also a strong impediment to the acceptance of heliocentrism. These and other technical problems meant that Copernican heliocentrism was not immediately hailed as a self-evidently correct or superior astronomical system. But other issues loomed, too, including religious objections that heliocentrism seemingly contradicted passages in the Bible. Copernicus dedicated *De revolutionibus* to Pope Paul III, perhaps to stave off such objections. Pope Clement VII had earlier learned of Copernicus's views in the 1530s and did not object, and Catholic astronomers and churchmen did not take theological exception to the Copernican hypothesis in the second half of the sixteenth century. Some leading Protestants, on the other hand, including Luther and the Danish astronomer Tycho Brahe, did pose such objections, but only in the next century when Galileo fanned the flames of theological controversy did they flare up against Copernicanism.

A spurious prefatory letter attached to *De revolutionibus* explains why Copernicanism did not provoke more strenuous theological reactions. A Lutheran cleric, Andreas Osiander, saw Copernicus's book through the press and on his own authority added an anonymous foreword, "To the Reader Concerning the Hypotheses of this Work." Osiander-cum-Copernicus wrote that heliocentrism need not be true or even probable, merely that it provides a convenient mathematical device that permits astronomers to make more accurate calculations. Copernicus himself held heliocentrism to be a true description of the physical world, but based on Osiander's preface he was taken merely to have created a useful fiction. Paradoxically, Osiander qua Copernicus may have helped pave the way for acceptance of Copernicanism by making it superficially palatable for all concerned.

The idea of heliocentrism slowly diffused among astronomers after Copernicus. A new set of astronomical tables—the so-called Prutenic Tables calculated on Copernican principles by the astronomer Erasmus Reinhold and published in 1551—represents one practical result forthcoming from Copernicus's work. In 1582, based on these new tables, authorities finally effected calendar reform by instituting the Gregorian calendar in use today. (Named after Pope Gregory XIII, the Gregorian calendar suppresses leap years for centennial years, except those divisible by four.) By the same token, although Copernicus's book was reprinted in 1566 and again in 1617, only a handful of technical astronomers ever read him. A revolution in astronomy is barely discernible even in the second half of the sixteenth century. Not an abrupt transformation of contemporary astronomy or of worldview, the Copernican revolution was, at most, a revolution by degrees.

Tycho's Turn

The great Danish astronomer Tycho Brahe (1546–1601) added to the momentum of the revolution Copernicus had quietly begun. A haughty and arrogant aristocrat, for 20 years from the mid-1570s to the mid-

Fig. 11.4. Tycho Brahe and the mural quadrant. The remarkably accurate set of naked-eye astronomical observations compiled by the sixteenth-century Danish astronomer Tycho Brahe and his assistants depended on large instruments, such as the mural quadrant depicted here. This famous engraving also shows the other components of the research installations erected by Tycho, including an alchemical lab. No contemporary university or university professorship could have paid for Tycho's activities. He depended on major subsidies from the Danish crown.

1590s Tycho ruled over the Danish island of Hveen given to him as a fief by the Danish king Frederick II, along with its village, farms, and peasants. There he built and equipped two great astronomical palaces—Uraniborg, the castle of the heavens, and Stjerneborg, the castle of the stars—which together comprised the most magnificent scientific installation of its day. Tycho added his own printing press, a paper mill, a library, and several alchemical laboratories. A practicing alchemist and astrologer, he cast horoscopes for patrons and friends and gave away alchemical medicines. Having lost part of his nose in a duel,

Tycho sported a prosthetic replacement, and with his own jester, pets, and coteries of assistants the lord of Uraniborg may seem a virtual self-parody of a Renaissance magus. Tycho had a falling-out with a succeeding Danish king and left Denmark in 1597 to take up a court position as Imperial Mathematician to the Holy Roman Emperor, Rudolph II, in Prague. Kepler reports that Tycho died from a burst bladder from having drunk too much at the banquet table and being too polite to get up to relieve himself.

But Tycho was not merely an eccentric. He was also an adept astronomer who understood the needs of his science. Early in his career he became convinced that the perfection of astronomy depended on accurate and sustained observations of the heavens, and he made it his life's work to undertake those observations. To that end he built large and delicately calibrated naked-eye instruments, such as mural quadrants and armillary spheres—some twenty major instruments at Uraniborg and Stjeneborg. Indicative of the "big science" nature of his enterprise, Tycho received government support totaling some 1 percent of crown revenues, and he boasted that many of his instruments individually cost more than the annual salary of the highest-paid university professors. (Like Copernicus, Tycho's career developed outside the university.) Using these huge and expensive instruments, shielding them from wind stress, minimizing temperature variations, testing and correction for their intrinsic errors, and adjusting for atmospheric refraction, Tycho produced the most precise naked-eye observations ever, trustworthy down to five or ten seconds of arc in some cases, a minute or two in others, and four minutes of arch in all cases. (A minute of arc is 1/60 of a degree; a second of arc is 1/60 of an arc minute; and, of course, 360° span a circle.) This margin represents an exactitude double that of ancient astronomical observations and one not bested by telescopic observations for yet another century. But the beauty of Tycho's data derived not only from their intrinsic accuracy, but from the systematic character of the observations that Tycho and his assistants methodically compiled night after night over an extended period of years.

Two celestial events further shaped Tycho's astronomy. On the evening of November 11, 1572, as he left his alchemical laboratory, Tycho noticed a "new star" blazing as brightly as Venus in the constellation of Cassiopeia. He wrote about this "stella nova" giving rise to our term nova or supernova, meaning an exploding star. The nova in question shone for three months, and by executing exact parallax observations, Tycho showed that the new star was not located in the earth's atmosphere or in the region below the moon, but stood in the heavens above the sphere of Saturn. In other words, the "new star" really was a new star, even if a temporary one. Tycho thus demonstrated the mutability of the heavens and thereby issued a strong challenge to a central tenet of received dogma in Western cosmology.

Tycho's observations of the comet of 1577 likewise proved unsettling to traditional astronomical theory. Again based on parallax observations, Tycho showed not only that the comet moved in the regions above the moon, but he raised the possibility that it also cut through the crystalline spheres supposedly carrying the planets. In other words, the celestial spheres—those mainstays of Western cosmology and celestial dynamics from at least the fourth century BCE—were not real. After Tycho, the only spheres in the heavens were the observed spherical bodies of the sun, the moon, the earth, and the other planets.

Although his work challenged received doctrines, Tycho rejected Copernicus and heliocentrism on strong empirical grounds, especially the lack of observable stellar parallax and because of the consequence that in Tycho's calculations, given the heliocentric system, the fixed stars had to lie an inconceivable 7,850,000 Earth radii distant from the center. The daily or diurnal motion of the earth in the heliocentric system also seemed absurd, and, indicative of a new imagery stemming from the Military Revolution, Tycho introduced a new argument against a spinning earth: a cannon fired toward the west (and a rising horizon) ought to outdistance shots fired toward the east (and a sinking horizon), altogether against experience. Then, too, Tycho, a Protestant, voiced religious objections to the heliocentric system.

In response to deep problems affecting both Ptolemaic and Copernican astronomy Tycho proposed his own system in 1588. In the Tychonic geoheliocentric system the earth remains quiescent at the center of the cosmos, the planets revolve around the sun, and the sun revolves around the earth. This system possessed several advantages: it accounted for stations and retrogradations of the planets without using epicycles, it removed the absurdities of a moving earth, it maintained the traditional scale of the universe, it eliminated the crystalline spheres, and it was mathematically as accurate as its competitors. Holding the earth at rest, the Tychonic system was the equivalent of the Copernican system without the disadvantages of the latter. The Tychonic system represents good, even if conservative, science. But by 1600 with three competing systems and research programs in existence—Ptolemaic, Copernican, and Tychonic—a crisis in astronomy began to mount.

The Music of the Spheres

The case of Johannes Kepler (1571–1630) belies the notion that the internal logic of scientific discovery alone suffices to account for scientific change. Early in his intellectual career Kepler became obsessed with astrology and number mysticism, and more than anything else these obsessions drove his work, shaped his scientific accomplishments, and redirected the course of the Scientific Revolution. Coming from an impoverished and dysfunctional family—his father an errant soldier

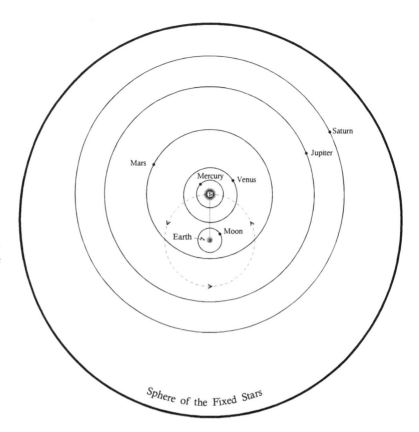

Fig. 11.5. The Tychonic system. In the model of the cosmos articulated by Tycho Brahe in 1588, the earth remains stationary at the center of the universe. The sun orbits the earth, while the other planets revolve around the sun. Tycho's system was good science in that it solved a number of difficult problems in contemporary astronomy, but it did not receive wide acceptance.

of fortune, his mother later in life tried as a witch—Kepler attended Lutheran schools and the university at Tübingen as a talented scholarship boy. An unhappy person with bad eyesight and a number of other physical afflictions, Kepler compared himself to a mangy dog. Although he disdained some aspects of astrology, he saw it as an ancient and valid science, and throughout his life he cast horoscopes and wrote up prognostications and calendars (like farmers' almanacs), from which he earned a regular income. On first learning of the Copernican system he became a convert, finding it, like Copernicus did, "pleasing to the mind" and revealing of the workings of the divine in nature.

Kepler did not set out to become an astronomer, but pursued higher studies in theology. However, before granting his degree, authorities at Tübingen nominated Kepler to fill a position as provincial calendar maker and teacher of mathematics in the Protestant high school at Graz in Austria, and Kepler accepted. He was a poor teacher, with so few math students that the school assigned him to teach history and ethics. One day—he tells us it was July 19, 1595, in front of his presumably bored geometry class—Kepler had a revelation. He was discussing the cube, the tetrahedron, the octahedron, the icosahedron, and the dodecahedron, which are the five regular solids with identical faces and angles between the faces. (The Greeks had proven that there cannot be more than these five.) Kepler's mystical insight consisted in imagining that these solids might somehow frame the universe, that is, establish the mathematical proportions that spaced the planetary orbits outward from the sun. Thus inspired, Kepler developed his views on the geometrical structure of the universe in a book, *Mysterium cosmographicum* (*The Mystery of the Universe*), that appeared in 1596. Kepler's *Mysterium* was the first overtly Copernican work since *De revolutionibus* more than a half a century before, and its origin in pedagogy is one of a handful of exceptions proving the historical rule that nothing of importance for science ever happens in classrooms.

Psychically driven to unpack divine mathematical harmonies of the universe, Kepler was physically driven from Graz in 1600 by the Catholic counter-Reformation and his refusal to convert to Catholicism. He managed to find his way to Prague and a position as Tycho's assistant for the last two years of Tycho's life. The noble, aging Dane assigned the pitiable, younger Kepler data for the planet Mars, in the hope that Kepler could reconcile Tycho's extraordinarily accurate ob-

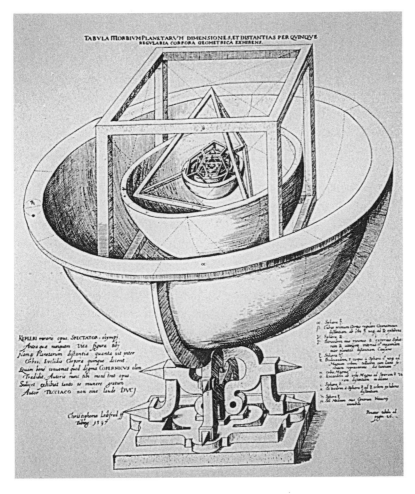

Fig. 11.6. The mystery of the cosmos. In his *Mysterium cosmographicum* (1596) Johannes Kepler conjectured that the spacing of the six known planetary orbits could be explained by nesting them in and around the five regular solids.

servations with Tychonic theory. The choice of Mars was fortuitous, in that the orbit of Mars is the most eccentric, or noncircular and off-centered, of all the planets. Kepler took up the problem with a vengeance, but only to rescue Mars for Copernican theory and his own intuitions of celestial harmonies. In an intellectual struggle of epic proportions, Kepler worked fervently on the problem for six years, leaving behind some 900 manuscript pages of calculations, testimony to a heroic endeavor and the fact that this exercise in curve fitting proceeded without benefit of mechanical or electronic calculators. In his published account Kepler leads his reader through every tortuous twist and turn. At one point, his circular model for Mars matched the observational data to within 8 minutes of arc, a tremendous achievement, but knowing that Tycho's data were good to 4 minutes, Kepler had to reject his own accomplishment. He made mistakes in his calculations, and then made other mistakes that corrected for them. He got the "right" answer and failed to see it. Then, on recognizing an obscure mathematical connection concerning the secant of an angle, Kepler had another flash of insight. "It seemed as if I awoke from sleep and saw a new light break on me," he wrote, and indeed he awoke in a new world.

Kepler concluded that the planets orbit the sun not in circles but in ellipses. This discovery, of course, marks an astonishing turn, in that circles had provided a physics and a metaphysics governing heavenly motion at least since Plato nearly 2,000 years earlier. In his *Astronomia Nova* of 1609 Kepler enunciated the first two of his celebrated three laws of planetary motion: 1) that the planets orbit in ellipses with the sun at one focus, and 2) in what amounts to a planetary speed law, that their radii sweep out equal areas in equal times. Kepler's second law carries the equally disturbing consequence that the planets do not move *uniformly*. As it turns out, Kepler developed his second law before the first, and himself never drew attention to his laws per se. Nevertheless, with the planets moving as Kepler described them and with the sun now indisputably central, Kepler's *Astronomia nova* truly represented a "new astronomy."

Kepler remained in Prague as Imperial Mathematician to Rudolph II until the latter's abdication in 1612. Thereafter he secured a position as provincial mathematician in Linz in Austria, which lasted to 1626, when he moved on to Ulm and Sagan. In the latter period of his life, while the disastrous Thirty Years' War swept over Germany and disrupted his very existence on more than one occasion, Kepler wrote an *Epitome of Copernican Astronomy* (1618–21), which presented his own elliptical vision of the solar system more than the Copernican one, and the *Rudolphine Tables*, a new and highly accurate set of astronomical tables based both on Tycho's data and on Copernican/Keplerian heliocentrism.

In 1619 Kepler issued his *Harmonice mundi*, or *Harmonies of the*

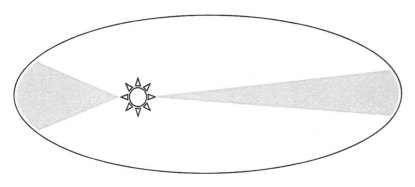

Fig. 11.7. Kepler's elliptical motion of the planets. Based on Tycho Brahe's data, Johannes Kepler broke with the age-old notion that the heavenly bodies move in circles. Kepler reformulated planetary motion in what came to be known as Kepler's three laws: (1) the planets orbit the sun in ellipses with the sun at one focus; (2) planets sweep out equal areas in equal times; and (3) the square of the time of an orbit equals the cube of the mean distance from the sun, or $t^2 \propto r^3$. Kepler reached these conclusions entirely on the basis of astronomical observations and geometrical modeling without providing a convincing physical explanation for why the planets move as they do.

World. This work culminated the effort that began with the *Mysterium cosmographicum* and Kepler's meditations and researches on the mathematical order underlying the structure of the cosmos. In the *Harmonice mundi* Kepler calculated astrological relations, correspondences of planets with metals, the music of the spheres—those unheard tones that he believed were generated by the planets in their motions—and like connections. Buried in this work lay Kepler's third law, that the square of the time of a planet's orbit is proportional to the cube of the mean radius—$t^2 \propto r^3$—at the time an oddly empirical law.

Astronomers and physicists preceding Kepler—even those few who held to heliocentrism—possessed a traditional dynamics for heavenly motion: the planets were understood to move uniformly in circles by their innate natures or carried by crystalline shells. But having dispensed with uniform circular motion, Kepler faced the problem of providing a dynamics and explaining why the planets move the way they do through space. Kepler was aware of this obligation and subtitled his *Nova Astronomia* of 1609 *The New Astronomy, Based on Causes, or Celestial Physics.* At an early stage Kepler believed that the sun possessed an *anima motrix*, a moving soul or spirit akin to the Holy Ghost, that propelled the planets through their courses. In his more mature formulations he substituted for this animate spirit a more inanimate moving force or *vis motrix*, a kind of magnetic power. Kepler derived this latter notion from William Gilbert's influential *De Magnete*, published in 1600, which showed the earth to be a huge magnet. For Kepler, then, the planets deviate from circular motion as the sun's and the planets' magnets alternately attract and repel. Kepler's celestial physics provided a plausible explanation for planetary motion, but not a compelling one, for problems remained. For example, Kepler does not explain how the force emanating from the sun acts tangentially, that is, how it "sweeps" the planets along, something like a broom, and can act at right angles to lines of force emanating from the sun. Also, paradoxically, he never treats this motive power with the same mathematical rigor and precision he used in determining the orbits of planets. After Kepler, the dynamics of heavenly motion remained an open question.

Kepler died of a fever in 1630 while traveling to petition for money owed him. Although he contributed to it mightily, Kepler did not culminate the Scientific Revolution. We extract Kepler's three laws all too easily from the corpus of his work because we know their historical role and significance for later science, but contemporaries did not and could not. Few astronomers actually read his works, and by and large Kepler did not win converts. Indeed, most scientists who became aware of Kepler's work, notably his great contemporary, Galileo, rejected his views. As an eccentric mystic, Kepler enjoyed the reputation, rather, of a great astronomer gone slightly mad.

The Crime and Punishment of Galileo Galilei

Galileo Galilei (1564–1642) looms as a pivotal figure in the Scientific Revolution and in the history of modern science. His great renown and importance for this story derive from several sources and accomplishments. His improvement of the telescope, his astronomical discoveries, and his research on motion and falling bodies brought him international fame and an enduring place in the annals of science. No less striking is Galileo's career as a Renaissance scientist, reflecting as it does deep changes in the social character of science in the sixteenth and seventeenth centuries. And his infamous trial and recantation of Copernicanism at the hands of the Catholic Inquisition is a notorious chapter in relations between faith and reason, which contributed to the slowly emerging recognition of the value of intellectual freedom.

Galileo, Court, and Telescope

Galileo's life and career unfolded in clearly demarcated stages. Born in Pisa and raised in Florence, throughout his life Galileo maintained his identity as a Tuscan. His father served the Medici court as a professional musician. Galileo attended the university at Pisa as a medical student, but he secretly studied mathematics, and eventually the father consented to his son's pursuit of the more socially vague career of mathematician. After a brief apprenticeship, and through patronage connections, at age 25 Galileo secured a term appointment at the University of Pisa in 1589.

At this stage Galileo followed the medieval model of a career in mathematics and natural philosophy institutionalized in universities. He toiled diligently as a university professor of mathematics for three years at Pisa and then, again through patronage connections, from 1592 at the University of Padua in the republic of Venice. As professor of mathematics at Padua, Galileo endured a lowly status in the university, earning only an eighth as much as the professor of theology. He lectured daily and disconsolately during university terms on a variety of subjects—astronomy, mathematics, fortifications, surveying—and came to feel that teaching interfered with his ambitions to conduct research: "It is a hindrance not a help to my work." To make ends meet, Galileo boarded foreign students and took in others for private tutoring. He employed an artisan to manufacture a "geometric and military compass"—the proportional dividers that he had invented and sold to engineers and architects. In Padua he took on a long-term mistress, Marina Gamba, with whom he had three children. Galileo was a devoted father. He had red hair, a short temper, a mastery of language, a gift for mockery in debate, and a liking for wine. All in all, until he stumbled onto the telescope—or better, until the telescope stumbled onto him—Galileo was a hardworking, low-paid, disgruntled, relatively undistinguished professor at a second-rate university. He was already 45 years old in 1609 when he suddenly achieved fame and immortality.

A Dutchman named Hans Lipperhey invented the telescope in Holland in 1608. In Padua Galileo heard a report of the "toy" and that was evidently enough for him to understand the concept of the telescope and to craft one. Galileo ground his own lenses of Venetian glass. His first attempt resulted in an eight-power telescope, which he soon bettered with models of 20 and 30 magnifications. Galileo's renown stems from his boldly turning his improved spyglass to the heavens and discovering a fabulous new celestial world. He rushed into print in 1610 with the forty-page pamphlet *Sidereus nuncius* (*Starry Messenger*) and, with a job-seeker's instincts, he dedicated it to Cosimo II de' Medici, the grand duke of Tuscany. In the *Starry Messenger* Galileo announced the existence of myriads of new stars, never before seen by anyone, making up the Milky Way. He showed that the moon, far from being a perfect sphere, was deformed by huge mountains, craters and valleys, and that it may well have an atmosphere. (See figure 12.1.) Most spectacularly, Galileo revealed that four moons circled Jupiter. The previously unknown moons of Jupiter indicated that other centers besides the earth or the sun existed around which bodies could orbit. Recognizing their importance and their potential for his career ambitions, Galileo unashamedly named these four moons the Medicean Stars.

The first telescopic discoveries did not involve simply pointing the telescope to the heavens and instantaneously seeing what Galileo reported. We should not underestimate the difficulties or the early disputes that arose over interpreting the visual images presented by the telescope, conceptualizing new astronomical entities, and the acceptance of the telescope as a legitimate tool in astronomy. Galileo could "see" the mountains on the moon only by interpreting the changing shadows they cast over a period of weeks; and he could "see" the

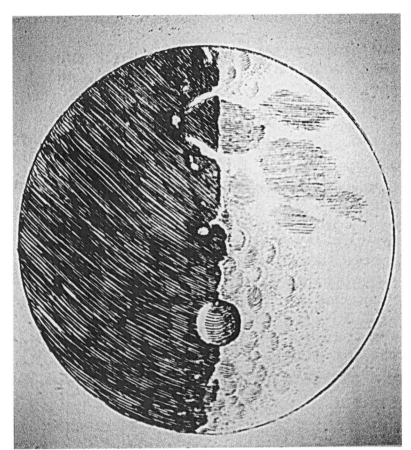

Fig. 12.1. Galileo's *Starry Messenger*. In his famous pamphlet published in 1610, Galileo depicted the results of the observations he made with the telescope. He showed, contrary to accepted dogma, that the moon was not perfectly smooth and had mountains.

moons of Jupiter only by similarly observing their changing positions by careful and protracted observations. As a result, Galileo's marvelous discoveries soon became incontrovertible, and they brought to the fore the question of the true system of the world.

Galileo craftily parlayed his new fame as an international celebrity into a career move from the University of Padua to the position of Chief Mathematician and Philosopher (with a handsome salary) at the Medici court in Florence back in his native Tuscany. He had long wooed the Medicis, tutoring the crown prince (turned Grand Duke Cosimo II) for a number of summers. The appointment culminated a ritualized negotiation between client and patron. Having presented his telescope to the Venetian Senate, receiving a salary increase and life-tenure at the university in Padua in turn, Galileo burnt his bridges to Venice by accepting the court post in Florence. He wanted the more prestigious position and the time to pursue his own work, free from the burden of teaching undergraduates. For its part the Medici court added another star to the heavens of its reputation, and someone with nominal engineering expertise who might do useful work. Galileo's title of philosopher elevated his status, putting him on par with his professorial adversaries, and it raised the status of his mathematical natural philosophy. With his move from Padua to the court at Florence, Galileo fashioned himself as a scientific courtier.

Galileo's career displays a new pattern for the organization and pursuit of science at the turn of the seventeenth century. He was a Renaissance scientist, albeit a late one, the equivalent in science of Michelangelo in art, and like his Renaissance compeers, Galileo Galilei is known universally by his first name. That Galileo left traditional university life underscores how much this new "Renaissance" mode differed from medieval university-based precedents. His Aristotelian opponents remained intellectually and institutionally rooted in universities, while his own science found its arena among the public and in the courts of the great.

Universities were not active seats of change in the Scientific Revolution, and Renaissance courts and court life provided the key new setting for science. One can usefully distinguish between the court itself, where the ruler and his retainers did business, and full-fledged states that later developed more elaborate bureaucratic apparatus, of which the court formed only a part. An entire patronage system arose, especially in the courts of Renaissance Italy, which provided a new, historically significant means of social support for science. Medici court patronage shaped Galileo's career and his science, while patronage forthcoming from the Holy Roman Emperor, Rudolph II, had supported Tycho and then Kepler at Prague as Imperial Mathematicians. European courts employed varieties of experts: artists, physicians, surgeons, alchemists, astronomers, astrologers, mathematicians, engineers, architects, projectors, surveyors, and cartographers. In offering court-based patronage and support, patrons certainly sought useful services and were motivated mainly by the hope of gaining practical results. But Renaissance patronage represents a social and cultural system that involved much more than patrons "purchasing" useful services. Such exchange constituted a small part of a patronage system that flourished in aristocratic cultures and that entailed hierarchies of "clients" and patrons. For example, patrons accrued glory and enhanced their reputations by supporting glorious clients. Patrons also stirred up controversies, and Galileo became embroiled in several that began as arguments at court. As a social institution, the court patronage system legitimated and helped define the social role of science and scientists in the seventeenth century. The pattern of Renaissance court patronage and court-science seen in Galileo's career did not die out in the seventeenth century, but continued into the eighteenth. Even Isaac Newton became drawn into a controversy (over historical chronology) through his role as a courtier attending the Hanoverian Princess Caroline.

Renaissance academies complemented courts as new institutions for science. Appearing in the fifteenth century as outgrowths of the humanist movement and the spread of printing, hundreds, indeed thousands, of literary and fine arts societies sprang up wherever the educated gathered across Europe over the next three centuries. Private salons and informal associations of amateurs proliferated not only in opposition to university Aristotelianism, but also because of a limited number of university positions. Renaissance-type academies characteristically operated with formal constitutions, but they usually lacked official charters certified by the state. The patron often played the essential role, and many Renaissance academies proved unable to survive without active promotion by the patron.

Renaissance science academies represent a late manifestation of the humanist academy movement. Two early anti-Copernican academies, dating to 1550 and 1560 and both called Accademia degli Immobili (the Academy of the Unmoved), may have been the first to concern themselves directly with science or natural philosophy. Naples was a center of esoteric knowledge, and the Neapolitan magus Giambattista Della Porta (1535–1615) organized an early experimental society, the Academy of the Secrets of Nature (Academia Secretorum Naturae or Accademia dei Secreti) in Naples in the 1560s. A volume of curiosities and wonders, *Magia naturalis* (*Natural Magic*), published by Della Porta in 1558 and again in 1589, probably reflects the interests and activities of Della Porta's academy. (More than fifty editions and translations of this influential work appeared in the sixteenth and seventeenth centuries.) Although Della Porta's dabbling in magic brought the scrutiny of the Inquisition, which forced him to disband his academy, he enjoyed a sizable reputation for his learning, and he received patronage offers from numerous courts.

The next Renaissance academy to concern itself with science was the Accademia dei Lincei (1603–30). The Accademia dei Lincei (or Academy of the Lynx-Eyed) appeared in Rome in 1603, patronized by Roman aristocrat Federico Cesi. Della Porta was an early member, but after 1610 the Lincei, led by Cesi, shifted to Galileo's more open approach and program for science. Galileo became a member of the Accademia as part of a triumphal trip to Rome in 1611, and afterwards he valued his title of Lincean academician and used it proudly in print. In turn, Cesi and the academy published several of Galileo's works, including his *Letters on Sunspots* (1613) and *The Assayer* (1623). The Accademia provided a key extra-university institutional prop upon which Galileo erected his career as Renaissance courtier and scientist, and Cesi's untimely death in 1630 and the collapse of the Accademia dei Lincei left Galileo without important support when he came to trial in 1633. In the long run the Renaissance-style social pattern exemplified by Galileo's career gave way to one centered on the nation-state and on national scientific academies. In the meantime, particularly in Italy, the Renaissance court provided a notable home for scientists.

Galileo, Copernicus, and the Church

Galileo's was always a contentious personality, and as soon as he moved from university to court he became embroiled in disputes with adversaries in Florence. He brought with him controversies surrounding the telescope and what it revealed, and he promptly quarreled with Aristotelian opponents over the physics of floating bodies. In these controversies university academics were Galileo's main opponents, and at issue were points of science and Aristotelian natural philosophy. The telescopic discoveries soon put the Copernican question at center stage, and all too quickly theological objections emerged, bringing a whole new class of enemies for Galileo—the theologians. Already in 1611 his name had come up in Inquisition proceedings. Some Dominicans publicly preached against him in 1614, and in 1615 zealous souls actively denounced Galileo to the Inquisition. Although nothing came of these first accusations, the bureaucracy of the Inquisition opened a file with his name on it. Well before his trial and conviction in 1633, Galileo in some sense had already become a prisoner of the Inquisition.

As his star began to shine more brightly at the Medicean court, Galileo preached Copernicanism evermore strongly. The *Starry Messenger* promised that a later book on the system of the world would "prove the Earth to be a wandering body." His *Letters on Sunspots* followed in 1613 with further telescopic novelties concerning spots at or near the surface of the sun, the changing shape of Venus as it makes its orbit, and curious news about the planet Saturn. The discovery of sunspots in particular challenged the purported incorruptibility of the sun, and Galileo would later use the phases of Venus as definitive evidence against the Ptolemaic system. In the *Letters on Sunspots* Galileo asserted that his observations "verified" Copernicus's *De revolutionibus*. A dispute arose at the Medici dinner table in late 1613 over the religious implications of Copernicanism and apparent conflicts with literal interpretations of the Bible. Out of this dispute came Galileo's courtly but provocative "Letter to the Grand Duchess Christina"— the mother of his patron—"Concerning the Use of Biblical Quotations in Matters of Science" (1615). In this work Galileo took the position that faith and reason cannot be in contradiction since the Bible is the word of God and nature is the work of God. However, in instances where there *appears* to be a contradiction science supersedes theology in questions concerning nature, for, as he put it, the Bible was written to be understood by the common people and can readily be reinterpreted, but nature possesses a reality that cannot be altered. For Galileo, if scientists demonstrate some truth of nature that seems to contradict statements found in the Bible, theologians must then articulate reinterpretations of the literal sense of Holy Writ. (This is essentially the position of the Catholic Church today.) Galileo's postulate that science and the human study of nature should take priority over traditional theology represents a radical step, much removed from the medieval role of science as handmaiden to theology, and one almost calculated to provoke the animosity of theologians. Especially offensive was his insufferable arrogance in counseling theologians in the conduct of their business.

Galileo actively defended Copernicanism and led a vigorous campaign to persuade church authorities to accept Copernican heliocentrism over the Aristotelian-Ptolemaic worldview to which religious and scientific thought had long been wedded. In 1616 Galileo's position lost out when the Inquisition ruled Copernicus's opinion erroneous and formally heretical, and the Congregation of the Index placed Copernicus's *De revolutionibus* on the list of books banned by the church. Robert Bellarmine (1542–1621), ranking cardinal of the Inquisition and elderly defender of the faith, wrote that he had understood (from Osiander's preface) that Copernicus posited heliocentrism merely as a mathematical conceit to facilitate astronomical calculations. Since God in his omnipotence could make the heavens go around in any of a thousand ways He pleased, it would be dangerous, said Bellarmine, to set human reason above divine potency and clearly stated biblical language unless clear proof existed that the Bible was mistaken.

On the surface Galileo managed to keep his name out of the proceedings that condemned Copernicanism in 1616, and indeed Bellarmine and the Inquisition afforded Galileo the honor of prior notification of the decision. Different versions have been proposed for what actually transpired at Galileo's meeting with Bellarmine on February 26, 1616. The Inquisition secretly specified steps to be taken (up to and including imprisonment) if Galileo refused to accept the verdict of church authorities as communicated to him not to hold or defend Copernican views. It seems evident that Galileo must have immediately acquiesced, but nevertheless—either in 1616 or at some later time—an irregular, possibly forged, notarial document, one that would come back to haunt Galileo, found its way into the Inquisition's files indicating that a specific personal injunction had been delivered to him "not to hold, teach, or defend [Copernicanism] in any way whatsoever, either orally or in writing." For his part, Galileo obtained a written certificate from Bellarmine later in 1616 confirming that he, Galileo, merely could not "hold or defend" Copernicanism, leaving the option in Galileo's mind to "teach" the disputed doctrine, even while not exactly holding or defending it.

Galileo lost in 1616, but life went on. He was over 50, but still famous and still a Medici luminary. Copernicanism remained off limits, but other scientific subjects engaged his interest and other quarrels developed. In 1618 a controversy erupted over three comets observed that year, with Galileo drawn into a bitter intellectual brawl with a powerful Jesuit, Orazio Grassi. Their pamphlet war culminated in

Galileo's *The Assayer* of 1623, sometimes labeled Galileo's manifesto for the new science. The timing for the publication of *The Assayer* proved propitious, for in 1623 the old pope, Gregory XV, died and a new pope was elected, Urban VIII—Maffeo Barberini, a Florentine himself and a longtime friend of Galileo's. Prospects looked bright, and *The Assayer* seemed to be the perfect vehicle for Galileo to establish patronage ties at the highest level in Rome. The Accademia dei Lincei published *The Assayer* with an effusive dedication to Urban, who had the book read to him at mealtimes. The new pope was delighted. He showered presents on Galileo, who was sojourning in Rome for six weeks in 1624, and invited him for promenades and discussions in the Vatican gardens. Apparently in that context Galileo asked for and received permission to revisit Copernicanism. Regarding the projected work, Urban insisted, however, that Galileo provide an impartial treatment of the Ptolemaic and Copernican systems and that he stress the point that God could move the heavenly bodies in numberless ways regardless of appearances, and, hence, humans cannot detect the true causes of observed events. Urban's involvement in shaping Galileo's opus extended even to its title. Galileo had wanted to call it *On the Tides*, to highlight his theory that the earth's motion causes the tides and that, hence, the tides confirm that motion. Urban immortalized the work with the title, *Dialogue on the Two Chief World Systems*, a title meant to imply an impartial review of the two planetary theories.

Eight years followed before the *Two Chief World Systems* appeared in print in 1632. By then in his 60s, Galileo was ill during much of the time it took to write the work. Delays ensued in securing the required approvals and licenses, as anxious censors and officials in Rome and Florence gingerly processed the manuscript. When it finally issued from the press in Florence in late February of 1632 Galileo's book was a bombshell for more reasons than one.

First and foremost, the *Dialogue on the Two Chief World Systems* made the clearest, fullest, and most persuasive presentation yet of arguments in favor of Copernicanism and against traditional Aristotelian/Ptolemaic astronomy and natural philosophy. Galileo wrote the work in Italian for the largest popular audience he could reach, and he cast it as a literary dialogue among three interlocutors, Salviati (who in essence spoke for Galileo), Sagredo (who represented the interested, intelligent amateur), and Simplicio (the "simpleton" who staunchly voiced Aristotelian views). A spirited and accessible work of literature, the "action" unfolds over four "days." In day 1, using evidence concerning the moon and other new telescopic discoveries, Galileo developed a devastating critique of traditional Aristotelian notions of place, motion, up and down, and the venerable distinction between the celestial heavens and the earth. In day 2 he treats the earth's daily rotation on its axis and deals with apparent conundrums, such as why objects do not go flying off a spinning earth, why we do not experi-

Fig. 12.2. The crime of Galileo. In his Two Chief World Systems (1632) Galileo presented strong arguments in favor of Copernicanism. Written in Italian as a set of informal conversations, the work was and is accessible to readers not trained in astronomy. The book led to Galileo's arrest by the Roman Inquisition and his subsequent condemnation and imprisonment.

ence winds constantly out of the east as the earth spins, why birds or butterflies have no more difficulty flying west than east, why a dropped ball falls at the base of a tower on a moving earth, and why a cannonball flies the same distance east and west. His explanations hinged on the idea that earthbound bodies share a common motion and seem to move only relative to one another. In day 3 Galileo moves on to consider the heliocentric system and the annual motion of the earth around the sun. Among an array of arguments in favor of Copernicanism and heliocentrism, Galileo introduces his "smoking gun" against Ptolemaic astronomy, the phases of Venus. Seen through a telescope, Venus changes its shape like the earth's moon: a new Venus, a quarter Venus, "horned" Venus. The upshot of Galileo's technical point is that the observed phases of Venus are incompatible with the geocentric

Ptolemaic system. The phases did not prove Copernicanism, however, for the observations are consistent with the Tychonic system, too; but Galileo brushed Tycho aside. Last, in day 4 of the *Dialogue*, Galileo offers what in his own mind represents positive proof of the Copernican system, his idiosyncratic account of the tides. His explanation: a spinning, revolving earth induces sloshing motions in the seas and oceans and thereby causes the tides. He gives an elegant mathematical explanation for seasonal variations in tides, and he introduces William Gilbert's work concerning the earth as a giant magnet.

In assessing the merits of the astronomical systems of Ptolemy and Copernicus, Galileo adopted, at least superficially, the impartial posture urged upon him by Urban VIII, pretending to treat indeterminately "now the reasons for one side, and now for the other." He sprinkled his text with the appropriate nonpartisan caveats, saying here and there that he had "not decided" and that he merely wore Copernicus's "mask." But the sharply partisan character of the work could not be denied. Not only was the superiority of Copernicus argued at every turn, Galileo repeatedly refuted Aristotle and made Simplicio seem ignorant and foolish throughout. To make matters worse, Galileo inserted the language dictated to him by the pope concerning God's divine omnipotence and the limits of human reason, but only at the very end of the work and, most provocatively, in Simplicio's voice as "heard from a most eminent and learned person." If Urban had winked at Galileo back in 1624 when they discussed an impartial treatment of Copernicanism and the Ptolemaic system of the world, then Galileo had duly maintained the appearances. If Urban was, or later became, serious about a balanced treatment, then Galileo mocked him and was in deep trouble.

Once the *Dialogue* began to circulate, the reaction was immediate and harsh. In the summer of 1632, on the pope's orders, sales stopped, copies were retrieved, and materials were confiscated from the printer. In an unusual move designed either to shield Galileo or simply to frame his downfall, Urban convened a high-ranking special committee to evaluate the situation. The matter then passed formally to the Inquisition, an institution dangerous, even fatal, to dissidents. The Dominican priest Giodarno Bruno (1548–1600) had been burnt at the stake for his heretical views including his heretical cosmology. The Inquisition treated Galileo with kid gloves, but in the fall of 1632 it called Galileo to Rome to answer charges. At first Galileo, now 68, resisted the summons, at one point pathetically going so far as to send a "doctor's note" testifying to his inability to travel. But the Inquisition remained adamant; Galileo would be brought in chains if necessary. But it was not; Galileo made the trip to Rome on a litter.

The raw facts concerning Galileo's trial have been well known for more than a century, but radically differing interpretations and explanations of the trial continue to be offered down to the present.

An early view—now dismissed—envisioned the episode as a great battle in the supposed war between science and religion, with Galileo as hero-scientist beaten down by theological obscurantism for having discovered scientific truth. Another interpretation, reflecting an awareness of twentieth-century totalitarian regimes, uncovered the essence of Galileo's trial in the bureaucratic state apparatus of the Inquisition. From another perspective, Galileo was simply insubordinate and questioned biblical authority at a time when the church was locked in a deadly embrace with the Reformation. A further documented but more conspiratorial version sees the formal charges stemming from Galileo's Copernicanism as a ruse by church authorities to hide a bitter dispute with the Jesuits as well as other more serious theological accusations having to do with Galileo's atomism and with difficulties reconciling atomism with the miracle of transubstantiation in the Catholic mass. Others argue that Urban's papacy had lost much of its luster and was in political trouble by 1632, and recent accounts evoke a patronage crisis for Galileo and his fall as a courtier. Galileo's trial is a tablet on which historians write and rewrite improved versions, similar to the way legal scholars review and argue old court cases.

Galileo's trial did not proceed in a wholly straightforward manner. Galileo first stayed at the Medicean embassy in Rome, and although treated preferentially, he then had to enter the Inquisition prison like all persons prior to questioning. He appeared before the Inquisition for the first time on April 12, 1633, but he was not told of any charges against him. He rather cockily defended his *Dialogue* by claiming that, in fact, he had not sided with Copernicus against Ptolemy, but had instead shown Copernicus's reasoning to be "invalid and inconclusive." The Inquisitors possessed the tainted notarial minute of 1616, so they thought they had caught Galileo violating a personal injunction not to deal with Copernicanism in any way. When confronted with this document, Galileo produced the certificate he had from the late Cardinal Bellarmine that only prohibited him from "holding or defending" Copernicanism, not from teaching it or treating it altogether.

The Inquisition procured expert opinions that in fact Galileo did defend and hold Copernicanism in his book, but Galileo's genuine certificate from Bellarmine remained an obstacle to resolving the neat case the Inquisition initially envisioned. To avoid a potentially embarrassing outcome, an Inquisition official visited Galileo extrajudicially in his cell to discuss a compromise: Galileo would be led to see and admit the error of his ways, for which some slap on the wrist would be imposed. Appearing next before the Inquisition, Galileo duly confessed to inadvertence and "vainglorious ambition." Shamefully, after he was dismissed he returned to the Inquisition's chambers to volunteer to write yet another "day" for his *Dialogue* to truly set the matter right.

Alas for Galileo, the compromise fell through, and having obtained his confession, Urban rejected the Inquisition's solution and insisted that formal charges of heresy be pressed against Galileo. Hauled before the Inquisition once again, Galileo had nothing to say. Threatened with torture, he said only that he was not a Copernican and had abandoned the opinion in 1616. "For the rest, here I am in your hands; do as you please."

The Inquisition found Galileo guilty of "vehement suspicion of heresy," just a notch below conviction for heresy itself, for which punishment was immediate burning at the stake. Galileo's *Two Chief World Systems* was put on the *Index of Prohibited Books*, and on June 22, 1633, this once-proud Michelangelo of Italian science, 69 years old, kneeling in a public ceremony dressed in the white gown of the penitent and with a candle in hand, was forced to "abjure, curse, and detest" the Copernican heresy and to promise to denounce all such heretics. Galileo remained a formal prisoner of the Inquisition under house arrest for life. Arriving for confinement at Siena in July of 1633, myth has it that, stepping out of his coach, Galileo touched his finger to the earth and uttered the words, "Eppur si muove"—"And yet it moves." The History of Science Museum in Florence displays the bones of Galileo's middle finger, a scientific relic gesturing defiantly at us today.

The trial and punishment of Galileo are sometimes invoked to support the claim that science functions best in a democracy. This claim is demonstrably false, as some of the least democratic societies have been and continue to be successful in the pursuit of science and technological advance. The issue, rather, is the independence of scientific communities, regardless of the political context. Where it has occurred, the intervention of political authorities—whether the Catholic Church or the Communist Party—has inhibited scientific development. Fortunately, political authorities rarely have any interest in the abstractions of theoretical science. In the Christian tradition only the movement of the earth and the origin of species pitted natural philosophy against biblical sovereignty. Whether science takes place in democratic or antidemocratic societies has little to do with its development.

Galileo, Falling Bodies, and Experiment

Transferred to his home outside of Florence in December of 1633 and attended by his daughter Virginia, Galileo turned 70, a prisoner of the Inquisition, humiliated by his recantation, and already half blind. Remarkably, he did not simply give up the ghost. Instead, Galileo turned to producing what many judge to be his scientific masterpiece, *Discourses on Two New Sciences* (1638). In his *Two New Sciences* or the *Discorsi*, as it is known, Galileo published two remarkable discoveries—his mathematical analysis of a loaded beam or cantilever and his law of falling bodies. The work represents Galileo's greatest positive contribution to physical science and to the ongoing course of the Scientific Revolution of the sixteenth and seventeenth centuries. The *Two New Sciences* also shows other dimensions of Galileo's genius as an expert mathematician and experimenter.

In writing the *Two New Sciences* Galileo did not suddenly begin a new research program. As might be expected, he went back to his old notes and to scientific work he had done before 1610, before his first telescope, before international scientific fame, and before the astronomy, the polemics, and the condemnation. The technical and esoteric topics Galileo deals with in the *Two New Sciences* are politically safe and theologically noncontroversial subjects involving how beams break and how balls roll down inclined planes.

The *Two New Sciences* was published somewhat covertly in Protestant Holland by the Elsevier press in 1638. Like its companion masterpiece, *Dialogue on the Two Chief World Systems*, the *Two New Sciences* is set in dialogue form and divided into four "days." The same three interlocutors appear as in the *Dialogue*: Salviati, Sagredo, and Simplicio, although this time they act less antagonistically and play somewhat different roles. Whereas previously Salviati clearly represented Galileo, Simplicio Aristotle, and Sagredo the interested amateur, the three characters in the *Two New Sciences* more likely represent the chronological stages Galileo himself went through in arriving at his mature views concerning mechanics. Simplicio represents his initial, Aristotelian phase, Sagredo an Archimedean middle period, and Salviati his late views. In the *Two New Sciences* the Aristotelian Simplicio, in particular, seems much more flexible, even admitting, "If I were to begin my studies over again, I should try to follow the advice of Plato and commence from mathematics."

The *Two New Sciences* was far more of a mathematical treatise than the previous *Two Chief World Systems*. At one point Salviati reads from a Latin mathematical text written by "our Author," Galileo himself. The book opens with a discussion among Salviati and his two friends at the Arsenal of Venice, that famous center of technology, the largest and most advanced industrial enterprise in Europe, where craftsmen and artisans built ships, cast cannon, twisted rope, poured tar, melted glass, and worked at a hundred other technological and industrial activities in the service of the Venetian Republic. Galileo set the scene at the Arsenal of Venice in order to juxtapose, rhetorically and self-consciously, the enterprises we designate as science and technology. While the Arsenal was clearly an important device for Galileo as an extra-university setting for his new sciences, experts disagree over whether Galileo came there to teach or to learn.

Historians of ideas have tended to skip over the first two "days" of the *Two New Sciences* where Galileo treats the mundane field of strength of materials. They prefer to concentrate on days 3 and 4, where Galileo presents his original findings on the more abstract study of motion, of such high importance for Newton and the Scientific Revolution. But the strength of materials is directly relevant to engineering and to the connection of science and technology.

In days 1 and 2 of the *Two New Sciences* Galileo explores the general topics of cohesion of bodies and the breaking strength of materials. In day 1 he considers a potpourri of technical and theoretical problems, some old and some new. He wonders, for example, about size effect (the theory of scaling) and why one cannot build a wooden boat weighing a million tons. He asks what makes a marble column hold together. He presents an extraordinary matter theory (involving infinities of infinitesimals), and he ingeniously tackles the coherence of bodies, surface tension, the nature of fluids, condensation and rarefaction, gilding, the explosion of gunpowder, the weight of air, the propagation of light, geometrical propositions about cylinders, mathematical paradoxes concerning infinity, and discoveries about the constancy of the swing of the pendulum. (Galileo supposedly discovered the isochronism of the pendulum back in the 1580s while at Pisa.) The discussion is always brilliant and entertaining.

In the second "day" of his *Discourse* Galileo extends the range of ancient mechanics in considering the loaded beam. He mathematically determines the effects of the internal stresses induced by external loads and by the intrinsic heaviness of the beam itself. This problem had previously received little theoretical attention and might be of interest to architects, stonemasons, carpenters, shipwrights, millwrights, and engineers. Galileo conducted no experiments. Instead, he applied theoretical statics to the problem and, despite a misguided assumption about the distribution of internal stresses in the beam, he arrived at the fundamentally correct conclusion that the strength of the beam (its flexural strength) is proportional to the *square* of the cross-sectional depth. (*AB* in fig. 12.3.) In the end, however, artisans and craftsmen greeted Galileo's results with indifference. Contemporary engineers were fully capable of solving their problems using traditional and hard-won rules of thumb without adopting the meager theoretical principles that contemporary science could offer.

In the more closely studied days 3 and 4 Galileo unveiled the second of his *Two New Sciences*, the study of local motion, that is, motion in the neighborhood of the earth. In a word, Galileo overthrew the traditional Aristotelian conception held by almost all contemporary scientists that the rate at which a body falls is proportional to its *weight*. The medium through which bodies fall, rather than playing an essential resisting role per Aristotelian interpretations, became merely an accidental "impediment" to ideal fall that would occur in a vacuum. To so reconceptualize motion and fall was to strike at the core of Aristotle's physics, and Galileo's work in these areas proved central to breaking down the Aristotelian worldview.

In examining the historical development of Galileo's thought, one

Fig. 12.3. The strength of materials. While under house arrest in 1638 Galileo, at the age of 74, published the *Two New Sciences* in which he formulated the law of falling bodies and deduced that the strength of a loaded beam is proportional to the square of the depth of its cross-section (AB).

must recognize a "process of discovery" as Galileo fought his way through the maze of factors involved in motion and fall. How to conceptualize the factors, much less how to relate them, was not clear to Galileo at the outset. As a student he adhered to Aristotle's views. Early on he became convinced that a rigid Aristotelian interpretation of fall was false, and his skepticism may have led to a demonstration concerning falling bodies at the Leaning Tower of Pisa. At one point in the long and complex process of working out his ideas, Galileo thought that the density of the medium through which bodies fell (such as air) was a key factor in determining how fast they fall. By 1604 he had come to believe that all bodies would fall at the same speed in a vacuum and that the distance a body covers in free fall is measured by the square of the time. But in 1604 he envisioned the right law for the wrong reason, thinking (mistakenly) that the velocity of fall is proportional to the distance covered, rather than time passed (as he finally concluded). Only after returning to his old work in 1633 did Galileo arrive at his mature view that velocity is proportional to time elapsed, not distance covered; the distance a body falls in free fall remains proportional to the *square* of the time of fall (distance $s \propto t^2$). This is Galileo's law of falling bodies. For Galileo, all bodies (independent of their weight) fall at the same accelerated rate in a vacuum.

Our knowledge of the "correct" answers may obscure Galileo's intellectual achievement. Galileo's position ran counter to common sense and everyday experience. Heavy objects do seem to fall faster than light ones, as Aristotelian theory would predict—a heavy book reaches the ground before a light sheet of paper, for example. Many factors are involved in fall: the weight of a body, or, as we would say, its "mass" and "momentum," the latter measured in several ways; the medium a body moves in, the density or specific gravity of an object, the buoyancy of the medium, the shape of the falling body, the resistance it might offer (and different measurements of same), the distance covered, time elapsed, initial velocity (or speed), average velocity, terminal velocity, and accelerations of various types. Which factors are essential? Galileo faced formidable conceptual problems in coming to terms with free fall.

Two further points need to be made concerning Galileo's fundamental contribution to theoretical mechanics. First, his law is a *kinematical* law, that is, it renders a *description* of motion, not an account of the causes of motion. Galileo's law *describes how* bodies fall; it does not *explain why* they fall. It's predictive, not explanatory. In this way Galileo self-consciously avoided any discussion of *cause*. As a methodological maneuver, the beauty and power of the move derive from what can be gleaned from kinematics alone without inquiring into causes. In effect, Galileo is saying, look what we can accomplish by concentrating on a mathematical description of phenomena without confusing matters by arguing over causes of phenomena.

The second point concerns the fact that all of the kinematical rules Galileo announces in the *Two New Sciences*, including the germ of his law of falling bodies, were, as previously noted in chapter 10, discovered and enunciated three centuries earlier by Nicole Oresme and a group of late medieval scholastics at Oxford University known as the Mertonians or the Calculators. There were differences, however; the most important was, as Galileo himself was quick to point out, that while the Mertonians speculated about abstract possibilities for motion, Galileo believed what he had discovered applied to the real world and to the way bodies actually fall here on earth.

Days 3 and 4 of the *Two New Sciences* also speak to the question of the role of experiment in Galileo's science and about how he satisfied himself that his mathematical formulations concerning motion apply in nature while the Mertonian-like ideas remained merely speculative. Galileo is often referred to as the "father of experimental science," and indeed, doesn't the popular stereotype portray Galileo "experimenting" by dropping balls from the Leaning Tower of Pisa? Alas, it has proven all too easy to (mis)cast Galileo as the father of the "experimental method" or the "Scientific Method." The advent of experimental science marked an important development in science in the seventeenth century to which Galileo contributed mightily, but simplistic and uncritical readings of Galileo's work misplace the role of experiment in his science and they reinforce myths about how science works. Galileo decidedly did not operate according to some cliché of the "Scientific Method" that, stereotypically, has scientists formulating hypotheses, testing them through experiment, and deciding their truth or falsity on the basis, simply, of experimental results. The reality of Galilean experiment is more interesting, more complex, and historically more important.

With regard to the Leaning Tower experiment, as a junior professor at the University of Pisa between 1589 and 1592, Galileo supposedly dropped balls from the Tower before an audience of students and professors in order to demonstrate the falsity of Aristotle's theories about heavy bodies falling faster than light bodies. One can question whether such an experiment actually took place because the first written record that Galileo performed any demonstration at Pisa dates only from 1657, 15 years after Galileo's death. The popular image is surely wrong that Galileo was trying to prove his law of falling bodies experimentally. We know that Galileo did not arrive at a formulation of his law until 1604, so he could not have been "proving" it in an experiment a decade before in Pisa. But he may well have performed a demonstration to illustrate problems with the Aristotelian analysis of fall.

The precise role of experiment in Galileo's science and the reality of *how* Galileo came to reject Aristotle are much more complex and nuanced than the cliché of Galileo dropping balls from the Leaning Tower of Pisa would suggest. The way Galileo arrived at his law of falling bodies and associated kinematics did indeed involve considerable experimentation in the sense that he repeatedly made all manner of tests and trials in coming to grips with the phenomena. He left voluminous manuscript records of such experiments. Galileo was a deft experimenter, manually dexterous, and expert with equipment, as his facility for building telescopes also demonstrates. Experiment in this sense figures prominently in Galileo's approaches to research. But formally, Galileo reserved experiment not to test his propositions, as we might think retrospectively, but to confirm and illustrate his principles. In a word, Galileo's experiments do not confirm hypotheses, they demonstrate conclusions previously arrived at through analytical reasoning.

In a key passage in the *Two New Sciences*, after laying out his rules concerning motion, Galileo has Simplicio inquire, "But I am still doubtful whether this is the acceleration employed by nature in the motion of her falling heavy bodies [and I ask you to present] some experiment . . . that agree[s] in various cases with the demonstrated conclusions." To which Salviati replies, "Like a true scientist, you make a very reasonable demand, for this is usual and necessary in those sciences which apply mathematical demonstrations to physical conclusions, as may be seen among writers on optics, astronomers, mechanics, musicians, and others who *confirm their principles with sensory experiences.*"

With that Galileo turns to his celebrated inclined plane experiment. He first describes his experimental equipment: a wooden beam 24 feet long, three inches thick, with a channel chiseled in one edge, smoothed, and lined with parchment. One end of the beam is raised two to four feet, and a rounded bronze ball allowed to roll down the channel. Two paragraphs later he describes his timing method: he collected water running from a container and measured its weight to determine a time interval. It hardly needs to be pointed out that a not-perfectly-spherical or uniform bronze ball rolling and rattling down and over a vellum-lined channel no matter how smooth could not exactly produce the predicted results. Too many "impediments" were at play in the experiment. Under the control of the human eye and hand, Galileo's ingenious water device doubtless slipped a drop or two here and there, and inaccuracies stemming from measurement even in a delicate balance had to throw off the results still further. Yet Galileo goes on unashamedly to claim that "by experiments repeated a full hundred times, the spaces were always found to be to one another as the squares of the times . . . these operations repeated time and again never differed by any notable amount." But he presents no data, and he does not tell us what "any notable amount" means. The community of scientists in France later had cause to doubt the validity of Galileo's claimed results when they tried to replicate his experiment based on his text. Galileo, however, took his own report of the experiment as sufficient evidence to confirm what mathematical analysis previously demonstrated as true. Beyond indicating the distinctive role of experiment in Galileo's science, the inclined plane experiment illustrates the complexities of how experiments work in practice rather than according to some abstract theory of "scientific method."

Finally, in day 4 of the *Two New Sciences*, Galileo extends his considerations of motion to cover projectiles and projectile motion. (See figure 12.4.) The figure shows the conceptual model Galileo used to analyze projectile motion. For Galileo the motion of a thrown or shot object is compounded of two other motions. On the one hand, the projected body falls downward according to the law of falling bodies presented in day 3. On the other hand, it moves *inertially* along a horizontal line, meaning that it moves of its own accord without any separate mover acting on it. Galileo's concept of inertia, first presented in his *Sunspot Letters* in 1613, requires further clarification, but its revolutionary implications should already be clear. As we recall, for "violent" motion Aristotle required a mover. How to provide such a mover in the case of a projectile after it separates from its launcher had been a nagging problem for Aristotelian mechanics for 2,000 years. Galileo

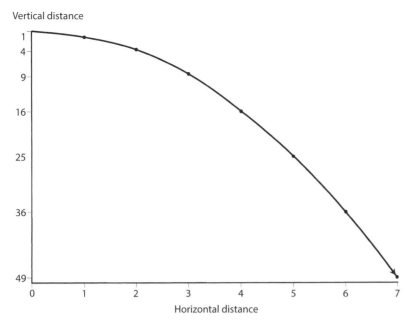

Vertical distance

Horizontal distance

Fig. 12.4. Parabolic motion of projectiles. By separating projectile motion into two components, Galileo could show that projected bodies follow parabolic paths. Along the vertical axis a body falls with uniformly accelerated motion; it moves inertially with a uniform, constant speed on the horizontal axis; when these two motions are combined, the body traces the curve of a parabola. Galileo thus reconceptualized the centuries-old problem of projectile motion, recasting it in something like its modern form.

offered a revolutionary reformulation that eliminated the problem altogether. For Galileo, and later for Descartes and Newton, no mover is required, for there is nothing to explain in accounting for the natural inertial motion of bodies. Such is the stuff of scientific revolutions.

A crucial and revealing difference separates Galileo's view of inertia from the view later adopted by Descartes and Newton. Whereas the latter (and modern science generally) adopts *rectilinear* or straight-line inertia, Galileo held to horizontal or so-called *circular* inertia. He believed that bodies moving inertially would travel, not in straight lines, but in curves following the horizon, in fact, in circles around the earth. For Galileo the horizontal line is not straight, but a segment of a circle around the earth's center, that is, the "horizon." Galileo's revolutionary "discovery" (or "invention," if you prefer) of inertia removed a major objection to Copernicanism, for if objects move inertially they will not appear to be left behind as the earth moves, and the concept contributed forcefully to the overthrow of Aristotle and the Aristotelian worldview. That Galileo held to circular inertia would be an oddity of history, except that it provides yet another instance of the continuing power of circles on the imagination of scientists well into the seventeenth century.

Galileo drew the corollary from his analysis of the compound motion of projectiles that the resulting curve is a parabola—at least in theory. (Ironically, it would be a parabola only if the earth were flat.) Galileo's discovery of the parabolic motion of projectiles represents another notable achievement, and this one offered obvious practical possibilities for artillery and ballistics. He recognized such applied uses, and in day 4 he published detailed mathematical tables of elevation and range for gunners derived wholly from theory. The case would seem a golden instance of theoretical science being turned to an applied end but, alas, Galileo's theoretical understanding had no impact on gunnery in practice. By the time Galileo published his tables, cannons and artillery had reshaped the face of Europe for over 300 years. Skilled gunners and military engineers had long since worked out artillery "rules," tables, and procedures for hitting their targets. The technology of cannonry may have been more influential on Galileo's science than the other way around.

Galileo knew that problems remained in the realm of mechanics. He understood very well, for example, that a ball of cotton behaves differently from a ball of lead, and he groped his way toward notions of what we would call "force" and "momentum." But Galileo never got beyond initial speculations about these matters, and at one point in day 4 of the *Two New Sciences* he says, almost nostalgically, "I should still like to find a way of measuring this force of impact." That work—the measurement of force—would come after Galileo, with Isaac Newton. Galileo went completely blind, and doctors withheld his beloved wine. He died in 1642, the very year Newton was born. His erstwhile friend, Pope Urban VIII, prohibited any monument to his memory.

Post-Galileo

The trial and punishment of Galileo did not end scientific activity in Italy in the second half of the seventeenth century, but the case severely dampened the level and quality of contemporary Italian science. The atmosphere in Italy remained repressive, and church authorities were vigilant. Copernicanism and grand cosmological theorizing remained off limits, and Italian scientists avoided them in favor of safer endeavors like strictly observational astronomy. A tolerance toward Galileo emerged only a hundred years after his death with an Italian edition of his works sanctioned by the liberal Pope Benedict XIV. The Catholic Church authorized the teaching of Copernicus only in 1822, and Copernicus was finally taken off the Index in 1835. Galileo himself was not fully rehabilitated until the 1990s.

Galileo and the patronage system are partly to blame for the failure of a Galilean school to take hold in Italy. Especially in the initial controversies in the 1610s, Galileo had followers and, as something of a patron in his own right, he succeeded in placing a number of them, including Benedetto Castelli (1578–1643) as professor of mathematics at the University of Pisa. Because he was a courtier, however, he did not train students. Vincenzio Viviani (1622–1703) and Evangelista Torricelli (1608–47) joined the master as copyists and assistants only in the last few years of his life. The younger mathematician Francesco Bonaventura Cavalieri (1598–1647) was a true pupil, and Galileo's son, Vincenzio Galilei (1606–49), also carried on his father's work, especially in the development of the pendulum clock. But Galileo's few direct scientific descendants, excepting Viviani, had passed from the scene by 1650. The patronage system robbed him of offspring.

With the waning of Italian science after 1633, a characteristic feature of the period of the Scientific Revolution is the geographical movement of scientific activity northward out of Italy and into the Atlantic states—France, Holland, and England. An active scientific community of independent amateurs arose in France and included such luminaries as Pierre Gassendi (1592–1655), Pierre Fermat (1601–65), Blaise Pascal (1623–62), and René Descartes (1596–1650). Although they were insulated from the direct power of the church in Rome, Galileo's trial by the Inquisition produced a chilling effect on French scientific intellectuals. Descartes, for example, suspended publication of his Copernican treatise, *Le Monde*, in 1633.

René Descartes inherited the mantel of intellectual leadership for the new science. Trained by the Jesuits, Descartes was a polymath genius and soldier of fortune who retired at the age of 32 to take up a contemplative life devoted to philosophy and science. Part of Descartes's fame stems from his achievements in algebra and analytical geometry, including the introduction of "Cartesian" x-y-z coordinates. He also produced original work in optics and meteorology, and he demonstrated his concern for how scientific knowledge is produced in his historic *Discourse on Method* (1637). Descartes likewise wrote on theology and metaphysics and is often hailed as the father of modern philosophy. For our purposes Descartes's importance derives from the fact that he developed a complete cosmology and world system to replace Aristotle's and competing alternatives that were at play in the early decades of the seventeenth century.

In coming to grips with the state of science and philosophy in his day, Descartes proposed a completely mechanical view of the world, and his mechanization of the universe represented a radical departure. For Descartes, the world and everything in it functions as a great machine linked and governed by the laws of mechanics and of impact. In this sense Descartes provided explanations for motion and the functioning of the universe, not just descriptions that Kepler and Galileo had been able to provide. On a cosmological scale Descartes pictured moons carried around planets and planets carried around the sun in great whirlpools of aetherial matter. His *Principles of Philosophy* (1644) elaborated this heliocentric vortex theory. (See figure 12.5.) In the realm of physiology and medicine Descartes provided a rational, mechanical alternative to traditional Aristotelian-Galenic accounts. Although his system was mathematically vague and open to critical challenges, Descartes may be fairly said to have capped the Scientific Revolution, in that Cartesian natural philosophy subsumed all the controversies raised over the century since Copernicus and encompassed all the discoveries of the new science. Even more, Descartes

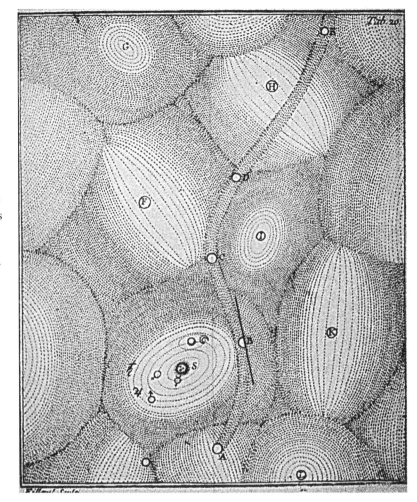

Fig. 12.5. Descartes's system of the world. The constitution of the universe was an open question. The great seventeenth-century French philosopher and mathematician René Descartes responded by envisioning a universe filled with an aetherial fluid sweeping the planets and other heavenly bodies around the vortices. The figure depicts a comet crossing the vortex of our solar system.

provided a comprehensive explanatory alternative to Aristotle and all other competing systems. Whether Descartes was right was the only issue in science after his death in 1650.

Descartes lived and worked for two decades in the Netherlands, that Protestant republic famous for its social and intellectual tolerance. The Dutch republic produced its own outstanding contributors as part of the northward movement of the Scientific Revolution, including the mathematician and engineer Simon Stevin (1548–1620), the atomist Isaac Beeckman (1588–1637), and most notably Christiaan Huygens (1629–95), perhaps the foremost Cartesian and spokesman for the new mechanical science in the second half of the seventeenth century.

The Low Countries likewise became the locus of pioneering work using the microscope. The drapier-turned-scientist Anton van Leeuwenhoek (1632–1723) achieved international renown for uncovering a heretofore hidden and unknown "world of the very small." The novelties he discovered included observing blood corpuscles, spermatozoa, and other minute "animalcules." His countryman Jan Swammerdam

(1637–80) likewise pushed the microscope to new limits, particularly in his delicate dissections and preparations of plants and insects. These trailblazing Dutch microscopists were joined by the Italian Marcello Malpighi (1628–94) and the Englishman Robert Hooke (1635–1703), whose *Micrographia* appeared in London in 1665. All of these early investigators used single-lens beads for their microscopes, and technique proved crucial for success. But unlike its cousin instrument, the telescope, which became universally accepted and an essential tool in astronomy, the microscope raised more questions than it answered for seventeenth-century observers and theorists. What one "sees" through the microscope represents a complex interplay of ideas and images, and agreement over what was seen and what the images said about insect anatomy, capillary circulation, or embryology, for example, was not forthcoming. The diverging fates of the microscope and the telescope in the seventeenth century suggest that shared intellectual frameworks are required for establishing research traditions, not the instruments themselves. Only in the nineteenth century, and under a different set of conditions, did the compound microscope become a standard piece of laboratory equipment.

England was another Protestant maritime state that fostered a community of men who pursued science after Galileo. We have already mentioned the court physician William Gilbert (1544–1603) and his influential work on the magnet. We saw, too, that the English physician William Harvey (1578–1657) made the revolutionary discovery of the circulation of the blood in 1618. One can likewise point to Francis Bacon (1561–1626), the lord chancellor of England who proved such an effective spokesman for the new science; the aristocrat and experimental chemist Robert Boyle (1627–91); and, of course, Isaac Newton (1642–1727), just to name a few of the luminaries of seventeenth-century English science. Institutions facilitated the growth of science in contemporary England, notably, the Royal College of Physicians (1518), Gresham College (a new institute with salaried professors founded in 1598), and, later in the seventeenth century, the Royal Society of London (1662) and the Royal Observatory at Greenwich (1675). Royal funding for new scientific chairs at Oxford (geometry-astronomy in 1619 and natural philosophy in 1621) and later at Cambridge (1663) likewise helps explain the flourishing of English science in the later seventeenth century.

Ideology and Utility

Although not absolutely new, claims for the social utility of science began to be widely asserted in the seventeenth century, the conviction that science and scientific activities can promote human welfare and should therefore be encouraged. The ideology was activist and contrasted with the Hellenic view of the practical irrelevance of natural philosophy and the medieval view of science as the subservient hand-maiden to theology.

The ideology for the social utility of science sprang from more than one historical source. Renaissance magic and Hermeticism, with their belief in the possibility of controlling forces that permeate the universe, represent one route from which emerged the doctrine that knowledge can and should be made useful and practical. Alchemy, in both its medicinal and metallurgical forms, exemplifies another. The Neopla-tonist and humanist Pico della Mirandola (1463–94), for example, saw magic as simply the practical part of natural science. Although he walked a tightrope in these regards, Giambattista Della Porta also favored the idea that natural magic held powers useful to princes and governments. Astrology and the occult formed a notable part of the re-ward systems of patrons: Tycho turned out astrological forecasts, and Kepler made a career as a court astrologer. Philip II, the level-headed ruler of Spain and the Spanish Empire from 1556 to 1598, became deeply involved in the occult. He patronized numerous alchemists, and he built a huge alchemical laboratory capable of manufacturing al-chemical medicines in volume. Charles II of England possessed his own alchemical laboratory. Reliable reports circulated of the alchemical multiplication of gold, and the utility of occult studies remained widely accepted through the 1600s.

The perception of science as useful knowledge found its foremost ideologue in Francis Bacon. Bacon pointed to gunpowder, the com-pass, silk, and the printing press as examples of the kind of worthwhile inventions potentially forthcoming from systematic investigation and discovery. (Bacon neglected to say that these technologies arose inde-pendently of natural philosophy but, no matter, for future scientific research promised similarly useful devices and techniques.) Among the castes of laborers Bacon envisioned for a scientific utopia, he set aside one group, the "Dowry men," especially to search for practical benefits. In categorizing different types of experiments, Bacon likewise specified that "experiments of fruit" must be combined with "experi-ments of light" to produce practical outcomes. His influence in the world of science was largely posthumous, but it proved no less power-ful for that.

Descartes, too, was influential in advocating what he called a "prac-tical philosophy" and the idea that knowledge should be put to use "for the general good of all men." Descartes considered medicine a principal arena where useful advances and practical applications of theory might be found. Later in the seventeenth century Robert Boyle likewise enunciated the goal of new medical therapies derived from experimental philosophy. Notwithstanding the fact that ties between contemporary scientific theory and effective medical technique were tenuous at best and largely remained so until the twentieth century, the seventeenth-century ideologues of the new science were quick to hitch their wagons to the draft horse of medical practice.

Deep in Book II of the *Principia* even Isaac Newton made an argument for utility. After completing a complex demonstration concerning hydrodynamics and the shape of bodies producing the least resistance while moving in fluids, he commented dryly that "this proposition I conceive may be of use in the building of ships." Newtonian theory—the result of pure science—stood far removed from economic reality and any practical application, but the case brings home the distinction between what the new ideology claimed and what it could deliver.

As part of their ideology, seventeenth-century thinkers likewise expressed new attitudes about nature and the exploitation of nature. Bacon and Descartes separately voiced the view that humans should be the master and possessor of nature, that nature and the world's natural resources should be vigorously exploited for the benefit of humankind—that is, those who own or control knowledge. The notion that nature was subject to human dominion possessed biblical authority and was already operative in the Middle Ages. But a distinctive imagery of the violent rape and torture of nature as an aspect of scientific practice came to the fore in seventeenth-century thought on these matters. Bacon, for example, asserted bluntly that "Nature must be taken by the forelock."

The notions that science is useful, that science is a public good, and that knowledge is power have ruled as cultural leitmotifs in the West since the seventeenth century and everywhere since the nineteenth. The further implication was twofold: science and scientists deserved support, and the power they brought should be used for the commonweal. The status of the old ideas that natural philosophy was for natural philosophers or subservient to theology diminished. Evidently the new notions were more consistent with the interests of the new centralized states and the development of merchant capitalism in Europe.

"God said, '*Let Newton be!*'"

The Scientific Revolution was a complex social and intellectual affair that was much more than a collection of individual scientific biographies. Nevertheless, Isaac Newton (1642–1727) so dominates the intellectual landscape of the later seventeenth and early eighteenth centuries that his life and works compel close scrutiny. Newton's career speaks volumes about the end of one era and the beginning of another, and in following the trajectory of Newton's biography, we can simultaneously survey the thematics of the Scientific Revolution raised in previous chapters.

Newton's *Principia Mathematica Philosophia Naturalis* or *Mathematical Principles of Natural Philosophy* (1687) effectively concluded a line of theoretical inquiry into cosmology and the underlying physics of the world that stretched back through Descartes, Galileo, Kepler, and Copernicus and ultimately back to Aristotle. With universal gravitation and his laws of motion, Newton's physics finally unified the celestial and terrestrial realms separated since Aristotle. Newton buried not only the moribund Aristotelian world, but also the newer mechanical alternative worked out by Descartes. He is also acclaimed as a mathematician and the co-inventor (along with Leibniz) of the calculus; and he did fundamental work in optics as well. The renown of Newton's genius stems from the many domains of his accomplishment as well as from the profundity of his contributions to each.

But Newton's overriding importance in the history of modern science derives not simply from what he did for contemporary science, but also from his abiding role in molding scientific traditions that came afterward. Newton culminated the Scientific Revolution and at the same time set the agenda for scientific research in astronomy, mechanics, optics, and a range of other sciences. In so doing he profoundly, indelibly shaped the history of science over the two centuries following his death.

Personally delicate and largely disagreeable, Newton was a man of unparalleled scientific genius whose influence is felt still today. Newton's professional career likewise is of general import and repays examination. That he metamorphosed from the reclusive Lucasian Professor of Mathematics at Cambridge University into a bureaucratic civil servant at the Royal Mint, president of the Royal Society of London, and finally Sir Isaac is revealing of the social history of science and reflects larger patterns of social and institutional changes in European science in the seventeenth century.

Newton also comes to us as a popular stereotype. Who has not absorbed the cliché of Newton discovering gravity by being hit on the head by a falling apple? Newton the egghead genius. In his own day and variously afterward Newton and Newtonian science have been the objects of mythological development and political exploitation. One especially influential image crafted in the eighteenth century held Newton to be the epitome of the rational scientist. Recent work—the product of a substantial scholarly industry—now portrays the great man not only as a brilliant natural scientist but also as a magus, deeply involved in alchemy, religious fanaticism, and investigating occult forces at play in nature and in history. The different Newtons revealed in the historical record not only raise the perennial question about historical truth, but the case highlights no-less-pointed issues concerning the cultural meaning of Newtonian science and the social uses to which it was put in his own day and afterward.

From Lincolnshire to Cambridge

Isaac Newton was born on the family farm in Woolsthorpe, Lincolnshire, England, on Christmas day, 1642. He was a posthumous child, his father having died earlier that fall. (The absence of the father and the coincidence of his birthday with Jesus's may have confirmed for Newton his sense of having a special relationship with God.) On his father's side the Newtons were a modestly rising yeoman family of English farmers and herders; he was better connected on his mother's side through a wealthier and more educated family of rural gentility and clergy. Underweight and premature, the newborn was not expected to survive. He reportedly fit in a quart jar and was too weak to lift his head to the breast.

Newton experienced an awkward and unpleasant childhood. His mother remarried when he was only 3 and left him at Woolsthorpe with his grandparents. All readings of Newton now acknowledge that he was an alienated and tortured neurotic, emotionally disfigured early in life. As a widow with other children, Newton's mother returned to her son when he was 10, and at 12 Newton began grammar school in nearby Grantham. Leaving the Grantham school at 17, he appeared headed for an unhappy life as a rural farmer and property manager, but the obviously clever lad so failed at farming and so rebelled against his fate that the only alternative was to ship him off to university. And so, through a family tie and after a bit of brushing up back at Grantham, Newton headed up to Cambridge in 1661.

England had entered a new period of its history just the previous year when it restored the monarchy under Charles II after two decades of civil and religious strife. Although not directly affected by the turmoil, Newton's Lincolnshire youth had coincided with civil war in England (1642–49) and the subsequent republican period of the English Commonwealth (1649–60) which had overthrown the monarchy and the official Anglican church. The peaceful installation of Charles II as Anglican king of England in 1660 reinstituted both monarchy and the state church, but on more constitutional terms and with a measure of religious and political toleration. Further changes lay ahead, and tensions ran high in Restoration England as Newton proceeded to Cambridge.

Newton matriculated at Trinity College as a subsizar or scholarship boy, and for his first years he suffered the indignity of serving the senior students. (Newton resented—but always retained—his somewhat peasant roots.) Cambridge University was a sleepy intellectual backwater that continued to embrace Aristotle even in the 1660s. Fortunately for Newton, its relaxed rules and lack of oversight left him free to study on his own. He soon made his way to the cutting edge of contemporary mathematics, mechanics, and natural philosophy. Symptomatic of the changing state of science in the later seventeenth century, Newton mastered Descartes's work before he studied Euclid. After he took his B.A. degree at the university in 1665, he stayed on at Trinity and was soon elected a permanent fellow of the college. In 1669, on the retirement of Isaac Barrow (his own professor), Isaac Newton, M.A.—26 years old—became the second Lucasian Professor of Mathematics, the new chair for science established in 1663 to invigorate the university. In the space of eight years, but as part of a medieval career pattern that had gone on for four centuries, Newton slipped into the role of university don.

In 1665, while Newton was still a student, the plague struck at Cambridge. The university closed for almost two years, and Newton returned to Woolsthorpe. Earlier biographical accounts made much of the so-called Miracle Year that followed in 1666, wherein Newton supposedly discovered gravity, invented the calculus, and developed his theory of light and colors. Historians today take a more nuanced view of Newton's development at this stage, setting 1666 in the context of the preceding period of intense study of science and mathematics and the leisure at Woolsthorpe of thinking through problems on his own. Unknown to the rest of the world, in 1666 Newton was in fact the world's leading mathematician and was as knowledgeable as anyone about science or natural philosophy (new or old). He thought about gravity and calculated in a rough way the effects of gravity extending to the moon. Using prisms, he investigated light and colors, discovered new phenomena, and toyed with new explanations. He also gained his fundamental insight into the calculus by seeing a relation between tangents to curves and areas under curves (what we call the derivative and the integral). What emerged in 1666 was not fully formulated and completed work, as legend has it. Instead, Newton's initial insights into mechanics, optics, and mathematics developed further after he returned to Cambridge in 1667, and indeed these subjects occupied much of his life's work in science.

Newton published first on optics. In a paper appearing in 1672, he offered his views that light is composed of rays, that different rays are refracted to different degrees through lenses or prisms, that each ray corresponds to a different color, and that white light represents an amalgamation of all rays and colors. Newton adduced these views using carefully controlled experiments. In his famous "crucial experiment," for example, he passed a beam of light through a prism to create a spectrum. Refracting portions of that spectrum through a second prism failed to produce another spectrum or other changes, thus demonstrating to his satisfaction that colors are properties of light and not produced by refraction. The conclusions Newton drew embodied radical new concepts of the nature of light and colors, contrary to both Aristotelian and Cartesian theory, but he naively believed he was simply making public straightforward facts about nature. Newton noted a practical, technological spinoff of his discoveries: to avoid chromatic aberration from light refracting through lenses he designed a *reflecting* telescope that used a mirror to focus light. Newton presented his reflecting telescope to the Royal Society of London, which in turn elected him a Fellow in 1672.

Newton's 1672 paper on light proved a tour de force, but disputes soon arose. Aristotelian and Cartesian opponents attacked Newton's findings, and he became entangled in arguments that continued doggedly for decades over precise details of experimental procedure and the interpretation of results. These events led Newton to shun a public life in science. Some of his mathematical work circulated privately in manuscript, but after his initial public notice Newton retreated as much as possible into his own world at Cambridge. There, spurred by the requirement of his professorship to become a priest in the Anglican Church, he took up serious study of theology and biblical prophecy in the 1670s and early 1680s. Religion fervently occupied Newton all his life, but he rejected Christian orthodoxy, seeing, for example, the Christian trinity as a hoax foisted on the early church. He developed other heretical theological views (known as Arianism, something like a fiery Unitarianism) which put him dangerously out of touch with English society around him. Newton's fanaticism extended to the belief that there existed bodies of secret, pristine, and arcane knowledge concerning matters religious and scientific that God had initially conveyed to Noah, which then passed to Moses and Pythagoras, and which had come down to Newton's day in an esoteric oral tradition of prophets and the elect few like Newton himself, or so he believed, who could read hidden codes in nature and the Bible. He seems to have taken seriously the poetic line later written about him by Edmond Halley, "Nearer the gods no mortal may approach." But Newton kept these dissident views very private, and in 1675 he was spared, one might say miraculously, having to resign the Lucasian professorship when the requirement for taking religious orders was dropped.

In the quest after secret knowledge, alchemy occupied the major portion of Newton's time and attention from the mid-1670s through the mid-1680s. His alchemical investigations represent a continuation and extension of his natural philosophical researches into mechanics, optics, and mathematics. Newton was a serious, practicing alchemist—not some sort of protochemist. He kept his alchemical furnaces burning for weeks at a time, and he mastered the difficult occult literature. He did not try to transmute lead into gold; instead, using alchemical science, he pried as hard as he could into forces and powers at work in nature. He stayed in touch with an alchemical underground, and he exchanged alchemical secrets with Robert Boyle and John Locke. The largest part of Newton's manuscripts and papers concern alchemy, and the influence of alchemy reverberates throughout Newton's published opus. This was not the Enlightenment's Newton.

Science Reorganized

Newton's 1672 paper on light was published in the *Philosophical Transactions* of the Royal Society as a letter to its secretary, Henry Oldenburg. The sponsoring organization, the Royal Society of London for Improving Natural Knowledge, was a new scientific institution present on the English scene. Founded in 1660 and granted a royal charter by Charles II in 1662, the Royal Society was a state scientific society that maintained itself through the dues of its members.

The Royal Society (1662) and the Paris Academy of Sciences (1666) were the flagships of an organizational revolution of the seventeenth century. They created a new institutional base for science and scientists, and they ushered in a new age of academies characteristic of organized science in the following century. Major national academies of science subsequently arose in Prussia, Russia, and Sweden, and the model of a state academy or society of science spread throughout Europe and to its colonies around the world. Scientific academies and societies coordinated a variety of scientific activities on several levels: they offered paid positions, sponsored prizes and expeditions, maintained a publishing program, superintended expeditions and surveys, and rendered a diversity of special functions in the service of the state and society. These institutions, incorporating a broad array of scien-

Fig. 13.1. Louis XIV visits the Paris Academy of Sciences. This fanciful seventeenth-century engraving, showing King Louis XIV of France visiting the Paris Academy of Sciences, illustrates the growing role of state support for the sciences in seventeenth-century Europe.

tific interests, dominated organized science until the coming of specialized scientific societies and a renewed scientific vitality in universities in the nineteenth century.

Growing out of Renaissance and courtly precedents, these new learned societies of the seventeenth and eighteenth centuries were creations of nation-states and ruling governments. State-supported scientific societies possessed a more permanent character than their Renaissance cousins in that they received official charters from government powers incorporating them as legal institutions and permanent corporations. Given the increasing separation of government operations from royal households, official state scientific societies became detached from court activity and integrated into government bureaucracies. Society members tended to act less as scientific courtiers and more as expert functionaries in the service of the state. The state academies and societies were also institutions specifically concerned with the natural sciences; they were not subservient to other missions, they largely governed themselves, and, unlike universities, they did no teaching. The growth and maturation of state academies and societies of science in the eighteenth century provide impressive evidence of the greater social assimilation of science after the Scientific Revolution.

Whereas upper-class women reigned in the literary and intellectual *salons* of the seventeenth and eighteenth centuries, the world of science and the learned society remained largely a man's world. The French marquise Madame du Châtelet (1706–49) is a notable exception, being a member of several academies, a contributor of original science, and Newton's underrated French translator. Several Italian towns with universities and academies produced a string of female scientific luminaries—the experimental Newtonian Laura Bassi (d. 1778), for example. Bassi and her fellow female scientists brought credit on themselves and their gender, on their towns, and on the contemporary world of learning.

Other extra-university patterns and alternatives developed for organizing and communicating science in the seventeenth century. Up until that point personal travel, the private letter, and the printed book represented the mainstays of scientific communications. As the Scientific Revolution unfolded, informal circles of correspondence began to effect new lines of communication among people interested in the new science. Then, in the second half of the seventeenth century, coincident with the new state-supported learned societies, the periodical journal, the main form for the publication of scientific research ever since, made its appearance. The *Philosophical Transactions* of the Royal Society of London and the *Journal des Sçavans* in France both appeared in 1666 and other influential early scientific journals followed. Journals provided a new mode of communicating and disseminating scientific knowledge and research. They allowed for comparatively speedy publication, and they created the scientific paper, which became the unit of production in the scientific world.

To better administer commerce at home and abroad, European states, following the precedent established in Islam, also founded royal and national astronomical observatories in France (1667), England (1675), Prussia (1700), Russia (1724), and Sweden (1747). Similarly, states began to sponsor and maintain national botanical gardens: the Jardin du Roi in Paris dates from 1635 and the Royal Gardens at Kew from 1753. These state gardens and hundreds like them arose either by fiat or by central governments coopting older university pharmacy gardens. The state botanical gardens of Europe became centers for scientific study, even as mercantilistic policies spread networks of Dutch, English, and French botanical gardens worldwide.

From the sixteenth century onward, European courts and governments institutionalized science, and, in turn, scientific specialists assisted governments in botanical gardens, in observatories, in the scientific societies, in specialized professorships, and in various nooks and crannies of government bureaucracies all across Europe. The River Commission of Medici Florence, for example, possessed a technical staff of fifteen, and the court called on Galileo for his opinion in engineering matters. Newton later served the English crown at the Mint. Like their predecessors in hydraulic civilizations, European scientists increasingly became state functionaries. Scientific experts and the institutions they represented performed useful services for the state. The Paris Academy of Sciences, for example, administered patent applications as an arm of the royal government of France, and, like the Astronomical Bureau in China, it controlled publication of official astronomical tables.

The ideology of utility cemented new relations between science and government in early modern Europe. European governments struck a firm deal with science and natural philosophy, a reciprocal exchange of useful services and powers in return for recognition, support, and self-government. Science actively sold itself to government, and, to a degree at least, the state—first the courts of absolute princes, then the bureaucracies of nation-states—began to buy into science. The historical significance of the new contract established between science and the state in early modern Europe is that European governments began to imitate ancient hydraulic civilizations in the use of scientific experts.

One should not exaggerate the extent of government support for science in Europe which developed by the end of the Scientific Revolution. Mathematicians, scientists, and technical experts did not generally enjoy a high social status. Charles II famously ridiculed his own Royal Society for the useless "weighing of air" with its air-pump experiments. Often, too, receiving payment for technical services or promised stipends proved difficult for scientific and technical personnel in court or state employ. Even the great Paris Academy of Sciences received meager funds in its early decades and at a level far lower than its sister fine arts academies. Tellingly, Louis XIV visited his science academy only once and then reluctantly. European governments insisted that science be cheap as well as useful. Until the twentieth century they saw to it that the first condition obtained. As for the second, they were only partially successful.

Reframing the Universe

In August of 1684 Edmond Halley traveled to Cambridge to ask Isaac Newton a question. Earlier that year at the Royal Society in London, Halley, Robert Hooke, and Christopher Wren toyed with connections between Kepler's law of elliptical planetary motion and a force of attraction emanating from the sun—clearly the idea of such a force was in the air. Newton and Hooke, curator of experiments at the Royal

Society, had earlier corresponded over the idea in 1679 and 1680. Wren had even proposed a prize for a mathematical demonstration. Halley visited the reclusive Newton and asked him about the motion of a planet orbiting the sun according to a $1/r^2$ attractive force. Harking back to 1666, Newton immediately replied that the shape of the orbit would be an ellipse and that he had calculated it. After fumbling through his papers, Newton promised the awestruck Halley that he would send him the calculation. Three months later Halley received a nine-page manuscript, "On the Motion of Orbiting Bodies," that outlined the basic principles of celestial mechanics. Everyone who saw it immediately recognized the significance of Newton's little tract.

The *Principia* remained to be written, however. When Newton returned to his initial calculations of 1666, he discovered errors and conceptual ambiguities. His initial response to Halley in 1684 merely sketched the new physics, and an intense two-year period followed as he struggled with the conceptual, mathematical, and authorial obstacles in expanding his insight into a rigorous, complete treatment. Halley shepherded the great work through the press, nominally under the aegis of the Royal Society. The *Principia* appeared in 1687 with the imprimatur of the Royal Society.

The *Principia* is a highly mathematical or, better, geometrical text, and Newton begins it with definitions and axioms. He defines his terms (e.g., mass, force) and states his historic three laws of motion: 1) his inertial law that bodies in motion remain at rest or in straight-line motion unless acted upon by an outside force; 2) that force is measured by change in motion (although he never wrote $F = ma$); and 3) that for every action there is an equal and opposite reaction. In his front matter he introduces his ideas about absolute space and time; and in a scholium—a mere footnote—he shows that Galileo's hard-won law of falling bodies ($s \propto t^2$) follows as a consequence of his, Newton's, laws. At the outset it is evident that Newtonian dynamics, in explaining motion with forces, subsumes Galilean descriptive kinematics.

The body of the *Principia* consists of three books. The first provides an abstract treatment of the motion of bodies in free space. Section 1 actually continues the preceding prefatory remarks in presenting the analytical techniques of the calculus (integration and differentiation) used in the rest of the *Principia*. Newton states these techniques in the language of geometry because virtually he alone knew the calculus while virtually all his potential readers understood only geometry.

Newton gets down to cases in section 2 of Book I, "The Determination of Centripetal Forces." There he proves that a body orbiting an attracting central force obeys Kepler's second law and sweeps out equal areas in equal times. (See figure 13.2.) That is, if a body at point *A* orbits *S* and is drawn toward *S* by a centripetal or gravitating force of some variety, then the line *AS* will sweep out equal areas in equal times. Newton also shows the converse: if an orbiting body obeys Kep-

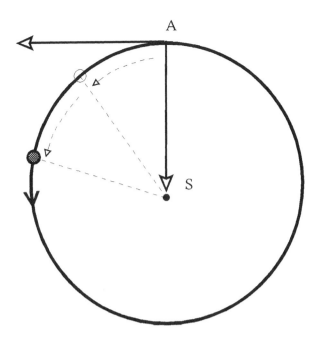

A

S

Fig. 13.2. Newton's *Principia*. In Book I of his *Mathematical Principles of Natural Philosophy* (1687) Isaac Newton linked gravity to Kepler's laws of planetary motion. In the proposition shown, Newton demonstrates that a body moving inertially and affected by a gravitational force will obey Kepler's second law and sweep out equal areas in equal times. That is, given Kepler's second law, orbiting bodies can be understood to be moving under the influence of some kind of gravitational force.

ler's second law, then an attracting or gravitational force of some sort may be seen to be operating.

Newton then turns to a deceptive Proposition IV, Theorem 4. In this proposition he considers the following simplified situation. (See figure 13.3.) Acted on by a centripetal (or attracting) force, a body sweeps out an arc. The illustration shows a body (*b*) moving with a velocity of (*v*) from A to A′ along arc (*a*) in a certain time (*t*) and at a radius (*r*) from the center of attracting force at *F*. Newton is interested in the abstract mathematical relations between and among these parameters: (*a*), (*r*), (*F*), (*t*), and (*v*). In the proposition and then in a staccato of corollaries, he reels off those relations. In the sixth corollary Newton slips in: if $t^2 \propto r^3$, then $F \propto 1/r^2$. This seemingly meek proposition masks a deep insight into nature, for in it Newton is saying that if Kepler's third law holds true, then bodies are maintained in their orbits by a gravitational force that varies inversely as the square of the distance, which is Newton's law of gravity. And conversely, Newton shows that bodies orbiting under a $1/r^2$ law of gravity must obey Kepler's third law. In other words, Kepler's third law demonstrates Newtonian gravitation and vice versa. In a footnote Newton owns up to the explosive significance of what he has done, saying, "The case of the sixth Corollary obtains in the celestial bodies."

The remainder of Book I develops the full mechanics implied in these initial propositions. Newton extends the analysis to all the conic sections; he proves that the attractive force of an extended solid body (like the earth) can be mathematically reduced to attraction from its center; he shows off his virtuosity in discussions of pendulums; he explores the abstract mathematics of universal gravitation; and he pro-

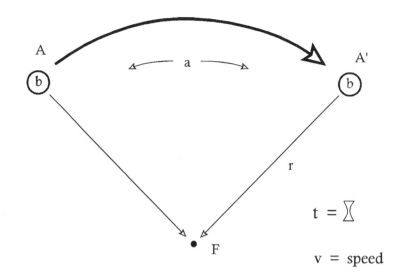

Fig. 13.3. Proposition linking gravitation and Kepler's third law. In this proposition from Book I of the *Principia* Newton analyzed various factors affecting the motion of a body, *b*, while it moves along the arc (*a*) from A to A′ while under the influence of an attracting gravitational force *F*. He was able to show that if the force of gravity decreases as the square of the distance, then $t^2 \propto r^3$, or Kepler's third law applies, and vice versa. In other words, given Kepler's third law, then Newton's $1/r^2$ law of gravity necessarily follows, and given Newton's law of gravity, then Kepler's third law must also apply. Newton thus provided a new physics to explain the motion of the heavenly bodies.

vides the mathematical tools for determining orbits from observations and vice versa.

Rather than pressing the astronomical implications of Book I, like the second act of a three-act play or the slow second movement of a concerto, the second book of the *Principia* pauses to consider the motion of bodies, not in free space, but in resisting media. In essence, in Book II Newton provides a mathematical treatise on hydrostatics and hydrodynamics. At first glance this diversion away from the main theme of gravitational celestial mechanics seems peculiar, until one recalls that in the Cartesian system a thick aether fills the cosmos and that, for Descartes, planets are carried around in vortices or whirlpools which are, in effect, hydrodynamical systems. In exploring the physics of these systems, Newton seeks to smash the Cartesian system, and Book II concludes with the devastating remark, "Hence it is manifest that the planets are not carried round in corporeal vortices. . . . Let philosophers then see how [Kepler's third law] can be accounted for by vortices."

The *Principia*'s denouement occurs in Book III, "The System of the World." Newton first presents the "phenomena" of a heliocentric solar system where orbiting bodies obey Kepler's three laws. In particular, he provides reliable observational data coupling Kepler's third law ($t^2 \propto r^3$) with the motions of the moon around the earth, the planets around the sun, and satellites around Jupiter and Saturn, respectively. Geocentrism is shown to be absurd and inconsistent with the known facts. Using Kepler's third law and Book I of the *Principia*, Newton then proposes that the forces holding the world's planets and moons in their orbits are $1/r^2$ attracting forces and, in particular, "that the Moon gravitates towards the Earth."

Newton's celestial mechanics hinges on the case of the earth's moon. This case and the case of the great comet of 1680 were the only ones that Newton used to back up his celestial mechanics, for they were the only instances where he had adequate data. With regard to the moon, Newton knew the rough distance between it and the earth (60 Earth radii). He knew the time of its orbit (one month). From that he could calculate the force holding the moon in orbit. In an elegant bit of calculation, using Galileo's law of falling bodies, Newton demonstrated conclusively that the force responsible for the fall of bodies at the surface of the earth—the earth's gravity—is the very same force holding the moon in its orbit and that gravity varies inversely as the square of the distance from the center of the earth. In proving this one exquisite case Newton united the heavens and the earth and closed the door on now-stale cosmological debates going back to Copernicus and Aristotle. In proving this and the comet case, Newton simultaneously opened the door on a whole new world of problems to solve.

The remainder of Book III spelled out areas for research in the new era to follow the *Principia*: the moon's precise orbit, astronomical perturbations, the shape of the earth, gravity and the tides, and especially comets. With regard to comets, Newton introduced his other smoking gun: observations of the comet of 1680 and calculations proving its Keplerian solar orbit under the influence of the sun's $1/r^2$ gravity. Newton's would be the model solution for every comet to follow.

The first edition of the *Principia* in 1687 tailed off in an alchemical allusion to cometary vapors transmuting into terrestrial substances through a "slow heat." Twenty-five years later in 1713—a very different period in Newton's life—the author added a General Scholium to the second edition of the *Principia*. Commensurate with the grandeur of the subject, in the General Scholium Newton offered a disquisition on God, the "intelligent and powerful Being" evident in "this most beautiful system of the Sun, planets, and comets." Newton's natural philosophy led him to natural theology and the conclusion that God can be known through His contrivances in nature. Newton's God becomes the great Clockmaker who watches his great machine tick away according to His established laws of nature. In the General Scholium Newton also explains his approach to treating gravity mathematically without inquiring into its cause; about the cause of gravity, Newton noted, in his famous phrase, that he "feigns no hypotheses" (*hypotheses non fingo*), and in this Newton fell back on the methodological cut made by Galileo that an understanding of phenomena that gives us predictive power offers something significant for science if not something fully explaining why bodies fall or the nature of force or gravity, for example. The Scholium itself ends with an incongruous last paragraph about a "subtle spirit" correlated with light, heat, cohesion, electricity, physiology, and sensation. These matters, indeed, as Newton wrote, were "not to be explained in a few words."

Virtually overnight the *Principia* made Newton, at 44, a famous man. But in the short run, it made little difference in his reclusive life at

Cambridge. After the *Principia* Newton returned to his private meditations and alchemical fires. In 1693 he suffered a serious mental breakdown and remained incommunicado for four months. He went without sleep for days on end; he wrote paranoid letters to acquaintances, such as John Locke, whom he accused of "endeavoring to embroil me with women." Newton himself later spoke of the "distemper" that seized him at that time. He remained what we would call clinically depressed for at least 18 months. The year 1693 also marks the end of the period of creative scientific work in Newton's life.

Several factors combined to precipitate Newton's breakdown. He had exhausted himself in his alchemical program and possibly became disillusioned with it. Perhaps he had poisoned himself with heavy metals (like mercury vapor) and for a while became "mad as a hatter." A priority dispute with the German philosopher G. W. Leibniz over the invention of the calculus was beginning to heat up, and his early efforts to secure a post-*Principia* appointment in London had come to naught. Still and all, historians are loath to say directly that another reason behind Newton's disintegration in 1693 stems from the breakup of his relationship with the 25-year-old Swiss mathematician Nicolas Fatio de Duillier. The climax of their presumably chaste affair in June of 1693 proved the proximate cause of Newton's psychic collapse.

Newton's mental disorders of 1693 underscore an oft-noted affinity between creativity and madness. His salvation from the abyss came from an unexpected quarter: politics. The Glorious Revolution in England in 1688 deposed the Catholic King James II and installed the Protestants William and Mary of Orange as joint king and queen. The period prior to this revolution was tense and affected the university at Cambridge where James tried to impose Catholic Fellows contrary to university rights and privileges. Somewhat out of character, Newton stood firm publicly on this issue, and the university elected him one of its representatives to the king. His election can be explained in part by Newton's eccentricity, which served as a front for those dons who had more to lose, and doubtless also by Newton's abiding hatred of Catholicism as the institution of the Antichrist. Newton went on to represent the university at the Convention Parliament that greeted William and Mary. He had shown himself courageous politically and had sided with the winning party. He stood to be rewarded.

After some disappointment and delay, Newton succeeded in obtaining an appointment. He left Cambridge in 1696 and moved permanently to London to become Warden of the English Mint, and in 1699 he succeeded to the top post of Master of the Mint. Newton applied himself assiduously at the Mint in recoining efforts and in the grisly task of prosecuting counterfeiters. His expertise in alchemy provided the requisite knowledge in scientific matters pertaining to the Mint. Most of his income derived from his position at the Mint, which he held until his death in 1727.

Newton had spent 30 years sequestered at Cambridge, and his passage from the university to the government, from don to functionary, deserves to be highlighted. Indicative of the changing social circumstances of science in the seventeenth century, Newton's employment as a government civil servant—and notably one with technical expertise—illustrates again that contemporary universities were not active scientific centers and that new careers in science were pursued elsewhere. As the century went on individual princely patronage waned, and the new social setting for science in Newton's day saw scientists more and more institutionalized in the bureaucracies of central governments and in national scientific societies. Christopher Wren, for example, was president of the Royal Society before Newton, became the king's chief architect, and, after the fire of London, rebuilt St. Paul's Cathedral.

That Newton became president of the Royal Society of London comes as no surprise. Newton's association with the Royal Society was longstanding, but he stood for president only in 1703 after the death of his longtime enemy, Hooke. Although not without opposition, Newton was reelected annually, and he remained sitting president until his death. With one foot at the Mint and the other at the Royal Society, Newton thus stood well planted at the center of the small but growing world of official English science. (That world also included the Royal Observatory at Greenwich.) In his new role Newton played the autocrat of English science. At the Royal Society he packed the Fellowship and stacked the governing Council. He maneuvered to control the observatory at Greenwich, and his voice was decisive in scientific appointments of protégés throughout Great Britain. Not for nothing, he became Sir Isaac, knighted by Queen Anne in 1705.

Newton shamefully abused his authority and position at the Royal Society in the notorious priority dispute with Leibniz over the invention of the calculus. Both prodigies seem to have independently invented the fundamental theorem of the calculus, Newton in 1665–66 and Leibniz around 1676, but they clashed mightily over who should get the credit. Leibniz published first (in 1684) and was recognized first, but in claiming sole credit he was not candid about what he learned of Newton's mathematics on a visit to London in 1676. Newton kept his mathematical accomplishments more or less private, but a paper trail existed, and as word leaked out and as material began to be published about Newton's priority in the discovery of the calculus, Newton's minions leveled veiled charges of thievery, fraud, and plagiarism against Leibniz. In 1711 Leibniz made the mistake of appealing to the Royal Society for justice in the matter, thereby falling into Newton's hands. Newton created a kangaroo court within the Society, and he himself furiously penned the inquisitorial report, the *Commercium epistolicum*, that "reckoned Mr. Newton the first Inventor." Newton further pursued Leibniz in 1714, writing and anonymously publishing a long "Account of the *Commercium epistolicum*" in the *Philosophical Transactions* of the Royal Society. Even after Leibniz's death in 1716 Newton actively pushed his claims for sole credit. The unfortunate episode sparked some worthwhile philosophical discussion about the worldviews of Newton and Leibniz, but otherwise it proved a disaster. The priority dispute over the calculus says much about Newton's character, about simultaneous discovery in science, and about the still-nascent state of modern social norms of science, such as establishing priority and sharing credit for discoveries. Ironically, Newton's version of the calculus became a dead end, while Leibniz's notation and concepts of d (for derivative) and $\int$ (for integral) prevailed.

Coincident with his ascent to the presidency of the Royal Society, Newton published the first edition of his *Opticks* in 1704 in English. The *Opticks* ranks alongside the *Principia* as the other major work through which Newton and Newtonian science became widely known. The *Opticks* did not represent any new scientific work. Rather, the book set out completed work stretching back to 1666. The *Opticks* was more readable and accessible than the *Principia*, being experimental in approach and without complicated mathematics. A notable aspect of Newton's experimentalism in the *Opticks*, likewise differentiating it from the more deductive *Principia*, was Newton's offering "proof by experiments," by which he presented whole collections of experiments to argue about the phenomena they revealed.

Newton wrote the *Opticks* in three books. He began by summarizing the body of optical knowledge developed from antiquity to his own day. He then recapitulated in nearly 200 pages his short paper of 1672, presenting experiments concerning the refraction of light, the heterogeneity of white light, the nature of colors, colored bodies, and the rainbow, and ways to improve the telescope. In Book I and elsewhere in the *Opticks* Newton revealed a Pythagorean and Neoplatonic side, by linking the seven colors of the rainbow with the seven notes in the Western musical scale and with underlying mathematical ratios. In Book II Newton moved on to consider thin-film phenomena such as the colored patterns that arise on soap bubbles or when a lens and a plate of glass are pushed together, phenomena now known as Newton's rings. Newton had difficulty reconciling the phenomena of the rings with his particle theory of light, and he had to introduce awkward explanations to get his rays to behave as required. Also in Book II, Newton speculated on matter theory and the atomic constitution of matter, expressing the hope that microscopes may be improved to where we could see atoms but, using the language of alchemy, he feared "the more secret and noble Works of Nature within corpuscles" must remain hidden. In Book III Newton took up the newly discovered phenomenon of diffraction, or the bending of light around edges and the appearance of fringes in shadows, but he abruptly broke off discussion, saying that since he first performed the experiments he had become "interrupted and cannot now think of taking these things into farther consideration."

Rather than a grand theoretical conclusion analogous to the General Scholium of the *Principia*, Newton concluded the *Opticks* by proposing a series of general scientific research questions, his famous Queries. Newton's queries were rhetorical, in that even if they ended with a question mark they expressed his views on a variety of scientific subjects. He explicitly intended that they spark "a farther search by others," and they had that effect. In three different editions of the *Opticks* Newton posed a total of thirty-one queries. In the 1704 edition he offered sixteen queries concerning light, heat, vision, and sound. He added seven more (later numbered 25–31) for the Latin edition of 1706 dealing with his particle theory of light, the polarization of light, and an attack on the wave theory of Huygens (and Descartes). In the last two queries of this set Newton ventured a frank discussion of (al)chemical and biological change ("very conformable to the course of Nature, which seems delighted with Transmutations"), and he set forth a lengthy disquisition on chemical phenomena. In these Queries his model for interpreting phenomena consisted of "massy, hard, impenetrable" atoms endowed with attractive and repulsive powers operating over a distance, atoms and powers which he called upon to explain gravity, inertia, electricity, magnetism, capillary action, cohesion, vegetable fermentation, combustion, and many other phenomena. Over a decade later, in the second English edition of 1717, Newton inserted eight more queries between the first and second sets. In these he retreated from the idea of atoms invested with active principles, and instead deployed various superfine, self-repulsive ethers to account for optical phenomena, gravity, electricity, magnetism, heat, and physiology. The collection of queries as a whole may have lacked intellectual coherence, but it set up research traditions and provided something of a general conceptual umbrella, particularly for the less well-developed sciences mentioned.

As in the *Principia*, Newton concluded the *Opticks* on a note of natural theology, by finding God in nature and claiming that it appears "from phenomena that there is a Being incorporeal, living, intelligent [and] omnipresent." Studying nature can reveal God's providence or design in nature and "our duty towards him." For Newton, natural philosophy again leads to natural theology.

Natural theology in Newtonian science resonated with the surrounding culture in ways that need to be emphasized. In particular, the content of his natural theology fit the larger interests and designs of English religious and political moderates (Latitudinarians) at and after 1688. Newton's Latitudinarian followers extracted from his science theological and sociopolitical points concerning God's existence, his providence, the sacredness of property, and the legitimacy of social hierarchy, duty, and enlightened self-interest. Newtonian cosmology and natural philosophy, in other words, supported, indeed stood at the center of, the reigning social and political ideology in England.

Nowhere is this coincidence of science and ideology more plain than in the noted series of Boyle lectures given from 1692 "for proving the Christian Religion against Infidels." Newton's followers controlled the administration of the lectures, and the chosen lecturers were all Newtonians. Newton himself had a hand in the appointment of the first lecturer, Richard Bentley, who pontificated on *The Folly and Unreasonableness of Atheism*, and Newton actively advised Bentley over points of science and natural theology. A not-untypical offering was that by William Derham, Boyle lecturer in 1711–12: *Physico-Theology: or, A Demonstration of the Being and Attributes of God from the World of Creation*. The Boyle lectures provided an influential means of spreading the word not only about God but about the Newtonian world picture, too. The case of the Boyle lectures epitomizes the social use of scientific knowledge, but it does not exhaust Newton's great influence in extrascientific, social realms.

Newton grew old and fat. He superintended the Mint and ruled at the Royal Society. With the help of younger disciples he attended to new editions of his *Opticks* in 1717 and 1721 and the *Principia* in 1713 and 1726. In his latter years Newton turned again to theology and biblical prophecy. His heretical views had not changed in the decades since Cambridge, but he continued to hide them. (A sanitized version of his observations on prophecy saw a posthumous publication.) Late in life and through his connections to the royal English court, Newton became drawn into a controversy over systems of historical chronology, which brought the aging giant into conflict with the learned scholars at the Académie des Inscriptions et Belles-Lettres in Paris. Circulated in manuscript and published posthumously in 1728, Newton's *Chronology of Ancient Kingdoms Amended* proposed an astronomically based dating system, but the French academicians bested Newton in erudition and showed up the shortcomings of his chronology. In the last years of his life Newton became incontinent and stuck more and more to his rooms, which he had decorated in crimson. Finally true to his religious beliefs, he refused the rites of the Anglican Church and died on March 20, 1727, at the age of 84.

The Newton who died refusing the comfort of the church was not the same Newton England buried in Westminster Abbey—the heart and soul of English Anglican rule. Newton, the closet Arian, the last of the magicians, the emotionally warped and scarred human being in life, underwent a resurrection of sorts as a national hero and the model of the rational scientist. Alexander Pope immortalized this glory to England in his apotheosis:

Nature, and Nature's Laws lay hid in Night.
God said, *Let Newton be*! and All was *Light*.

Voltaire attended Newton's funeral and brought back the famous anecdote of having left France and a universe filled with the Cartesian aether to arrive in England and a universe of empty Newtonian space. By the mid-eighteenth century, aided by Voltaire and Madame du Châtelet, Newton's science conquered France and won out over Descartes among French intellectuals and scientists worldwide.

Theory and Practice

The new ideology of the seventeenth century strengthened the claim that science should be useful and applied. What, then, were the real connections between science and technology in the period of the Scientific Revolution? By and large no technological or industrial revolution paralleled the Scientific Revolution in sixteenth- and seventeenth-century Europe and, although inventions like the printing press, cannons, and the gunned ship had epoch-making impact, their development proceeded without the applications of science or natural philosophy. With the possible exception of cartography, no technological application or spinoff from science produced a significant economic, medical, or military impact in the early modern period. Overall, European science and technology remained the largely separate enterprises, intellectually and sociologically, they had been since antiquity.

Cannonry and ballistics again prove revealing. As already noted, cannon technology became perfected independently of any science or theory. Governments created artillery schools, but instruction was almost entirely practical, and all commentators agree that the science of the day was irrelevant to gunnery. Experience alone counted. The same might be said about fortification and building techniques in the seventeenth century; theory mattered little. Indeed, as we saw, Galileo's ballistic theory and published tables of range *followed* the maturation of early artillery techniques. In this case, once again reversing the usual logic of applied science, technology and engineering posed problems for scientific research and so shaped the history of science.

Cartography—the art and applied science of making maps—may well have been the first modern scientific technology. Spurred by the voyages of discovery, the printing press, and the humanist recovery of Ptolemy's *Geographia*, sixteenth-century European cartographers quickly surpassed Ptolemy and all ancient and medieval predecessors. Cartography and mathematical geography were certainly scientific in that practitioners had to know trigonometry, spherical geometry, the gnomon, cosmography, and the practical mathematics associated with geodesy, surveying, and drafting. Gerardus Mercator (1512–94) was a Flemish professor of cosmography and court cosmographer, and notably his Mercator projection can only be constructed mathematically. Early Portuguese and Spanish support for cartography has been mentioned. In France an ambitious project to map the kingdom scientifically began in 1669. Conducted by the Cassini family from the Paris Observatory and the Paris Academy, the project lapsed from time to time for want of adequate funding, but for over a century the French cartographical establishment—also in coordination with the French navy—turned out whole series of highly accurate maps of France, Europe, and the overseas colonies, many unsurpassed until the twentieth century. Accurate maps formed a nontrivial adjunct to the pursuit of trade and the government of nations and empires. Maps and mapping projects not only concerned navigation and surveying, but they also connected with resource identification and economic development. In these ways cartography as an applied science made a difference in the development and expansion of early modern Europe and indicated what might be forthcoming from state support of science.

Other efforts to exploit or somehow apply science were less fruitful than the case of cartography. The practical problem of finding one's location at sea—the famous problem of longitude—illustrates the more characteristic distance between the promise of early modern science and its practical payoff in technology. For nearly 300 years European maritime activity proceeded in the face of this unsolved navigational problem. In the Northern hemisphere the determination of *latitude* is relatively simple: it requires only the measurement of the angle to the North Star, a procedure that presents no great problem even on a moving ship. But the determination of *longitude* was an intractable problem for ships' captains, trading houses, and Atlantic maritime nations that had to rely on fallible experience and navigators' reports for east-west voyages. Already in 1598 Philip III of Spain offered a monetary prize for a practical solution to the problem of longitude; the Dutch republic put up 25,000 florins in 1626; the Royal Observatory at Greenwich in England was created with the specific charge of "perfecting navigation"; in 1714—propelled by merchant interests—the specially created British Board of Longitude offered the extraordinary sum of £20,000 for the solution of the longitude problem; and in 1716 the French government responded with a substantial prize of its own.

In principle, contemporary astronomy offered several means of solving the problem. The key to the solution lay in determining the differences in time between a known locale such as Greenwich or Paris and the unknown locale. In 1612 Galileo himself proposed that his newly discovered satellites of Jupiter could function as a celestial clock against which the required times could be measured, and he worked (unsuccessfully) to develop practical procedures for the Spanish government. In 1668 the French-Italian astronomer J. D. Cassini published tables of Jupiter's satellites for longitude observation purposes, and other such tables followed in the eighteenth century. Astronomers also tried simultaneous lunar observations. Successful for determining land-based longitude, lunar observations were impractical for observations at sea.

The ultimate solution to the problem of longitude finally came not from the world of science but from the world of the crafts—the improved technologies of clockmaking. In the early 1760s the English clockmaker John Harrison perfected a chronometer (a time-measuring instrument) which employed counteracting balances to compensate for the pitching and rolling of a ship and thermocouples to compensate for temperature variations. With Harrison's chronometer sailors could carry Greenwich time on their voyages, and by comparing it with easily ascertained local time they could at last determine their longitude at sea. After considerable wrangling, the clockmaker finally received the prize from the Board of Longitude. It must be said, though, that Harrison shared the prize with the German astronomer Tobias Mayer for his astronomically derived lunar tables which at least in principle could serve the same purpose. Science split the prize with technology.

Practical efforts undertaken in the early Royal Society likewise typify the general failure of practical science in the seventeenth century. The early Royal Society expressed its Baconian commitments through committees it established to pursue useful inquiries into navigation, shipbuilding, reforestation, and the history of trades, but very little came from this extended collective effort. In response to a request by the Royal Navy, for example, the Royal Society conducted experiments on the strength of beams. Fellows tested specimens of various types of wood and different cross-sectional shapes and arrived at a conclusion at odds with Galileo's theoretical determination. A decade later one of them, William Petty, recognized their error and described how Galileo's result could be used by builders to size beams, but here again practicing engineers, in possession of reliable rules of thumb, had no need of Petty's reassurance.

The case of scientific instruments in general and the telescope in particular provides a minor but revealing example of more complex interactions possible between science and technology which began to emerge in the seventeenth century. Scientific instruments, especially the telescope and the microscope, became more common in seventeenth-century research, and they demonstrate once again the historical impact of technology on contemporary science. The development of the telescope, however, shows a further pattern of sequential and reciprocal interactions over time, back and forth, between scientific theory and technological practice. The first telescopes were crafted entirely without the benefit of any theory, Galileo's claims to the contrary notwithstanding. Once the telescope existed, new discoveries were made in optics, notably chromatic and spherical aberration, distortions introduced in the passage of light through lenses. These discoveries in turn posed practical issues for improving telescopes and theoretical ones for the science of optics in accounting for the phenomena. On the practical front scientists and lens grinders responded by trying to grind nonspherical lenses. Newton in his 1672 paper on light provided a theoretical explanation for chromatic aberration and, based on this discovery, he went on to propose a remedy with his reflecting telescope. Ultimately, the solution to the problem of chromatic aberration—the creation of compound lenses with compensating indices of refraction—emerged only after the 1730s and came from the world of technology and the glassmaking craft. All the while, of course, European astronomers made groundbreaking astronomical discoveries with technologically improved telescopes, a reaffirmation of the principle that science often follows, rather than leads, technology.

By the same token, as notable as the telescope case was—and it certainly was for optics and astronomy—in general science and technology did not interact strongly in the era of the Scientific Revolution. Where scientific insights had (at least potentially) practical import, natural philosophers and theoreticians were ignored in favor of engineers, builders, architects, craftsmen, and others with practical empirical experience. Indeed, contemporary technology seems to have had a greater effect on science than the other way around, and one needs to be wary of concluding that the modern alliance of science and technology appeared with the Scientific Revolution. Not until the nineteenth century did ideological promise bear substantial fruit.

Matter and Methods

Competing visions of science and nature jostled among different groups of European intellectuals in the sixteenth and seventeenth centuries: old-fashioned Aristotelianism, various occult traditions, and the "new science," also known as the mechanical or experimental philosophy. The new science reintroduced doctrines, dormant since antiquity, of atoms and corpuscles, and it offered mechanical explanations for nature's actions in terms of force, matter, and motion. The new natural philosophy tended to sustain scientific claims through experiments and, unlike occult learning, it espoused the open communication of knowledge. By 1700, the new science—especially in its new institutional settings—had won out over alternatives whose validity was unquestioned a century earlier. Such "open" systems of knowledge proved of broader appeal to princes and states, and that fact may help account for the relative decline of "closed," magical, or secret doctrines which could possibly be politically dangerous.

Proponents of the new science defended atomism and the mechanical philosophy aggressively against charges of atheism and determinism. Descartes handled these issues with his well-known dualism between matter and spirit that preserved all the formalisms of official religion. We have seen how natural theology formed so important a part of Newtonian science. As a standard operating procedure, the new academies and societies refused all discussion of politics and religion. Scientific research was directed at nature, not society or divinity.

The seventeenth century witnessed the rise and spread of experimental science. Gilbert played with magnets; Galileo rolled balls down inclined planes; Torricelli toyed with tubes of mercury to arrive at the principle of air pressure; Pascal sent a barometer to the top of a mountain to confirm the conjecture that the atmosphere forms a great sea of air; William Harvey dissected and vivisected countless animals in his quest to understand the heart; Newton ran beams of light through prisms and lenses. A certain degree of playfulness runs through the work of experimental sciences of this era. Although seventeenth-century scientists experimented, they used a variety of approaches and they did not agree on an "experimental method" or the precise role of experiment in generating new knowledge. In particular, what is usually thought of as the hypothetico-deductive method was not historically prominent in seventeenth-century science. That is, one finds little of the procedure later adduced by philosophers of science of formally posing a hypothesis, testing it through experiment, and accepting or rejecting the initial hypothesis based on the results of experiment. Galileo, as we saw, utilized experiments to confirm points already established; Newton used experiment one way in his "crucial experiment" of the 1672 paper and in another way in his "proof by experiment" in his *Opticks*. Harvey undertook systematic experimentation in his dissections, but only as an old-style Aristotelian concerned with "animality" and not to demonstrate the circulation of the blood. "Thought experiments"— experiments impossible to carry out in practice—likewise formed part of seventeenth-century science; consider, for example, Galileo's proposal to view the earth from the moon. Given these different approaches to experiment, including the hypothetico-deductive, experiment has to be seen not as a fixed entity or technique, but as historically contingent: various approaches to experiment and experimental science emerged, and from heterogeneous roots.

Alchemy doubtless formed one route to modern experimental science. So, too, did handbooks of "secrets" that mushroomed along with printing. These sixteenth- and seventeenth-century "how-to" texts contained practical empirical formulas and recipes for making inks, paints, dyes, alloys, wine, perfumes, soaps, and the like. In their Academy of Secrets Della Porta and friends did not undertake experiments to test scientific hypotheses, but to verify the effectiveness of the practical "secrets" of which they had heard. The Renaissance model of experimentation has been likened to a noble "hunt" for secrets of nature with successful hunters sometimes receiving big rewards from patrons.

As a component of his inductive program, Francis Bacon argued for a multifaceted role for experiments. In the first instance he intended that they produce new phenomena artificially, to "twist the lion's tail," as the expression has it. Such experiments were not to test anything, but to add facts and examples for later inductive analysis. A classic example of a Baconian experiment is Boyle's sustained observations of a rotting piece of meat that glowed. So, too, is the story of Bacon himself stuffing a chicken with snow to observe the outcome. (He died from the resulting chill he suffered.) But Bacon also envisioned a role, something like theory testing, for experiments conducted at a higher level in the inductive process, "experiments of light," experiments designed to question nature more specifically and in the manner of an experimental test as we commonly envision it. Then, Bacon also combined "experiments of light" with practical applied research in his "experiments of fruit."

Descartes, on the other hand, rejected Baconian induction and Bacon's approach to experiment. Descartes's deductive method of philosophizing from first, mechanical principles led him to downplay the role of experiment altogether. Instead he wanted to deduce effects from causes and not to derive causes experimentally through effects. The masses of experiments performed by others confused rather than clarified matters for Descartes, although he did allow a restricted role for experiments at an advanced stage of deduction as a means of testing not a single theory but between and among plausible theoretical alternatives.

As experimental science matured and developed, experiment came to be used as a refined tool to test theories or hypotheses and advance scientific discourse. Newton's "crucial experiment" proving light to be composed of rays of differing degrees of refraction had this character. Robert Hooke arrived at the generalization that stress and strain are proportional—Hooke's Law—by tests conducted with springs. Robert Boyle is also widely credited for promoting experimental science as a powerful new technique. With increasing, if diverse, attention given to experiment in the seventeenth century, the scientific enterprise naturally became more instrument based and technology dependent. Generations of amateur and professional inquirers into nature applied themselves with telescopes, microscopes, thermometers, beakers, scales, barometers, clocks, inclined planes, prisms, lenses, mirrors, and, later, static electricity machines. (The increasing demand for instruments also provided employment for a growing number of expert instrument makers.) Some of these technological implements extended the range of the senses, and instruments became essential tools for the production of knowledge in a number of arenas of science.

To conclude with one prominent example, Boyle, assisted by Hooke, invented the air pump or vacuum pump in 1658–59. With an evacuated receiver providing a new "experimental space," Boyle's pneumatic pump added a spectacular new instrument to the armamentarium of seventeenth-century experimental science. Boyle used the apparatus to

Fig. 13.4. The air pump. This invention led to many experiments in which objects were observed under conditions of a near vacuum. The realities of such experiments were more complex and difficult than apparent in the reported results.

investigate the elasticity or "spring of the air," and he came to discover the law correlating pressure and volume of confined gases. Textbook accounts of this discovery make it seem misleadingly simple and the straightforward product of the experimental apparatus. Newer research has revealed the social and instrumental complexities of producing new facts about nature through the use of instruments and experiments. In the decade after Boyle's invention, for example, a mere handful of costly pumps existed in just a few locales. They all leaked and were difficult to maintain in working order. Replicating experiments and effects proved elusive. Only a practiced operator could demonstrate claims, and claims were regularly contested. A demonstration setting like that of the Royal Society became a public yet restricted forum where effects were tried and where social standing and prestige mattered a great deal. Word of experiments spread through written reports of trustworthy observers allowing others to "witness" the experiment at the removes of time and space. The intricate ways in which these factors played themselves out in the discovery of Boyle's Law embodied subtle social conventions for establishing matters of fact, and the case underscores the fundamentally social nature of the process that is the production of new knowledge.

Science and the Enlightenment

The Scientific Revolution had a profound impact on larger social, cultural, and intellectual contexts in contemporary Europe and America. As the seventeenth century rolled into the eighteenth, the Western world woke up to what Newton, Descartes, and their predecessors had accomplished. Not only had these giants given us a new world to live in, they had given us new ways to know about that world, ones that stood in stark contrast to traditional sources of knowledge and authority. At least in retrospect, contemporaries saw in Newton and the heroes of the Scientific Revolution a model of rationality and knowledge-making that validated human reason and that could be extended to a range of subjects beyond the world of science and natural philosophy. In other words, the Enlightenment of the eighteenth century originated as a game-changing response to the Scientific Revolution and an extension to the human and social sciences of what the natural sciences had so impressively achieved the century before.

Along with Bacon and Descartes, Newton may be fairly called a founding father of the Enlightenment. Indeed, the first Enlightenment document may well be the ode penned by Edmond Halley to preface the first edition of Newton's *Principia* in 1687, wherein Halley wrote of understanding the world: "In reason's light, the clouds of ignorance/ Dispelled at last by science." Newton's influence on his slightly older contemporary, John Locke (1632–1704), is especially to be noted. Locke, the influential English philosopher and author of the foundational Enlightenment texts *Two Treatises of Government* (1689) and *Essay Concerning Human Understanding* (1690), made a point of meeting Newton. He took Newton and the Newtonian worldview to heart, after satisfying himself through the great Cartesian physicist Christian Huygens that the mathematics in the *Principia* was solid. Locke even prefaced his *Essay Concerning Human Understanding* with an encomium to "the incomparable Mr. Newton" as one of the "master-builders, whose mighty designs, in advancing the sciences, will leave everlasting monuments to the admiration of posterity." Locke, like Newton, shaped English natural theology and the Enlightenment campaign of reason that followed.

The Enlightenment was a largely secular movement. Activists railed against what they perceived as superstition, obscurantism, and irrationality especially as embodied in the Catholic Church and traditional social divisions. In the Enlightenment, science and rationality became new sources of authority previously monopolized by tradition and religion. The Enlightenment spread from Great Britain and France across eighteenth-century Europe and America. Participants formed the cosmopolitan commonwealth of intellectuals known as the Republic of Letters.

Inspired by an idealized Newton, Newton's clockwork God, and the success of the Scientific Revolution generally, influential Enlightenment *philosophes* and thinkers sought to extend to the social and human sciences the progress exemplified in the natural sciences in the previous century. Locke, Voltaire, Montesquieu, David Hume, Adam Smith, Immanuel Kant, and others wrote critically in a new vein and with lasting impact about politics and government, religion, law and justice, the economy, psychology, epistemology, and moral theory. As Kant expressed it in 1784, the Enlightenment was humanity's coming of age.

A key tenet of Enlightenment thought is the idea of progress—the notion that humanity is moving toward a better tomorrow. This belief, too, was one inspired by the Scientific Revolution. To contemporaries, men of science like Descartes and Newton exemplified what human intelligence might achieve, and they became exemplars who inspired the possibility of progress in others. In his *Sketch of the Progress of the Human Mind* (1794), for example, the late Enlightenment *philosophe*, mathematician, and man of science, the marquis de Condorcet (1743–94), traced humankind's technical and intellectual ascendancy in progressive steps from primitive, even brutish origins through the invention of agriculture, writing, the printing press, and on "to the Period when the Sciences and Philosophy threw off the Yoke of Authority." Humankind had demonstrably progressed, and Condorcet's Tenth Epoch, "Future Progress of Mankind," prognosticated the end of slavery, increasing equality among peoples and nations, greater liberty and freedom, economic fairness, moral improvement, universal education, better health, increased longevity, cheaper goods, the perfection of language, gender and racial equality, the repudiation of war, more harmonious societies, continued advances in the sciences, a boundless future for human understanding, and ultimate human perfectibility.

The forces associated with Newtonian science and the European Enlightenment were liberal, universalist, progressive, reformist, and even revolutionary, and they continue to echo today. Enlightenment thought underlies the American Revolution of 1776 and the political philosophy embodied in the 1787 Constitution of the United States. The French Revolution of 1789 is unthinkable without the Enlightenment and the background of the Scientific Revolution that preceded it. As evidenced in the Declaration of Independence, for example, with its proposition that "all men are created equal," the political realm can be represented as a Newtonian system of politically equal citizen-atoms moving in law-like patterns under the influence of a universal political gravity and a democratic impulse toward civic association.

The sciences enjoyed great prestige. Co-editors Denis Diderot and Jean le Rond d'Alembert enshrined science and rationality in their great and influential *Encyclopédie* (26 volumes, 1751–72). The English captain James Cook undertook daring voyages to the Pacific, stopping on behalf of the Royal Society of London to observe the Transit of Venus of 1769 from Tahiti. The French explorer La Perouse similarly sailed the Pacific in two floating scientific laboratories, the *Astrolabe* and the *Compass*, and an anxious public back in Europe awaited news of the doomed expedition. But Cook and La Perouse were merely the best known as European scientists fanned out across the globe in the eighteenth century to the applause and attention of those at home. Benjamin Franklin was a hero domestically and abroad, not least because of fame garnered as a scientific investigator of electricity, and Franklin was welcomed at the Academy of Sciences in Paris and elsewhere around the world. The first Montgolfier lighter-than-air balloon lifted off from Annonay in France in 1783, and soon these testimonies to human ingenuity rose everywhere across Europe and America to the applause of the witnessing multitudes. Science and humankind had achieved the conquest of the air. Science had its critics, but at no previous point in history had the scientific enterprise been as well institutionalized nor so well received by a larger public socially.

The enterprise of science itself developed after the Scientific Revolution with a post-revolutionary Newtonian worldview and a host of research questions that occupied inquirers. As a social entity, science found itself with a new institutional base in the learned scientific societies and related technical organizations of the day (such as navies), most directly or indirectly supported by governments. For most individuals involved in contemporary science, the Hellenistic model of state-supported experts in institutions that encompassed pure *and* applied research prevailed; other individuals inclined toward science and natural philosophy managed a more Hellenic style of self-supported disinterested inquiry. As the Enlightenment movement indicates, science had shown itself to be powerfully influential in the wider world. Then, too, the ideology of the utility of scientific knowledge had at long last been articulated by Bacon and Descartes, and that transformative notion diffused through levels of European societies. Although not without being contested by traditionalists and fringe elements, science occupied a much more public and respected place in the eighteenth century than ever before in history.

As well ensconced as science was socially, something major contemporary science did *not* do in the century following the Scientific Revolution was to transform the wider world of technology in any major way. Traditional technologies and craft practices employed in traditional settings by traditional artisans continued as the order of the day, well away from the more-refined urban world of the scientists and their sponsors. The Industrial Revolution that unfolded in the eighteenth and nineteen centuries owed its origins not to science or the Scientific Revolution directly, but to the wider world of the crafts and the particularities of the economy and society of eighteenth-century England. Part IV takes up the Industrial Revolution, the coming of industrial civilization, and the story of how science ultimately became connected to technology as today's applied science.

Part IV

Science, Technology, and Industrial Civilization

Back in the mists of prehistory, two great technological revolutions utterly transformed human existence: the Neolithic and the Urban Bronze Age revolutions—the transitions from food-collecting to food-producing, and then to complex societies that began some 12,000 and 6,000 years ago, respectively. A third great technological revolution—the Industrial Revolution and the coming into being of industrial civilization—has unfolded only within the last three hundred or so years, a comparative blink of the eye. The result has been yet another world-historical transformation, as industrialization has irreversibly altered the course of history and fundamentally refashioned human societies and ways of living. The social and economic consequences of industrialization on a world scale have been more far-reaching than the earlier Neolithic and Urban revolutions.

The Industrial Revolution was a seismic technological and socio-cultural transformation that began to rumble quietly in England in the eighteenth century. In essence, industrialization entailed a shift away from agriculture as the primary object of human labor and the main means for the creation of wealth to the mechanization of production of goods in factories. As a result, a new mode of human existence began to take shape, industrial civilization, and with industrial civilization the human species entered a new historical era. Today, we grow our food, raise our children, organize our economies, and plan and carry out our activities as individuals and as societies in ways that are vastly different from life prior to the onset of the Industrial Revolution in eighteenth-century England. An imposing list of these differences is not hard to compile: Production became mechanized and powered by engines burning fossil fuels; industrialization revolutionized agriculture in many regions; a wholesale restructuring of societies brought vast new wealth and entire new classes of owners, workers, managers, political leaders, and consumers. Populations skyrocketed. Interlocking technological systems of great complexity fused in the technological supersystem that undergirds industrial civilization today. Industrialization opened the door to a brave new world we have yet to fully grasp or master.

Paralleling developments in the realm of technology, dramatic intellectual and social changes unfolded in the world of science and natural philosophy since the seventeenth century. From Newton to Einstein and beyond, from Darwin to DNA and beyond, the roster of great scientists and great scientific accomplishments has grown exponentially, and we know the world in different and vastly more sophisticated ways than ever before. In Newton's day, the scientific enterprise was organizationally not that different from what it had been in ancient Alexandria. Today, supported to unprecedented levels by governments and by new players on the historical stage—multinational corporations and high-tech industry—the enterprise of science has moved from the periphery of society to the very heart of the economies and priorities of industrial civilization today.

As part of this dramatic socio-technological revolution, science and technology have forged new connections, and their full merger in the modern era represents another defining element of contemporary industrial civilization. We have argued that historically science and technology were overwhelmingly separate enterprises through the eighteenth century. The great novelty is the historical marriage of science and industry in the nineteenth century, as somewhat slowly and grudgingly governments and an increasing number of industries came to recognize the full possibilities of applying research in science to technology and industry. In the twentieth and now twenty-first century government and industry support for pure and applied research has only accelerated. Led by key science-based industries, the result has been a dramatic expansion in the applications of science to technology. In short, from their separate origins and historically occasional contact, thinking and toolmaking—science and technology—combined to give us the world we know today.

Industrialization brought a potent combination of new technologies in power production, transportation, military hardware, communication, and entertainment that spread tentacles across the globe, that replaced many traditional technologies, and that effaced many of the barriers separating nations and peoples of the world. Modern science, particularly in the twentieth century and increasingly in the twenty-first, switched from being a narrowly European and "Western" institution to becoming a dynamic and defining element of world culture. Unlike other cultural traditions, science today is fully ecumenical, a treasured part of the patrimony of world civilization, and high testimony to the achievement of humankind as a whole. In these ways science and technology have played essential roles in the historical transformations that go under the rubrics of modernization and globalization.

Part IV pursues these themes and follows the evolution of the new world order wrought by industrialization. We use the term "industrial civilization" to characterize the new mode of existence that arose coincident with the underlying socio-technical revolution. While it is worthwhile to discriminate episodes in the history of industrialization, it is nonetheless essential to conceive of industrialization as a single world-historical phenomenon that originated in the eighteenth century, that gained extraordinary momentum, and that continues to transform the world in unexpected ways. To think of this revolution otherwise misses its essential unity and the grand historical sweep that puts it alone in a class with the Neolithic and Urban Bronze Age revolutions of earlier eras. Because we are still so close to it and because we experience its dynamism in our daily lives, industrialization might seem more disjointed than it does when one takes in the long sweep of human history, and from this point of view how industrial civilization will play itself out is the fundamental question facing humanity today. This wide-angle view is essential if we are to understand and deal rationally with the consequences of industrialization that we presently face.

Textiles, Timber, Coal, and Steam

And was Jerusalem builded here,
Among these dark Satanic Mills?

—William Blake (1804)

The Industrial Revolution and the advent of industrial civilization began in England in the eighteenth century. But industrialization did not suddenly commence then. Far from it; and the world in 1750 was decidedly not a world without industrial-scale production or machines large and small. The contemporary international wool trade and cloth manufacture, for example, were major and technologically intricate economic activities. Shipyards and arsenals were bustling hubs of technological industry. Mining and quarry operations across Europe were industrial scale. Humanity's acquired set of technologies had multiplied in manifold and ingenious ways since the first civilizations and the invention of the wheel. As advanced and technologically sophisticated as contemporary peoples had become, however, their worlds can only be characterized as traditional in almost every way.

At the beginning of the eighteenth century, as the Scientific Revolution receded into history, Europe remained a continent filled with agrarian societies. The bulk of the population, more than 90 percent, lived in rural settings and engaged directly in animal husbandry and agricultural activities. Of the urban dwellers, few worked in what we would consider factories. As robust and important as certain sectors had become, the overall scope of industrial-scale manufacture remained limited. Manufactured goods were for the most part the products of what is known as the domestic or cottage or putting-out system of production that was ideally suited to rural and farm communities. This form of production was decentralized and household-based where merchants would farm out raw materials to workers who would work, usually part time at home, using their own equipment. Local artisans such as blacksmiths or masons provided necessary services everywhere, and in cities skilled craftsmen in artisanal shops fashioned other goods needed by society. Wind, flowing water, and the labor of animals powered these traditional societies. Wood fueled the fires of the kitchen and the forge.

But then, here and there and in small ways hardly noticeable at first, one thing led to another and something radically new began to emerge. That something was the epoch-making socio-technological transformation we know as the Industrial Revolution that took off in eighteenth-century England and gave birth to a whole new mode of human existence, industrial civilization. The coming of industrial civilization entailed a demographic shift away from traditional agriculture and trade to the mechanization of production and the elaboration of the centralized factory system. Previously unheard of machines tapped different sources of power, notably steam power. A new class of workers emerged to labor in factories. Novel forms of transportation and modern financial institutions arose alongside and in support of these developments. A wholesale reorganization of societies and cultures ensued that set humanity on a fresh and unforeseen historical path. Iron, coal, and steam power became the emblematic resources.

Today, the transformation wrought by industrialization is familiar to us in everything that surrounds us: our infrastructure, our technologies, our way of living. Cars, airplanes, smart phones, our cities, dental floss, the list goes on and on. Nothing is as it was before the Industrial Revolution.

The changes wrought by the Industrial Revolution are of a scale and magnitude not seen since the Neolithic Revolution 12,000 years ago when humans first turned from foraging to food-producing, or the great Urban Revolution that occurred with the rise of cities and fully civilized life in the pristine societies at the dawn of written history 5,000 years ago. Largely as the result of the Industrial Revolution, the technical, economic, political, and social bases of life have become transformed virtually everywhere over the last 250 years. The point applies not only to strictly industrial societies, but also to traditional agrarian societies, remaining groups of pastoral nomads, and surviving hunter-gatherers—all humanity has been affected by the arrival of industrial civilization. The scale of this transformation from traditional agrarian societies to intensified industrial production that characterizes industrial civilization today is profound and world-historic. We need to explain what happened and why.

The Mill

We first see the mechanization of production in the textile industry in England in the eighteenth century. Mechanization in this modern sense means having machines make a product as opposed to machines being

Fig. 14.1. The textile
mill. This idealized en-
graving from 1819 shows
the stages of "carding,
drawing, and roving" that
prepared cotton for spin-
ning. Mechanical power
to run the machines
depicted is delivered overhead by rotating shafts;
these are driven by steam
engines (not shown).
Light comes in on the
right from the outside.
Women and younger girls
tend the machines. The
Industrial Revolution
and the factory brought
a whole new way of life
and a whole new way
of working compared to
traditional rural life.

tools that humans use to produce this or that. With industrialization
machines *are* the means of production, which more than suggests that
with mechanization human workers become auxiliaries who tend machines. (See figure 14.1.) Textile manufacture in England was the first
industry to become mechanized in this way, and the early textile mill
exemplifies all that mechanization and the factory system of production came to entail.

Making cloth is intrinsically a multistage process. Raw materials
(wool, cotton, flax) have to be gathered, cleaned, and otherwise processed in preparation for spinning. (Making silk is only somewhat different.) Thread has to be spun, and cloth then woven, not to mention
further processes of dying and finishing. The multistage production of
textiles represents a case where bottlenecks in production drove technical innovation and ultimately the mechanization of the entire industry.
Alternating technical innovations in spinning and weaving machinery propelled development. In 1733 John Kay, initially a clockmaker,
invented the "flying shuttle," which improved weaving but thereby
caused a lag in the spinning of thread. A combination of technical
developments elaborated by a series of artisans and engineers in the
1760s and 1770s then mechanized spinning in the spinning jenny and
the "mule." These improvements left weaving as the constriction point
in the production process, an imbalance intensified by the invention
of cylinder carding in 1775, which made spinning even more efficient.
The mechanically powered loom appeared in 1785, and with weaving

now mechanized, textile production became a process where machines made cloth from start to finish.

A series of power-driven textile mills created by Richard Arkwright in the 1770s and 1780s, employing hundreds of workers, represents the appearance of the first factories in the modern sense. These factories were powered initially by water and waterwheels. Water power (and wind power) had been used in Europe since the Middle Ages in milling grain, for example, or forging. But technical improvements in hydraulic engineering and waterwheel design provided the new factories with increased power. Early manufacturing sites such as Lowell, Massachusetts, or Paterson, New Jersey, in the United States show what could be done by tapping the mechanical energy of flowing and falling water. But soon the new steam engine became the dominant independent motor supplying power to manufactories, and the steam engine freed mills from having to locate on or near watercourses. In 1813, 2,400 power looms operated in Britain; by 1833, the number had skyrocketed to 100,000. (The power loom replaced not only the hand loom but also hand-loom weavers.) Not unexpectedly, between 1764 and 1812 worker productivity in the cotton industry increased by a factor of 200. The mechanization and industrialization of textile production marked the arrival of industrial civilization in England.

A new mode of production came into being, and more than one feature of the new factory system stands out beyond the fact that powered machines were doing the actual manufacturing. Factories arose in cities, not in London, but in new, crowded, and smoky ones like Manchester and Birmingham. Production became centralized and was no longer farmed out. By dint of that, raw materials had to be transported to factories and finished goods shipped to markets, and new and improved transportation systems arose to accommodate these needs, notably first with shipping canals and then railroads. England was becoming the bustling workshop of the world, and the need to transport bulky goods was rapidly increasing. Despite contemporary efforts to build improved "turnpikes," animal-drawn wagons on the primitive road system proved inadequate, especially for moving large quantities of coal on inland routes. At first, the solution was river and canal transportation, and in 1757 and 1764 the first two canals were constructed to link coal fields with Manchester via the River Mersey. Canal mileage and the associated number of locks and crossings increased dramatically thereafter, and for several decades canals remained the primary means of transporting raw materials and finished goods. The railroad did not begin to have an impact until the 1830s.

A new work force had to be called into being to tend the machines. An industrial, urban-based labor force formed a new working class that rivaled the traditional rural peasantry. This novel laboring class and army of wage earning industrial workers came to be called the proletariat to distinguish it from the owning-class and those who profited from the new industrial system. Wage labor and a money economy replaced traditional exchanges of goods and services. The factory created a new relationship between workers and the goods they produced; it imposed an unprecedented alienation from the products of their labor, meaning that they had no control over the goods they produced, much less being paid enough to afford to buy them. But the factory produced alienation from home and family life also. Bosses were a novelty, and life inside of factories involved rigid hierarchies of supervisors governing workers. And the clock, the medieval device to tell time, became the industrial master that governed the workplace. Working at the mill was worlds away from the outdoor life of the field and the plow. The formal assembly line would come only much later, but factory labor was indoor drudgery governed by the clock, and workers were not free to labor as they might as more independent tillers of the soil. The mill demanded punctuality, sobriety, and reliability, values that had to be instilled. Especially in its early phases in England, the factory system entailed a severe exploitation of labor. Two-thirds of Arkwright's 1,150 workers in 1789, for example, were children. The Parliamentary Act of 1799 made it illegal to unionize and prescribed a three-month prison sentence for anyone seeking to ameliorate working conditions by means of organized action. The Bill of 1825 recognized worker "combinations" but severely restricted union activity; comparable restrictions on business organizations and price-fixing were not enforced. The doleful effects on labor are one of the traditional themes in the study of the Industrial Revolution. Quite apart from the political and moral issues, the exploitation of labor formed a systemic component of industrialization. During the second quarter of the nineteenth century, for example, as cotton cloth production quadrupled and profits doubled, wages remained almost unchanged.

Industrialization unleashed processes of fundamental social change as well as technological innovation and economic growth. People migrated from the countryside to cities, expanding urban populations of low-paid factory workers; factory labor increased and class conflict intensified; new coercive institutions such as public schools and well-regulated prisons came into being as agents of social control; the family ceased to be a center of production, and a new division of labor took hold—typically men secured employment in dreary factories while women were mainly restricted to domestic duties. A further demographic upheaval accompanied industrialization as mass migration saw millions of Europeans head westward across the Atlantic and eastward into the expanding Russian empire. The process of industrialization surged and became a global tidal wave that continues to bring change to every corner of the world, often with unsettling results.

Observers early on began to sense this change in history's direction, not always approvingly. In 1804 the English poet William Blake (1757–1827) lashed out against the "dark Satanic Mills" that had begun to dot England, and we can readily call up images of grimy factories belching smoke and soot, of ceaseless machines tended for long hours in dangerous conditions by downtrodden and underpaid women, girls, and scraggly boys. It's not hard to imagine the cramped and dirty neighborhoods surrounding factories. As Charles Dickens would later, Blake deplored the new circumstances he witnessed in England, and he condemned the disenchantment of the world brought on by industrialization and, in his mind, by Newton and science!

Ecological Stimulus, Technological Response

How can we explain this remarkable transformation? The history of industry in Europe since the rise of European civilization in the tenth century is the history of long-term economic development and technological innovation against a background of environmental constraints and pressures. Indeed, the industrialization of Europe generally marched in step with a growth of population that produced a constant threat of scarcity. From a low of 2 million in the middle of the fifteenth century, after a hundred years of repeated ravages of the Black Death, the population of England and Wales rose to about 5.5 million by the end of the seventeenth century and, with increasing rapidity, to 9 million by the end of the eighteenth. (This increase resulted from a lowering of mortality rates, probably through improved hygiene, and from changing agricultural practices.) As the population quintupled during a span of 350 years, pressure on resources increased and in some cases became severe.

Perhaps the most serious shortage that developed was in land itself. In England, in many ways a typical agrarian society, land was put to many uses—as cropland, as pasture for cattle, horses, and sheep, as forest for timber, and, increasingly, for expanding towns and cities as the burgeoning population sought nonagricultural means of subsistence in urban centers. During the sixteenth and early seventeenth centuries, English towns, trade, and industry had been growing, but by the mid-seventeenth century growth began to be checked as critical bottlenecks formed. Because economic life is a process of interlocked activities, a shortage or restriction in one area can disrupt the entire system. During the eighteenth century several of these constraints were successively broken by technological innovations, and the British economy began to grow again at an unprecedented, rapid pace. The textile industry, iron-making, mining, and transportation were all improved by the application of new techniques. Although most improvements were incremental, many of the innovations were radical and ingenious departures from traditional methods. The main effect of these technological novelties, however, was to expand the economy in the face of a population that was outpacing limited resources and inadequate methods. Behind the process of industrialization a growing population pressed against its economic and ecological limits.

Indicative of these pressures and constraints, the emergence of a set of new farming techniques known as the Norfolk system provided the necessary agricultural surplus to support the coming of industrialization to England. The new practices replaced the medieval three-field system with a new four-field system of crop rotation; the added production of turnips and clover permitted the over-wintering of greater numbers of cattle, with subsequent and significant rises in meat production and agricultural production generally. The Norfolk system succeeded in part by enclosing public land (the Commons) and making it subject to cultivation and private ownership. The enclosure movement increased agricultural productivity, but it also rendered landless sizable numbers of marginal villagers and farmers who were then "freed" for an industrial labor pool.

The English "timber famine" may also serve as an instructive case study of ecological and economic tensions that propelled change in eighteenth-century England. The British Isles were never heavily endowed with forest, and with the advent of Neolithic farming thousands of years earlier timber reserves became further depleted by the expansion of cropland and pasturage. In the early modern era military and naval requirements along with the beginnings of industrial intensification placed increasing strains on a dwindling supply of timber. Shipbuilding, for example, a major industry in a maritime nation, consumed vast quantities of timber. By the beginning of the eighteenth century construction of a large man-of-war devoured 4,000 oak trees. And just prior to the American War of Independence one-third of the British merchant marine had to be built in the American colonies where timber remained plentiful. The smelting of iron ore, another major industry, depleted whole forests, with each furnace annually consuming the equivalent of four square kilometers of woodlands. And, like the smelting and refining of iron, the making of bread, beer, and glass likewise depended on wood as fuel in the form of charcoal (charred wood). In none of these production processes could coal be substituted since, using contemporary techniques, the fuel or its fumes came into direct contact with the product, which would then be ruined by coal's impurities, notably sulfur. Also, in the heating and lighting of buildings wood was preferred as fuel since coal fires produced noxious fumes. As the scarcity of timber spread, its price inevitably rose. From 1500 to 1700 while general prices rose fivefold in England the price of firewood rose tenfold. As a result of this energy crisis, by the beginning of the eighteenth century British iron production actually declined, owing to a shortage of fuel. Under these conditions the incentive to conserve or find substitutes for timber was a response to a threatening decline in the standard of living. It sprang from a problem whose roots lay in a wasting ecological imbalance, intensified by population growth and the conversion of woodland to other uses.

The increasing scarcity of wood created a bottleneck in several industries. While iron-making consumed extravagant quantities of timber, it was an industry that seemed to lend itself to the substitution of plentiful coal for scarce wood. During the seventeenth century many attempts were made to smelt iron ore using coal as fuel, but all of them proved unsuccessful. In processes like cooking, for example, where the product could be separated from the fuel by the use of pots the substitution of coal for wood caused no problems. But in the traditional method of smelting iron the fuel and the ore had to be physically mixed for them to react chemically. In the eleventh century Chinese ironmasters had already developed methods of smelting that employed coal instead of wood as fuel, but in Europe a comparable development did not take place until the eighteenth century. In 1709 Abraham Darby, a Quaker ironmaster, succeeded in using coke (charred coal) instead of charcoal in the blast furnace, although not until mid-century did the new process come into common use.

Darby arrived at his discovery strictly through tinkering with contemporary methods. Neither scientific theory nor organized or institutionalized science played any role in the process. Applicable theoretical principles of metallurgy had not come into being, and even "carbon" and "oxygen" were entities yet to be defined. As a typical artisan-engineer, Darby left no record of his experiments or, rather, tinkering, and we can only guess at how he might have achieved his success. As both the size of the blast furnace traditionally used for smelting iron ore and the strength of the blast slowly and incrementally increased, higher temperatures may have proved capable of burning off the impurities in the coal which had ruined the iron in earlier attempts.

In 1784 the English inventor Henry Cort developed the "puddling" process for converting pig (or cast) iron to wrought iron using coal, a technique that involved stirring the melt with an iron bar and so burn off any impurities. These changes rendered English iron production geographically and materially independent of the forest. With the lag in iron production thus relieved, the world entered a new Iron Age. In the course of the eighteenth century British iron production multiplied more than tenfold from a low point of fewer than 25,000 tons per year. And from 1788 to the middle of the nineteenth century, with railroad construction booming, iron production increased another fortyfold.

Another key industry of the Industrial Revolution, coal mining, displayed a similar pattern of development. It too had been growing in step with population growth, and it too encountered a production bottleneck. As superficial deposits became depleted, mine shafts were sunk deeper and therefore filled with groundwater at a more rapid rate. Traditional methods of removing water from the mines employed pumps of various designs driven by animal power or water power at the pithead. By the end of the seventeenth century it became clear that a more effective source of power was required to drive the pumps. "Fire engines" soon resulted—devices that would employ fire in one way or another to raise water. In 1712 an obscure English ironmonger, Thomas Newcomen, invented the first practical steam engine.

The steam engine was a technological innovation that changed the course of industrial development. The origin of the steam engine lies in independent craft traditions. Through intuition, tinkering, and a stroke of luck, Newcomen and his plumber assistant, John Cawley (or Calley), hit on the method of condensing steam in a cylinder and creating a partial vacuum so that atmospheric pressure would drive the piston. Even if the idea of atmospheric pressure as a potential motor was, as it were, "in the air," the actual design of the steam engine owed nothing to science. The complicated valve mechanisms, the technique of injecting cold water into the cylinder to condense the steam, and the mechanical linkages to the pump evolved through trial and error. The alternate heating and cooling of the large cylinder made the engine highly inefficient and a profligate consumer of coal; but, nonetheless, Newcomen engines proved sufficiently economical and were widely adopted because they operated primarily at coal mines where coal was cheap. The heavy consumption of coal, however, remained a short-coming of the design, especially when employed in other applications, and it provoked attempts to increase the efficiency of the engine.

Around the mid-eighteenth century two English craftsmen, John Smeaton and James Watt, using entirely different approaches improved the Newcomen engine. Their work also remained essentially within the craft tradition without applying scientific abstractions. Smeaton, who later became president of the Society of Civil Engineers (1771, "the Smeatonians"), employed strictly empirical methods and systemati-cally tested model steam engines, varying the dimensions of the parts without modifying the basic design; he thereby achieved a doubling of efficiency of the Newcomen engine. Watt, on the other hand, intro-duced a fundamental novelty that resulted in a radical improvement in efficiency. In a flash of insight during a Sunday stroll in 1765 he ar-rived at the idea of condensing the steam in a separate vessel kept cold outside of the cylinder, thereby leaving the cylinder hot throughout the cycle. By eliminating the alternate heating and cooling of the cyl-inder, the saving in coal was considerable. Constructed and sold in a famous partnership with the Birmingham manufacturer Matthew Boulton, Watt engines were widely adopted and were soon used in industries other than coal mining. The success of the Watt engine stemmed in part from the commercial strategy of leasing the machines and charging only for a percentage of the savings in the cost of coal over traditional Newcomen engines. Freed from a dependence on cheap coal at the coal mouth, the Watt engine could be set up and used

Fig. 14.2. Steam power. As mine shafts went deeper it became increasingly difficult to remove ground water. The problem was initially solved by Thomas Newcomen's invention of the atmospheric steam engine in 1712. Because it alternatively heated and chilled the cylinder Newcomen's engine was inherently inefficient and was only cost-effective when deployed near coal mines where fuel was inexpensive.

to drive mills virtually anywhere, thereby promoting the expansion of manufacture in urban centers. Five hundred steam engines puffed away in Britain in 1800, and the number rose rapidly after that.

Since the early steam engine relied on atmospheric pressure it was necessarily a large machine (sometimes called a "stationary" engine). By the end of the eighteenth century stationary engines were used to drive machinery in factories, and since boats were sufficiently large to hold them, early steamboats were also powered by atmospheric engines. But the steam locomotive and the railroad had to await the invention of a compact, high-pressure design, developed in 1800 by another Englishman, Richard Trevithick. By employing pressures much higher than atmospheric, Trevithick's engine was much smaller and could thus be placed on a carriage of reasonable size. At first Trevithick intended it as a replacement for the atmospheric engines used primarily in mines and mills, but he found that mine operators and manufacturers were unwilling to scrap their atmospheric engines for the still-questionable benefits of the new design, especially since the high-pressure engine was reputed to be unsafe (a reputation publicized by Watt who attempted to fend off the competition of the new design). Trevithick then took his invention to Peru where he hoped that at the reduced atmospheric pressures of high-altitude mining his engine would have a competitive advantage. When that venture failed he returned to London and did what inventors have often done before and since—he turned his invention into a novelty by running a locomotive on a circular track as an amusement for which he charged admission.

But the railroad could not be denied, and Trevithick's high-pressure steam engine altered the economics of transportation by making the railroad possible. In 1814 British engineer George Stephenson unveiled his first steam locomotive. The railroads initially hauled coal for short runs from the mines, but the railroad age truly dawned with the first public line, which opened between Liverpool and Manchester in 1830. Indeed, a railroad mania ensued, as Britain's and then the world's land areas began to be encrusted with a dense network of iron rails. Britain in 1830 ran 95 miles of track; that number rose to 1,500 miles in

Fig. 14.3. Watt's steam engine. In 1765 James Watt hit upon a way to improve the efficiency of steam engines: condense the steam in a condenser separated from the main cylinder. That way the cylinder could remain hot through the entire cycle, thereby increasing efficiency and lowering operating costs. As important as Watt's technical innovation was, the success of his steam engine depended as much on the manufacturing partnership he established with the early industrialist Matthew Boulton and the marketing strategies they devised.

1840 and then to 6,600 miles by 1850 and 20,000 by 1890. As rail-road transportation developed, it interacted with the iron industry in a pattern of mutual stimulation: the rapid growth of railways was made possible by the availability of cheap iron, and, in turn, rail transporta-tion facilitated—indeed demanded—the further acceleration of iron production.

The railroad revolutionized transportation and changed society. It brought economic development wherever it extended. It cut travel time dramatically and it reduced freight costs substantially. The railroad tied the countryside to cities in new ways, and it extended the city out-ward to new communities along the railroad line. The railroad gave rise to a new style of tourism and travel, so now folks could "take the waters" at spas or vacation by the seaside. Even more, in the cases of the transcontinental railroad in the United States, completed in 1869, or the Trans-Siberian railway in Russia, completed in 1916, the rail-road opened up vast areas of wilderness for the expansion of industrial civilization.

The process that began in Britain in the 1780s—what economic historians have called the "takeoff into sustained growth"—can be at-tributed to the mutually reinforcing effects of major industries as they developed separately and jointly. Thus, as iron began to be smelted with coal, it stimulated the growth of the coal industry, which led to the use of steam engines to clear the mines, while the need to transport large quantities of coal led to the railroad, which then led back to an enormous increase of iron production in an upward-spiraling, symbi-otic process. The result was a nation, and eventually a world, trans-formed. A population of rural farmers became a population of urban factory workers. Locomotives and iron tracks replaced horses and dirt roads. Iron increasingly replaced wood and stone as structural mate-rials, and the steamship replaced the sailing ship. The actors involved did not see clearly where all these changes were leading, but as with the Neolithic and Urban Revolutions previously, once these fundamental processes of change were under way, there was no going back to earlier modes of social or economic life.

Industrial Civilization

Not every aspect of emerging industrial civilization can be treated here, but several further features require attention: new energy sources fuel-ing the Industrial Revolution, new means of financing industrial devel-opment, and ideological changes that accompanied industrialization.

Premodern societies depended overwhelmingly on human and ani-mal muscle power and to some extent on wind and water power. For fuels they used renewable resources, notably wood. In turning to the steam engine and nonrenewable fossil fuels like coal and, later, oil, the advent of industrial civilization brought about significant changes in available energy resources and patterns of energy consumption. The production of coal and then oil rose exponentially from the eighteenth century, to the point where per capita energy consumption in industrialized societies today is 5 to 10 times greater than in traditional pre-industrial cultures. With its sharply increased consumption of energy, industrialization produced not merely a reordering of traditional society but, rather, a new type of society in which energy-intensive industrial production represents the major economic and human activity.

Like labor, capital and new means of financing were equally essential for industrialization. The history of European capitalism extends back to the late Middle Ages, and the development of the new industrial capitalism emerged out of the background of merchant capitalism and the success of overseas trade in commodity products through the eighteenth century. Profits from British colonial trade in sugar and slaves in large measure provided the accumulated capital needed to fund industrial development. Although authorities created the Bank of England in 1694, the official state bank did little to promote industry. Rather, hundreds of private banks arose in the English Midlands to handle capital requirements of nascent industries. Interest rates fell steadily through the seventeenth and eighteenth centuries, reaching a low of 3 percent in 1757 and remaining low thereafter, despite some ups and downs (as during the American war). A low interest rate made available large amounts of cheap money, without which the capital requirements of early factories could not have been met. Growing out of prior groups of commodity brokers and insurance agents, the London Stock Exchange opened in 1773, and in 1803 it offered its first list of traded stocks.

The ideological effects of industrialization proved no less powerful. Mercantilism, or the idea of a state-controlled economy restricting free trade in the interests of bolstering exports and accumulating gold and silver for the state's coffers, had stood as the reigning economic theory and the basis of the economic policies of contemporary European governments. But the Industrial Revolution killed mercantilism. Not coincidentally, new ideas about open markets and free enterprise or "laissez-faire" capitalism emerged along with the Industrial Revolution. Adam Smith's history-making work, *The Wealth of Nations*, appeared in 1776 and it signaled this new ideology of profit and the marketplace. As the Industrial Revolution gained momentum, however, and as anticipated labor strife and social costs emerged more clearly, other voices began to speak in opposition to free-market capitalism. In particular, the work of Karl Marx (1818–1883), *Das Kapital* (three volumes from 1867), provided an analytical critique of the new economic relations. Marx underscored the inevitable exploitation of labor for profit by the owners of factories and the means of production. Class warfare between workers and owners, according to Marx, was intrinsic to the organization of capitalist society and was bound to result in a transformation of society in favor of the proletariat, just as the conflict between the land-owning aristocracy and merchant capitalist resulted in the transition from feudalism to capitalism. In this way Marx provided not only a trenchant analysis of capitalism that has endured to this day but also the ideological underpinnings of socialism and communism as political doctrines. What Marx didn't foresee clearly was the rise of a middle class—a technical-managerial class—composed not of proletarian laborers nor of owners of the means of production, but an in-between class of experts with knowledge and skills to keep the machine running. This component of modern industrial civilization became clearly manifest only after Marx's death.

The later eighteenth and nineteenth centuries also witnessed the flowering of the Romantic movement, of which William Blake is an early representative. In poetry, literature, music, and other of the fine arts artists turned away from classical styles of the preceding decades and toward themes associated with the simplicity and beauty of nature, with sensibility, the value of the family, and matters of the human heart. This blossoming of romanticism needs to be seen first and foremost as a reactionary response to the ravages of industrialization.

The processes of industrialization continued to gain momentum in England in the nineteenth century. Worker productivity doubled from 1830 to 1850. Iron production jumped from 700,000 tons in 1830 to 4 million tons in 1860. Coal production soared from 24 million tons in 1830 to 110 million tons in 1870. In 1850 for the first time anywhere the urban population of England topped 50 percent. The first "world's fair"—the Great International Exhibition—opened in London in 1851. The machines on display within the magnificent iron and glass "Crystal Palace" vividly exemplified the power of industrialization and the new technologies then transforming the world. Britain, at least, was a very different country than it had been a century earlier.

Science and the Early Industrial Revolution

All of the technical innovations that formed the basis of the Industrial Revolution of the eighteenth and the first half of the nineteenth centuries were made by men who can best be described as craftsmen, artisans, or engineers. Few of them were university educated, and all of them achieved their results without the benefit of scientific theory. Nonetheless, given the technical nature of the inventions, a persistent legend arose that the originators must have been counseled by the great figures of the Scientific Revolution. During the eighteenth century John Robison, a professor at the University of Edinburgh, publicized the myths that Thomas Newcomen, the inventor of the steam engine, had been instructed by Robert Hooke, one of the leading savants of seventeenth-century English science, and that James Watt had applied

Fig. 14.4. The Industrial Age. The Crystal Palace, erected in 1851 for an international exposition in London, was a splendid cast-iron and glass structure that heralded the new industrial era.

Joseph Black's theory of latent heat in arriving at the separate condenser. These claims have been discredited by historical research. In fact, the French physicist Sadi Carnot, for example, produced the first scientific analysis of the operation of the steam engine in his work *Reflections on the Motive Power of Fire* published in 1824, long after steam engines had become commonplace. And Watt's cleverly designed parallel motion could not even begin to be studied scientifically until kinematic synthesis developed the appropriate analytical techniques in the last quarter of the nineteenth century, partly in an attempt, in fact, to analyze Watt's device. These examples rank among the many instances where, contrary to the claims that eighteenth-century engineers benefited from scientific theory, technical developments provoked the interests of scientists and led to theoretical advances.

The myth that the theoretical innovations of the Scientific Revolution account for the technical inventions of the Industrial Revolution found reinforcement in the common belief, which has been challenged repeatedly in these pages, that technology is inherently applied science, a belief only partially true even today when research and development are indeed often conducted in close contact. In the eighteenth and early nineteenth centuries it was almost never the case. This is not to say that science played no social or ideological role in promoting industrialization. On the contrary, as the Industrial Revolution unfolded in England, science permeated the social and cultural fabric of European civilization. Dozens of learned societies and academies dotted the European map, such as the famous Lunar Society of Birmingham where scientists and literate engineers occasionally rubbed shoulders. Public lectures alerted large, nonspecialist audiences to the triumphs of scientific discovery and the analytical potency of experiment and the scientific method. Natural theology, the doctrine that the study of nature is an act of piety, strengthened the concordance between science and religion and reinforced the notion of the useful exploitation of nature. Science raised the status of the reasoned life and was honored as a cultural and intellectual enterprise. The rational sciences offered a new outlook and worldview. In this sense, scientific culture was important and perhaps essential to the Industrial Revolution. But the scientific enterprise itself continued to be shaped in a Hellenic mold or Hellenistic government employment, largely divorced from practical applications, and technologists and engineers proceeded without tapping bodies of scientific knowledge.

Although technology developed along traditional lines without the benefits of scientific theory, several eminent craftsmen made social contact with the world of science in eighteenth-century Europe, and the lines do get blurred. In England, engineers James Watt and John Smeaton and the potter Josiah Wedgwood became members of the Royal Society of London and contributed to its *Philosophical Transactions*. But, in fact, their publications had little or nothing to do with their contributions to industry. Watt published letters and articles on the composition of water and on the "medicinal use of factitious airs" in terms of phlogiston chemistry. None of these scientific contributions had any bearing on his steam engineering. Wedgwood became keenly interested in chemistry, performed chemical experiments, discovered that clay shrinks when heated, and in 1782 invented the pyrometer based on that discovery. This last certainly was a case of technology as applied science, and it foreshadows further connections between science and industry. Wedgwood also corresponded with eminent chemists, including Joseph Priestley and Antoine Lavoisier. But his novel and successful ceramic ware, which still goes under the name of Wedgwood, came *before* his entrance into the world of chemistry. His father and brother were potters, and Wedgwood himself was unschooled. He came to pottery through a craft apprenticeship in his brother's factory, and his career as a potter led him to an interest in chemistry, rather than the other way around. And Smeaton, who coined the term *civil engineer* to distinguish civilian consultants from the military engineers graduating from the newly founded Royal Military Academy at Woolwich, distinguished himself as a builder of large-scale public works. He won the Copley Medal of the Royal Society for a paper he published in the *Philosophical Transactions*. The paper, a report on his empirical demonstration that overshot waterwheels are more efficient than undershot, contained a keen insight, and he applied it in his own hydraulic projects where he consistently preferred the overshot wheel.

(Other engineers continued to use undershot wheels because they were less costly to build.)

In 1742, the English government, recognizing the need for formally trained artillery (and, soon afterwards, engineering) officers, established the military academy at Woolwich. There, cadets were schooled in the "fluxions" (Newton's form of the calculus) and the elements of statics, among other subjects. However, the scant knowledge of the graduates combined with their lack of craft experience disqualified them as working engineers, and during the eighteenth century only civilian engineers, unschooled in any science and untrained in any school, contributed to the program of industrialization. The period of the Industrial Revolution did indeed see a revision in the traditional affiliation between technology and the crafts in favor of new sociological links between technology and science: the perceived rational methods of experimental science began to be applied in industry, and some of the leading engineers drew closer socially to the world of science. But the gulf between practical applications and theoretical research remained to be bridged.

An episode in Watt's career reflects the dimensions of that gulf. During the 1780s he performed experiments on the commercial application of the chlorine process for bleaching textiles, a process discovered by the French chemist C. L. Berthollet. For Berthollet the research was purely scientific, and he published his results without regard for commercial possibilities or financial gain. But Watt's father-in-law, James MacGregor, was in the bleaching business, and Watt hoped that the three of them—Watt, MacGregor, and Berthollet—could, by keeping their improvements secret, acquire a patent and reap substantial profits. When Watt complained to Berthollet that by "making his discoveries . . . publick" he undermined the possibility, Berthollet replied, "If one loves science one has little need of wealth." Berthollet stood almost alone among scientists of the eighteenth century called upon to defend the ethos of pure science. For the others the occasion to apply theoretical research to practical problems almost never arose.

Almost never, but not quite. A curious instance occurred at the turn of the nineteenth century when English scientists were called upon to apply their knowledge to industry. The resulting failure points conclusively to the features of applied science that were still lacking. The Port of London determined that another bridge across the Thames was required to serve the growing metropolis. Proposals were solicited, and among them Thomas Telford (who later became the first president of the Institution of Civil Engineers) submitted a spectacular design in the form of a single cast iron arch with a span of 600 feet. Since cast iron bridges were still a novelty, with only three or four having been built, no established tradition or rules of thumb existed to guide the design. Nor was there any theoretical science that could be brought to bear on the project, or any professorships of what would nowadays be called engineering science at universities. The parliamentary committee considering plans for improvement of the port recognized the difficulty and proposed to consult "the Persons most eminent in Great Britain for their Theoretic as well as Practical Knowledge of such Subjects."

Since such knowledge, theoretical and practical, was not combined in any individuals in 1800, Parliament created two committees, one of mathematicians and natural scientists and the other of practicing builders. Each was asked to respond to a questionnaire about Telford's design in the hope that useful points would somehow emerge from the combined answers. The results illustrate the futility of the proceedings, based on the misconception that practical outcomes could be derived from the combined expertise of mathematicians who possessed little knowledge of building and builders who were largely ignorant of mathematics and theoretical mechanics. Among the answers provided by the "practitioners" a few sensible suggestions emerged, but engineers were still defeated by their lack of any theory of structures (which had not yet been formulated) and by the complexity of the design, which would present a formidable theoretical problem even today. But the inability to bring contemporary scientific knowledge to bear on the solution of practical problems was especially evident in the replies of the "theoreticians," who included the Astronomer Royal and the Professor of Geometry at Oxford. The Astronomer Royal's views on mechanical engineering were ridiculed not long afterwards

Fig. 14.5. Thomas Telford's iron bridge across the Menai Straits (completed 1824). In the eighteenth century, iron was adapted to structural engineering. The cast iron arch bridge was followed by wrought iron suspension bridges and wrought iron tubular bridges. Organizational and managerial skills proved no less important for the success of bridge building than the technical mastery of iron as a construction material.

as "a sentence from the lofty tribunal of refined science, which the simplest workman must feel to be erroneous." The astronomer's incompetent testimony was enriched only by his knowledge of heavenly phenomena: he suggested that "the Bridge be painted White, as it will thereby be least affected by the Rays of the Sun," and that "it be secured against Lightning." The contribution of the Savilian Professor of Geometry was equally silly: he calculated the length of the bridge to ten-millionths of an inch and its weight to thousandths of an ounce.

To their credit, some of the theoreticians on the committee recognized that science was not yet prepared to minister to technology. Isaac Milner, the Lucasian Professor of Mathematics at Cambridge (Newton's professorship), observed that theory would be useless in such applications until it was combined with practical knowledge. The theoretician, he observed, "may . . . appear learned, by producing long and intricate Calculations, founded upon imaginary Hypotheses, and both the Symbols and the Numbers may be all perfectly right to the smallest Fraction, and *the Bridge be still unsafe*." And John Playfair, professor of mathematics at Edinburgh, ended his report by noting that theoretical mechanics "aided by all the Resources of the higher Geometry, [has] not gone farther than to determine the Equilibrium of a Set of smooth Wedges." At the beginning of the nineteenth century the design of a structure as complex as Telford's fixed arch was still to be left to the intuition and experience of craftsmen—"Men bred in the School of daily Practice and Experience." It would be another half century before John Rankine's *Manual of Applied Mechanics* and similar works would show the way toward the engineering sciences.

Whatever cultural influence the Scientific Revolution exerted on the Industrial Revolution, it did not extend to the application of scientific theory to technical invention. Although the governments of Europe were rational in their hope that science would assist society, their interests were more narrowly focused on governance, while the technical dimension of the Industrial Revolution was left to be crafted by the ingenuity of unschooled artisans and entrepreneurs working without benefit of theoretical knowledge or any training in science. That knowledge was not yet compacted into textbooks. The universities had no programs or even courses in the engineering sciences. There were no professional engineering societies until the Smeatonians came on the scene in 1771. The physical constants and practical tables that would convert abstract mathematical principles into engineering formulas had not yet been determined and compiled. And no research laboratories had yet come into being. These developments and applied science awaited a later day.

Legacies of Revolution: From Newton to Einstein

The movements that historians label the Scientific Revolution and the Industrial Revolution were epochal in shaping the modern world, but it took many decades for their consequences, separately and jointly, to fully unfold and for science, technology, and society to assume their completely modern characters. Modern science did not emerge full-blown from the Scientific Revolution or full-fledged from the work and worldview of Isaac Newton. Global industrial civilization did not arise overnight in the aftermath of the Industrial Revolution or developments sparked by James Watt and the steam engine. And, despite the ideology of the useful application of knowledge espoused by Francis Bacon and René Descartes in the seventeenth century, theoretical science did not immediately find applications in industry.

The close sequence of the Industrial Revolution following hard on the heels of the Scientific Revolution can easily confuse historical understanding. In fact, in the eighteenth century and most of the nineteenth century theoretical science and technology (the crafts) continued to go their traditional, separate ways. And when they began to merge in the nineteenth century it was as much institutional factors, along with intellectual and technical developments, that shaped the partnership. In this chapter we trace the intellectual trajectory of the physical sciences from Newton to Einstein, and we likewise follow the course of industrialization through the nineteenth century. We then pick up the story of how engineering transformed itself into a profession and began to direct the traffic of history. As science and industry forged crucial new ties that established the pattern for applied science today, the application of scientific knowledge can, at long last, be clearly seen in the world of the toolmaker.

In the Wake of Bacon and Newton

The exponential growth of knowledge since the seventeenth century presents a problem in thinking about the history of science after the Scientific Revolution. The scale of the scientific enterprise and its output in the form of publications have increased by orders of magnitude since then, making it hard to review comprehensively the content of the sciences of the past three centuries. One way around this difficulty involves modeling the phenomena, that is, reducing a complex reality to a simpler model by identifying essential elements, linking them conceptually, and seeing how they interact.

It is thus possible to envision two distinct scientific traditions developing in the aftermath of the Scientific Revolution. One, often labeled the "Classical sciences," includes astronomy, mechanics, mathematics, and optics. These fields originated in antiquity; they matured as research endeavors in the ancient world; and they were, in fact, the sciences revolutionized in the Scientific Revolution. Prior to their reformulation in the early-modern period, these sciences were already highly theoretical with research aimed at solving specific problems. All things considered, the Classical sciences were not experimental in approach. Instead, they built on mathematical and theoretical foundations, and they were clearly the province of the trained expert.

The other group of sciences, the "Baconian sciences," developed parallel to but largely separate from the Classical sciences during and after the Scientific Revolution. The name derives from the style of science advocated by Sir Francis Bacon. The Baconian sciences—primarily the systematic study of electricity, magnetism, and heat—were without roots as formal sciences in antiquity, but sprang into existence as domains of empirical investigation more or less as a result of the ferment surrounding the Scientific Revolution. That is, as the Classical sciences became transformed in the Scientific Revolution, the Baconian sciences arose and took shape in the general intellectual excitement of the times. In contrast to the theory-dependent and more mathematical Classical sciences, the Baconian sciences were generally more qualitative in character and experimental in approach, and they therefore depended on instruments to a much greater degree than their Classical counterparts. The Baconian approach was more empirical and only loosely guided by theory.

Newton's *Principia* provided the exemplar for the Classical sciences in the technical details of its physics, in its general approach, and in dictating problems that the community of mathematical scientists worked on over the course of the eighteenth century. The predicted return of Halley's comet, for example, in 1758–59 showed the awesome power of Newtonian mathematical science. In another confirmation and extension of Newtonian physics, other scientists measured the earth's curvature. Then, in 1761 and again in 1769 international teams of observers clocked the rare passage of Venus across the face of the sun and calculated the earth-sun distance reasonably well for the first time. On the Continent French and Swiss mathematicians extended research in theoretical mechanics into highly technical fields like hydrodynamics, the mathematics of vibrating strings, and elastic deformation.

Technical work of this sort continued into the nineteenth century. A famous example is the discovery of the planet Neptune in 1846. Based on observed irregularities in the orbit of the planet Uranus—itself discovered in 1781 by William Herschel—British and French astronomical theorists predicted the existence of Neptune, and German astronomers in Berlin duly observed the planet the night after they received the prediction. This tradition of the Classical sciences may be fairly said to have culminated, at least conceptually, in the work, *Celestial Mechanics*, by P. S. Laplace (1749–1837). Newton's *Principia* with its abstruse geometrical diagrams seems quaint and antiquated in comparison with Laplace's magisterial work (five volumes, 1799–1825) written wholly in the language of the calculus. And where Newton saw God's presence in his physics, Laplace saw His absence. In a famous exchange, the emperor of France, Napoleon Bonaparte, remarked that he found no mention of God in Laplace's opus. "But, Sire," Laplace reportedly replied, "I have no need of that hypothesis." The Classical sciences had progressed so far that Laplace could formulate a mathematically complete and ordered universe based on the fundamental laws of mechanics established by Newton and elaborated by his successors.

Newton's *Opticks* (1704), on the other hand, provided the conceptual umbrella beneath which the Baconian sciences developed in the eighteenth century. In a set of research questions that Newton appended to the *Opticks* the great man at one point posited a series of superfine, self-repulsive substances to account for phenomena. How can the heat of a warm room penetrate the glass of an evacuated receiver, Newton asked, except by means of such an ether? Similarly, Newton invoked various imponderable ethers and atomic powers to explain electrical, magnetic, certain optical, and even physiological phenomena.

Developments in eighteenth-century electricity illustrate the character of the Baconian sciences in the era following Newton. Static electricity was, of course, known at least since antiquity. Its investigation took off in the eighteenth century, as new instruments were developed to generate and store static electricity and as scientists applied themselves to studying a host of new facts relative to electrical conduction, insulation, attraction, and repulsion. (Before the invention of the battery in 1800 current electricity did not yet exist in the world of science.) Benjamin Franklin's kite experiment—first performed in 1752—that identified lightning as a static electrical phenomenon—seems entirely typical of this manner of experimental, qualitative research. On the theoretical side, in the spirit of the *Opticks* Franklin offered a single electrical ether to account, not entirely successfully, for the phenomena. Other theorists countered with a two-ether theory. Notably, there was no agreement. Much the same points emerge in considering the scientific study of several other fields, all of which took their lead from the *Opticks*. With regard to magnetism, for example, the German-Russian scientist F.U.T. Aepinus (1724–1802), working at the imperial science academy in Saint Petersburg, articulated an ether theory to account for magnetic attraction and repulsion. The English physiologist, Stephen Hales (1677–1761), deployed a "vegetable" ether in experiments on plants. While Franz Anton Mesmer's work on "animal magnetism" and early hypnotism may strike readers today as beyond the pale of science, quite the contrary is the case, because Mesmer (1734–1815) operated with the full authority of Newton's *Opticks* and the tradition of ether-based research and scientific explanations. Mesmer's sin was not in invoking a magnetic ether to explain the seemingly miraculous cures he produced in his medical patients; it was, rather, his refusal to share the secrets of his discoveries with other members of the scientific and medical communities. The refusal led to his downfall and the repudiation of his mesmerizing ether.

One can extend the concept of the Baconian sciences to include research in eighteenth-century meteorology, natural history, botany, and geology, all of which were more observational and empirical than theoretical. Regarding meteorology, the scientific societies served as central depots and published reports sent in by individuals, and they independently sponsored several large-scale meteorological data-collecting projects. Obviously instruments (such as thermometers and barometers) were required, and collecting weather data allowed a provincial amateur, say, to feel as if he (or in a few cases, she) participated in the grand enterprise of European science in the eighteenth century. Analogous circumstances prevailed in botany and natural history, where the primary activity consisted in collecting specimens, often from far-flung corners of the world. Specimens found their way to central depots in London, Paris, or Uppsala in Sweden, where theorists such as the Count de Buffon (1707–88), Sir Joseph Banks (1743–1820), or Carolus Linnaeus (1707–78) attempted to develop rational systems of classification. "Botanizing," indeed, became something of a fad in the eighteenth century, where, armed with simple manuals for identifying plants (and perhaps a bottle of wine), individuals would pass the time in scientific communion with nature. Progress in geology in the eighteenth century likewise depended on systematic collections of data. In all these instances research proceeded without sophisticated theory and the other trappings characteristic of the *Principia*-based Classical sciences.

In considering the different traditions represented by the Classical

Fig. 15.1. Early electrical equipment. The scientific investigation of static electricity blossomed in the eighteenth century aided in essential ways by the development of new experimental equipment, such as glass or sulfur ball generators, shown here. This engraving is taken from the Abbé Nollet's *Lessons on Experimental Physics* (1765). It illustrates something of the parlor-game character that often accompanied demonstrations concerning static electricity in the eighteenth century.

and Baconian sciences in the eighteenth century, chemistry seems the odd science out. With deep roots in traditional alchemy, chemistry underwent no revolutionary restructuring in the Scientific Revolution of the sixteenth and seventeenth centuries, and chemistry in the eighteenth century does not fit well into either the empirical Baconian sciences or the problem-oriented research of the Classical sciences. Contemporary chemistry was highly experimental and instrument-dependent, yet in the early decades of the eighteenth century it developed an agreed-upon theoretical structure known as phlogiston chemistry, and at the end of the century chemistry underwent an independent conceptual revolution.

The history of the Chemical Revolution fits the pattern previously seen for scientific revolutions in general. The theoretical framework prevailing through the 1770s was that of phlogiston chemistry. Phlogiston was envisioned as the principle of combustion, vaguely akin to the ancient Greek concept of "fire" that is active and released in combustion. Thus, for example, according to the theory, a burning candle releases phlogiston; the candle goes out when covered by a jar because the air contained within becomes saturated with phlogiston, a circumstance that prevents further combustion. (Note that this is the very opposite of the view of combustion after the Chemical Revolution which postulated that the candle goes out because the air in the jar becomes exhausted rather than saturated.) As illustrated in figure 15.2, phlogiston theory gave a coherent account of a diverse range of phenomena—combustion, plant growth, digestion, respiration, smelting—and thus provided a solid theoretical framework within which research in eighteenth-century chemistry unfolded.

Several factors led to the downfall of phlogiston chemistry and its replacement by Antoine Lavoisier's oxygen theory of chemistry and combustion. The discovery of "fixed air" (what we know as carbon dioxide) by Joseph Black in 1756 and its identification as a distinct gas represents a major milestone. With its own specific properties, "fixed air" helped break down the traditional notion of "air" as a single element or entity, and with improved equipment chemists soon identified a series of other new "airs." Finally, a string of anomalies, which at first appeared minor, posed problems for phlogiston theory, notably that mercury seemed to *gain* weight in combustion (under certain conditions), whereas with the release of phlogiston, according to

Fig. 15.2. Phlogiston chemistry. Phlogiston was thought to be the element or principle of fire given off during combustion. Early modern chemists had ample evidence supporting the existence of phlogiston. The explanatory scope of phlogiston theory added to its authority by linking combustion with plant growth, human physiology, and metallurgical phenomena in cycles wherein phlogiston is variously given off and absorbed. The French chemist Antoine Lavoisier initiated the eighteenth-century revolution in chemistry by proposing that oxygen is taken out of the air in combustion, rather than phlogiston given off, and with that phlogiston ceased to exist as an entity in the world.

φ = Phlogiston

Wood

Food

Plant Growth
(Phlogiston absorbed)

Respiration
(Phlogiston released)

Combustion
(Phlogiston released)

Digestion
(Phlogiston absorbed
from food)

Body Heat
(Phlogiston released)

Metals = [Ore + φ]

Smelting: Ore + Charcoal + Heat ==> Metal
(Rich in
Phlogiston)

Corrosion (rust)
(Phlogiston released)

the theory, it ought to have *lost* weight. Such problems became more and more acute for theoretical chemists. Lavoisier (1743–94), a young chemist, knew phlogiston theory well, yet he began his theoretical researches with the radical notion that in combustion something was taken out of the air rather than released into it.

That "something" proved to be oxygen, although we hasten to add that even Lavoisier's mature views of oxygen were not the same as those taught in chemistry classes today. Nevertheless, with the discovery of oxygen gas and the identification of its role in combustion, Lavoisier effected a revolutionary reconceptualization of chemistry, aided in significant measure by his careful accounting for the inputs and outputs of the reactions he produced. Typical of revolutionary transformations in science, other chemists did not immediately subscribe to Lavoisier's radical new views. Indeed, led by the older English chemist Joseph Priestley (1733–1804), they modified phlogiston theory in ways that preserved satisfactory, rational accounts of chemical phenomena. Well into the 1780s chemists could still reasonably hold phlogistonic views, if only because the latter seemed more familiar. Indeed, Priestley never converted to the new chemistry and went to his grave in 1804 as virtually the last adherent of phlogiston chemistry.

If not through an irrefutable test or piece of evidence, why, then, did European chemists shift their allegiance to the new chemistry? Rhetoric and persuasion played crucial roles in the dynamic of the Chemical Revolution. Not only did Lavoisier and a band of like-minded colleagues make new discoveries and publish exciting experimental results, in 1787 they formulated an entirely new system of chemical nomenclature. In Lavoisier's new system, "inflammable air" became hydrogen, "sugar of Saturn" became lead acetate, "vitriol of Venus" became copper sulfate, and so on. The proponents of the new system wished to have language rationally reflect chemical realities. But, as a result, students schooled in the new chemistry could only speak the new language, and phlogiston chemists were left behind. This shift was compounded by a related step, the publication of Lavoisier's textbook, the *Elementary Treatise of Chemistry* in 1789, which taught only his new chemistry. In this work, phlogiston is banished. As a result of the revolution effected by Lavoisier, from being a central element of theory, phlogiston ceased to exist as an entity in the world, becoming a mere historical oddity.

One feature of Lavoisier's book deserves particular notice in the present context. In the opening section of his *Elementary Treatise*, Lavoisier is careful to separate heat phenomena from truly chemical phenomena. Thus, for Lavoisier water remains chemically water even though it may change its physical state from ice to liquid water to water vapor. To account for changes in state and other thermal phenomena, Lavoisier introduced a new ether, caloric. Like the other ethers we have encountered, caloric was a self-repulsive fluid-like material, much more fine than ordinary matter. Thus, caloric penetrated a block of ice, pushing its particles apart and melting the ice into water, the addition of more caloric transforming the water into water vapor. By introducing caloric Lavoisier normalized chemistry within the intellectual framework laid down by Newton in the *Opticks*, and so chemistry after all comes to fit the Classical-Baconian sciences model under discussion here.

The Second Scientific Revolution

A second Scientific Revolution began to unfold at the turn of the nineteenth century. Two closely connected trends characterize this pivotal historical transformation in the sciences: the *mathematization* of the previously more qualitative Baconian sciences, and the theoretical and conceptual *unification* of the Classical and Baconian sciences. That is, previously separate traditions became conjoined in a new scientific synthesis familiar to us today as "physics." As the second Scientific Revolution unfolded and these processes of mathematization and unification proceeded, a single set of universal laws and a powerfully coherent scientific world picture began to emerge. By the last decades of the nineteenth century that world picture, known as the Classical World View, seemed at once to integrate all the domains of the physical sciences and to promise a complete understanding of the physical world and thereby the end of physics itself.

The pattern of mathematization and unification can be seen in many different specialties and areas of research in nineteenth-century science. Developments in electricity and their ramifications for magnetism and chemistry present a compelling example. Through the eighteenth century, the scientific study of electrical phenomena involved static electricity only. The accidental discovery of current electricity opened the door to a whole new area of research. In experiments conducted with frogs' legs in the 1780s, the Italian scientist, Luigi Galvani (1737–98), did not set out to extend the range of electrical science but, rather, in the tradition of the *Opticks* he sought to investigate the etherial "animal electricity" that seemed to "flow" in an animal's body. His compatriot, Alessandro Volta (1745–1827), built on Galvani's work and in 1800 announced the invention of the pile, or battery, which could produce flowing electricity. Volta's battery and the ever-larger ones that soon followed manifested profound new connections between electricity and chemistry. The battery—layers of metals and cardboard in salt (later acid) baths—was itself a chemically based instrument, and so the generation of current electricity was self-evidently associated in fundamental ways with chemistry. More than that, through electrolysis or using a battery to run electricity through chemical solutions, scientists, notably Humphry Davy (1778–1829), soon discovered new chemical elements, such as sodium and potassium, appearing at the poles of the battery. As a result, an electrical theory of chemical combination—that chemical elements were bound by electrical charges—predominated in chemistry during the first decades of the nineteenth century.

These discoveries in electrochemistry generally supported atomistic interpretations that had gained ground since the early nineteenth century. Following Boyle's lead, Lavoisier had been content to describe chemical elements as merely the last products of chemical analysis without saying anything about the constitution—atomic or otherwise—of these elements. In 1803, out of a background in meteorology and pneumatic chemistry, John Dalton (1766–1844) noticed that the proportions of elements entering into reactions were often ratios of small integers, suggesting that chemical elements are in fact discrete particles. He thus became the first modern scientist to propose chemical atoms—or true indivisibles—in place of the more vague concept of chemical elements. Atomism was not immediately accepted in all quarters, but by mid-century the doctrine had become a fundamental component of contemporary chemistry. In advocating chemical atomism Dalton and his successors established a link with "philosophical" atomism that had been so prominent a feature of the new science of the seventeenth century.

While scientists suspected some unity between electricity and magnetism, only in 1820 did the Danish professor of natural philosophy, Hans Christian Oersted (1777–1851), accidentally demonstrate the sought-after link. Repositioning an electric circuit and a compass after a classroom lecture, Oersted discovered that opening and closing the circuit produced a magnetic effect if the wire stood parallel to the compass needle (rather than perpendicular to it, as might have been expected). By showing the magnetic effect of current electricity to be motion, Oersted unveiled the principle later applied to the electric motor. New discoveries followed, including the electromagnet and the attraction and repulsion of current-carrying wires.

These developments culminated in 1831 with the discovery of electromagnetic induction (or the creation of electricity by magnetism) by Michael Faraday (1791–1867), the self-educated experimentalist at the Royal Institution in England. Faraday generated an electrical current by plunging a magnet through a closed coil of wire. The significance of Faraday's discovery stemmed only in part from its technical potential in the form of the dynamo (or electric generator) or its being the sought-after analog to Oersted's production of magnetic effects through electricity. On a deeper philosophical level Faraday proved the interconnection of electricity, magnetism, and mechanical motion. After Faraday, given two of these three forces of nature, scientists could readily produce the third.

Faraday's explanations of electromagnetic phenomena, although initially idiosyncratic, proved highly influential in the long run. Lacking

Fig. 15.3. Faraday's lines of force. Michael Faraday postulated that lines of force occur in a field surrounding a magnet. He illustrated this proposition by observing patterns that iron filings form in the space around a magnet.

mathematical expertise, Faraday conceived the effects of electricity and magnetism visually as mechanical distortions in space. The case of iron filings distributing themselves around a magnet convinced Faraday of the reality of electromagnetic *fields* and "lines of force" emanating from magnets and electric currents. Faraday thus shifted attention from magnets and wires to the surrounding space and initiated field theory. Unmistakable in all this work is an incipient merger of scientific theory and technological applications, as the new science of electricity and novel electrical devices grew hand in hand.

Complementary developments in optics constitute a central component of the Second Scientific Revolution. In the eighteenth century the authority of Newton was such that his particle theory of light predominated, even though scientists knew of an alternate wave theory espoused by Newton's contemporary, Huygens. In this case Newton's influence was oppressive, and significant work in optics did not take place in the eighteenth century. That situation changed radically with the work of Thomas Young (1773–1829) and Augustin Fresnel (1788–1827). Dissatisfied with accounts of diffraction phenomena (or the slight bending of light around the edges of objects) Young proposed a wave interpretation in 1800. He envisioned light as a longitudinal pressure wave, something like sound. In France, the young Fresnel upset the established scientific community by proposing that light consisted of transverse waves, like waves in the ocean. Fresnel's interpretation better accounted for the full range of optical phenomena, including interference phenomena where one set of waves interacts with a second set. A peculiar consequence of Fresnel's theory soon emerged, that under appropriate experimental conditions a white spot should appear at the center of the shadow cast by a circular disk. In a

LEGACIES OF REVOLUTION 329

dramatic experiment Fresnel demonstrated that the predicted phenomenon in fact proved to be the case. He won the prize competition of the Paris Academy of Science in 1819.

With the gradual acceptance of the wave theory of light, old problems, such as polarization, had to be reevaluated, and new research areas opened up, such as determining wavelengths and analyzing the optical spectrum through spectroscopy. (The latter study, incidentally, showed unexpected ties between light and chemistry, with each chemical element emitting a distinctive light spectrum.) The wave theory of light also posed a major theoretical problem: what was the medium in which light waves propagated? The answer emerged that light consisted of waves in an all-pervasive cosmic ether. In the Second Scientific Revolution all the particular subtle fluids of the eighteenth-century Baconian tradition collapsed into this one unitary world ether.

The study of heat added to the conceptual innovations transforming the intellectual landscape of science in the nineteenth century. With his notion of caloric as a material substance Lavoisier had initiated a fruitful line of research in the measurement of heat. In his *Analytical Theory of Heat* (1822), Joseph Fourier (1768–1830) applied the calculus to the investigation of various modes of heat flow, but without pronouncing on the nature of heat. In 1824 the young French theorist Sadi Carnot (1796–1832) published his landmark tract, *Reflections on the Motive Power of Fire*. In this work Carnot analyzed the workings of the steam engine and elaborated what we know as the Carnot cycle, which describes what happens in the cylinders of all heat engines. For us, his *Motive Power of Fire* has another telling importance. Carnot's became the first scientific investigation of the steam engine. By the time Carnot wrote, steam engines had been in use for more than 100 years, and the Industrial Revolution, propelled in large part by the application of steam power as we have seen, was well under way in Europe. The case is precisely the opposite of the cliché of technology as applied science. Carnot's analysis of the steam engine provides the paradigm case, rather, of technology setting the agenda for scientific research.

In the wake of the Industrial Revolution, Romanticism and the Romantic movement were felt in science and natural philosophy no less than in the arts and sensibility. A distinctive train of thought, *Naturphilosophie,* or romantic science, blossomed in the nineteenth century and had notable consequences. Not just a critic of industrialization, in his art and in his poetry the English artist William Blake (1757–1827) expressed his feeling that Newton and the mechanism and emotionless rationality stemming from the Scientific Revolution and the Enlightenment eviscerated the heart and soul of nature. The famous German poet and author Goethe (1749–1832), for example, also wrote on botany, anatomy, and, particularly, colors, penning a landmark treatise, *Theory of Colors* (1810) wherein he repudiated the Newtonian, mechanistic approach. In a like vein, Hans Christian Oersted's sense of a connection between electricity and magnetism came from his almost spiritual belief in an underlying unity to the forces of nature, and this notion of "force," or *Kraft* in German, proved a key concept leading to the discovery of thermodynamics in the 1840s.

The most remarkable development in heat studies—and, indeed, in all of the physical sciences in the nineteenth century—was the creation of an entirely new theoretical discipline, thermodynamics, which unified the sciences of heat and motion. In the years leading up to 1847, the recognition arose from a variety of sources that the forces of nature— heat, light, chemistry, electricity, magnetism, and motion—might not simply interact with each other but might be mutually interconvertible and manifestations of some underlying force. In the 1840s several individuals independently enunciated the first law of thermodynamics, the conservation of energy—the principle that the various forces of nature can change from one form to another and that an indestructible entity called *energy* is conserved in the transformations. In the steam locomotive, for example, chemical energy stored in coal is released, some of which is transformed into heat, light, and the mechanical motion that drives the piston that propels the train. What saves the first law of thermodynamics from being merely a metaphysical principle (although metaphysical principles concerning the unity of nature proved instrumental in its formulation) is that, quantitatively, the transmutation of energy from one form to another takes place according to fixed exchange rates. The English experimentalist, James Prescott Joule (1818–89), worked out the mechanical equivalent of heat to a high degree of exactitude, with the fall of a standard weight exactly equal to a specific rise in temperature of a given weight of water. Building on the work of the pioneers, in a series of fundamental papers in the 1850s and 1860s the German physicist, Rudolf Clausius (1822–88), formulated the second law of thermodynamics. This law concerns the behavior of energy over time; specifically, it postulates that in a closed, undisturbed system energy peaks and valleys will even out until no temperature differences exist in the system. The second law implies that energy, like water, naturally "runs downhill" and that, without additional work, reactions are not naturally reversible.

Thermodynamics was one of two entirely new scientific fields to emerge in the nineteenth century that fundamentally transformed our outlook on the natural world, the other being the theory of evolution. The concept of energy and the laws of thermodynamics united the physical sciences on a deeper and entirely unprecedented level and provided the basis for the conceptually unified worldview that coalesced at the end of the century.

For the physical sciences, at least, the Classical World View (or Classical Synthesis) of the latter half of the nineteenth century offered a comprehensive vision and a unified intellectual picture of the physical world unmatched historically since the Middle Ages and the heyday of the Aristotelian outlook. The unity of the Classical World View crystallized around the work of James Clerk Maxwell (1831–79). Maxwell mathematicized Faraday's more qualitative notions of the electromagnetic field and gave the world the elegant mathematical expressions that describe the field in the form of wave equations, known as Maxwell's equations. Two aspects of Maxwell's achievement proved instrumental in confirming the Classical World View. Electromagnetic waves possessed a finite velocity, and a constant, c, appearing in Maxwell's equations proved to be identical to the speed of light. When this understanding emerged it seemed to confirm a deep connection between electromagnetism (via Faraday-Maxwell) and optics (via Fresnel). Secondly, Maxwell's equations seem to imply that under appropriate conditions electromagnetic waves might be generated and transmitted. When Heinrich Hertz (1854–94) proved the existence of these electric waves in 1887–88—we know them as radio waves—Maxwell's equations and an integrated view of electricity, magnetism, light, and radiant heat seemed abundantly confirmed.

Taking its fundamental parameters from Newton and German philosopher Immanuel Kant (1724–1804), the Classical World View thus began with notions of absolute space and time—space being uniform and Euclidean and time flowing inexorably and constantly. The vision that came into focus around Maxwell's work then posited three entities in the world: matter, a universal ether, and energy. Matter consisted of chemical atoms that had no interior parts, that were individually identical and distinct from one another; in this view all oxygen atoms, for example, were identical and categorically dissimilar from all hydrogen atoms. The Russian chemist, D. I. Mendeleev (1834–1907), arranged the elements into a chemical table of families, each atom with its own atomic number and weight. Although atoms could combine chemically, thus giving rise to the wealth of chemical substances evident in the world, the nature of the chemical bond remained obscure. Atoms, molecules, and larger bodies are endowed with mechanical energy and move; the dynamical motion of atoms and molecules determine their degree of heat; for gases, that motion came to be analyzed by the new discipline of statistical mechanics.

The material stuff of the world—its atoms and molecules—also possessed an intrinsic attractive force that allowed particles to agglomerate into larger and larger bodies. The force of gravity, therefore, provided a bridge between the invisible atomic world and the macroscopic world known to mechanics and astronomy. (Scientists no more understood the nature of gravitational force at the end of the nineteenth century than they had in Newton's day, or do in ours, but all experience confirmed such a force.) On the cosmic scale moving bodies like the earth, moon, planets, and comets obeyed the laws of classical physics, and in this way the Classical World View incorporated the tradition of the Classical sciences initiated by Newton and perfected by two centuries of problem-solving research.

The world ether complemented ordinary matter. This universal ether, as we have seen, provided the substratum for radiation: light, radiant heat, and the electromagnetic field, all of which embodied energy. Regular matter, the ether, and energy were all interconnected and obeyed the laws of thermodynamics. Thus, mechanical, chemical, electrical, magnetic, and light energy can become transformed one into another. The second law of thermodynamics, in particular, by introducing what has been called the "arrow of time," provided a major pillar on which the Classical World View rested. For seventeenth-century mechanics, for example, the laws of impact, say, were perfectly reversible—theoretically billiard shots work the same in one direction or its reverse; in contrast, the second law of thermodynamics posited an irreversible direction to time and the behavior of energy. Although couched in abstract and highly mathematical terms, the second law envisioned the ultimate "heat death" of the universe with all energy equally diffused throughout the cosmos, with atoms and molecules ultimately vibrating with a uniform degree of heat just above absolute zero.

The Classical World View outlined here came together in the 1880s. It represented a powerful, coherent, and mathematically exact understanding of the physical aspects of the cosmos and the interconnectedness of natural phenomena. With the coalescence of the Classical World View, the enterprise of natural philosophy that had begun so long ago in ancient Greece seemed almost to draw to a close, at least as far as the physical sciences were concerned. Yet it would be a mistake to overemphasize the consensus that developed around the Classical World View or how much agreement the elaborate and esoteric claims of contemporary science actually attracted. In point of fact, serious and lively debates arose among scientists and philosophers as to whether the Classical World View represented the true and underlying reality of nature. Furthermore, a series of unexpected discoveries soon dispelled any overconfident sense of closure and set the stage for yet another revolution, the twentieth-century revolution in physics launched by Albert Einstein.

The focus on the physical sciences in the foregoing analysis should not obscure important developments to be noted in the history of the life sciences in the nineteenth century. That the term *biology* (or science of life) was coined only in 1802 signals how much biology was a characteristically nineteenth-century field, particularly in the coming-into-being of laboratory-based and experimental approaches to investigating the chemistry and physiology of life. Although Robert Hooke had coined the term *cell* in the seventeenth century, cell theory did not emerge until the 1830s when German scientists M. J. Schleiden (1804–81) and Theodor Schwann (1810–82) looked through their microscopes and identified the cell as the basic unit of plant and animal tissue and metabolism. Claude Bernard's *Lessons in Experimental Physiology* (1855) and his *Introduction to the Study of Experimental Medicine* (1865) provided exemplars for the new style of research. Out of this background emerged the germ theory of disease, articulated notably by the German physician Robert Koch (1843–1910) and the great French chemist and experimentalist Louis Pasteur (1822–95). Koch identified the bacteria responsible for anthrax, cholera, and tuberculosis, and articulated key rules for proving the connections between bacterial pathogens and disease. Koch held academic and administrative positions in Berlin, ultimately as director of a new Institute for Infectious Diseases; he was awarded the 1905 Nobel Prize in Physiology or Medicine. Ideas about *microbes* in the 1850s led Pasteur to his studies of fermentation, and as a microbiologist Pasteur achieved world fame first through solving practical, applied science problems in the dairy, wine, silk, vinegar, and beer brewing industries. He did so sometimes flamboyantly, but his efforts led to pasteurization and improvements in production and in understanding bacteria. Built on his exquisite experimental techniques, Pasteur's further work in medicine, notably developing vaccines against anthrax and then rabies, cemented his fame and led to the establishment of the first Pasteur Institute in 1888 and the spread of Pasteur Institutes throughout the world thereafter. The germ theory of disease and the work of Koch and Pasteur marked the advent of truly scientific medicine and strongly reinforced the claims and status of science-based medicine and medical research. This work eliminated once and for all the ancient view of humoral pathology as well as competing environmental explanations for the causes of disease. Experimental medicine and biology, like so much else of contemporary science, evolved under changed institutional and professional circumstances. We turn next to this new institutional basis for science in the nineteenth century.

Science Reorganized, Again

The professionalization of science and careers in science represents a milestone along the route to today's scientific culture. Who is a scientist and how does one become a scientist? Historically, individuals who pursued investigations into nature have occupied many disparate social roles: the priests and anonymous scribes of the first civilizations; Greek natural philosophers; Islamic doctors and astronomers; Chinese mandarins and officials; medieval European university professors; Renaissance artists, engineers, and magi; and Enlightenment academicians. What needs emphasis here is that the modern social role of the scientist first appeared in the nineteenth century coincident with the Second Scientific Revolution.

A major element involved in the creation of the modern scientist as a recognizable social type was the establishment of a new institutional base for science in the nineteenth century, a second "organizational revolution" comparable to the first that formed part of the initial Scientific Revolution. The mainstays of organized science in the eighteenth century—state-sponsored learned societies—continued in the nineteenth, but less as centers for original research and more as honorary organizations rewarding past scientific achievement. In their place a complementary set of more vital institutions emerged for the practice of science. Established in 1794 in revolutionary France, the École Polytechnique provided a key institutional setting where France's leading scientists taught advanced theory to an exceptional generation of students who made France the leading scientific nation and French the leading scientific language through the 1830s. (The École Polytechnique also became the model for West Point and polytechnic colleges in the United States and elsewhere.)

In England the Royal Institution, founded in 1799, provided a home for such luminaries as Davy and Faraday. A revealing institutional novelty and sign of the times was the creation of hundreds of Mechanics Institutes throughout Great Britain and North America which at their height in the nineteenth century provided instruction in the sciences to over 100,000 artisans and interested middle-class amateurs.

The reform of the German university system represents the most thoroughgoing manifestation of this new organizational basis for nineteenth-century science. Beginning with the foundation of the University of Berlin in 1810, the natural sciences gradually gained a powerful new position within the network of universities in the German-speaking states. Nineteenth-century German universities became secular state institutions, and science instruction fulfilled a service function for the state in helping to train secondary-school teachers, physicians, pharmacists, bureaucrats, and other professionals.

An unprecedented emphasis on scientific research distinguished science education in this new context. That is, the role of a science professor was not merely to transmit old knowledge to students, but to lead the way in the production and dissemination of new knowledge. Several new pedagogical modes emerged to facilitate the new research agenda within universities, including the now-familiar teaching laboratory (inaugurated in 1826 by Justus von Liebig's chemistry lab at Geissen), graduate-level scientific seminars and colloquia, and specialized institutes within universities equipped for advanced research. The formal textbook as a medium for instruction in the sciences emerged for the first time as part of these developments, and the Ph.D. became a requirement for a career in science. The decentralized nature of the German university system spurred competition among the separate German states for scientific talent and thereby raised the level of research in science. The advent of polytechnic schools in later nineteenth-century Germany—the *Technische Hochschulen*—strengthened these trends by elevating pure science research in the university above applied-science training in the *Technische Hochschulen*. And the importance of science and scientists within the context of German higher education grew further as connections to technology and industry developed in the second half of the century, notably in the chemical industry, electrotechnology, and precision optics. German was the international language of science at the time, and Americans who wanted to study science went to Germany. The model of the research university soon spread outside of Germany; the foundation of Johns Hopkins University in 1876 brought the model of a research university producing Ph.D.s to the United States.

A characteristic feature of the first Scientific Revolution had been the social and intellectual turning away from the medieval university by the vanguard of science. After two centuries in the background the university once again became the leading institution for the natural sciences as part of the second Scientific Revolution of the nineteenth century. Even in England, where the universities were slow to respond to the upsurge in scientific research, by the third quarter of the century the two oldest universities, Oxford and Cambridge, along with newly founded ones in London (1826) and elsewhere in Great Britain, established new scientific professorships and sponsored research, some of it in fields bordering on technology and industry. None of these developments, however, changed the essentially male-dominated character of nineteenth-century science. As in the past, a mere handful of women became directly involved in science, usually exercising subsidiary functions. Some nontrivial exceptions include the American astronomer Maria Mitchell (1818–89) and the Russian mathematician Sonya Kovalevsky (1850–91), who received a Ph.D. from the University of Göttingen in 1874. The American geographer Ellen Churchill Semple (1863–1932) became president of the Association of American Geographers but was not permitted to matriculate when she studied at the University of Leipzig in the 1890s and, when attending lectures, was required to sit alone in an adjoining room.

In a separate set of developments the hospital likewise became reorganized as a locus of medico-scientific research as well as practice. Particularly large urban hospitals provided cadres of patients for systematic clinical trials, autopsies, and statistical compilations. The hospital has provided an institutional mainstay for medical science and scientific research since the nineteenth century.

The professionalization of science also came to entail specialized institutions. While the traditional learned societies, such as the Royal Society in London or the French Academy of Sciences, generally represented all of the sciences, in the nineteenth century specialized organizations devoted to single disciplines gradually supplanted the former as the primary seats for the identity of scientists and the presentation of their research. England led the way in new institutions of this sort with the Linnean Society (1788), the Geological Society of London (1807), the Zoological Society of London (1826), the Royal Astronomical Society (1831), and the Chemical Society of London (1841). New publication patterns also emerged, wherein specialized journals competed with the general journals of the traditional scientific societies as the loci of original publication in the sciences. Among the noted early specialized journals one might mention Lorenz Crell's *Chemische Journal* (1778), Curtis's *Botanical Magazine* (1787), the *Annales de Chemie* (1789), and the *Annalen der Physik* (1790), with similar specialized journals proliferating as the nineteenth century progressed. Finally in this connection, the nineteenth century saw the appearance of societies representing the professional interests of scientists, including the Association of German Researchers (the *Deutsche Naturforscher Versammlung* from 1822), the British Association for the Advancement of Science (from 1831), and the American Association for the Advancement of Science (1847).

The English philosopher and historian William Whewell (1794–1866) coined the English word *scientist* in 1840 as "a cultivator of science." That he did so is powerful testimony to the profound social changes surrounding science and scientific investigators at the time. The modern profession of science, the modern social role of the scientist, and modern beliefs about the nature and utility of science fully emerged in the nineteenth century and are recognizable today. Obviously, science and its practitioners had been part of the learned world at least since the dawn of civilization in ancient Mesopotamia. Yet it is telling of the changed circumstances of organized science in the nineteenth century that only then did the "scientist" emerge full blown as a social and professional entity.

The Spread of Industrial Civilization

Coincident with these developments in the world of science and natural philosophy in the nineteenth century, industrialization accelerated its impact on the landscape of human existence. What had begun haphazardly in the coal, iron, and textile industries in England in the eighteenth century coalesced into a mighty new mode of production, consumption, and social organization. Yet, at the beginning of the nineteenth century industrialization had still made only limited impact even in England, and its effects had hardly been felt elsewhere in the world. But from that point in time industrialization spread outward from England in waves across the globe, and by 1900 the world became a very different place than it was in 1800, as industrialization worked its transforming effects. But this deep technological and social metamorphosis occurred unevenly and in stages down to today. Furthermore, unlike the Neolithic and Urban Bronze Age revolutions of prehistory, industrialization has unfolded with amazing rapidity.

Fig. 15.4. Industrial civilization and pollution. In developed societies today scenes such as this one from the 1920s are less and less common, as industries and nations begin to take care to control their emissions and preserve a healthy environment. Such concerns are less strongly felt in developing regions where the expense of pollution controls is less easily met. It remains to be seen if industrial civilization can achieve a sustainable balance with the earth's ecology.

The eruption of industrialization on the world scale may be said to have begun in the period 1820–40. Belgium, Germany, France, and the United States were the first countries outside of England to be affected. Germany reached the landmark of 50 percent urban population in 1900, France and the United States by 1920. While the Netherlands, Scandinavia, parts of Spain, and northern Italy joined the core area of industrial civilization, the spread of industrialization over the European continent and North America was not uniform. Most of Eastern Europe, for example, and pockets within industrialized countries (such as Ireland and the American South and Southwest) remained overwhelmingly agrarian and pastoralist well into the twentieth century, as did much of the rest of the world into the twenty-first century.

Although notable national and regional differences characterize the process, industrialization in each case centered primarily on textiles, the iron industry, the railroad, and later, electrification. As industrial intensification continued, new areas of activity arose in construction, food processing, farming, and housework. By the end of the nineteenth century, whole new science-based industries began to emerge coincident with advances in electricity and chemistry. An expanded service sector complemented the core industries, often employing women as clerks, schoolteachers, nurses, secretaries, and, following the invention of the telephone by Alexander Graham Bell in 1876, telephone operators. And where industrial development took root, higher education eventually reflected its progress. Engineering, nursing, teaching, and architecture became professions in the nineteenth century through the establishment of university programs. In the case of engineering such programs leaned heavily (and increasingly) on the basic sciences, even when they were not fully applicable. Today's knowledge economy was off the ground.

By the 1870s these developments transformed Europe and the United States into the dominant powers in the world. A new era of colonial development unfolded, and the so-called new imperialism of the nineteenth century ensued with the solidifying of English rule in India, the creation of a new French colonial empire in Southeast Asia and Africa, the spread of Russian power eastward across Asia, Western incursions into China, the forcible opening of Japan to the United States in 1853–54, and the "scramble for Africa" by European powers after 1870. The point that deserves emphasis is that Europe and the United States dominated other countries and peoples in large measure because they monopolized the process of industrial intensification. Other countries did not possess the resources or technical capabilities to match the nations of Europe or America in output of armaments, railroad track, electricity, or shipping. By the nineteenth century Europeans had been making themselves felt on the world scene for three centuries, but at least the traditional high civilizations in India, China, and Japan had more or less successfully resisted succumbing. That changed as the nineteenth century unfolded, and by 1914 the global empires of the Western powers enveloped 84 percent of the world.

The spread of industrialization is thus intimately linked to the history of European colonialism and imperialism. Imperialism fostered industrialization by spurring production in Europe, by securing cheap sources of raw materials overseas, and by creating new markets for the products of industrial production. India provides a relevant example. Prior to the Industrial Revolution in Europe and the formal takeover of India by the British government in 1858, India ranked as one of the most technically advanced regions of the world. With the loss of its political independence, what followed amounted to the collapse of a flourishing and technically advanced traditional economy. In 1750, for example, India accounted for one-quarter of the world's manufacturing output, but by 1900 that figure had fallen to just 2 percent. In a word, British rule brought Western-style industrialization but undermined and restructured the traditional economy. Railroads, of course, proved essential to effective British control over India, and railroads spread rapidly over the subcontinent. The first railroad began operations in India in 1853. By 1870 engineers had laid over 4,500 miles of track; by 1936 mileage had risen to 43,000 miles, creating the fourth largest railroad system in the world. Hardly a decade after Morse's first telegraph line in America, a telegraph system arose in India literally alongside the railroad. By 1857 4,500 miles of wire had been strung, and in 1865 India became connected to Britain by telegraph cable. But notably, these technologies did not spur industrial development in India itself. Rather, they served as instruments of colonial control and as transportation and communications conduits to extract raw materials and commodity products and to bring British manufactured goods to Indian markets. Thus, for example, India went from being the world's largest exporter of textiles to a net importer. The traditional shipbuilding industries that flourished in India through the nineteenth century—often fulfilling European contracts—became completely outmoded and outdated once the shift to steamships began. Although indigenous Western-style industrial activity developed in India later in the nineteenth and twentieth centuries, the initial effect of contact with industrial Europe resulted in increasing poverty through technological displacement, unemployment, and neglect of the traditional agricultural infrastructure. The pattern of colonial and imperialist exploitation seen in India repeated itself in virtually every other European encounter with the nonindustrialized world. In a word, the success and power of the Western industrial economies virtually precluded the independent development of industrialization elsewhere. Poorer and less-developed nations stood hardly a chance against the technical and economic power of the European onslaught, soon followed by an American second act.

A constellation of new technological practices provided the basis for further European dominance in the nineteenth and twentieth centuries and for the imposition of European imperialism on an even greater worldwide scale, particularly in Africa and Asia. At the heart of the matter is what has been labeled the "industrialization of war," that is, the application of industrial methods to the production of war materiel. Consider, for example, the steamship. Developed in the first decade of the nineteenth century, the steamship quickly and reliably moved goods and people and furthered the penetration of river systems around the world. As a military vessel, the flat-bottomed steamship was introduced into the Royal Navy in 1823, and it secured British victories as early as the 1840s in the opium wars in China. Development of the steamship culminated in the steel-hulled coal-fired ship armed with breech-loading rifled artillery firing high-explosive shells. Floating citadels, like the massive HMS *Dreadnought* launched in 1906, allowed the projection of Western power to virtually any coastal area of the world. Ultimately, with 15-inch guns and elaborate sighting, loading, and recoil systems, they could strike targets within a 20-mile range. Explosive shells, in place of cannonballs, spelled the end of the wooden warship. Only those nations with the resources and know-how to produce steel could play the deadly new game.

A word needs be said about steel. Although the term "steel" was used loosely prior to the nineteenth century, the material, an alloy of iron and carbon, was first produced in quantity after the spectacular invention in the 1840s of an inexpensive method of producing wrought iron. What became known as the Bessemer process was hailed at the time as "the greatest invention ever made." The product of the traditional furnace is cast, or pig, iron, a material with the brittle characteristics of stone, but with the advantage of being able to be cast in intricate shapes. Chemically, it is iron with around 2 percent of carbon. To produce wrought, or malleable, iron, molten cast iron was laboriously stirred, whereby the carbon was reduced and a one- to two-hundred–pound "bloom" of wrought iron was produced. It was the blacksmith's iron, malleable under the hammer. That laborious and time-consuming process was replaced through a totally unexpected effect. First an American ironmaster, Thomas Kelly, and soon afterward a British experimental entrepreneur, Henry Bessemer, accidentally hit on the surprising discovery that if air is blown through the molten cast iron in a blast furnace a violent, explosive reaction occurs and in less than half an hour a large vat of cast iron is converted to wrought iron. (The carbon in the cast iron combines with oxygen in the air and passes off as a gas.) The Age of Wrought Iron had arisen, symbolized by the Eiffel Tower (1889) and numerous other structures, including the armature of the Statue of Liberty in New York City. But this age was short-lived and soon succeeded by the Age of Steel as British ironmasters and chemists learned to produce true steel, with a carbon content intermediate between that of cast and wrought iron. The new material was malleable like wrought iron but with a tensile strength much higher than wrought iron or, especially, cast iron. If the Eiffel

Fig. 15.5. The industrialization of war. Launched in 1906, the HMS *Dreadnought* was the most formidable armored battleship of its day. With a main battery of ten 12-inch guns, driven by a stream turbine power-plant, and capable of a speed of 21 knots, this revolutionary vessel surpassed every previous warship in design and was testimony to the might of industrial civilization in Britain.

Tower in Paris is the symbol of the Age of Wrought Iron, the Empire State Building in New York City—built in 1930—is the symbol of the Age of Steel.

Steel provided the raw material for the armaments of industrialized warfare. The arms race that developed in the second half of the nineteenth century in Europe among England, Germany, and France escalated military developments and led to the creation of machine guns, rapid-firing bolt-action rifles, new types of bullets, submarines, self-propelled torpedoes, destroyer battle groups, and the full armamentarium of modern warfare, including "gunboat diplomacy" so characteristic of Western relations with the non-Western world. In the 1860s the American Civil War, mainly through the use of railroads to move military units, presaged the effects of the industrialization of land warfare, and the victory of the United States over Spain in the Spanish-American War in 1898 signaled the effectiveness of the new style of gunboat and the beginnings of extraterritorial American imperialism. The industrialization of war expanded in World War I, with combatants dependent on heavy industry, submarines, tanks, railroads, nascent air power, and poison gas, the last clearly indicating the participation of scientists, science, and the chemical industry in the war effort. Indeed, World War I has been called the "chemists' war." And finally, along these lines, in the interwar period the rapid development of the warplane and motorized military vehicles gave European imperialism and colonial powers new weapons with which to enforce their rule and to compete among themselves. Military technologies and industrial development stimulated each other at the end of the nineteenth century, in a manner akin to the effects of the steam engine and railroad-building over the previous century.

The industrialization of Russia extended industrial civilization eastward from its European base. Only partly a European nation, nineteenth-century Russia preserved a traditional, agrarian economy. The Tsarist government prevented the outright colonization of the country by foreign powers, but industrialization in Russia depended on outside expertise and capital (mostly from England). At the same time, railroad building provided a mechanism for Russia's own imperialist expansion across Asia. The Moscow–Saint Petersburg line opened in 1851, and the number of railroad miles mushroomed from 700 in 1860, to 12,500 in 1878, to 21,000 in 1894, and over 36,000 in 1900. The first leg of the Trans-Siberian railroad opened in 1903. By 1913, with over 43,000 miles of railroad, Russia possessed the fifth largest industrial economy in the world.

Following the Russian Revolution of 1917, the Soviet government speeded the industrialization of the nation, achieving during the 1930s the most rapid rate of industrial development in history. By the late 1940s, despite the heavy toll of World War II, the Soviet Union emerged as the second largest manufacturing economy in the world (the United

States was first), with leading heavy industries in coal, iron, and steel production, chemicals, oil production, and electric power generation, and, as always, with increased urbanization. Officials achieved the transformation of the Soviet economy through heavy-handed state planning, extraordinary inefficiencies, and at great human cost, but also with more attention to worker issues, including concern for education, medical and child care, and recreation. From the point of view of the history of industrialization, the case of the Soviet Union stands out, especially insofar as developments there took place largely independently of industrialization elsewhere and in the face of hostile reactions in the West, including an American embargo on trade that lasted, at least officially, until 1933. The future of industrial development in the former Soviet Union is a question that historians will be keen to judge.

But, inevitably, industrialization spread beyond its European and America homeland. Japan presents a notable case in the elaboration of the world industrial economy today. Through 1870, industrial intensification remained confined to the West, and Japan became the first to break the European stranglehold. Forcibly opened to outside trade in 1854, Japan, unlike India, managed to maintain its political independence and never became subjected to outright foreign control. The Meiji Restoration of 1868 signaled the end of feudal government in Japan, and a combination of traditional merchant classes and progressive government civil servants began to promote industrial development. Created in 1870, a Ministry of Industry set an enduring pattern for state planning and the financing of Japanese industry. Initial investments went into railroads, and the first line linking Tokyo and Yokohama opened in 1872. Japanese shipbuilding also received early government support to promote trade and in order to compensate for the lack of raw materials in Japan itself. Mechanizing silk production was another arena of early industrial policy by the government. A demographic spurt from 30 million in 1868 to 45 million in 1900 (and to 73 million in 1940) provided a population base for an industrial working class. In stark contrast to Europe, a large proportion of women in the workforce (above 50%) was a unique feature of early industrialization in Japan. Paternalism and the group-identity characteristic of Japanese culture generally helped ease the transition to an industrial economy, with less social and political strife than seen in the West. In the twentieth century Japan, too, became an imperial power, with the victory of the Japanese in their war with the Russians in 1904–5 and expansion in East Asia and the Pacific in the 1930s and 1940s. Regarding the scientific underpinnings of Japanese industrial development, the first university in Japan was founded only in the 1880s.

Japan led the way in the twentieth century in breaking the industrial monopoly of Europe and America, and especially after World War II industrialization proceeded on a truly multinational and global basis.

We will return to this point, but in the meantime the scientific and industrial developments surveyed thus far in this chapter brought about another transformation of great historical moment: the modern merger of science and technology in the context of industrial civilization.

Applying Science in Industry

Science and industry and the cultures of science and technology generally began their historical unification in the nineteenth century. The main thesis of this book has concerned the historically limited degree of applied or practically oriented science prior to the nineteenth century. In the earliest civilizations and state-level societies thereafter, governments patronized useful knowledge and science in the service of administration. In Europe, state support for what were deemed useful sciences appeared slowly after the Middle Ages, as noted in the case of cartography, for example, or somewhat later as an outgrowth of the Scientific Revolution in the creation of state scientific societies. The conviction that natural philosophy ought to be turned to public utility became an ideological commonplace in the seventeenth century. More readily apparent, however, are the intellectual and sociological disjunctions between the sciences and the vast body of technology as these have developed throughout history. One measure alone illustrates the scope of the separation—in Europe scientists were university educated, engineers and craftsmen were not. In the Industrial Revolution in eighteenth-century England the worlds of science and technology drew closer together, but we were hard-pressed to find historical evidence to support the view of contemporary technology as applied science. In the nineteenth century, however, several important novelties appeared that began to recast the age-old separation of science and technology that originated in Hellenic Greece. Firm connections came to link theoretical science and the key new player on the scene, large-scale industry. To be sure, much of science and technology remained separate, but the new dimensions of applied science that arose in the nineteenth century in the context of industrialization represent historical departures of great consequence that solidified in the twentieth and twenty-first centuries on a global scale.

The new nineteenth-century science of current electricity spawned several applied-science industries, of which the telegraph represents a prime example. Following the discovery of electromagnetic induction by Michael Faraday in 1831, the scientist Charles Wheatstone and a collaborator invented the first electric telegraph in 1837. Wheatstone and other European and American scientists and inventors worked to create a telegraph industry, spurred in part by the utility of the telegraph as an adjunct to railroad development. These efforts quickly culminated in the system patented by Samuel F. B. Morse in 1837 and field-tested in 1844, incorporating Morse's renowned alphabetic code using dots and dashes. London and Paris became connected by telegraph in 1854, the first trans-Atlantic telegraph cable was laid in 1857–58, the first transcontinental telegraph in North America linked New York and San Francisco in 1861, and the telegraph and the railroad spread together worldwide thereafter. The upshot was something of a communications revolution.

The telegraph tapped a body of preexisting scientific knowledge, but the development of the new technology of telegraphy involved the solution of a myriad of problems—technical, commercial, and social— that had little or nothing to do with contemporary scientific research or theory. In other words, the coming into being of a science-based technology usually involves the creation of a complex technological system to the point where it is misleading to think of such systems as merely "applied science."

The telephone was a potent new technological system that emerged out of this same complex of science and industry. Alexander Graham Bell invented the telephone in 1876, but it took some time before telephony challenged the telegraph as an effective communications medium. An infrastructure of wires, central switching stations, manufacturing operations, telephone operators (mostly women), and social behaviors had to evolve. The first commercial exchange opened in 1878, telephone lines linked Boston and New York in 1884, and Bell himself placed the first transcontinental phone call in 1915. The diffusion of the dial telephone and automatic switching after 1905 proved keys to the elaboration of the telephone system, as were government subsidies for extending telephones lines into rural areas.

Much the same point regarding technological systems is evident in considering the electric lighting industry that also arose in the last quarter of the nineteenth century. Such an industry clearly derived from prior work in the new science of electricity. As celebrated in traditional biographies of great inventors, Thomas Alva Edison (1847–1931) in New Jersey and Joseph Swan (1828–1914) in England independently created the incandescent light bulb through elaborate empirical trials in 1879. By the 1880s and 1890s the science involved in the developing electric lighting industry was hardly new. Some of it, with regard to insulators, for example, harked back to the eighteenth century. This example brings home that in considering applied science it can be analytically fruitful to distinguish between whether the science involved is, say, "boiled down" science or whether it represents the application of more recent and cutting-edge theory. Furthermore, the light bulb itself hardly constitutes the establishment of a practical electric lighting industry. As we will see further in chapter 17, a large and complex technological system had to be brought into being before an electric lighting industry could be said to have existed, a system involving gen-

Fig. 15.6. Invention without theory. Thomas Edison was a prolific inventor (with more than 1,000 patents) who had little education or theoretical knowledge. His ability to tap "boiled down" science and his organizational skills in institutionalizing invention in his laboratory in Menlo Park, New Jersey, proved essential to his success.

erators, power lines to distribute electricity, appliances, meters to measure consumption, and methods of billing customers, to name only a few of its many elements.

Another of the early instances where science and up-to-date scientific theory became applied to technology and industry was the case of radio communications, where practical application followed closely on the heels of theoretical innovation. Seeking to confirm Maxwell's theory of electromagnetism, Heinrich Hertz demonstrated the reality of radio waves in 1887. Hertz worked exclusively within the tradition of nineteenth-century theoretical and experimental physics, but when the young Italian Guglielmo Marconi (1874–1937) first learned of Hertzian waves in 1894, he immediately began to exploit them for a practical wireless telegraphy, and by the following year he produced a technology that could communicate over a distance of a mile. Marconi, who went on to build larger and more powerful systems, received his first patent in England in 1896 and formed a company to exploit his inventions commercially. In 1899 he sent his first signal across the English Channel, and in a historic demonstration in 1901 he succeeded with the first radio transmission across the Atlantic. The creation of this new technology involved much more than the application of scientific theory, however direct, and, although Marconi's contribution was essentially technical, in this instance the line between science and technology became so blurred that in 1909 Marconi received a Nobel Prize in *physics* for his work on wireless telegraphy. The case is also noteworthy because it illustrates that the outcome of scientific research and technological change often cannot be foreseen. What drove Marconi and his research was the dream of ship-to-shore communications. He had no prior notion of what we know as radio or the incredible social ramifications that would follow the first commercial radio broadcasts in the 1920s.

The growth of applied science in the nineteenth century was not limited to physics or industries connected solely to the physical sciences. In the realm of scientific medicine, for example, the introduction in the 1840s of anesthesia in dentistry and surgery, and the antiseptic measures developed in the 1860s by Joseph Lister (1827–1912) proved boons for humanity. As mentioned, the scientific and medical researches of Louis Pasteur held practical and economically important consequences for a variety of industries, and his pioneering work in developing vaccines likewise brought applied science to the service of the public good.

Chemistry was another domain where important practical applications in nineteenth-century industry were forthcoming from science. Through the middle of the century the dye industry in Europe remained a traditional craft activity with no contact whatsoever with the world of science. Then, in 1856, out of the background of German advances in organic chemistry, the English chemist William Perkin discovered an artificial dye producing a purple color. The economic value of bright, synthetic dyes became immediately apparent, and mastery of the chemistry of dyes and dyestuffs derived from coal tar became essential to the textile industries. Until the various German states adopted a uniform patent code in 1876 competition between firms amounted largely to raiding one another's expert chemists. After 1876, however, with patent rights assured, the emphasis shifted to research and development of new dyes. A new institution uniting science and technology—the industrial research laboratory—emerged as a result. The Friedrich Bayer Company, makers of Bayer aspirin, created a research division and hired its first Ph.D. chemist in 1874. In 1896 the number of its salaried staff scientists reached 104.

The style of applied research undertaken at the Bayer research laboratory deserves emphasis. The story hinges on the fact that the German chemical industry established close contacts with research universities. Industry supplied universities not only with materials and equipment for advanced studies in chemistry, but also with students and with opportunities for consulting. Universities reciprocally offered industry trained graduates and prospects for scientific cooperation. In a further division of labor, fundamental research became the province of the universities, while industry undertook mainly empirical research and routine experiments to test dyes and their fastness on different materials in the hope of developing commercial products. For example, in 1896 the Bayer Company subjected 2,378 colors to an assortment of tests but marketed only 37. As this case indicates, even where theory was applicable, "research" often still took the form of trial and error, albeit conducted by scientists in a research environment. The reality of applied science was (and is) often far removed from the assumption that technology is merely the translation of scientific theory into practice.

The model of the research laboratory spread widely in late nineteenth- and early twentieth-century industry. Thomas Edison's laboratory at Menlo Park, New Jersey, established in 1876, represents an early example. Others include those of Standard Oil (1880), General Electric (1901), DuPont (1902), Parke-Davis (1902), Corning Glass (1908), Bell Labs (1911), Eastman Kodak (1913), and General Motors (1919). Today thousands of such labs operate in the United States alone. The advent of industrial research has been hailed as the "invention of invention," although this view is somewhat misleading in that research labs are generally not innovators of new technologies. For the most part research labs concern themselves with the development and extension of existing technologies, and they often function as part of a business strategy to develop and control patents to ward off competitors. But even today, as we will see further, pathbreaking inventions, like xerography and the personal computer, may still be the work of independent inventors rather than established scientists or engineers working in industry.

But these theoretical and technical developments were not alone in transforming the landscape of European science and culture in the nineteenth century. As Britain expanded its empire and spread its economic tentacles across the globe, it sent ships to far corners of the world to map lands, to trace wind and ocean currents, to search out resources and markets, and to collect specimens of plants and animals. An utterly unexpected result of these efforts was the theory of evolution.

Life Itself

Science usually changes incrementally, and the increments are usually small. A new molecule might be discovered now and then; a new fossil might be found, or a new star. But these incremental discoveries leave the theoretical framework of a science undamaged. Indeed, they generally strengthen it. But occasionally a major upheaval occurs, and when the dust settles a new theoretical framework has arisen in place of the old. These upheavals—scientific revolutions—not only replace old and long-held ideas with sharply revised concepts, they also modify research boundaries and create new problems for research that the old science could not even formulate. The Copernican Revolution was one such revolution that unfolded in the sixteenth and seventeenth centuries. In the nineteenth and twentieth the Darwinian Revolution similarly refashioned the intellectual landscape of science.

Charles Darwin published *The Origin of Species* in 1859. The year and the book represent another watershed in the history of science. On the one side of the Darwinian divide stood what might be labeled the traditional Christian worldview. This view, sanctioned by biblical authority and apparently confirmed by both folk wisdom and scientific observations, held that the species of plants and animals were created separately and remain fixed—like producing like. This worldview was largely static, not admitting of significant change. It included the notions that each distinct species was separately created at a not-very-distant point in time, possibly only 6,000 years ago; that various catastrophes, notably Noah's flood, explain our observed geological and biological surroundings; and that humans occupy a special place in the cosmos, one created by God who is the architect of all the apparent design in the world and who plays an active, divine role in its historical unfolding.

On the other side of the landmark year of 1859 radically contrary views, springing from Darwin's seminal ideas, gained ground to the effect that the species are not fixed, that there were no separate creations, that the life forms we observe around us evolved through the process of natural selection, that biological evolution and geological change have been unfolding gradually over eons of time, that humankind is no more than an artifact of the history of nature, and that the study of nature offers no evidence of miracles or any divine plan.

The Copernican and Darwinian Revolutions display parallel characteristics. The Copernican Revolution departed from a set of astronomical beliefs that in their essential form had been held for 2,000 years, and because of their seemingly self-evident truth—the stability of the earth and the motion of the sun—had been taken for granted in both astronomy and religion. The Darwinian Revolution departed from the age-old belief, also preserved in the biblical tradition, of the fixity of the species. Both Copernicus and Darwin held back publication of their new ideas, not out of fear of religious or political authorities, but rather out of fear of ridicule for far-fetched theories that in their day they could not prove. And both of them were vindicated by subsequent research. The result of these two revolutions is the scientific worldview in which heaven and earth obey the same physical laws while man and beast share the same biological origins.

Natural Theology and the Background to Darwin

Consistent with Galileo's advice, a consensus now reigns in scientific and most theological circles that in the study of nature biblical authority must defer to scientific investigation. Historians of science long ago dismissed the idea of an innate and abiding conflict between science and religion. Indeed, one outcome of the Scientific Revolution of the seventeenth century was precisely to strengthen ties between science and the traditional Christian worldview. Natural theology or the idea that one can gain insight into the divine plan by examining God's handiwork in nature gained significant ground, particularly in England. The conviction was that, by investigating apparent design in nature, we could better understand the Great Designer and his providential provisions for humans and their needs. In other words, as evidenced by such tomes as John Ray's natural history, *Wisdom of God in the Creation*, and Thomas Burnet's geology, *Sacred History of the Earth* (both 1691), seventeenth-century religious sensibilities stimulated scientific research toward a harmonious reconciliation of genesis and geology.

Empirical research in botany, natural history, and geology exploded in the eighteenth century, and by 1800 scientists knew much more about the world than a century earlier. Yet the strand of natural theology that sought to find God in His handiwork remained no less an element, particularly of English science at the turn of the nineteenth century, than it had been in the seventeenth. The first edition of William Paley's argument that design requires a Designer, *Natural Theology, or*

Evidences of the Existence and Attributes of the Deity Collected from the Appearances of Nature, appeared in 1802, and it brought up to date for a new generation of Englishmen, including the young Charles Darwin, the idea that the natural sciences and received religion were two faces of the same coin. The argument, embraced by the young Darwin, was that just as a pocketwatch found by the roadside implies the existence of a watchmaker, a beetle or a butterfly, which are infinitely more complex and purposeful, imply a Creator. The tradition of natural theology continued strongly in England into the 1830s when the earl of Bridgewater commissioned the so-called Bridgewater Treatises, eight scientifically based studies designed to show the "power, wisdom and goodness of God in the works of creation."

In the eighteenth century, Swedish botanist Carolus Linnaeus (1707–78) brought order to the explosion of knowledge in the organic realm—what he thought of as the plan of God—by arranging plants and animals in the "binomial" system of classification still in use today. (Each plant or animal receives two Latin names, the first, the *genus*, places it in a general category, while the second, the *species*, refers to its specific characteristics.) Although his strict classifications seemed to confirm the fixity of species, late in life Linnaeus began to wonder about the distinction between species and varieties within species and hinted vaguely that both might be, in his words, "daughters of time."

Just as suggestions that the planetary system may be sun-centered predated Copernicus, a series of precursors to Darwin circled around the principle of the transformation of species. French naturalist and superintendent of the royal gardens in Paris, the Count de Buffon (1707–88) held that the species evolved. Buffon believed not in progressive evolution but that the plants and animals we observe around us today devolved or degenerated from more robust ancestors that existed at earlier eras in time. He never provided a mechanism to explain how such transformations were possible, however, and he later recanted his belief when confronted with religious criticism as to the antiquity of the earth.

Another French naturalist, Jean-Baptiste Lamarck (1744–1829) went further. He postulated a *mechanism* whereby evolution could occur—what has become known as the inheritance of acquired characteristics. Essentially, Lamarck proposed that, through use and disuse of parts, an individual organism modifies itself by striving to adapt to its changing environment. Moreover, the modifications are supposedly inherited by the organism's offspring, establishing a line of descent that from time to time produces new species. Thus, for example, the giraffe has a long neck because its ancestors continually stretched theirs to reach foliage high up in trees, and longer and longer necks were then inherited. The inheritance of acquired characteristics was an appealing and influential idea, and Darwin himself found refuge in it when attempting to account for the source of variation within species. Lamarckian inheritance of acquired characteristics intermittently served as an anchor for those who clung to a hope that evolution contains an element of guidance or purpose, or at least of direct environmental (read social) influence and is not merely the result of random, long-term processes. Lamarckianism became empirically discredited but survived into the twentieth century, notably in the case of Soviet agronomist Trofim Lysenko and his disastrous program under Stalin for improving agricultural production in the former Soviet Union.

Around the time that Lamarck formulated his ideas, Charles Darwin's own grandfather, Erasmus Darwin (1731–1802), put forth similar notions in a series of scientific poems. The elder Darwin believed that useful traits are passed on through biological inheritance and their slow accumulation results in diverse forms. To start the process life emerged from inanimate matter:

Hence without parents, by spontaneous birth,
Rise the first specks of animated earth.

Some of these new ideas were consistent with old ones. It had long been assumed that the natural world formed what has been termed the Great Chain of Being from the simplest globule of life to the highest and most divine form, all in accordance with a law of continuity that stated that the links in the chain are infinitely small and there are no jumps or gaps between them. Both the prevailing theory of separate creation and the new theories of evolution embraced the idea of an organic continuum unbroken by any missing links. But, apart from a few such points of intersection, the two approaches necessarily diverged in the face of the evidence of natural history. With regard to another telling point—the age of the earth—traditional belief was content to accept either the biblical chronology or at most a modest extension backward in time. But the new view seemed to call for a much greater antiquity of the earth since the postulated evolutionary changes were necessarily small and incremental and required long stretches of time to produce the diversity that the natural world displays. Only slowly did some naturalists by the end of the eighteenth century become convinced that the earth is substantially older than the few thousand years granted by biblical authority and the story of Noah's ark and the flood.

The accumulating evidence from the fossil record and geological strata proved inconsistent with traditional beliefs about the age of the earth. Some fossil impressions were embedded in what appeared to be extremely old rocks. And some were found in puzzling strata. For example, marine fossils found at high elevations away from any body of water suggested great upheavals, evidently over long periods of time, in the earth's surface. Traditionally fossils were thought to be geological accidents or "sports of nature" and not the truly fossilized remains of organic beings. Around 1800, however, the discovery and recognition of large vertebrate fossils ("elephant bones") dramati-

Fig. 16.1. The Megatherium. In the first decades of the nineteenth century the discovery and reconstruction of the fossils of extinct large animals revealed a "lost world of prehistory" and heightened the problem of how to account for apparent biological change over time.

cally changed the debate. Based on recent developments in the science of comparative anatomy of living creatures, experts could reassemble newly discovered fossil remains. The stunning revelation of the reality of biological extinction and a lost world of prehistory emerged with shocking clarity. By 1830 there was no denying that strange large creatures, now extinct, once roamed the earth.

Squaring the mounting new evidence with biblical convictions and traditional chronology tested the imaginations of many naturalists. A conservative theoretical synthesis known as *catastrophism* coalesced around the work of the eminent French naturalist, Baron Georges Cuvier (1769–1832), the successor to Buffon at the Museum of Natural History in Paris. Cuvier concluded that seemingly great changes occurred through a few catastrophic events spaced over a relatively brief past. The apparently sudden breaks observed in the geological and fossil records could then be accounted for by this series of catastrophes (floods, fires, volcanoes) and new life forms. Furthermore, the apparent sequence of life forms shown in the fossil record (of reptiles and fish followed by birds and mammals, of extinct forms preceding living ones) would seem naturally to point to the coming of Man. In this manner the Cuvierian synthesis preserved the (relatively) brief history of the world required in the traditional view. It also admitted the reality of extinction and of progressive biological change over time, but *without* the necessity of any organic transformations or violation of the principle of the fixity of species. Cuvier's was a great intellectual achievement.

At the end of the eighteenth and beginning of the nineteenth centuries, "catastrophist" doctrines in geology were challenged by what became known as *uniformitarianism* or gradualism—the claim that the physical processes now operating on the surface of the earth have been operating uniformly, rather than catastrophically, over long periods of time and have slowly produced the changes that the geological record reveals. In 1795 Scottish geologist James Hutton (1726–97) published his landmark work, *Theory of the Earth*, in which he attributed the geological features of the earth to the combined actions of two opposing forces—a leveling tendency induced by gravitation and a lifting tendency produced by the heat within the earth. These forces, which as we observe them at present produce their effects slowly, would have required immense stretches of time to account for present geological conditions. They would have acted uniformly, as they evidently do today, and would have produced a world with, as Hutton put it in his majestic aphorism, "no vestige of a beginning, no prospect of an end." After Hutton, no rational hope existed for reconciling the limitless time span of uniformitarian geology with the traditional biblical account.

In his *Principles of Geology* (three volumes, 1830–33), Charles Lyell (1797–1875) revived and strengthened Hutton's uniformitarian argument that the physical features of the earth result from processes we observe today working slowly and continuously over long periods of time, in opposition to catastrophism which granted little time to geological processes even though it recognized extinction and biological change. While uniformitarian geology admitted infinitudes of time, it hesitated over the issue of biological change. Publicly at least, Lyell himself, the most eminent geologist in Britain, remained adamant in denying the possibility of the transformation of species: "There are fixed limits beyond which the descendants from common parents can never deviate from a certain type." In this intellectual environment the question of how to account for biological change on uniformitarian principles awaited its prophet.

Darwin

Charles Darwin (1809–82) was born into a well-to-do family of rural English gentility. His father, like his grandfather Erasmus, was a successful country physician. His mother was the daughter of Josiah Wedgwood, the potter-entrepreneur, but she died when Charles was only eight years old, and Darwin's older sisters largely saw to his upbringing. In his privileged youth Darwin showed little passion for study, although something of the future naturalist might be read into his love of the outdoors, of dogs, and especially in his fascination with beetles. His corpulent father demeaned the young Darwin, saying, "You care for nothing but shooting, dogs, & rat catching, & you will be a disgrace to yourself and all your family."

Darwin entered the University of Edinburgh to study medicine but, sickened by the sight of blood, he soon disappointed his father by dropping out. He then transferred to Christ's College of the University of Cambridge with the intention of studying to become an Anglican clergyman. Darwin was a middling student who enjoyed cards and hunting, but his naturalist's interests led him to make contact with the eminent science professors at Cambridge, the botanist John Henslow (1796–1861) and the geologist Adam Sedgwick (1785–1873). Darwin received his degree in 1831 at age 22, still with a vague intention to become a clergyman. Indeed, that lack of firm commitment settled his fate, for shortly after graduation he was presented with the opportunity to sail as the naturalist on board the HMS *Beagle* on a projected two-year mapping project in South America.

In the end, the voyage lasted five years, from September 1831 to October 1836, and it transformed both Darwin and ultimately the world of science. Darwin proved to be an acute observer, an avid collector of plant and animal specimens, and a proficient writer of letters to his former teacher, John Henslow, who transmitted them to the larger community of English naturalists.

Darwin's mind was prepared for his conversion by the doctrine of uniformitarianism that he received from Lyell's *Principles of Geology*, the first volume of which he carried on the *Beagle* and later volumes of which he received en route. In the face of his Christian roots and his respect for Lyell's teaching, only with the greatest intellectual effort did Darwin eventually arrive at his belief that as the physical features of the earth changed, plants and animals well-adapted at one point in time became ill-adapted—unless they, too, became modified in step with the geological sequence.

Everywhere the *Beagle* landed Darwin saw evidence of change. Traveling along the coast of South America and across the pampas, he noted biological change over space as related species of birds abruptly gave way to others. Discovering the fossils of extinct giant armadillos where their miniature descendants scurried about in his day, he confronted the question of biological change over time. But nowhere was the puzzle of biological diversity and change more pronounced for Darwin than on oceanic islands. Two weeks after leaving England the *Beagle* arrived at the Cape Verde Islands off the west coast of Africa where Darwin observed animals similar but not identical to animals on the nearby mainland. And later, when he explored the Galapagos Islands off the west coast of South America he recorded a strikingly similar pattern—animals comparable to those on the coast of Ecuador but not quite the same. Since the Cape Verde and Galapagos Islands seem environmentally similar, why were their animal populations similar to their respective mainlands rather than to each other? He also observed a variety of birds on the Galapagos Islands—since known as "Darwin's finches"—and he noted their different features island to island. But Darwin was not yet a Darwinist. What he later recognized as separate species, on the voyage of the *Beagle* he generally

took to be merely varieties. Only once at this stage did he speculate that if varieties diverged sufficiently it "would undermine the stability of Species." And only after he returned to England and consulted with zoologists about the specimens that he had collected did it dawn on Darwin that these patterns he observed on the Galapagos of minor variations in restricted geographical locales could not be explained by the doctrine of divine creation, but rather could only be accounted for through some sort of evolutionary process.

Poor Darwin was seasick the whole time the *Beagle* was at sea. After his return he did not travel much and never left England again. In 1837, a few months after the *Beagle*'s circumnavigation of the earth, Darwin began a notebook on "the species question." He had become convinced of the fact of biological evolution, but he still groped for a mechanism that would drive the process. He found that mechanism in the writing of an English clergyman.

In his *Essay on the Principle of Population* (1798/1803) the Reverend Thomas Robert Malthus reached the conclusion that "population, when unchecked, increases in a geometrical ratio. Subsistence [food] increases only in arithmetical ratio. A slight acquaintance with numbers will shew the immensity of the first power in comparison with the second." In other words, as population necessarily outpaces food supplies, competition for resources becomes increasingly fierce, a competition in which the swifter, stronger, hardier, and more cunning are most likely to survive and reproduce. That insight into the pressure of population against resources gave Darwin the decisive piece of the puzzle he needed. Malthus had applied the principle to social change; Darwin redirected it to plants and animals. He now possessed, in 1838, the theory of evolution by natural selection: despite high rates of reproduction, populations of species are kept in check by high mortality as individuals compete for resources, and those best adapted to the habitat are more likely to survive and produce offspring. That process, he believed, given enough time, could account for the patterns of variation he had observed on the voyage of the *Beagle* and which he was already convinced could only be explained by descent from a common ancestor.

Darwin had formulated his theory of evolution, but he had told no one. In 1838 he was the only person in the world who grasped the reality of evolution. He alone recognized the mechanisms driving evolutionary change, and he was the first to confront the deep ramifications of modern evolutionary thought. Darwin bore a heavy burden. He was fully aware of the importance of his discovery, and like all scientists he had every incentive to publish in order to secure his priority. In 1842 he wrote a short sketch of his theory, which he expanded in 1844 into a substantial manuscript of 231 folio pages. He left a letter with instructions that the manuscript should be published if he died prematurely. But he still held back from announcing the great work publicly. Al-

Map 16.1. (opposite) The voyage of the *Beagle*. Charles Darwin sailed as naturalist aboard the HMS *Beagle* on its five-year voyage to map the coasts of South America (1831–36). A keen observer, Darwin became exquisitely sensitive to the distribution of plants and animals, particularly on the Galapagos Islands off the coast of Ecuador which he visited toward the end of his voyage. He returned to England convinced of the fact of evolution but struggled to articulate a theory or mechanism to account for evolutionary change.

Fig. 16.2. Darwin's finches. The variations in beak formation of the birds that Darwin collected on the Galapagos Islands were later interpreted in terms of evolution from an ancestral stock that arrived on the islands from the west coast of South America. Darwin saw that different finches with different styles of beak survived better in different ecological niches.

though he was convinced in his own mind, he knew that he could not prove definitively that one species can be transformed into another in accordance with his theory. Just as Copernicus had refrained from publishing out of fear of being "hissed off the stage" for a theory that he could not prove, Darwin remained reluctant to publicize a theory that was at least as unsubstantiated and even more provocative.

Darwin settled first in London and began to make his way around the Linnean Society and the Geological Society of London as a junior member of the growing British scientific establishment. In 1839 he married his first cousin, Emma Wedgwood, with whom he was to have ten children. Darwin's father was wealthy enough, but her side of the family had real money, and the marriage meant that Darwin did not need to work to support himself or his family. He hated London, calling it an "odious dirty smokey town," and in 1842 he purchased an 18-acre estate in Down, Kent, a convenient 16 miles from London. There, with his independent means, he settled into the life of a Victorian country squire. Gradually over the succeeding years, Darwin went from a diffident Christian to an outright scientific materialist who attributed all being and all changes in the natural world to the action of natural processes.

Soon after his return to England Darwin began to suffer from the debilitating ailments that plagued him the rest of his life. Symptoms included chronic fatigue, violent gastrointestinal disruptions, and assorted dermatological ailments. Scholars still debate the causes of his medical condition: perhaps he caught a blood disease in South America; very likely he became psychologically crippled by panic attacks caused by the knowledge that he alone possessed the secret of organic evolution, a secret so much at odds with the prevailing Victorian culture he represented. In any event, Darwin remained frequently "unwell" for days, and when he did work at his science at Down it was only for a few hours in the morning.

Nevertheless, over a span of more than 40 years, undistracted by any other duties, his scientific output was prodigious. During the 1840s and early 1850s Darwin published more than half a dozen books and became recognized as one of Britain's most eminent naturalists. He spent years in painstaking research on barnacles, in large measure to establish his scientific credentials as a rigorous systematist and not merely a field naturalist.

Darwin received a cautionary lesson in the highly negative reception of Robert Chambers's *Vestiges of Creation*, which first appeared in 1844. Published anonymously it popularized the idea of transmutation of species as an innate part of design and God's plan, what Chambers called the "law of development." The book was very poorly received, and was severely criticized on scientific, philosophical, and religious grounds. The continuing debate brought about by Chambers's replies in subsequent editions of the *Vestiges* did little to promote serious scientific consideration of evolution and the species question. It also did nothing to expedite Darwin's own publication. He continued to mull over his "discovery," and he hoped eventually to produce a massive work that would carry the argument by the sheer weight of factual evidence. But in 1858—twenty years after he first conceived of evolution through natural selection—Darwin's reticence was cut short by an event every scientist dreads: he had been scooped! A letter arrived from the South Seas, the celebrated letter from Wallace, showing that Darwin's ideas had been arrived at independently.

Alfred Russel Wallace (1823–1913) was, like Darwin, a scientific maverick, but with a very different class background. He received only an elementary education and became a naturalist through self-study, eventually producing a body of research and publications which, even without his independent discovery of the principle of natural selection, would have established him as one of the most eminent naturalists of the nineteenth century. His spectacular collections of specimens were matched, almost from the beginning of his investigations, by a theoretical curiosity about the origin of species. By the 1840s, when he traveled to South America to explore regions of the Amazon basin, he was already convinced that species develop naturally and not by the intervention of a divine power, although he, like Darwin, did not initially possess any mechanism for evolutionary change. Ten years later he explored the Malay Archipelago, and there he arrived independently at the principle of natural selection as the driving force of evolution.

When Wallace's letter describing his thoughts on evolution arrived in June of 1858 Darwin feared that he might lose credit for the theory of evolution, and he allowed his friends Charles Lyell and Joseph Hooker (1817–1911) to arrange a simultaneous announcement at the

Linnean Society through which Darwin and Wallace were recognized as codiscoverers of the theory. Yet, perhaps because of Darwin's superior social and scientific status, Wallace still receives only token credit for his independent discovery of natural selection. Both men remained on amicable terms until Darwin's death and often exchanged views on evolutionary theory. While Darwin remained a strict selectionist, Wallace retreated on the question of human origins and adopted a spiritualist account in which a divine power, a "Higher Intelligence," played a role.

The Origin of Species

Darwin could no longer linger over a definitive work that would clinch the argument for evolution by natural selection. Within eighteen months he produced an "abstract" which he titled *On the Origin of Species by Means of Natural Selection, or the Preservation of Favoured Races in the Struggle for Life*. The *Origin* is, as Darwin put it, "one long argument," and he marshaled an impressive mass of evidence to demonstrate the greater plausibility of evolution over special creation. The book was written to be read like a novel. He uses many examples from his readers' world in making the work entertaining and persuasive. In the fourteen chapters of the original edition Darwin's argument unfolds in three phases, something like three acts of a highly dramatic play.

In the first act, he outlined his model for evolution. Since he could provide no direct evidence for what he knew was a highly controversial theory, he began by introducing a grand analogy in discussing the achievements in selective breeding or what plant and animal breeders had done in producing domesticated varieties. Darwin coined the term *artificial selection* to describe this process. He wanted to suggest that the great variety among domesticated dogs and pigeons, for example, warrants the view that natural varieties may, over much longer periods, become sufficiently extreme to result in separate species and not merely races of a single species.

Darwin then presented the fundamental elements of his own model: variation, struggle for existence, and natural selection leading to species change. His guiding principles were that individuals of each species will display variations for each characteristic, that Malthusian population pressures inevitably lead to competition and a struggle for existence with often the most fierce competition between individuals of the same species. Through a process of "natural selection" the best-equipped individuals will reproduce at a higher rate, thus diverging from the ancestral stock and in the long run producing species change. Natural selection is nature's version of selective breeding, and compared to what humans have wrought by artificial selection over a comparatively short period of time, how much greater will have been the effects of the "struggle for existence" acting through natural selection over "whole geological periods"?

In the middle chapters of the *Origin*—the second act of this great intellectual drama—Darwin raised what he called "difficulties" with the theory. He had become his own best critic, and he set out in advance to explain away what seemed to be the gravest problems besetting his views. Through small incremental changes, how could he account for organs of extreme perfection, such as the eye of an eagle, without the notions of design or Providence? How can natural selection explain the existence of animals with extraordinary habits or instincts, such as the cuckoo bird, which lays its eggs in other birds' nests, the honey bee with its exquisite hive-making abilities, ant colonies with sterile castes, or ant species that capture slaves? In each instance Darwin strained to show that the incremental addition of small variations could ultimately account for the extraordinary behaviors and traits observed in nature. With regard to the geological and fossil records, which in Darwin's day showed little evidence of the transitional forms between species that the theory required, Darwin reasoned that the geological record was simply imperfect and preserved only random samples from the vast history of life on Earth.

In the final chapters—the last act of the play—Darwin turned the tables and raised issues explained only with great difficulty by special creation yet readily accounted for by his theory of evolution. Why does the fossil record, even though imperfect, provide evidence of extinction and species change over time? How can we most reasonably account for the geographical distribution of plants and animals? How, Darwin asks, should taxonomists classify the obviously related varieties, species, and orders of living forms observed in nature, except through notions of descent with modification? How are the striking similarities of embryos of wildly divergent species to be explained except by evolutionary principles? Darwin also cited the many known cases of useless organs—"rudimentary or atrophied organs . . . imperfect and useless"—such as "the stump of a tail in tailless breeds." For the creationist doctrine rudimentary traits present an unbending difficulty. What deity would produce a kingdom of animals with sets of useless organs? For the evolutionist such traits actually strengthen the hypothesis—they represent "records of a former state of things" as organisms become transformed from one species to another:

> On the view of descent with modification, we may conclude that the existence of organs in a rudimentary, imperfect, and useless condition, or quite aborted, far from presenting a strange difficulty, as they assuredly do on the old doctrine of creation, might even have been anticipated in accordance with the views here explained.

For nearly 500 pages Darwin piled detail upon detail and argument upon argument, creating a preponderance of evidence for his theory of evolution through natural selection. *The Origin of Species* was an immediate bestseller among scientists and educated amateurs alike, and, in the end, Darwin left his readers with a haunting sense that the ground had shifted, that the biblical account of the origin of species was implausible and would be replaced by a naturalistic account.

Through almost the entire book Darwin refrained from implicating humankind in his great revision, but in the third paragraph from the end of his long monograph he opened the debate on humanity's place in nature with perhaps the most fateful statement in the history of science: "Light will be thrown on the origin of man and his history." Darwin hoped that others would develop this line of inquiry, but, dissatisfied with what he saw as the tendency, even among his supporters, to envision humankind in a different light, Darwin himself took up the cudgels in his *The Descent of Man* (1871). In this work he made explicit the human connection to an ape-like ancestor, even insisting that evolution had not only produced humankind's physical features but also played a role in the development of instinct, behavior, intellect, emotions, and morality. It needs to be said that Darwin's unalloyed racist views come across plainly in this work. In *The Descent of Man*, too, he discusses gender differences in the animal world and proposes another form of selection, sexual selection or competition for mates, as the mechanism to account for gender differences, and notably Darwin insists on the role of female choice in sexual selection.

The degree to which Darwin's theory has been confirmed in our day testifies to the power of his theoretical sagacity. At the time he wrote nothing was known about genetics, or, in general, about the biochemistry of inheritance. No convincing evidence existed that the earth was more than a few hundred thousand years old. And, since no human-like fossils had yet been found (the bones of Neanderthal man were at first assumed to be those of a modern European) there seemed to be a yawning gulf between humans and apes. Over the next century the pieces fell into place with a regularity and precision that transformed theory into accepted fact. In Darwin's lifetime, however, his theory did not win universal acceptance in the scientific community.

Darwin's understandings of how inheritance works and how variations arise and become stabilized in a breeding population—"the various, quite unknown, or dimly seen laws of variation," as he put it, remained weak links in his argument. Along with his contemporaries he assumed that inheritance embodies a blending process akin to mixing two pigments and producing an intermediate color. But if inheritance occurs through blending how can a more fit individual spread its fitness to an entire population? On the contrary, the improved trait will, in a few generations, be first diluted and then completely washed out through successive blending. At one point Darwin had thought that species originate on islands or in other small and isolated groups where the effects of blending would be minimized and, by the laws of chance, an improved variety might then spread over the whole population. But even in small breeding groups, it was shown, blending would prevent the formation of new species. Based on the same reasoning and with equally compelling force, critics likewise posed grave difficulties concerning the evolution of mimicry in nature—or how two different species of butterfly, for example, can come to have the same outward appearance.

Another devastating source of criticism sprang from an unexpected and authoritative quarter: the community of physicists. That evolution requires long periods of time stood as a fundamental tenet of Darwin's theory. In the first edition of *The Origin of Species* Darwin mentioned a figure of 300 million years for a single geological formation, the implication being that life probably evolved over hundreds of millions or even billions of years. But just as Darwin's views became known, the discipline of physics, based in particular on the first and second laws of thermodynamics, began to assume an unprecedented intellectual coherence and mantle of authority, as we saw. Given what was known of the radiational cooling of the earth and the presumed combustion of the sun (thought to be something like a burning lump of coal), physics—notably in the august personage of William Thomson, Lord Kelvin—pronounced against the extraordinary amounts of time required by Darwin's entirely qualitative theory.

Over the next two decades Darwin responded to his critics. In the process he ceased to be a true Darwinist and acquired the characteristics of a Lamarckian. He maintained an emphasis on his beloved natural selection, but he now admitted that natural selection was perhaps not the exclusive mechanism for species change. But he refused to capitulate to the authority of the physicists regarding the span of time available for evolution, saying that we do not "know enough of the constitution of the universe and of the interior of our globe to speculate with safety on its past duration," and he suggested that perhaps the *rate* of evolutionary change was more rapid at an earlier point in time. He even allowed for the possibility that environmental factors may affect whole groups of plants and animals in a Lamarckian fashion, inducing wholesale changes and thereby minimizing the significance of individual variations. In a special section of his 1867 book on *Variation of Animals and Plants under Domestication* Darwin outlined a possible mechanism he called *pangenesis* for the inheritance of acquired characteristics; his explanation involved "gemmules" shedded by the working parts of the body that circulated and collected in the reproductive organs to be transmitted to the next generation. In Darwin's case the first edition of his great work gives clearer insight into his doctrines than the intellectually troubled later versions.

The Origin of Species provoked immediate and strong reactions from a conservative Victorian world forced to confront Darwin's views on evolution. The tone was set in 1860, just months after the publication of the *Origin* at the famous meeting of the British Association for the Advancement of Science at Oxford. There, in a staged debate the Anglican bishop of Oxford, Samuel Wilberforce, denounced Darwin and evolution for equating humans and apes; T. H. Huxley, Darwin's ardent supporter who earned the nickname of "Darwin's Bulldog," retaliated with devastating sarcasm.

While his friends and followers pushed the fight in the public arena, the retiring Darwin continued to work and tend to his gardens and family at Down. As the years unfolded he published substantial volumes on the emotions of animals, fertilization and cross-fertilization in plants, insectivorous plants, climbing plants, and the actions of worms. All of these works embodied evolutionary perspectives and were framed by the principle of natural selection, but none carried the day in fully persuading the contemporary world of science that the history of life was to be understood in evolutionary terms. Darwin died in 1882. Even though his ideas threatened the foundations of the Victorian social order and he never became Sir Charles Darwin, the kindly old man himself clearly stood with the greats of British science and culture. He was buried in Westminster Abbey alongside Sir Isaac Newton.

Ironically, one of Darwin's main problems was solved only six years after the publication of the *Origin* when Gregor Mendel (1822–84), an Austrian monk, described his experiments on plant hybridization which showed that heredity is not a blending process but instead maintains traits in discrete units. These units—later designated *genes*— pass from generation to generation, and since they are never blended out they may preserve a useful trait and eventually spread it through the breeding pool. Darwin may have read Mendel's paper, but it did not impact him. In the larger world Mendel's publication was almost wholly neglected, and by 1900 Darwin's theory of evolution through natural selection remained far from established in the scientific community. Indeed, at the turn of the twentieth century it would be hard to speak of a Darwinian revolution anything like what came to guide science and human thought in the twentieth century and today.

Toolmakers Take Command

Science, industry, and engineering had begun to find common ground by the turn of the twentieth century, and business executives, managers, and engineers now took command in transforming the socio-industrial landscape on a global scale. Several paradigmatic industries—first and foremost electrification, but also the automobile, aviation, domestic technologies, and entertainment—illustrate how new technologies came to form intricate and interlocked systems that spread across the world from their European and North American roots. These complicated and impressive new technologies, the major ones examined in this chapter, utterly transformed how people around the world live in the twenty-first century. Together, this set of interlocked technological systems grew to constitute a world-encompassing technological super-system and to furnish the primary components of the infrastructures of virtually all twenty-first-century societies.

Electrical devices, which were nonexistent prior to the last quarter of the nineteenth century, are an utter commonplace now and integrated into every need and activity in everyday life from lighting to communications, transportation, and entertainment. Aviation has made the world functionally smaller and has effected global interconnections, while the automobile industry has driven economies and facilitated suburban sprawl and a mobile lifestyle. The household has been transformed, and today some people amuse themselves splendidly with on-demand movies, music in infinite variety, the vastness of the Internet and social media, video games, radio, and yes, even TV or actually going to the movies.

In only a few short centuries humans have created an entirely new mode of existence for themselves—industrial civilization. This revolutionary transformation of how humanity lives originated only in the eighteenth century with the onset of the Industrial Revolution, as we have seen, and it has snowballed in unexpected and unforeseen ways into what we know today. The socio-technological revolution can be equated only with two earlier great leaps in human history, first to the Neolithic and then to Bronze Age civilizations so many centuries and millennia ago. Industrial civilization has come a long way since the early steam engine and the first railroad. While it has vastly expanded human potential and improved the quality of life for many—but far from all—of the world's peoples, the modern merger of science and technology has displayed a dark and deadly side, notably in the industrialization of warfare.

Electricity and the Supersystem of Industrial Civilization

At exactly 3 p.m. on Monday, September 4, 1882, an operator threw a switch at the generating station of the Edison Illuminating Company at 255-257 Pearl Street in lower Manhattan. At that very same moment at the offices of the financier J. Pierpont Morgan nearby on Wall Street, Thomas Alva Edison himself turned on the lights in the room. In so doing, Edison inaugurated not only the first commercial electrical utility but a new age.

Electricity is an especially versatile energy source, and the development of technologies to generate and transport electricity is deeply significant and may be equated to human mastery of fire so many millennia ago. The electric power industry inaugurated by Edison that day in 1882 was transformative because so many other technologies and technological systems came to use electricity to power whatever it might be . . . lighting, heating, refrigeration, and so forth. There are other sources of energy to be considered in this connection (water, coal, oil, natural gas, nuclear), but electricity represents the fundamental technological system on which modern industrial civilization depends, and electric power can stand as a surrogate for industrial civilization as a whole. The reason is because so many other technologies piggyback on top of electrical systems and are designed to connect to a source of electrical current. Refrigeration, for example, eats up 15 percent of the world's electricity, and without electricity there would hardly be any air conditioning or refrigeration. In this way a complex mix of technological systems—a supersystem of technologies powered by electricity—reveals itself as the true face of industrial civilization.

As it turns out, current electricity was not a century old by the time Edison inaugurated the Pearl Street Station in 1882. *Static* electricity was known from antiquity, and in the seventeenth and eighteenth centuries static electricity and its properties were investigated by Franklin and other scientists, as we saw earlier in chapter 15. But a continuously flowing *electric current* is another matter; it is not altogether an exaggeration to say that the very phenomenon of current electricity did not exist before 1800 when the Italian scientist Alessandro Volta (1745–1827) built the first battery and created a chemical source for producing a continuous current. That first battery and its ever-improved successors drove early-nineteenth-century chemistry in a certain direction, as we likewise saw, but more to the point here the stunning new applied-science technology of the telegraph, which emerged in the later 1830s and early 1840s, was built upon batteries and electrical circuits. In 1831 Michael Faraday announced the principle of electromagnetic induction (that of the generator or dynamo), whereby given mechanical motion and a magnetic field, one can produce an electrical current. But it was not until 1867 that the first practical, industrial dynamo with its own electromagnets was independently unveiled by the German inventor and industrialist Werner Siemens (1816–92) and the English scientist Charles Wheatstone (1802–1875). Only then did these great electricity-generating machines, themselves powered by ever more sophisticated and efficient steam engines, come to play their part in introducing electricity into the veins and sinews of society and industry. In the 1870s and 1880s this new technology involving current electricity was deployed in arc light street lighting in Paris and New York, but bright arc lamps are not suitable for indoor lighting.

This set of circumstances led Edison and others to search for a practical incandescent lamp. After extensive experimentation in 1879 Edison filed for a patent for a long-lasting and commercially feasible incandescent bulb with a carbon filament. (The British physicist and inventor Joseph Swan independently developed an incandescent bulb also in 1878–1879.) Demonstration illumination projects quickly followed, and a few private residences of the rich came to have self-contained systems and be lit with electricity. But the Pearl Street central generating station that Edison inaugurated in 1882 is rightly famous for being the first commercial electric generating plant in the world and the mother of all that followed.

At an earlier time, when historians of technology focused more on artifacts, the invention of the light bulb stood out as a landmark achievement. But the light bulb is a red herring that leads us astray from Edison's real accomplishment, which was the implementation of an entire technological system for providing electricity, first for lighting and then for so much else. Since the 1980s historians and students of technology have recognized that "systems thinking," as it were, needs to be applied to understanding technology, and no case better exemplifies the notion of a technological system than electric lighting.

In this light Edison (1847–1931) is not so much a great inventor but a transformative system builder. At the Pearl Street Station power plant he deployed six 200-horsepower, 27-ton, 110-volt generators. Each of the behemoths, dubbed Jumbos by Edison after a popular circus elephant, was powered by its own coal-fired steam engine. Edison's crews snaked 80,000 feet of underground mains to customers through a square mile of lower Manhattan. He took the idea of underground mains from contemporary gas lighting companies, but otherwise he had few models to go by as he engineered his system for generating, distributing, and metering electricity. He already had his "horse" of a light bulb before the "cart" of his electrification system, and although he relied on existing technologies where he could (notably the copper wire industry), he was left to develop many key parts himself. The strictly electrical part of the system came to include generators, steam engines to power the generators, mains and transmission lines, voltage regulators, meters, switches, fuses, standardized sockets and fixtures, as well as the bulbs themselves, which were manufactured separately. Edison had to train his own cadres of workers to run and maintain the system.

The Edison Illuminating Company, headed by Edison and backed by the New York industrialist J. P. Morgan and the wealthy Vanderbilt family, was a business that sold electricity, after all, and so the no-less-essential business side of the system had to be brought into place, too: the legal, corporate/managerial, and back-office requirements of keeping the system going as well as recruiting paying customers and ultimately paying investors and stockholders. Incidentally, as a corollary of the coming into being of electric lighting systems like the one Edison created, the light bulb, which initially was just a curious demonstration item, became a commodity.

Edison's Pearl Street Station was a remarkable success. The number of customers rose to 200 in just three months after the plant opened and to over 500 customers a year later with a system of some 10,000 lamps. The Pearl Street Station continued to generate electricity and illuminate homes and offices until 1890, when a fire destroyed the facility. But an unstoppable technological force had been unleashed on the world, as the Pearl Street Station became the prototype for the electric power industry and the rapid electrification of America and the world. By 1900 some thirty different companies were in the business of generating electricity in and around New York City alone. From there, the numbers escalate exponentially.

Edison's technological system produced a direct current (DC), and he and George Westinghouse (1846–1914) waged an ugly public war ("The War of Currents") through the 1880s and 1890s over the merits of Edison's system versus the system involving alternating current (AC) that Westinghouse engineered based on the patents of the pioneering genius Nikola Tesla (1856–1943). In the end because of its advantages in transmitting electricity over longer distances Westinghouse's AC won out, but at stake, too, were the differing social and technological alternatives of generating electricity locally at essentially neighborhood stations or at larger-scale regional power plants that transmitted electricity over high tension lines. The latter vision won out, and by 1917 95 percent of electricity generated in the United States was through an AC system. AC current powers most of the world today.

Electrification was electrifying, or at least it seems so in retrospect. Public and private power grids arose quickly first in cities, and by the 1930s, 90 percent of urban dwellers in America had access to electricity. In 1935, as part of the New Deal, the Roosevelt administration created the Rural Electric Administration that, as an initiative of the federal government, dramatically extended electrification to the hinterland of America. By the same token, demand had to be created, particularly as the load on electrical systems varied over the course of the day, and early consumers had to be convinced to adopt electricity. The electrification of the home and the introduction of electric home appliances, as we will see further, was a mechanism for turning the public's initial unfamiliarity with the nature or need for electricity into a necessity of everyday life. Across the country and worldwide, electrical power plants form vast regional networks, which occasionally collapse causing blackouts and power failures. The successor to Edison's

Fig. 17.1. (opposite) Electricity and industrial civilization. Modern life depends completely on the generation and distribution of electrical power. Thomas Alva Edison put into place the first commercial generating station in New York City in 1882. The figure depicts the elaborate system that begins with a generating plant (liquefied natural gas in this case, but that could also be hydroelectric, coal, oil, wind, or nuclear) and that across powerful transmission lines brings electricity into households and businesses.

Pearl Street Station, Consolidated Edison or Con Ed in New York, for example, now employs over 15,000 people and provides electricity to more than 3 million billable accounts through 130,000 miles (218,000 km) of wires and cable. The production and consumption of electricity have grown spectacularly in tandem with the rest of industrial civilization. The rough numbers are striking. World production of electric power quadrupled from 1900 to 1950, quadrupled again by 1975, doubled again by 2000, and increased another 50 percent through 2014. The growth rate in the production of electricity worldwide has tapered off to about 10 percent per annum today.

The history of electrification reveals two great facts about industrial civilization: the dramatic increase in scale over previous modes of human organization and a great divergence between the haves and the have-nots of nations and peoples in the world today. The increase in scale is evident in the numbers just presented. A hundred times more energy is produced every year today than in 1850, and from the point of view of simple thermodynamics, people living in economically developed societies today have and use more energy per capita than any human group ever in history. The big story in recent years is energy growth in Asia, notably in China. Beginning in the 1980s the production and consumption of electricity exploded in China, and China has now raced ahead of the United States as the world's leading producer of electricity. China's accelerated growth is predicted to continue for the next several decades.

But this overall growth obscures lingering energy imbalances where nations in the developed world with comparatively small populations produce and consume a disproportionate share of the world's energy. This skewed pattern is evident in the United States, for example, which is per capita by far the world's leading energy consumer. The West still commands the lion's share of commercial energy resources, and the top 10 nations (including notably Japan) consume nearly 40 percent of the world's energy. At the same time, the whole continent of Africa with approximately 15 percent of the world's population produced only 3 percent of the world's electricity in 2014, and that figure would fall to 1 percent for the remaining 13 percent of the world's population on the African continent if Egypt and South Africa were taken out of the equation. Bangladesh, the seventh most populous nation in the world with 157 million people in 2014 (over 2% of the world's total) and one of the poorest countries in the world, generates under 0.2 percent of the world's electricity. At present the average person in the United States taps nearly 35 times more energy than someone in Bangladesh. Energy-rich and energy-poor nations jostle side by side as industrial civilization continues its global expansion.

Electricity is fundamental to industrial civilization, and the technological system that supplies electricity is the fundamental technological system underlying industrial civilization. True, electricity has to be generated, and it takes other sources of energy to do that. So, systems designed to produce and distribute electricity are not fundamental in the sense that they somehow come before or are independent of other systems and the rest of industrial civilization. Indeed, industrial civilization heavily exploits other rich sources of energy, notably oil, coal, and natural gas. Cars, trucks, and airplanes burn tons of fuel, and coal is still widely used in various industries, as is natural gas in a variety of ways, including for domestic heating and cooking. And, be it noted, these other energy sources come to us via complex commercial technological systems. But lots of those natural resources also go toward generating electricity! And so, a feedback loop from the growing demand for electricity has helped spur the enormous expansion of the gas, oil, coal, and nuclear industries. Two-thirds of all electric power plants in the United States are fueled by coal or natural gas, for example. Hydroelectricity provides another example, where the mechanical energy of falling water is converted into electricity. Nuclear power plants generate electricity, as do solar- and wind-power installations. So, the electrical grid is enmeshed with other essential technological systems. Just the same, pull the plug on electricity and civilization as we know it shuts down.

The centrality of electricity to industrial civilization worldwide today makes plain the existence of the aforementioned supersystem of technologies without which we cannot live. Shall we name some examples: The granddaddy of electric lighting? The sump pump in the basement? Hair dryers? The air traffic control system? Your device that needs recharging? A further laundry list won't do more to make the point that industrial civilization is to be understood not only in terms of technological systems but as an integrated supersystem of technologies that emerged on the world stage in the twentieth century. This technological supersystem itself is dynamic and continues to unfold on that stage just as it continues to shape our lives.

The history of electrification and electrical systems raises the hoary question of technological determinism, or whether technological systems impose themselves and effect social change independently of human volition or politics. Are systems actors on the historical scene? Do artifacts have agency or will? Is technology social destiny? The cases we have seen of the stirrup or the deep plow in the early Middle Ages or cannon technology in the Renaissance provide food for thought in this connection, but electrification pointedly illustrates how the unavoidable demands of a new technology can transcend established social and political orders. A revealing case in point concerns hydroelectricity in the United States and in the former Soviet Union, the USSR.

At first, in the United States electric power was generated under the traditional auspices of capitalism—private ownership of electric power production and distribution. During World War I, the American Congress, as a strictly wartime improvisation to provide electricity for the manufacture of gunpowder, authorized the construction of a hydro-electric station at Muscle Shoals in Alabama—as the authorizing legislation put it, "To be operated solely by the Government and not in conjunction with any other industry or enterprise carried on by private capital." Once the war ended, however, the unfinished Wilson Dam became a bone of political contention; although it was still owned by the government, its output would now be sold in competition with private electric power companies. As the engineering work proceeded, the production of hydroelectricity in a society devoted to private enterprise brought into focus conflicting political ideas. Private electric power producers, along with *laissez-faire* defenders of capitalism, charged that the government's ownership of Wilson Dam amounted to "communism."

In 1921, Henry Ford, the automobile tycoon, stepped forward with a proposal to turn the project over to private capital. He offered to "buy" Muscle Shoals. Ford enlisted the endorsement of Thomas Edison, and the popularity of the two technical wizards produced strong support in Congress. But in 1924 Ford abruptly withdrew his offer. The technical realities proved too great and could not be overcome by ideology, politics, or money. Hydroelectric stations cannot be built only to produce and sell power and light for profit. Large dams inevitably affect the public interest: shorelines are altered, land is inundated, and dams must also be designed for use in ways that cannot turn a profit for private investors without government intervention—irrigation, flood control, and recreation.

In 1933, when Franklin Roosevelt became president, one of his first acts was the establishment, through Congress, of the Tennessee Valley Authority (TVA) to develop the vast water-power potential of the southern states. Roosevelt's initiative was not motivated by any socialistic bias. Instead, it recognized the technological imperatives dictated by the nature of hydroelectric production. Indeed, Roosevelt's politically conservative initiative was embodied in an "Authority," which meant that the TVA was required to appeal to Congress for financing every year, thus restricting the government's ability to compete aggressively with private electric power interests.

Another clear indication that large-scale technology, when it impinges on the public interest, overwhelms ideology and politics was Vladimir Lenin's dictum, promulgated shortly after the Russian Revolution of 1917, that "Communism is Soviet power plus electrification." This was no mere rhetorical flourish. The communist leader Joseph Stalin and his acolytes took it seriously, and in a remarkable instance of technology transfer between 1927 and 1932 the world's largest hydroelectric power station was constructed on the Dnieper River in Russia under the supervision of American engineers and with American turbines and generators. Within the frameworks of both American capitalism and Soviet communism hydroelectric technology wrote its own script. The technical and economic realities of electric power production weighed more heavily than ideology and politics on the balance of social change. Older technologies could be confined within the limits of private enterprise—indeed, they nurtured the principles of private enterprise. Hydroelectric power production defied those limits. The Three Gorges Dam in China today is merely the latest example of this historical phenomenon. To engineer, build, and maintain, the toolmakers necessarily took command.

"See the USA in Your Chevrolet"

The steam engine remained the prime mover in industry and in transportation systems—railroads and steamships—to the end of the nineteenth century. Then, with the development of effective diesel and gasoline internal combustion engines by German engineers a new prime mover appeared that, mated to the wagon and the plow, created the automobile and the tractor, supplanting the omnipresent horse. Cars powered by internal combustion engines were first developed in the 1880s, and the "horseless carriage" began to make a public impact. Auto sales in the United States in 1900 reached 4,000. By 1911, 600,000 automobiles puttered along on U.S. roads. By 1915, that number had risen to 895,000, and by 1927 it had soared to 3.7 million. Some of the early automobiles were electric or steam-driven, but the internal-combustion engine soon won out as the most successful power source.

In more heroic versions of the history of American technology, Henry Ford (1863–1947) stands as a lone visionary personally responsible for the creation of the automobile industry in America. A self-trained mechanic with a lifelong disdain of experts with university degrees, Ford built his first automobile in 1893, and a decade later he founded the Ford Motor Company. He intended to produce "the car for the great multitude," and to do so he had to harmonize mass production with mass consumption. Ford was not the first manufacturer to use interchangeable parts or to run an assembly line, but in his quest to produce an inexpensive and standardized product he perfected assembly-line production techniques; and the results proved dramatic. In 1908, before he introduced the assembly line, Ford made 10,607 Model Ts—the "Tin Lizzie"—which he sold for $850 each. He shifted to an assembly line in 1913, and production quickly rose to 300,000 cars a year. In 1916 he sold 730,041 Model Ts for $360 each, and in 1924 he produced 2 million of the cars retailing at $290 each. A total of 15 million Model Ts rolled out of Ford plants before production ceased in 1927. Prior to Ford, it took over twelve hours to assemble a car. By contrast, his first assembly line turned out a Model T every 93 minutes, and by 1927 Ford made a Model T every 24 seconds! The Ford Motor Company became not only the world's largest automobile manufacturer, but the world's largest industrial enterprise.

Fig. 17.2. The assembly line. Improved by Henry Ford, the assembly line greatly increased production. The case of the automobile illustrates well the nature of a technological system that must combine many subsystems.

Ford indeed created a car for the masses. By the 1920s the automobile was no longer an esoteric toy for enthusiasts or the rich and idle, but was well on its way to becoming a necessity in industrialized societies and a mainstay of the modern global economy. In less than a century the automobile industry resulted in 10 percent of the American landscape being covered with paved roads. The model of mass production and mass consumption originated by Ford with the Model T would come to be applied to many other "necessities" of modern life.

It is too easy to see Ford as the heroic inventor single-handedly changing the landscape of American industry and culture. Our understanding of his accomplishment is enriched if we think of him, like Edison, as a system builder who orchestrated the efforts of thousands of others and who oversaw the coming-into-being of a diverse and self-sustaining technological conglomeration. Within his own company, Ford headed a sizable team of talented and enthusiastic young engineers, foundrymen, and toolmakers to bring his first assembly line together (and other companies quickly hired some of them away). He also created an elaborate organizational structure to secure materials, assemble cars, and then market them, and the automobile industry provides another clear example of what historians of technology mean when they speak of a technological system. A piece of technology does not exist in a social vacuum but is connected with makers, users, and other technologies in often complex ways. The automobile itself can be thought of as a technological system, one composed of thousands of parts that integrate fuel, an engine, transmission and power train, brakes, suspension, lights and electrical subsystems, to name only some of the major components. The invention of the electric starter in 1912 and the introduction of balloon tires in 1921 were innovations that significantly improved the automobile considered as a collection of subsystems. The starter replaced the hand crank and, incidentally, brought women into the automobile market.

Be it noted that the administration and management of technology represent essential, if intangible, aspects of the many new technological systems that arose in the twentieth century. Frederick Winslow Taylor (1856–1915), a mechanical engineer with a degree from Stevens Institute of Technology, pioneered "scientific management" in the later nineteenth and early twentieth centuries, and his influence on the course of industrial development and manufacturing around the world cannot be overemphasized. With his stopwatch and clipboard Taylor performed elaborate "time and motion" studies precisely measuring the movements and activities of shop workers to determine "the one best way"—that is, the exact speed and the exact motions workers should use to perform a task. His *Principles of Scientific Management* appeared in 1911, and as the first management consultant himself and through a set of disciples Taylor's notions of industrial efficiency and the rational analysis of manufacture became universally adopted in commercial and technological sectors. They transformed business practices and put a focus on employee training and close supervision by management, a hallmark for better or worse of much of manufacturing and industrialized life today. The fields known as operations research (OR) and decision sciences are other notable outgrowths of Taylor's work. The embrace of Taylorism in the communist USSR, not to mention in manufacturing around the world today, is testimony again to how industrial civilization has a momentum of its own that transcends other barriers. In the end, the great technological achievements of the twentieth century, such as Henry Ford's assembly line or NASA's landing humans on the moon, were as much managerial triumphs as they were purely technological ones. Today, technology management and systems engineering have become integral elements of innovation and development more narrowly conceived.

Technological systems were not wholly imposed from above but developed dynamically as users adopted *and adapted* new technologies and as feedback loops operated between and among whole classes of agents to shape technological systems. Those groups include manufacturers, distributors, salespeople, actual users, people who can be identified as modifiers and maintainers of systems, and categories of non-users who opt out of a technology or who are technologically excluded. Down on the American farm, farmers transformed Ford cars into tractors and farm machinery before Ford itself made and sold them, and housewives on party lines battled the phone company's version of what they should be doing.

Another aspect of the automobile-as-technological-system emerges in focusing on manufacturing methods. Multifaceted production mechanisms had to come into place before the automobile could become a significant piece of technology in American life. The Ford factories at Highland Park and River Rouge, for example, became vast industrial sites that grouped together several subsidiary plants to service the main assembly lines, including a coking plant, a foundry, a steel plant, and a cement plant. Ford had his own coal mines, glass factories, rubber plantations, ships, and railroads. Labor comprised a key feature of this intricate manufacturing operation, and Ford gained fame for dramatically increasing the wages of his factory workers. While the prevailing rate was eleven dollars a week, Ford announced that he would pay workers five dollars a day for an eight-hour day, but with conditions attached, including inspections of workers' homes by the Social Department of Ford Motor Company to assess their family relations, morals, sobriety, cleanliness, and commitment to thrift. Otherwise no friend of organized labor, Ford took this bold step primarily in an effort to create a stable workforce, but it had the effect that Ford factory workers could now begin to be buyers of the very products they made.

Cars and car factories themselves form part of an even larger network of related technologies and social practices required for an effective technological system. In some measure, traditional techniques were combined—metalworking, glass making, and pumps. But the automobile industry also fostered technical innovations—the balloon tire, the carburetor, sealed-beam headlights, and, as an ancillary technology, timed traffic signals. And, around all of these subsidiary industries interwoven networks of technical and social enterprises formed and flourished. For example, cars ran on gasoline or diesel fuel, and so the explosive growth of the automobile industry both occasioned and would have been impossible without an equally massive expansion of the oil industry and improved techniques of "cracking" crude oil to refine gasoline. The local gas station became an essential element in the larger technological system, as did garages, repair shops, and replacement auto parts. The same point can be made about the necessity of reliable roads and road systems, standardized traffic signs and driving conventions, licensing, auto insurance, public road and bridge construction, and the bureaucracies of state motor vehicle departments. Similarly, without auto dealerships and advertising the automobile industry would be very different, to say the least. Along these lines, the car loan, introduced in 1915, and the "trade-in" were marketing innovations that, along with the car lease today, continue to play a large role in keeping the system going.

Mass-produced automobiles largely defined the culture of the twentieth century, and that culture rolls on today. Automobiles are aspirational, and automobile manufacture has spurred industrialization and grown economies around the world. From creating suburbia to transforming the sex lives of adolescent drivers, the social repercussions of the technological system of the automobile have been and continue to be immense.

The automobile today is a wonder of technology and an icon of industrial civilization. The complexity of cars as assemblages of technology is often taken for granted. Cars are powered by sophisticated, computer-controlled internal combustion, electric, or hybrid engines. Cars have "intelligent" braking, traction, and transmission systems, power steering, and highly sophisticated electrical, lighting, and suspension systems. The tires they run on are wonders of engineering and materials science. Air-conditioning systems automatically maintain cabin temperatures and supplies of fresh air. Entertainment systems incorporate not only AM, FM, and satellite radio, but TV, phone, and Internet connections, as well as digital electronic devices for playing music and movies and entertaining children in the back seats. Protective airbags deploy instantaneously in the event of an accident. Many vehicles come equipped with voice-activated satellite GPS systems that display maps and directions on color LCD screens. The self-driving car has arrived. The manufacturing standards are exquisitely high, and the interiors of cars can be of virtually unimaginable luxury. Out on the road, the car is a cocoon with the driver the master of his or her universe. For not a few drivers, their cars are more comfortable and technologically sophisticated than their homes.

What has been said of the passenger automobile needs to be extended to all the motorized beasts that transport goods and people over roads and seas everywhere around the world. Buses and trucks have grown in sophistication and efficiency, and trucking—now complemented by giant tankers and containerization shipping—is another crucial industry and technological system underpinning contemporary civilization.

The industrialization of agriculture powered by tractors and related mechanized farm equipment is another part of this same story. Although the mechanization of agricultural production occurred in the nineteenth century, horses and humans supplied the power. The tractor-driven plow, an offshoot of the automobile industry, changed matters significantly and, with other motorized farm equipment, accounted for a substantial increase in food production. During the 1920s, Ford-manufactured plows were being exported across the world. Distribution of food by road, rail, air, and steamship, along with the increased use of improved fertilizers, and, in the second half of the twentieth century, the introduction of more productive varieties of crops allowed food production and burgeoning populations to keep in tandem step. Indeed, in a clear and noteworthy instance of modern applied science, in the "Green Revolution" of the 1950s and 1960s science contributed significantly to increasing agricultural production worldwide and improving food security for millions of people. The agronomist Norman Borlaug (1914–2009) led the way in developing more hearty and productive varieties of wheat. Others did the same for rice, and they introduced these cultivars into Asia in particular. For his work Borlaug won the Nobel Peace Prize in 1970.

Today, we speak of agribusiness to mean large, industrial-scale food production by giant, often multinational corporations. The industrialization of food production has been a staggering transformation that has occurred in just the comparatively recent past, since the 1950s. Just 3 percent of the farms in the United States, for example, account for more than 60 percent of America's agricultural production. One side of what is known colloquially as Food, Inc., is monoculture in vast fields with genetically modified crops and planting with seed that has to be purchased from the Monsanto Corporation and the like; the other is the industrial-scale raising and processing of chickens, pigs, and cattle in huge barns and factories. A historically dramatic and unprecedented reduction in the numbers and percentage of the population directly involved in agriculture and food production has resulted. That figure stands at less than 5 percent in many industrialized countries today where most agricultural workers are underpaid and employed on large farms. In the United States the figure has fallen below 2 percent. In many countries small farmers persist in good measure only because of significant government subsidies.

The spectacular success of the American automobile industry led to the myth of "Yankee ingenuity" and "American know-how." But subsequent accomplishments, mainly after World War II, revealed the contingent nature of American success. Soviet heavy industry rivaled American production; Japanese automobile industry eclipsed the United States in both production and innovation; and East Asia, including China, has become an industrial powerhouse. Toolmakers everywhere, like their Paleolithic ancestors, are developing a universal industrial culture.

In many respects the automobile is synonymous with industrial civilization. Estimates vary, but as of 2014 the total number of passenger vehicles on the road worldwide was on the order of one billion or nearly one car for fewer than seven people. The United States leads the way with 253 million cars on its highways or one vehicle for every 1.3 people. Eclipsing Japan, China is now second with 78 million and one car for every 17 people, but China's car population is exploding. India's vehicle population is also rising rapidly. The total number of motorized vehicles goes up another three or four hundred million when one adds trucks and buses. Counting motorcycles and motorscooters, the numbers begin to approach 2 billion motorized vehicles of one sort or another driving trillions of miles and kilometers everywhere all over the world all the time. It's sobering to think how humanity has created a motorized auxiliary to itself.

Almost 85 million motorized vehicles were built worldwide in 2012.

Since 2008 China has been the leading manufacturer of cars and commercial vehicles, producing 20 million units, almost twice as many as the United States and Japan at roughly 10 million each; Germany, South Korea, and India follow, but the list of nations that build cars, trucks, and motor bikes does not stop there. What these figures mean for jobs and employment alone around the world, for example, underscores the high importance of the automobile and its manufacturing and servicing as economic engines driving industrial economies and societies.

Wings over the World

While electricity was being generated by coal- and oil-fired steam plants and wrung from the world's rivers by turbine-driven generators, and while rails, roads, and automobiles began to crisscross the landscape, humans conquered the air. The skies were invaded by unforeseen flying machines. The heavier-than-air vessel came out of the blue, as it were. In the eighteenth century, hot-air balloons had been invented in France, but their potential never went beyond recreation or limited military use as observation posts. Although human flight was an age-old dream, sustained powered flight became realizable only after compact and relatively light thermal engines, mainly internal combustion engines, were invented.

In 1903 the Wright brothers (Orville and Wilbur) flew the first airplane. On December 17 at Kitty Hawk on the Outer Banks of North Carolina it flew for twelve seconds and covered 120 feet. They made four flights that day, the longest of 852 feet. This was the first powered, controlled, and sustained flight of a human. The plane was a wood and fabric structure that embodied only a few recent technical developments—gasoline, the internal combustion engine, and techniques borrowed from the manufacture of bicycles (which, surprisingly, had been invented only twenty-five years earlier, long after the more complex locomotive). The Wright brothers stated that they could only foresee the use of airplanes in warfare, but they were soon proven, spectacularly, to be the "wrong brothers." Soon, airplanes were being used not only in warfare, but much more extensively in mail and passenger services. During World War I, less than fifteen years after the Wrights' primitive flyer, airplanes mounted with machine guns and carrying bombs were used in Europe, and development of the radial engine enabled Charles Lindbergh to fly across the Atlantic in 1927. By the end of the 1920s mail was being carried in rudimentary services, and in the early '30s passenger airlines got off the ground. The sleek aluminum-skinned DC-3 with its pressurized cabin became the workhorse of such systems. In a time frame of less than half a century the Wright flyer evolved into commercial airliners combining hundreds of techniques and materials, many of them newly invented. Passenger air travel expanded after World War II, eventually developing into the complex global air transportation system that exists today.

The social consequences of aviation were commensurate with its technical originality. Tourism and recreation were revolutionized and regions of the world that had been bypassed by the early Industrial Revolution were brought into the mainstream. And, as in the electric power industry, government became unavoidably entangled with private enterprise. Passenger terminals had to be constructed, inevitably by local and central governments. Mail delivery, now largely via airmail, was traditionally a state operation; and uniform operational and safety standards became a responsibility of government, even if aided by industry consultants and boards of professional experts. Fleets of military aircraft are of course owned by the government. Moreover, commercial aviation is a global industry; many nations have their own airline—either fully or partially government-owned and certainly government-regulated. A commercial airliner is now assembled from parts manufactured in different countries, and passenger service is staffed by individuals of different nationalities.

The invention of powered flight by the Wright brothers may have been a small step for them, but it was truly a "giant leap for mankind," to borrow the expression of Neil Armstrong when he landed on the moon. In only a century, humans have gone from that first flight of 120 feet to the organization of a huge technological system of air transportation that is another essential component of industrial civilization today. Ticketing is now largely handled online and electronically. Other transportation systems whisk us to airports, which are often small cities in themselves. There, we and our baggage are processed (factory-like!) through security and onto stupendous flying machines of extraordinary technological scale and complexity that are maintained and serviced by other elaborate components of the system. The Boeing 747, the inaugural "jumbo jet," first flew in 1970, and increasingly sophisticated aircraft with increasingly sophisticated electronic guidance and control systems have rolled out since. (The introduction of the first SST supersonic passenger plane, the Anglo-French *Concorde*, marked a turning point in 1976; but SST technology proved too costly and too environmentally unfriendly to be widely adopted in commercial aviation, and *Concorde* flights ceased in 2003.) The latest in this series, the European Airbus A380 passenger plane has two decks and four aisles and can transport 555 to 800 passengers along with baggage.

At any one time thousands of airplanes are in the air, seven thousand over the United States alone. That one can sip a Coke or an iced cocktail at 35,000 feet is a technological achievement of the highest order. Many vessels fly grand routes across the Atlantic and the Pacific Oceans, and it is now possible to get from virtually anywhere in the world to any of a thousand destinations in a day. Radar and air traffic control systems guide planes to their destinations and to exit

Fig. 17.3. A century of flight. The world's largest passenger plane, the Airbus A380, on its maiden flight, April 27th, 2005. Inset (to scale) The Wright brother's first flight at Kitty Hawk, North Carolina, on December 17, 1903. The A380 has a wingspan six times that of the Wright brothers' flyer and, fully loaded, weighs over two thousand times more.

points at other airports and hubs. Regional airports and secondary airlines extend the air-travel network to virtually everywhere in the world. Today, airplane travel has become common and comparatively inexpensive, and masses of passengers now funnel through this vast air transportation system.

The figures are staggering. Various air carriers around the world transported over 3 billion people in 2013, over 40 percent of the world's population. Air traffic is doubling every 15 years, and estimates are that the number of travelers will hit 4 billion by 2030. More than 28,000 commercial flights and some 87,000 total flights crisscross the United States every day. Because of its strategic location connecting to Europe, South America, and Africa as well as U.S. domestic traffic, Hartsfield-Jackson International Airport, home for Delta Airlines, in Atlanta, Georgia, is the world's busiest passenger airport with 94 million travelers passing through in 2013. Since 2010 Beijing Capital International Airport, hub for Air China, is next on the list with 83 million passengers. Passenger traffic in China is expected to grow to a phenomenal 415 million passengers by 2016, and not surprisingly the number of Asian airports among the top twenty worldwide rose from 2 to 8 between 2000 and 2010, while the United States, which counted 13 airports in the top 20, had only 6 in 2010. London Heathrow, Tokyo Haneda, and Chicago O'Hare rounded out the top five passenger airports in 2013. Over all, London area airports handle more passengers (135 million) than any other, with New York coming in second (112 million). Hong Kong, Memphis (Tennessee, USA), and Shanghai are the top three airports handling cargo traffic—Memphis as the main transit hub for the package-shipping giant FedEx and the others as centers for East Asia manufacturing. Situated at a global travel nexus of its own linking the Middle East, Africa, Asia, and Europe, the international airport in Dubai, United Arab Emirates, is well on its way to becoming the world's busiest airport; with over 67 million international travelers it overtook London Heathrow and Hong Kong in 2014. Emirate Airlines already is third in the world for carrying international traffic, and Dubai airport is on target to reach its capacity of 100 million passengers in 2019, at which point a second airport with a planned capacity of 120 million is scheduled to be built! The civil air transportation system is incredibly complex as a technological system and high testimony to human accomplishment. It never shuts down. It has shrunk our world.

In these considerations we must not overlook parallel military air transportation systems, and aviation in its military mode has rewritten the laws of war. The air transport of freight and mail is yet another highly sophisticated aspect of global transportation today, as companies like United Parcel Service (UPS) and Federal Express (FedEx) well illustrate.

Machines for Mother

A change has also taken place in family households as a result of the Industrial Revolution and the ongoing maturation of industrial civilization. This technological revolution of the household has affected billions of people in their everyday lives, particularly women. The kitchen has been transformed, the home reconstituted. What happened in the hearths of the family unit coincident with industrialization needs to be ranked with other particularly noteworthy technological systems that are fundamental to understanding industrial civilization today.

In 1900, even in the industrializing nations, the instruments of housework were the traditional washboard and hand wringer, the floor broom, the mechanical carpet sweeper, and the dish towel. Cast iron stoves for cooking and heating became common with the early Industrial Revolution, rendering obsolete the traditional fireplace; but coal and wood fires still needed to be tended in 1900. The iceman still made his rounds with his horse and wagon. Upper-class families had servants to mind the cooking, cleaning, heating, and other domestic chores. By 2000, circumstances had altered radically. The simple appliances and techniques traditional for running the home were replaced by a powerful and power-hungry army of machines: washing machines (1910) and dryers, refrigerators and freezers (with optional ice-makers and filtered water dispensers), dishwashers, vacuum cleaners (1901), gas and electric ovens and ranges, microwave ovens, toasters (1909), coffee makers, rice cookers, juicers, blenders, mixers, towel warmers, heating pads, garbage disposals, Jacuzzis, and a host of gadgets typical of millions upon millions of well-to-do family homes around the world today. And let us not forget frozen and prepared food and the differences these technologies have made for food preparation and meals. The modern middle-class home was the result, and there, servants are few.

This "domestic" revolution did more than relieve the drudgery of housework. Technological innovations played a role in the "women's liberation" movement. In the United States, in conjunction with increasing educational opportunities and militant action, it led, in 1920, to the Nineteenth Amendment to the Constitution granting women the right to vote. During World War II women entered the workplace in large numbers, confirming once again that sweeping technological innovations may result in social change. Critics have argued that labor-saving devices brought to the home by industrialization may have actually created more work for mother, not less, by imposing new sets of "needs" upon households and impossible standards of order and cleanliness that the homemaker is expected to maintain. But, for umpteen men, women, and children today, their modern lifestyle, not only as they live it at home, but in the world at large, would be impossible without the underpinning and support it receives from the technological arsenal of the home.

The twentieth-century transformation of homes into technological showcases bristling with machines and products to perform domestic chores was propelled by advertising that relentlessly sold the new way of life—and, conversely, by consumer demand. Today, homes are awash with cleaning fluids, sprays, soaps, powders, and like specialty products for a myriad of domestic cleaning purposes. Tile scum be gone, says one! Antibacterial wipes will cleanse your countertops, says another! New and sophisticated technologies for shaving and feminine hygiene are part of this story. And then, how babies are cared for and raised reflects yet another aspect of this domestic revolution, with industrially processed baby foods, specialty disposable diapers, the technologies of strollers, and other paraphernalia—all, again, heavily promoted by advertising and purchased by devoted parents.

Along these lines, the houses and apartments we live in also need to be seen as technological systems that have become refashioned through industrialization and the modern economy. Even modest homes today embody wonders that were rare or unimaginable in mansions prior to industrialization: running fresh, potable water (hot and cold); indoor showers and flushing toilets connected to reliable public sewage systems; fully automatic gas, electric, or fuel oil furnaces permitting central heating; air conditioning; security systems; electric lighting, of course, and then telephone and electric lines and cable and in some cases satellite connections in and out linking individual dwellings to the outside world of voice communications, television, the Internet, and reservoirs of electric power or natural gas. The home likewise connects with necessary garbage and recycling systems. The elevator made possible urban office and high-rise apartment buildings. The suburban

"It's all done - and no real work about it!"

"JUST an hour's interesting experience in operating a most wonderful machine that does both the washing and wringing. All the hand work required was feeding the wringer and hanging out the clothes.

"I'm glad all through that I can do my own washing, independent of unreliable servants, and get a lot of real pleasure out of it besides.

"There's just enough work in a Maytag washday to keep up healthy activity and high spirits."

For city homes and rural homes equipped with electric lighting plants—the Maytag Electric Washer. For village and farm homes without electrical convenience—the Maytag Multi-Motor Washer, whose power is supplied from a small, easily operated gasoline motor installed under the tub.

These two wonderful machines—identical in construction and operation, except as to the source of power—make available a labor-saving, self-contained power washer for every home laundry, in city, town and country.

Write for a Maytag Household Manual. It will be sent by mail gratis.

THE MAYTAG CO.
NEWTON, IOWA

BRANCHES:

Philadelphia, Indianapolis, Minneapolis,
Kansas City, Atlanta, Portland (Oregon),
Winnipeg

DISTRIBUTORS

BILLINGS, MONT. — Billings Hardware Company
BOISE, IDAHO — Stewart Wholesale Co.
CHICAGO — The Fair
HELENA, MONT. — A. M. Holter Hardware Company
LOS ANGELES — Woodill-Hulse Electric Company
NEWARK, N. J. — Newark Electrical Supply Company
OAKLAND, CALIF. — Creighton-Morris Company
SALT LAKE CITY — Utah Power & Light Company
SAN ANTONIO — Smith Bros. Hardware Sales Company
SEATTLE — Seattle Hardware Co.
SPOKANE — Holley-Mason Hardware Company
FOR UTAH AND IDAHO — Consolidated Wagon & Machine Company, Salt Lake City.

71

Maytag
Electric Washer

Maytag

house comes with its own range of ancillary equipment such as lawn movers, swimming pools, and barbecue grills. The system of the home very easily plugs into the world of the automobile, for in so many cases getting into their cars is the first thing people do when leaving their homes.

Innovative home appliances transformed housework and food preparation, and many new technological marvels of the twentieth and now twenty-first centuries, including frozen and prepared foods, made the home a much more comfortable and interesting place to eat, sleep, and base one's family and private life. It hardly needs to be said, however, that, comparatively speaking, this revolution in domestic technologies affected only a minority of households. In many less-developed parts of the world and among the world's poor everywhere, shelter is more primitive, if not precarious. Most women are still chained to the kitchen, and it's a kitchen with few amenities.

"Let Me Entertain You"

An extraordinary constellation of sociologically and economically significant new technologies developed in the twentieth century centered on personal and mass entertainment. The entertainment industries that arose are characteristic of the modern era. Radio was the first, a new science-based technology that arose quickly and more or less directly out of Heinrich Hertz's confirmation of electromagnetic waves in 1887 and Guglielmo Marconi's subsequent efforts to develop wireless telegraphy. Creating operational systems for radio required substantial technical refinements of transmission and reception of signals, and radio was improved by a series of inventors and radio amateurs in the early 1900s, improvements marked, notably, by the amplifying audion vacuum tube patented by Lee de Forest in 1906. In the fall of 1920 regular commercial radio broadcasts began over a 100-watt station in Pittsburgh, KDKA; and by 1922, 564 licensed stations dotted the country. Radio networks emerged with the BBC in 1922 and NBC in 1926. Radio quickly became a new and pleasant form of home entertainment, and the radio receiver an early piece of consumer electronics. It became the new family hearth. Radio was made feasible and profitable through advertising revenues, a development aided and abetted by the new science of applied psychology. The technology of radio broadcasting improved over the years, and the radio itself moved from hobbyist object to consumer product. In its first heyday in the 1930s and 1940s, radio became *the* medium of the Great Depression and World War II. Consumers purchased eight million radios, and three quarters of all American families had a radio in the home by 1936. Great clear-channel beacons like WLS in Chicago broadcast popular music and entertainment across America and through radio brought the wider world to small towns and the rural hinterland. In 1949 over

Fig. 17.4. More work for Mother? A 1919 advertisement for the Maytag Electric Washer and Wringer touting "an hour's interesting experience . . . independent of unreliable servants." New appliances eased the burden of housework for women, yet arguably enslaved them with more chores and higher standards for the modern home.

two thousand American stations were broadcasting, and in that year $628 million was spent on radio advertising. In 1950 there were two radios for every home.

The political applications of radio were less entertaining. Radio became a major medium for propaganda, often by repressive political authorities. And in warfare it gave rise to new espionage technologies of sending and detecting radio signals.

Music, both live and recorded, predominated in radio from early on. Sports broadcasting quickly became popular in the United States, especially baseball games and boxing matches; three hundred thousand people (!) in the New York region listened to the Dempsey-Carpentier fight in 1921, and that was just the beginning. Radios became the new hearths and the nexus of new forms of popular culture. Through the Depression and World War II families gathered around their radios to listen to the "fireside chats" of President Roosevelt. The American listening public panicked when Orson Welles broadcast his radio play about an invasion from Mars in 1938. Programs like *The Lone Ranger* or *The Shadow* inspired the imagination of millions of youths. The introduction of the transistor in the early 1950s made for truly portable radios and marked the end of vacuum tubes as a key technological component of radios. Engineers developed FM (Frequency Modulated) radio in 1939 that allowed for "high fidelity" reception, but the technology of FM radio spread only in the 1950s. Inevitably, the radio was soon married to the automobile, and already in 1935 over a million radios were installed in cars.

The invention of the phonograph and the rise of technologies of recorded music represent another powerful personal and home entertainment technology. (The capture of an ephemeral moment via a recording may be likened to the much earlier technology of writing.) Thomas Edison invented cylinder recording in 1877, but its initial use as an office dictating machine did little to change the world. Again, the creation of a complete technological system had to take place before "records" (patented by Emile Berliner in 1896) and record players could become standard items in American and European homes. This process was not fully stabilized until the introduction in 1906 of the "Victrola" record-playing machine that played recorded vinyl disks. On the other end, recording equipment and the technologies for recording music likewise continued to be perfected. There was much synergy between the radio and the music industry, obviously, as radio stations played recordings that listeners purchased to enjoy at home on their own record players. The 78 rpm was the original standard speed of recording and reproducing sound. "Long-playing" 33 1/3 records date from the 1930s, but they did not begin to predominate until the 1950s. The 45-rpm format originated in 1949 and proved very successful, particularly for popular music, and drove the 78 record to extinction. Stereo recordings followed in 1957. For better or worse, the phonograph record effectively ended traditions of making music in the home, and the technology changed the social and aesthetic experience of listening to music. Records themselves are long gone, except among a small breed of aficionados, replaced in the digital revolution by various devices and storage systems.

To this list we need to add moving pictures or "movies," of course. The redoubtable Edison patented the Kinetograph in 1888 for capturing motion and the Kinetoscope for viewing it. Kinetoscope parlors had machines for individual viewing of films that Edison filmed in his "Black Maria" studio in West Orange, New Jersey. In 1895, with their Cinématographe the French brothers Auguste and Louis Lumière first successfully brought together the requisite camera and projection technologies for mass viewing, and so launched the motion-picture era. With paying customers watching—sometimes stupefied, as with trains seeming about to hurtle off screens into theaters—movies immediately became a highly successful popular entertainment and industry. Not to be outdone, the Edison Manufacturing Company quickly adopted the new technology and produced 371 films, including *The Great Train Robbery* (1903), until the company ceased production in 1918. Sound movies—the talkies—arrived in 1927 with Al Jolson starring in *The Jazz Singer*, and by that time Hollywood was already the center of a vigorous film industry with its "stars" and with movie magazines on newsstands everywhere. The first color movie was made in 1922. With technical improvements made by the Kodak Company in the film itself, color movies came to the screen in the 1930s in such famous examples as *The Wizard of Oz* (1939). (Color did not become an industry standard, however, until the 1960s.) Elaborate production, marketing, and distribution systems brought movies, newsreels, and the glamor of industrial civilization to towns and villages large and small all over the world. The local movie house became the cultural and entertainment hub of communities. Television undercut the movie industry somewhat in the 1960s, but Hollywood soon began producing movies for television, sexually more explicit works, and films with dramatic special effects, e.g., *2001: A Space Odyssey* (1968) and *Star Wars* (1977). The introduction of low-cost videocassette recorders in 1969 created a subsidiary industry of movie rentals. Compact disk (CD) technology succeeded videocassette and tapes and dominated movie and music delivery systems for a while. The local Blockbuster movie rental store is decidedly a thing of the past, however, with CDs and DVDs now being replaced by streaming of movies, TV, and music via the Web. These trends have also made entertainment an evermore solitary activity. Different national traditions are notable in film, with the French, Italian, Japanese, and Swedish among the better known examples, while the film industry in India is the largest in the world. Movies made all but obsolete the traditional theater and concert hall that had long been centers of elite and popular entertainment in cities, certainly in the nineteenth century; and movies erased most residues of truly popular culture of music making, dancing, and like activities in rural settings. The opera house today is little more than a peculiar kind of art museum.

The technology of radio provided a natural model for the idea of sending pictures via electromagnetic waves. Against a background of several systems under development, including one by the pioneer Vladimir Zworykin (1889–1982) in 1927, Philo T. Farnsworth (1906–1971), a farm boy and then a college student, first patented the technology for electronic transmission and reception of moving images. In 1930, the first commercial television was already broadcasting. RCA, the Radio Corporation of America, famously promoted its television technology at the 1939 World's Fair in New York. World War II interrupted development, but television experienced explosive growth after the war and soon became a major industry, and the "box" a common household appliance. U.S. manufacturers produced four million TV sets in 1949, and in the same year ninety-eight television stations beamed programming into a rapidly growing number of homes; in 1952 there were 21 million TV sets in America. Color television emerged only after 1953, when the Federal Communications Commission settled on one of several incompatible technologies, but usage remained limited for technical reasons until NBC became the first all-color network in 1966. Beginning with Telstar in 1962, communication satellites allowed for global television broadcasts, which are now commonplace. The infrared "remote" appeared in the 1980s, changing viewers' habits and the experience of watching TV. Digital television transmission (HDTV) began in 1998 and has in many ways supplanted over-the-air transmission of television.

The power and importance of television as a key news and entertainment medium of industrial civilization is self-evident. Radio had already begun to change politics—reporting of election results and broadcasting from political conventions, but TV did even more, especially with the development of network television news, and from the 1950s onward television became a universal fixture in homes everywhere around the globe, a powerful outlet for incessant advertising and an omnipresent communicator of norms. Especially in the 1950s, television conveyed a monolithic set of social values, and only from the 1970s did TV in the United States come more to reflect America's racial, ethnic, gender, and class diversity.

Today, the technologies of radio, television, movies, and recorded music of one sort or another have indeed conquered the world. They have spread their roots deep into the bedrock of human cultures. Radio, TV, movies, the music industry, and in short, the modern electronic media have become intertwined as well through overlapping hierarchies of economic interests. Today's media are undeniably powerful instruments of cultural homogenization around the world. Television, for example, is known to have reduced regional accents all across America, and professional sports, for example, would not exist as they do at present without radio and television. Yet, ironically, even as giant media conglomerates become more global, centralized, and ever more rich and powerful, the technologies of the media have allowed for and even encouraged diversity and choice among consumers. In other words, the modern media, although monolithic in its ownership, has responded to federal regulations and to consumer and advertiser demand by breaking down markets and people into smaller subgroups. Radio, for example, is omnipresent and with over 15,000 licensed radio stations on the air in the United States alone. On average today each American listens to over two hours of radio a day, and radio's presence in our lives has only increased with the advent of satellite radio and Internet radio. Listeners can tune into a spectrum of specialized music stations that limit their repertoire to country music, rock, classical, jazz, easy listening, and so forth; others are national language stations or programs; some focus on news, sports, weather, and talk radio, yet others are devoted to religious broadcasting.

The same point holds true for television. In his 1964 landmark, *Understanding Media*, the prescient theorist Marshall McLuhan articulated the concept of the "global village" with television as the leading medium. Today's Internet and the World Wide Web constitute a new medium McLuhan could never have imagined for the global village, as we will emphasize in chapter 19, but for the moment television still today touches the lives of more people in the United States and around the world than the Internet or Web. Certainly no more powerful medium than television arose in the twentieth century for bringing peoples together. By creating a "you are there" feeling, television brings the world to viewers. The coverage of the moon landing in 1969 or the destruction of the World Trade Center in New York on September 11, 2001, riveted virtually all of humanity for one moment. The coverage of the long-ago war in Vietnam, to pick another example, brought the war into living rooms everywhere and was instrumental in the success of the contemporary antiwar movement.

The quadrennial World Cup soccer tournament is the world's most widely viewed sporting event, and it is emblematic of contemporary industrial civilization. Viewership is tricky to determine, given communal settings for viewing at home, at work, in bars, or in fanzones, but over 700 million people are estimated to have watched the 2010 World Cup final in South Africa, and perhaps as many as 3.2 billion people watched one or more of the 2010 World Cup games broadcast from there. The 2014 World Cup in Brazil shattered viewing records for a sporting event in the United States, a country not known for being enamored of soccer, and it was the most watched sporting event ever in many other countries; 42.9 million Brazilians, for example, watched the opening game between Brazil and Croatia, 34.1 million in faraway

Japan saw their team play Ivory Coast, and 34.7 million in Germany celebrated their national team's victory at the 2014 World Cup final against Argentina. Tellingly, that one game generated 3 billion posts on Facebook and 672 million tweets on Twitter.

For our purposes what is striking about the World Cup is the technology and worldwide reach of its broadcasts and what the event says about the world today. The 2010 event in South Africa was broadcast to 214 countries and territories, including the Arctic and Antarctic, on 245 different channels and in uncounted dozens of languages. Consider the globe-spanning technology required for that to happen. High-definition broadcasts of the 2014 competition in Brazil originated at twelve production centers at the various game sites. Then factor in the cameras, cable wires, transmission installations, satellites, studios, and commentators. Consider next the national and commercial channels that broadcast or streamed games over the Internet to the rest of world. Think of the broadcast rights and the advertisers who spent $1.5 billion to sell themselves before this vast audience. The event had a real-time immediacy that brought the world together and shrank the globe. World Cup soccer, no less than everything it takes to make it happen, is a stunning human accomplishment and testimony to the scale and character of contemporary world civilization.

And, thanks to communications satellites, on the Internet and every night on the nightly news, we see reporters live from hot spots around the world with video shot that day, or even live feeds for breaking news. Yet, diversity and fragmentation rule in the "cool" medium of television. Over 1,700 full-power television stations presently broadcast in the United States, and with the advent of cable and satellite TV viewers have literally hundreds of channels to choose from that cater to all kinds of tastes. TV stations now specialize in news, movies, business reporting, sports, weather, cartoons, children's subjects, music videos, hobbies, religion, pornography, and in blind self-reflection, TV itself. In many countries private television channels exist alongside state-sponsored ones, but the airwaves are regulated by state authority. In the United States the Public Broadcasting System (PBS) originated in 1967 with various missions in mind, including education, culture, and news presentation. The arrival of digital, high-definition TV (HDTV) has improved the quality of the images, but not the quality of the content of television.

Cable television, recording devices, and the Internet have radically changed the status of television. Families no longer sit around the "box" as they once did, too, with radio. The TV anchorman is no longer the nation's wise uncle. Cable television, as mentioned, expanded the number of channels available into the hundreds for paying customers, and so already diluted the power of the great broadcasting systems like ABC, CBS, and NBC and opened up the airways to a cacophony of voices. That cable channels began presenting their own original programming did the same. Recording capabilities freed viewers from the tyranny of the clock and the TV broadcast schedule. The Internet brought further, fundamental changes in not only becoming an asynchronous source of TV shows, but also of other, often self-generated programming. In any number of fascinating Websites, such as YouTube, hours and days can be whiled away clicking through one thing after another at one's leisure.

Movies and music, too, represent giant, converging industries and technologies. On often huge screens and with exquisite sound systems, multiplex cinema palaces show multimillion-dollar blockbuster films with their own spectacular computer-driven special effects. On the one hand, the same film will bind the world together by playing in theaters from Moscow to Sydney to Toronto. On the other hand, low-budget and art films persist, and on a national or regional basis at least, such as India's Bollywood, variation in the movie world exists. But movies can also be played on TVs and mobile personal devices at home in various formats or downloaded via the Internet, and here consumers are free to enjoy their favorite types of films, be they Kung-Fu movies, crime dramas, or the latest romantic comedy.

The recorded music industry today perhaps best represents the not-so-paradoxical paradox of technological homogeneity and global corporate reach alongside greatly increased diversity, individual choice, and concurrent social and cultural segregation of people into groups. Recorded music is essentially no longer played on records, except by a dwindling core of aficionados, and, after living through several varieties of magnetic tape, music technology today has moved over to new, digital platforms. The industry is worth billions of dollars, and today via iTunes and other services over the Internet consumers can download virtual cornucopias of music bursting with every conceivable genre, variety, and style, not to mention thousands of performers and performances. Music services such as Pandora available over the Internet offer an unending stream of music tailored to listener choices, and much music is posted and widely available on the Web. It is telling in terms of the history of music technology that megastores that only recently sold music discs have all but disappeared. In the 1960s music was shared collectively and a vehicle for social upheaval and social change; today, with the Internet and personal devices for storing and listening, music is more a personal statement with music available for every taste or mood. In this connection something might be said about video games, which are forms of entertainment produced by media giants that cost tens of millions of dollars to develop and that preoccupy significant sections of a generally younger population around the world. But digital music and video games are strongly tied up with computers and the digital world, a technological development of science of world historic proportions discussed further in chapter 19.

The distribution of these remarkable entertainment technologies is uneven, as would be expected, yet radio, TV, movies, and recorded music have penetrated well beyond the bounds of the industrialized world per se and have reached the most remote corners of the world. The new media of the twentieth century have indeed been powerful forces in spreading and unifying industrial civilization, and they continue to exercise powerful functions in shaping the global village today.

One Global System

The history of technology, particularly in the twentieth century, shows the rise of the great technological systems surveyed here, the ways they have come together to forge contemporary industrial civilization, and how they have spread to provide the technological infrastructure for globalization and global culture. The globalizing changes that continue to sweep the world are sustained by a supersystem of interlocked technological and social systems that cannot be confined to any region. The automobile industry became interconnected with aviation, another major system, and with rubber and glass production, domestic labor relations, foreign trade, communications, tourism, recreation facilities, and educational institutions. Automobiles are now manufactured in many countries, many beyond the precocious areas of Western Europe and North America; your Japanese brand car may well be made in America. Nor can the aviation industry be restricted to any country, or even any region of the world. Hydroelectric power production could not be contained within capitalism; electric power distribution knows no boundaries. Monoculture spreads beyond borders everywhere. A Coke or a McDonald's hamburger tastes more or less the same anywhere in the world. In countless ways, old and new technologies, embedded in diverse social and political settings, coalesced into a global culture by the turn of our third millennium of the common era.

To gain a sense of the scale of the global socio-industrial system that formed in the twentieth century and that continues to expand today, the automobile industry might be compared with the much smaller eighteenth-century system surrounding the steam engine. By the early nineteenth century steam engines powered the iron industry, coal mining, small-scale manufacturing, and on an even smaller scale, steam boats. Steam engines touched the lives of a small part of the population in only a few countries and reached only slightly beyond the borders of any nation. By contrast, the automobile today has directly altered the lives of everyone in the advanced industrial nations and indirectly everyone everywhere else. The parts of a single vehicle are manufactured in different parts of the world, and they are designed in conformity with international safety and environmental regulations. The road systems that automobiles have engendered are marked by internationally uniform icons, including, for example, the ubiquitous stop sign. The Toyota Motor Corporation, which sold nearly ten million vehicles in 2013, sells more cars than any other manufacturer in the world, and it sells them all over the world.

Did globalization begin with the prehistoric and historic spread of humanity to all habitable parts of the globe? Surely, in some sense; and human cultures have long been aware of each other with resulting exchanges. One thinks in this connection of historical relations between Japan and China, the appearance of Chinese silk in ancient Rome, or the spice trade. But globalization as we have come to recognize the phenomenon today may be said to have begun in the fifteenth century of the common era with Portuguese and Spanish expansion outside the borders of Europe. This process continued to unfold in the centuries that followed through the activities of merchant traders—including slave traders—and European colonists and imperialists. We noted earlier how the technological capabilities of the railroad, the steamboat, and the telegraph propelled European expansion and globalizing forces in the nineteenth century. These trends were dramatically accelerated by twentieth- and twenty-first-century developments that now include global media, global markets, and global corporations. Globalization has its own momentum, and the story of globalization has moved well beyond any earlier notions of Westernization.

Industrial civilization brought with it a historically unprecedented and potentially devastating population explosion. By 1750 world population had reached 760 million, and humanity crossed the threshold of one billion people only around 1800. That number rose exponentially to 1.6 billion in 1900 with more billions continuing to be added in shorter and shorter periods, especially after 1950. Another 1.6 billion came on board between 1990 and 2010. World population passed the threshold of 7 billion at the turn of 2012, and in 2014 it pushes 7.25 billion, adding 70 million new mouths to feed every year, 200,000 every day, 140 every minute.

Those who project world population foresee 8 billion people in 2024, 9.6 billion by 2050, and perhaps 11 billion by 2100. The upper end of a United Nations' estimate is 10.1 billion by 2050. China is the most populous nation in the world today at almost 1.4 billion people. India follows next with almost 1.3 billion; China and India surpass any other nation in population by an order of magnitude. China and India, especially India, are among the most densely populated countries in the world, and together they hold almost 40 percent of the world's people and a much larger percentage (60%) if the whole of Asia is included with them; the United States is third in population rank with 320 million, followed by Indonesia and Brazil.

Asia is also the fastest growing part of the world, and a demographic landmark should occur around 2025 when India will surpass China in total population. Chinese population itself is expected to peak around 2030 and Asian growth to top off in 2050. Africa is the demographic surprise in this race. In Africa, where industrialization lagged, the population nevertheless grew more than tenfold since 1750. Now, numbering 15 percent of the world's population, Africa's population is expected to explode in the upcoming decades, doubling by 2050 and perhaps even quadrupling to surpass Asia by 2100. Ten of the twenty fastest-growing economies through 2017 are expected to be in Africa. In particular, Nigeria, with a 2013 population of 174 million, is projected to grow to fully 440 million and surpass the population of the United States by 2050. A striking and perhaps ominous feature of these statistics is that essentially all this population growth is occurring in less-developed countries, which is only going to exacerbate tensions between the haves and the have-nots in the world.

By contrast, even as they consume most of the world's resources, populations in more developed countries today are growing only slowly, if at all, and in some cases, such as Japan and Italy, the rate at which children are being born is not sufficient to replace the existing populace. The population of Western Europe, for example, is currently 740 million or 10 percent of the world total, a figure expected to decline to 7 percent over the next several decades. The population of the United States at 320 million is expected to rise to 362 million by 2030. This rise is nontrivial, due significantly to immigration, yet it is a modest increase compared to the boom in the rest of the world.

These patterns of rapid, yet uneven, demographic growth coincide with the spread of industrial civilization, and they map well onto what we already know of rich and poor nations in the developed, developing, and underdeveloped world today. Population pressures manifest themselves intensely in many forms, notably surrounding immigration and controlling access to wealth and natural and social resources. Whether the natural world around us or industrial civilization as we know it can sustain a world population at these levels is an open question.

What has just been said about exponential population growth since the Industrial Revolution can equally well be said about another telling indicator of industrial civilization, urbanization. In 1800 Beijing was the only city in the world with a population larger than one million; by 1900 sixteen such world cities dotted the globe; Tokyo became the first city to surpass the 10 million mark, and today 28 megacities with populations greater than 10 million light up the night sky around the world. This urbanization is relatively recent and a concomitant element of industrialization. Urbanization on a world level stood at only 15 percent in 1900, but the figure doubled to 30 percent in 1950, it surged to 45 percent in 1990, and humanity as a whole crossed the landmark of being 50 percent urban dwellers in 2008. That figure rose further to 54 percent in 2014, and by 2050 two-thirds of the world's population will likely live in cities. Germany already reached 50 percent urban population in 1900, France and the United States by 1920, and by 2000 75 percent of the population of more developed countries were urban dwellers. The United States is now over 80 percent urbanized in terms of the number of people who live in cities. This remarkable historical and worldwide shift to cities was brought on by industrialization and is a defining feature of contemporary industrial civilization.

How to measure these things varies somewhat, but greater Tokyo remains the largest megacity in the world with a population of almost 38 million in 2014. Delhi, the capital of India, jumped to become the world's second most populous city in 2014 with almost 25 million inhabitants, more than doubling its population since 1990. Shanghai has dropped to third place with 23 million, followed by Mexico City, Sao Paulo, and Mumbai at about 21 million apiece. Beijing is the eighth largest megalopolis, New York the ninth. None of the top ten are in Europe. Lagos in Nigeria with a population above 17 million in 2014 ranks in the top ten of world cities and is the largest urban area in Africa. Virtually all projected population growth in the next thirty to fifty years is urban growth, with the three countries of India, China, and Nigeria alone expected to account for more than a third of that urban growth through 2050. Most of the newer megacities are rising in less-developed countries, a phenomenon that once again points to the key forces at play on the world stage today. Traffic congestion is a plague, and how increasingly packed cities are to be made livable is a great challenge facing industrial civilization. Related challenges have to do notably again with internal and external migration, that is, within countries as people move from the countryside to the cities and between more densely populated and poorer regions of the world toward richer, more urbanized countries.

As a result of the phenomenal growth of cities and the industrialization of food production, rural populations have declined accordingly worldwide and will continue to do so. In Asia rural populations are set to fall below the 50 percent mark soon, and the current goal in China is to move 100 million rural people into towns and cities by 2020. Urbanization in Latin America has skyrocketed since the 1970s and will continue to outpace a rural population at 20 percent and declining. Heavily urbanized Europe will only become more so. In 1950 about 12 percent of the U.S. labor force was involved in agriculture; in 1990 that figure, certainly a good proxy for rural life, fell to 2.6 percent, and today, depending on who is counted, that figure hovers just above 1 percent. This is an outstanding turnaround from the historical pattern that has dominated since the dawn of civilization, that of cities and urban dwellers being a small minority surrounded by seas of rustic tillers of the soil or engaged in animal husbandry.

The rapid industrialization of Japan and the Soviet Union in the twentieth century, and of China beginning in the 1980s, suggests that industrialization will attach itself to any system and ideological program. The Industrial Revolution that began in Britain in the eighteenth century and for the next two centuries spread itself wherever European populations were dominant induced the misunderstanding that it was inherently associated with European culture and bourgeois society. Now, the examples of Japan, China, South Korea, India, and Brazil show that rapid industrial development, along with its many earmarks that were displayed in its European expression, is unfolding on the world level in ways closely associated with global capitalism and the material conditions of industrial civilization.

But innovation generates not only technical progress, it also results in technical decay, and sometimes feeds a sense of nostalgia. The automobile replaced more than the horse. Wherever the culture of the automobile has taken hold it has often led to the decline of small communities and urban neighborhoods. The railroad industry that had encrusted the earth's surface with track during the nineteenth and twentieth centuries and fostered a habit of leisurely and refined travel, fell into decline in competition with hectic and tiresome road traffic. Sophisticated technological systems require a better educated and trained work force, but improvements often result in unemployment. For example, devices installed by water companies in customer's homes can transmit data automatically to a central computer, but this technology is rendering the old-fashioned meter reader extinct. The marvels of computers and the Web are myriad, but an unanticipated result was the abrupt demise of the inexpensive typewriter and the slide rule, that humble token of a more innocent engineering craft. The buggy whip went extinct, and the fact is underappreciated that technological systems not only come into being, but also pass away.

Industrial civilization has its decidedly dark side, too, notably in war. Just a small list includes military fighter jets, bombers, military and police helicopters, tanks and personnel carriers, aircraft carriers and destroyers at sea, nuclear bombs, chemical weapons, other weapons of mass destruction, rockets and missile systems capable of bringing down a passenger plane locally or destroying whole populations on the other side of the world, land mines, small arms, and destructive armaments of all sorts. The technologies of policing and national security infrastructures form part of the dark side of civilization. They are all inextricably interwoven with its marvels, benefits, and comforts.

But Not "One World"

By drawing attention away from individual artifacts, the concept of a technological system has proven an extremely useful one for thinking about technologies in general and about the particular bases of industrial civilization on a world scale today. The concept of a technological system has its limits, however. At the margin, it is hard to distinguish where one system ends and where another takes up. And then, there is variation in the implementation of the "same" technology in different social and cultural settings. In France, for example, Dijon mustard is served at McDonald's restaurants; in the 1950s and 1960s the sizes of cars in America and Japan were quite different because of social, not technological, factors. Then, too, we need to understand technology not only as a triumphal march of systems, but as subject to counter currents that resist the introduction and spread of new technologies and that operate to defeat systems. Resistance to farming genetically modified (GM) foods is a notable example, as are anarchist and anti-capitalism movements. The shift toward locally grown and organic products fits the point here.

A striking feature of industrialization in our day has been an increasing divergence of wealth on a global scale. Figures for per capita income are revealing. According to World Bank statistics, the median national per capita annual income for the whole world in 2013 was approximately $10,500 (or $14,000 factoring in purchasing power parity), meaning that half of the world's nations earned more or less on a per capita basis. In a handful of the richest nations, median per capita income pushes $100,000 per year. Low-income countries have a median income of roughly $1,000 a year, middle-income countries plus or minus $10,000, and high-income countries on the order of $40,000 a year personal income. The United States stood seventeenth in the world in 2013 with a median per capita income of $53,000. The median income in the top 10 percent of countries is over one hundred times that of the poorest 10 percent, and according to a United Nations report in 2014, the wealthiest 85 individuals in the world possessed as much wealth as the 3.5 billion poorest people in the world. The poverty of Africa is striking in this regard with twenty countries on that continent posting per capita incomes of less than $700 a year or less than $2 a day. Haiti is the poorest country in the Americas with its per capita income in 2013 having only risen to $810. Bangladesh, the South Asian country whose utterly insignificant energy consumption we noted earlier, has a per capita income of $900 a year. These and similar statistics paint a portrait of skewed wealth and uneven consumption of resources in the leading industrialized countries and regions compared to developing and third-world countries. Even within developed countries, of course, economic inequalities are striking. In the United States, for example, income has become distributed more unequally over the past 30 years, with those in the top quintile (20%) earning more than the bottom 80 percent combined, and the figures become only more skewed as one approaches the top one percent. Modern wonders like cell phones and computers are unavailable to large populations mired in poverty in nations inside and outside the developed world.

By the same token, although these disparities are distressingly real, a rising tide of economic activity on a world scale seems to have produced a more egalitarian world overall for more and more of the world's peoples. The scale of the U.S. economy, for example, has exploded in the last hundred years, some sixteen times larger in real terms today, and GDP per capita has risen by a factor of six. For those who have benefited, wealth has not necessarily translated into more leisure or diminished the stress of life in industrial civilization. Yet, although the answer is hardly certain, there may be grounds for thinking we're better off.

Most underdeveloped countries remain economically subservient to the industrialized nations. Decolonization after World War II brought political independence to many former colonies, but in most cases their dependent economic status remained largely unchanged. Nevertheless, a powerful set of new industrial or industrializing nations emerged after World War II, notably the Pacific rim countries—Japan, Hong Kong, South Korea, and Taiwan. A series of "emerging markets" now complement the more established industrial societies—markets like Malaysia, Singapore, Thailand, the Philippines, India, and countries of Latin America. In some cases, such as the development of the oil industry in the Persian Gulf region or diamond and gold mining in South Africa, only small sectors of a nation's economy have become industrialized. As apparent from what has already been reported, the current industrialization of China and to a lesser extent India, with their combined populations of 2.5 billion of the world's 7.25 billion people, are now surging toward the first rank, and they will write the next chapter in the ongoing history of industrial civilization. Europe and the United States are destined to remain potent regional powers for years to come, but they are being joined by Asia and a burgeoning Africa as humanity charts its course across the twenty-first century.

Several factors have combined to transform the more mature industrial economies, notably an increasing emphasis on service industries, information management, electronic industries, and biotechnology. The entry of substantial numbers of women into the workforce in the West and increasingly elsewhere has effected great social and economic change. The emergence of the multinational corporation has helped forge an interdependent global economic system. Multinational corporations are often more wealthy than some whole countries, and to a considerable extent they rival the importance of many nations and national economic policies. For example, the retailing giant Walmart is the largest business enterprise in the world in terms of revenue generated, $476 billion in 2014; it has 2.2 million employees and a market capitalization of $257 billion. The country of Slovenia, by contrast, in 2014 had roughly the equivalent population of Walmart employees (2.1 million) but a gross domestic product (GDP) of only $57.4 billion or eight times less than Walmart's revenues. Exxon Mobil Corporation, other oil companies, and a string of Chinese and international banks likewise are larger and more important entities than many nation-states today. The point is telling for understanding the history of industrial civilization.

New techniques and new industries called for new approaches to the solution of technical problems. Over the centuries traditional technologies had acquired refined rules of thumb. But for steel suspension bridges, electrical devices, airplanes, plastics, and the many innovations that followed in rapid succession in the nineteenth, twentieth, and now twenty-first centuries there were no traditional rules of thumb and no time for them to be derived from experience. Instead, rules derived from science increasingly took their place, and the merger of modern science and the new technologies created a scientific-industrial culture. The universities were brought into technology, and governments were brought into science. University-trained engineers, still a small minority in the nineteenth century, became the norm in the twentieth century and today. Government patronage of scientific research was necessitated by skyrocketing costs of "Big Science," and justified by the public benefits of its technological spin-off. Antibiotics and the atomic bomb are the emblems, beneficial and baleful, of the application of theoretical science to practical problems in the twentieth century. This key aspect of industrial civilization—applied science itself—receives deserved separate treatment in chapter 19.

The processes let loose by the Industrial Revolution continue to unfold across the globe. Industrial civilization has been accompanied by an expanded middle class and higher standards of living. Globalization has widened peoples' horizons, and the cultural impact of globalization has been huge in creating a single global culture. For a large number of people the results have been historically unprecedented lives with good health, comfort, and technological trinkets of amazing variety. By the same token, the material progress that many enjoy has not been achieved without heavy costs, not the least of which are the increasing stratification of rich and poor and a more hectic pace of life. Recent years have seen a drop in real wages in advanced countries particularly for the less well educated. Consumerism now represents the dominant values in many parts of the world. Current environmental problems with global warming/climate change, pollution, oil spills, lingering acid rain, the fragility of the ozone layer, waste disposal, loss of biodiversity, greenhouse gas emissions, extensive deforestation, and like concerns reflect the tremendous, and likely irreversible, ecological degradation accompanying industrialization. Population growth and increasing demands on limited resources such as fresh water, oil, and natural gas add to these pressures. The human impact on the planet has been so great since the onset of the Industrial Revolution that thinkers now designate a whole new geological era, the Anthropocene—

the Human Epoch—for the period that began with civilization and took off with industrialization. The ultimate outcome of events that began with the Industrial Revolution in England not so long ago is not clear, but it seems unlikely that the world can long sustain further industrial intensification. The toolmakers have done their work. Now the peacemakers and the stewards of the Earth must do theirs.

The New Aristotelians

The Hellenic tradition of natural philosophy—the disinterested inquiry into nature that sprang to life in ancient Greece—was a defining element of science in the twentieth century, and it continues to astonish with sparkling theoretical insights into nature in the twenty-first century. As a result, a host of conceptual novelties has emerged across a broad range of disciplines to transform our contemporary vision of the world. The significance of theoretical science in our day may ultimately stem less from its natural philosophical import than from the great practical impact it is having on society at large. Nevertheless, the tradition of pure science continues to shape our intellectual culture and the lives of thinking men and women everywhere.

Einstein, Relativity, and Quantum Physics

One of the most notable achievements of scientific thought in the twentieth century was the overthrow of the Classical World View of nineteenth-century physics and the substitution of revolutionary new parameters for understanding the physical world. This revolution is ofttimes labeled the Einsteinian Revolution, and it played a decisive role in shaping how we now think about the natural world around us.

Chapter 15 outlined the coming together of the grand intellectual synthesis known as the Classical World View. The reader will recall its major tenets: a Newtonian framework of absolute space and time; a world of immutable and indivisible atoms; an ethereal substrate that provided the basis for the electromagnetic field and for the undulatory propagation of light and radiant heat. The ether and matter mutually interacted, and the abstraction called energy regulated this interaction; energy manifested itself in several forms and obeyed the strict mathematical laws of thermodynamics. There was great unity, simplicity, and harmony to the worldview fabricated by physicists in the second half of the nineteenth century.

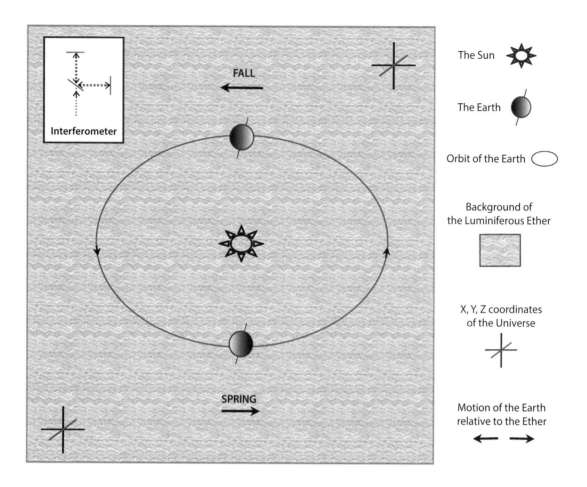

The Sun

The Earth

Orbit of the Earth

Background of
the Luminiferous Ether

X, Y, Z coordinates
of the Universe

Motion of the Earth
relative to the Ether

Interferometer

FALL

SPRING

Unlike Aristotle's scientific vision, which endured for 2,000 years, the Classical World View fell apart almost as quickly as it came together. In the last decade of the nineteenth century a series of nagging problems began to undermine the intellectual edifice of contemporary physics, and by the dawn of the twentieth century a serious crisis had developed in the physical sciences.

In a series of experiments beginning in 1887, American physicist Albert A. Michelson (1852–1931) failed to detect the motion of the earth relative to any ether. Theory predicted that the speed of light ought to change ever so slightly as measured from the earth moving relative to a stationary ether. That is, over a six-month period the orbiting earth moves in opposing directions around the sun, and so, from the frame of reference of a fixed ether, over that period of time the same beam of light ought to move first with the earth and then contrary to the earth's motion, and that difference in motion ought to be detectable. (See figure 18.1.) But through experiments of exquisite exactness Michelson and his colleague E. W. Morley consistently obtained a negative result. In retrospect, this "failure" can easily be portrayed as paving the way for Einstein's theory of relativity, in which the null result of the Michelson-Morley experiment is expected and easily explained by the constancy of the speed of light independent of relative motion. In point of fact, however, Einstein came to relativity through a different route. At first, contemporary physicists, rather than abandon classical principles on the basis of the Michelson-Morley experiment, characteristically rushed to "save the phenomena" and to patch up the existing system by improvising ad hoc explanations that preserved the ether.

Several other developments added to a mounting sense of crisis in contemporary physics. In the fall of 1895 the German experimentalist Wilhelm Roentgen (1845–1923), working with a standard piece of laboratory apparatus, uncovered X-rays, a new type of radiation. While not wholly outside the bounds of classical physics, this discovery extended the range of electromagnetic radiation well beyond its conventional limits and called into question traditional assumptions about the spectrum and about accepted laboratory procedures.

The discovery of the electron constituted a related and more serious problem. Already in the 1870s scientists recognized that an electric current passing through an evacuated receiver produced "rays" of some sort that they called cathode rays. In 1897 British physicist J. J. Thomson (1856–1940) demonstrated that cathode rays were particulate in nature; that is, they were distinct particles nearly 2,000 times smaller than an atom of hydrogen, the smallest atom. Thus, the traditional, indivisible atom was not the smallest unit of matter after all. How was that to be squared with received notions?

A year earlier, French physicist Antoine-Henri Becquerel (1852–1908) accidentally noticed that uranium ore clouded unexposed photographic plates, and he thus uncovered still another unexpected natural phenomenon. In 1898 the acclaimed Franco-Polish physicist Marie Curie (1867–1934) coined the term radioactivity. Marie Curie became the first female professor at the Sorbonne and the only woman ever to win two Nobel Prizes—the first in physics in 1903 shared with her husband, Pierre Curie, for their work on radioactivity, and the second in chemistry in 1911 on her own for her discovery of the radioactive element radium. After her seminal work it became clear that certain heavy elements spontaneously emitted several different sorts of radiation, including electrons, superenergetic electromagnetic waves (gamma rays), and subatomic particles called alpha particles (two protons and two neutrons, the nucleus of a helium atom). By 1901 the phenomenon of radioactive decay became apparent, wherein one element, say uranium, transformed itself through radioactive emissions into another element, say lead. Such transformations clearly violated the fundamental principle of the immutability of atoms. The fixity of atoms, like the fixity of species, broke down and provided yet another great riddle for scientific explanation.

Two other technical puzzles compounded this crisis: the photoelectric effect and the mathematics of "black body radiation." The photoelectric effect, discovered in 1887 by Heinrich Hertz, concerned the

Fig. 18.1. (opposite) The luminiferous ether. In the scientific worldview of the late nineteenth century a superfine ether permeated all of space and provided the physical substrate for electromagnetic radiation including visible light. Two American physicists, Albert Michelson and E. W. Morley, developed a highly sensitive instrument called the interferometer to measure the motion of the Earth relative to the universal ether. The interferometer split a beam of light and then recombined it to measure changes on the order of magnitude of a wavelength. But despite the satisfactory sensitivity of their instrument, Michelson and Morley never observed the sought-after evidence that would prove the existence of the ether. With the revolution in modern physics led by Albert Einstein in 1905 the ether simply vanished.

Fig. 18.2. Radioactivity. Pierre and Marie Curie in the "discovery shed" at the School of Industrial Physics and Chemistry, Paris, ca. 1900. Their discovery of radioactive elements and radioactive decay undermined received notions of the immutability of atoms.

apparent paradox that, only above a certain wavelength would light shining on certain materials induce an electric current. Even intense amounts of light below the requisite threshold would not trigger a current, whereas the least amount of light above it would. The black body problem concerned research that implied that an ideal system might emit more radiant energy than it received as input; that is, if the electromagnetic spectrum were truly continuous, then an initial wave of light or radiant heat might be redistributed into an infinite number of smaller waves and hence produce an infinite amount of energy. This was an obviously absurd finding that contradicted empirical experiments and the received laws of thermodynamics. The interpretations that began to emerge, notably those enunciated after 1901 by the German physicist Max Planck (1858–1947), involved suggestions that light (or radiation in general) came in discrete energy packets or quantum units and did not exist in the infinitely graded energy continuum required by classical physics. The great Dutch and French physicists Hendrik Lorentz (1853–1928) and Henri Poincaré (1854–1912) worked on these problems in the years leading up to 1905, introducing fudge factors that preserved existing certainties.

Albert Einstein (1879–1955) achieved intellectual maturity during this confusing period at the turn of the twentieth century. The son of an unsuccessful businessman, Einstein displayed no precocious talents as a child. At 16 he dropped out of the gymnasium (or high school) in Munich to join his family, which had moved to Italy. After some difficulties in being admitted, Einstein attended and in 1900 graduated from the Federal Polytechnic School in Zurich. Excluded because of his Jewish ancestry from becoming a schoolteacher, Einstein and his first wife moved to Berne, Switzerland, where he took a minor post in the Swiss patent office. He received a doctorate in physics from the University of Zurich in 1905, and he remained in Berne until 1909 pursuing physics in his spare time. Einstein was thus perfectly positioned to effect a revolution in contemporary physics: he was well educated technically in the central dogmas of the field, yet he was young enough and as an outsider professionally marginal enough not to be locked into established beliefs.

In what was indeed an *annus mirabilis*, Einstein published an extraordinary series of papers in 1905 which redirected modern physics. The most dramatic paper dealt with special relativity or the physics of bodies moving uniformly relative to one another. While highly technical, its conceptual novelties are easy to summarize. In essence, by positing that nothing can move faster than the speed of light, Einstein reformulated Newtonian mechanics, which contains no such restriction. The upshot was special relativity, an interpretation of motion that dispensed with the reference frames of absolute space and absolute time fundamental to Newtonian physics and the Classical World View. The slowing down and speeding up of time—time dilation—is predicted and explained by special relativity. In Einstein's interpretation the cosmos contains no privileged frames of reference, no master clock. All observations (such as when an event takes place, how long a ruler is, or how heavy an object) become relative and depend on the position and speed of the observer. As a feature of his new physics, Einstein posited his celebrated formula, $E = mc^2$, that equated mass (m) and energy (E), elements held scrupulously distinct in Classical physics, with the speed of light (c), a constant in the equation.

Sometimes, particularly in physics classes, Newtonian physics is portrayed as simply a special case of Einsteinian physics—slow-moving bodies supposedly obey Newton's laws while bodies approaching the speed of light follow Einstein's. Such a view, while it facilitates science teaching, distorts the historical record and obscures the revolutionary transformation effected by Einstein's 1905 papers. For Newton and Classical physics, space and time are absolute: an "Archimedean" point exists somewhere out there against which all motions may be measured; somewhere a standard pendulum beats out universal time; mass and energy are not interconvertible; objects can go faster than the speed of light. In every instance Einstein reached categorically opposite conclusions. Thus, simply because the same letter m (standing for mass) appears in the Newtonian formula, $F = ma$, and in the Einsteinian one, $E = mc^2$, one should not confuse the two fundamentally different concepts of mass or the different physics behind them.

Einstein's special relativity of 1905 concerned uniform motions. In 1915 he published on general relativity, or the physics of accelerated motions, wherein he equated gravity and acceleration, as in his famous thought experiment comparing an upward accelerating elevator with gravity. In a highly imaginative thought experiment, Einstein supposed that inside an elevator one could not distinguish gravitational forces induced by an imaginary planet pulling from below from forces produced by the upward acceleration of the elevator; both events would produce the same nervous flutter in a passenger. As a consequence of this equivalence, the nature of space underwent a profound transformation. Uniform, three-dimensional Euclidean space—yet another absolute of Newtonian physics and the Classical World View—became obsolete, replaced by the four-dimensional continuum of Einsteinian space-time. As a consequence of this reinterpretation, it came to be understood that bodies distort the shape of space. Gravity—a "force" in Newtonian mechanics—becomes only an apparent force in Einstein's general relativity, the result of the curvature of space warped by heavy bodies within it. Planets orbit the sun not because a gravitational force attracts them, but because they must follow the shortest path through curved space. Observations of a total solar eclipse in 1919 seemed to confirm Einstein's prediction that the mass of the sun should bend starlight; and similar, very precise calculations of Mercury's orbit around the sun brought like agreement with general relativity. By the 1920s Classical physics, with its absolutes and its ether, had become a thing of the past. Physicists led by Einstein had created a conceptually new world.

Another set of scientific developments—having to do with atomic theory and the physics of the very small—complemented relativity and likewise proved highly consequential for twentieth-century natural philosophy. Einstein was a major contributor to this line of work, too, especially with another paper that appeared in 1905 on the photoelectric effect and that supported the notion that light comes in discrete bundles (quanta) and not in continuous waves.

Once the discovery of the electron and radioactivity undermined the indivisible and immutable nature of the classical atom, atomic theory became a prime focus of experimental and theoretical research. J. J. Thomson, coincident with his discovery of the electron in 1897, proposed a model of the atom having negatively charged electrons sprinkled about like raisins in a cake. Using radioactive emissions as a tool to investigate the interior structure of the atom, in 1911 Ernest Rutherford (1871–1937) announced that atoms were composed mostly of empty space. Along with the Danish physicist Niels Bohr (1885–1962), Rutherford proposed a model for the atom that had electrons orbiting a solid nucleus, much like planets orbit the sun. In the 1920s problems with this solar-system model (e.g., why electrons maintained stable orbits or why, when excited, atoms radiated energy in a discontinuous manner) led to the formulation of the so-called new theory of quantum mechanics and yet another radical transformation of our understanding of nature.

The paradoxical principles of quantum mechanics were difficult to accept, yet empirical studies supported the theory, as did the social network that arose around Niels Bohr and the Institute for Theoretical Physics that he headed in Copenhagen from 1918 until his death in 1962. While of great mathematical and technical sophistication, the basic ideas behind the "Copenhagen interpretation" of quantum mechanics are not difficult to comprehend. Essentially, quantum theory replaces the deterministic mechanical model of the atom with one that sees atoms and, indeed, all material objects not as sharply delineated entities in the world but, rather, as having a dual wave-particle nature, the existence of which can be understood as a "probability equation." That is, quantum mechanical "waves" predict the likelihood of finding an object—an electron or an automobile—at a particular place within specified limits. Everything that is, is a probability wave.

The power of this counterintuitive analysis was greatly strengthened in 1926 when two distinct mathematical means of expressing these ideas—matrix mechanics developed by Werner Heisenberg (1901–76) and wave mechanics developed by Erwin Shrödinger (1887–1961)—were shown to be formally equivalent. In 1927 Heisenberg proposed his famous Uncertainty Principle that extended the conceptual grounds of quantum theory in surprising ways. In short, Heisenberg stated that in principle one cannot simultaneously determine the position of a body and its speed (or momentum) with equal accuracy. In other words, in contradistinction to the determinism of Classical physics, where in theory, given initial conditions, we should be able to predict all future behavior of all particles, quantum mechanics reveals an inherent indeterminacy built into nature and our understanding of it. With quantum mechanics, chance and randomness were shown to be an intrinsic part of nature. We can say nothing with certainty, but can only make probabilistic predictions. The Heisenberg Uncertainty Principle made it plain, furthermore, that the act of observing disturbs the subjects of observation. The key implications were that nondisturbing or "objective" observation is impossible, that the observed and observer form part of one system, and that probability waves "collapse" into observed reality with the act of observation. In other words, when we look within limits we find, but when we are not looking nothing exists but clouds of possibility.

Yet, when the implications of mature quantum theory came to the fore—implications concerning the impossibility of visualizing phenomena, an inherent uncertainty to knowing nature, and probabilistic limits to the behavior of particles—Einstein himself recoiled, objecting in a celebrated phrase that "God does not play dice with the universe." Nevertheless, quantum mechanics flourished without the master's blessing. More recent work in particle physics has only confirmed quantum-mechanical interpretations of natural phenomena.

By the 1930s, therefore, with such bizarre quantum possibilities gaining intellectual respectability, the Classical World View of the nineteenth century was a relic of the past. Thereafter, knowledge of subatomic particles developed rapidly. In 1930 Wolfgang Pauli postulated the existence of a virtually massless uncharged particle which he christened the neutrino. Despite the inordinate difficulties of observing the neutrino—detected only in 1954—it, too, soon entered the pantheon of new elementary particles. In 1932 the discovery of the neutron—a neutral body similar to the proton—complemented the electron and the proton. In the same year the detection of the positron—a positively charged electron—revealed the existence of antimatter, a special kind of matter that annihilates regular matter in bursts of pure energy. Based on the work of the flamboyant American physicist Richard Feynman (1918–88), quantum theory is now known as quantum electrodynamics or quantum field theory and, hand in hand with experimental high energy physics, it has revealed an unexpectedly complex world of elementary particles. With the use of particle accelerators of ever greater energy as research probes, nuclear physicists have created and identified over 200 different types of subatomic particles, most very short-lived.

Today, physicists classify elementary particles into three main groups—hadrons, leptons, and exchange particles—each with its matter and antimatter components. Heavy particles such as neutrons and protons, for example, are called hadrons and are now understood to be composed of various combinations of six even smaller units called quarks. In a remarkable confirmation of the theory behind quarks (called quantum chromodynamics), in 1995 physicists at the Fermi National Accelerator Laboratory in Batavia, Illinois, demonstrated the existence of the elusive "top quark." The electron is an example of the second class of particles known as leptons, which are generally lighter than neutrons and protons but which have mass and are not responsive to the strong nuclear force. In the year 2000 a team of Fermilab researchers uncovered direct evidence of the final kind of lepton to be discovered, the tau neutrino, thus strengthening the Standard Model of particle physics. Particles called exchange particles form the third group of elementary particles, which include photons or quanta of light and electromagnetism, and they lack rest mass. Exchange particles are thought to be vectors for carrying the four known forces of nature: electromagnetism, gravity, and the strong and weak forces—the latter two governing the binding of particles in the nuclei of atoms and radioactive decay. Discovered by scientists using the Large Hadron Collider at the European Center for Nuclear Research (CERN) in Geneva in 2012, the Higgs boson is the long-sought exchange particle that confers mass. The Nobel Prize committee awarded Peter Higgs (1929–) and fellow theorists the Nobel Prize in Physics in 2013 for their effort in laying the groundwork for this important discovery. Using large particle accelerators and other exquisite experimental apparatus, physicists continue to search for the elusive graviton, the particle that putatively conveys the force of gravity.

Einstein's spirit still haunts physics. A large and lingering problem in theoretical physics in the second decade of the twenty-first century remains how to harmonize the physics of the very large—general and special relativity—with the physics of the very small, particle physics and quantum mechanics. In particular, a quantum theory of gravity is presently the Holy Grail of theoretical physics. For a number of years physicists have been developing superstring theory as the conceptual basis for a quantum theory of gravity and as a tool to bridge the gap between the large and the small in physics. String theory originated in the 1970s, but remained a backwater of research until 1984 when it was shown that the particle that putatively carries the gravitational force—the graviton—could be described by string theory without the mathematically disagreeable side effects of the standard treatment. Touted as "The Theory of Everything," for a while string theory was disparaged by mainstream physicists as being utterly fantastic and untestable. In the 1990s, however, when several versions of string theory were seen to derive from a single, more general formulation and when string theory revealed new sorts of connections between types of particles (supersymmetry), the theory, dubbed superstring theory, took on greater seriousness.

Today, in superstring or M-theory, the fundamental particles of the universe are conceived of as unimaginably tiny one-dimensional "strings" that exist on the Planck scale of 10^{-33} centimeters, many orders of magnitude below our ability to experimentally confirm their existence. In the current version of the theory there are eleven dimensions of space-time, including the four with which we are familiar. (In another version there are twenty-six such dimensions.) Strings are thought of as compacting or curling up seven of these dimensions of space-time. According to theorists, strings form loops, and through their vibrations strings (and related "membranes") produce the particles and the forces in the common taxonomy of physics, different vibrations and resonances producing different particles and different forces. Superstring theory today has attracted considerable expert and popular attention, but it remains an abstruse and highly mathematical area of scientific research with many serious problems that threaten its credibility. A future history of science will tell us whether superstring theory proves as successful as its backers claim for uniting physics and our theory of the world.

The quest for understanding the material stuff of the world—a quest that began with the Milesian natural philosophers in the fifth century BCE—has achieved an unprecedented level of sophistication in our own time, and active research continues within the general theoretical frame established by contemporary quantum field theories. As part of this effort, theorists have worked to understand the forces of nature on a deeper, more unified level. They have achieved some success in this area, notably in the 1970s in conceptually unifying electromagnetism and the weak nuclear force into the so-called electro-weak force. An as-yet-unrealized Grand Unified Theory will perhaps add the strong nuclear force, but a Final Theory of nature uniting all the forces of the universe, including a quantum theory of gravity, remains a distant goal. Such theorizing, while of great philosophical import, remains utterly useless for practical purposes.

Cosmology

Cosmology represents another area where theorists forged profound conceptual novelties in the twentieth century that continue to underlie cosmological thinking today. In the eighteenth century, astronomers had broached the idea that the Milky Way was an "island universe," and observations of nebulous bodies in the nineteenth century suggested that many galaxies populated the cosmos outside the Milky Way. Also in the nineteenth century the spectroscope, an optical instrument to analyze the light spectrum of the sun and stars, indicated a single chemistry common to the entire universe. Of even greater intellectual consequence, lines in a spectrograph seemed to shift as a function of the motion of stars. This discovery opened the possibility after 1870 that the universe may be expanding. (The application of dry-plate photography to astronomical practice in the later 1870s greatly facilitated spectrographic and related work.) Only in the 1920s, however, due particularly to the work of the American astronomer Edwin Hubble (1889–1953), did the extragalactic nature of "nebulous" bodies, the immense distances involved, and the apparent expansion of the universe become established among cosmologists.

Beginning in the 1930s, relativity and particle physics greatly impacted cosmology. Einstein's equation of matter and energy, along with evolving understandings of nuclear processes, not only led to practical applications in the atomic and hydrogen bombs, but also to theoretical understandings of thermonuclear fusion as the energy source powering the sun and the stars. Such theorizing ultimately led to the recognition that chemical elements more complex than hydrogen and helium, including those necessary for life, were forged in stellar furnaces and scattered through the universe by exploding stars. In addition, the discovery of thermonuclear processes, by replacing ordinary combustion models, greatly extended the age of the sun and the solar system and provided a decisive response to Lord Kelvin and his critique of Darwinism premised on a comparatively short lifetime for the sun. A new generation of physicists, in other words, saved the Darwinian model by providing all the time needed for evolution to take place.

How to account for the apparent expansion of the universe remained a great debate among cosmologists through the 1950s. Essentially, two mutually exclusive views divided the allegiance of theorists. One, the "steady state" model, as articulated by the English astrono-

Fig. 18.3. Expanded horizons. The Whirlpool Galaxy (M51) and Companion Galaxy (NGC 5195) float through space 31 million light-years from earth. Photograph taken by NASA's Hubble Space Telescope in January 2005. Only in the last century did astronomers grasp the immense temporal and spatial scales of the universe.

mer Fred Hoyle (1915–2001) and others, held that new matter appeared as space expanded, with the density of the universe remaining constant. The alternative theory, first articulated by the Belgian abbé Georges Lemaître in 1931 and developed by the Russian émigré physicist George Gamow and colleagues in the 1940s and early 1950s, proposed that the universe originated in an incredibly hot and dense "Big Bang" and has continued to expand since.

Both camps had their supporters and arguments, although most cosmologists seemed predisposed to steady state models throughout the first half of the twentieth century. The debate was ultimately decided only after 1965 with the almost accidental discovery by two Bell Laboratory scientists, Arno Penzias and Robert Wilson, of cosmic microwave background radiation at a temperature of 3° Kelvin. The idea that explains their discovery is elegant. If the universe began as a hot fireball with a big bang, then it should have "cooled" over time, and calculations could predict what should be the residual temperature of the universe today. And it was this relic heat, measured by an omnipresent background radiation at roughly three degrees above absolute zero (2.73° K) that Penzias and Wilson stumbled upon. Amply confirmed by exquisite space-based instruments and observations, their discovery won them the Nobel Prize in 1979 and sounded the death knell for steady state cosmology.

The watershed discovery of the 3° background radiation showed the theoretical power forthcoming from the unification of particle physics and cosmology. In tandem, the science of the very small and the study of the largest astronomical systems proved extraordinarily fruitful in elaborating the coherent "Standard Model" of the universe. In the "inflationary" variant of that model generally accepted today, the universe began around 13.8 billion years ago in a violent explosion out of what theorists extrapolate to have been a virtually unimaginable "singularity" when the universe was infinitely small. The known laws of nature do not apply under such circumstances, but particle physics provides a reasoned account of the evolution of the cosmos from the merest fraction of a microsecond after the Big Bang: an initial union of energy, matter, and the forces of nature; a period of stupendous inflation as the universe ballooned seventy-five orders of magnitude in size coincident with a "phase shift" akin to a change of state, like boiling; the subsequent decoupling of energy and matter after a still comparatively brief period of 100,000 years as the primordial universe continued to expand and cool; and the ensuing evolution of galaxies and stars over the following billions of years, ultimately leading to the formation of our own nondescript solar system 5 billion or so years ago.

Many uncertainties affect this picture, and research continues in a number of key areas. Precisely how old is the universe? Experts seek a more exact value for the so-called Hubble constant to help settle the question. What will be the fate of the universe? Will it reach a limit and collapse back into itself in a "Big Crunch"? Or, will it keep on expanding to peter out in a "Big Chill"? Answers hinge in part on the amount of mass in the universe, and investigations are currently underway to discover "dark matter," the missing mass required to prevent the universe expanding forever. Complicating these inquiries, research announced in 1998 and confirmed in stunning studies in 2004 of X-rays emitted by distant galaxies suggests that a mysterious "dark energy" is in fact speeding up the expansion of the universe. These newer findings raise the possibility of the universe continuing to accelerate and ultimately exploding apart in a "Big Rip."

Then, while the cosmos is extraordinarily homogeneous on a large scale, how can we explain the obvious nonuniformities evident in galaxies and on even smaller scales? How did the universe begin out of nothing? Work in quantum cosmology concerning "imaginary time," "virtual particles," quantum "tunneling," and the chance coming-into-being of matter out of the quantum vacuum may prove useful in addressing these issues. Research into black holes—those extraordinarily dense celestial objects whose gravity prevents even the escape of light but that "evaporate" energy—may also illuminate these ultimate questions. Today, cosmologists entertain other esoteric theories and entities such as string theory, parallel universes (aka the multiverse), and equally provocative, but perhaps intellectually productive concepts. Such exciting research indicates the continuing vitality of theoretical science today. But to repeat a point made throughout this volume, the value of inquiries such as these derives principally from the intellectual and psychological satisfaction they provide and not from any potential utility. They represent the continuation of the venerable Hellenic tradition, alongside the increasingly practical bent of scientific research in the twenty-first century.

DNA and the Triumph of Darwinian Evolution

Physics and cosmology are of prime significance in providing the basic reference frames for comprehending the world around us. But explaining life has been no less momentous, and in this domain biologists in the twentieth century reformulated our view of living organisms and the history of life on Earth in ways that have fundamentally altered our way of seeing the world and the content and orientation of many disciplines in the life sciences and the physical sciences.

Readers will recall that at the end of Darwin's lifetime in 1882 his theory of evolution by natural selection was contested in the scientific world, and, while not neglected, Darwinian evolution had been relegated to a lesser role in defining life and the history of life on Earth. Several scientific trends combined after 1900 to produce a refurbished evolutionary theory known as the Neo-Darwinian synthesis. The discovery of radioactivity and Einstein's 1905 equation of matter and energy justified Darwin's skepticism about Lord Kelvin's estimate of the age of the earth. During the first half of the twentieth century estimates of the age of the solar system progressively increased first to millions and then to billions of years, and evolutionary theory ended up with all the time it needed.

Several researchers simultaneously rediscovered the Mendelian principle of particulate inheritance at the beginning of the twentieth century. Mendel's experiments on plant hybridization showed that heredity is not a blending process but maintains traits in discrete units. In 1900 Hugo de Vries complicated matters with a theory of macromutations that provided evolution with another mechanism of variation and also, by accelerating the process of evolutionary change, added credibility to explaining the origin of species within a limited time frame. Ironically, Mendel's laws of inheritance along with the mutation theory suggested to some scientists that these principles alone, without the operation of Darwinian natural selection, could account for evolution. Darwinism was finally adopted as the new paradigm of biological diversity during the 1940s with the recognition that genes, those discrete units of heredity, combine in extraordinarily complex patterns to produce the minute variations on which natural selection operates.

Prior to the turn of the twentieth century, improved staining techniques had already led to the discovery of chromosomes—thread-like structures in the nucleus of the cell—and in the 1880s August Weismann speculated that they contained the units of heredity, a speculation confirmed in the 1910s and 1920s. Working in the famed "fly room" at Columbia University, Thomas Hunt Morgan (1866–1945) and his colleagues experimented with the fruit fly, *Drosophila melanogaster*, confirming Mendel and the chromosomal basis of inheritance. In England R. A. Fisher and other specialists in population genetics demonstrated mathematically that a single variation conferring an advantage on an individual in the struggle for existence could in the fullness of time come to characterize an entire population. These trends converged in a series of scientific works appearing in the late 1930s and early 1940s, notably *Genetics and the Origin of Species* (1937) by Theodosius Dobzhansky, *Evolution: The Modern Synthesis* (1942) by Julian Huxley, and *Systematics and the Origin of Species* (1942) by Ernst Mayr. These and related works provided a full and compelling theory of evolution in line with the original doctrine outlined by Darwin in 1859.

In the 1930s and 1940s, scientists such as W. M. Stanley (1904–71) working with tobacco mosaic virus and Max Delbrück working with phage viruses, investigated proteins as the physical mechanism of genes, work for which they both won Nobel Prizes, but the specific material, biochemical basis of inheritance still eluded them. In the 1940s the key nucleic component seemed to be an organic acid, DNA—deoxyribonucleic acid, a white powder composed of carbon, hydrogen, nitrogen, oxygen, and phosphorus. In 1953 a maverick young American postdoc and an older British graduate student, James Watson and Francis Crick, decoded the molecular structure of DNA. Appropriating the X-ray diffraction data of the physical chemist Rosalind Franklin (1920–58), they showed the molecule to be in the form of a double helix, an elegant structure that provided a mechanism for inheritance and the appearance of new variations, a landmark discovery for which the two received a Nobel Prize in 1962. It represents the crowning achievement of a century of research—the code of life had become reduced to chemistry.

Practitioners in the biological sciences and a variety of related fields now accept Darwinian evolution, and its principles now guide research and scientific understanding. Still, in other sciences some of the implications of evolutionary theory have been and remain controversial.

From the very outset, Darwinism, through its substitution of natural for divine processes, courted a duel with the religious mentality and biblical authority, mainly in the United States. The Tennessee "monkey trial" of 1925, where public-school teacher John Scopes was convicted of violating a state law against teaching Darwinism, remains a notorious example. Controversies continue to flare up today over the teaching of biological sciences versus "scientific creationism" and "intelligent design" in public schools; and some publishers have simply removed evolutionary references from biology textbooks in an effort

Fig. 18.4. The double helix. The discovery of the double helical structure of DNA by James Watson and Francis Crick in 1953 demonstrated the material basis of inheritance and the substrate for how new variations can lead to species change. Their discovery marked the beginning of a new era in genetics and biochemistry and opened the way for the new fields of molecular biology and genomics.

to avoid the controversy. In the realm of theology, fundamentalism excepted, major religious authorities soon accommodated themselves to the general principle of evolution by adopting, *pace* Galileo, a non-literal interpretation of scripture and allowing divine intervention into the evolutionary process to infuse an immaterial soul into humankind at some moment during its physical evolution.

In the social sciences and philosophy, fierce conflicts have arisen over the application of Darwinian concepts to explanations of human nature. The first such conflict was provoked by what became known as Social Darwinism—the claim that the organization of society, and specifically the stratification into classes—can be accounted for biologically by the struggle for survival. In the United States, Social Darwinism tended to color social and psychological analyses by suggesting that race and ethnicity as well as class determined success in life. The general implication was that the status quo with all of its inequities is determined by hereditary biological characteristics and cannot readily be improved by social reform.

Various strains of Social Darwinism provided ideological support for an array of political, economic, and social positions. For example, in his book *Gospel of Wealth* (1900), the American industrialist Andrew Carnegie justified cutthroat laissez-faire capitalism, the sacredness of property and profit, and scientific opposition to socialism and communism on Social Darwinist grounds. Other apologists have put forward similar, and equally specious, arguments in favor of Nazi racial claims, class struggle, the eugenics movement, the sterilization of "inferior" persons, and the superiority of men over women. However, since the 1960s the mainstream of American social scientists have repudiated such applications of Darwinian thought, and they are no longer at play in social science research.

Fundamental discoveries in paleontology and the history of human evolution figure among developments in the twentieth-century life sciences with strong natural philosophical implications. Although Darwin had postulated the evolution of humans through various stages from simian ancestors, a coherent story of human evolution remains decidedly the product of twentieth- and now twenty-first-century scientific thought. As discussed in our opening chapter, fossils have now been found that represent at least three species of *australopithecines*—an extinct classification of bipeds linking humans with an ancestral ape living 5 million years ago—and at least two species of ancestral human beings who successively followed the *australopithecines* before the evolution of our species, *Homo sapiens*. The initial discoveries of the Neanderthal variety of *Homo sapiens* and of Paleolithic cave art (produced by anatomically modern humans) were not generally accepted until the turn of the twentieth century. The first *Homo erectus* fossil was unearthed in 1895, and the first *Australopithecus* surfaced only in 1925. The "discovery" of the notorious "Piltdown man" in 1908, with his big brain and robust, ape-like jaw, argued strongly against human evolution from more primitive, small-brained ancestors, and only in 1950 was the Piltdown artifact conclusively shown to be a fraud. Since that time, an increasingly clearer picture has emerged of the stages of human evolution, from *A. afarensis* ("Lucy" and her kin first uncovered by Donald Johanson in 1974), through *H. habilis*, *H. erectus*, and then varieties of *H. sapiens*. Exciting research is ongoing in this area, too. A treasure trove of *H. erectus* uncovered in 2013 in Dmanisi, Georgia, clarified the status of that species at the early date of 1.8 million years ago. Back in 2004 paleoanthropologists announced the discovery of an entirely new species of human, *Homo floresiensis* that lived on the island of Flores off of Indonesia and that coexisted with modern humans as recently as 12,000 years ago. Known as the "Hobbit," this diminutive species of human—one meter tall and weighing 25 kilograms (3'3" and 55 pounds)—had brains one-quarter the size of modern humans, yet possessed sophisticated toolkits, hunting patterns, and probably language skills. Of great intrinsic interest, the discovery of *H. floresiensis* has upset the standard history of human evolution. The discovery is contested, however, as other scientists think the specimens diseased or deformed, and the debate shows once again science as a dynamic interpretative activity.

While these discoveries can be seen as confirmations of Darwin's theory of evolution, consensus has emerged only with regard to the evolution of human physical traits. Human exceptionalism proved the rule, and resistance remained strong against extending to humans and human social behavior scientific findings taken from the animal world. Mental and behavioral traits were the province of the social sciences, where a medley of doctrines defended the sanctity of consciousness against the encroachment of the gene and natural selection. In reaction to Social Darwinism the sociologist Emile Durkheim enunciated the principle that social conditions must be explained by strictly social events, thereby erecting a barrier between the social and biological sciences. Because the scientific basis of Social Darwinism remained insecure, Durkheim's dictum held sway during the first half of the twentieth century. As a result biological approaches to the social and psychological sciences were treated with skepticism and as tainted by fascism and ideological extremism.

The situation began to change during the 1930s along with the clarification of Darwinism itself. Austrian zoologist and Nobel Prize winner Konrad Lorenz (1903–89) inaugurated the discipline of ethology, or the scientific study of animal behavior under natural conditions, and he and his followers drew attention to the genetic basis of mental and behavioral traits. Inevitably, some natural scientists focused their attention on the evolution of human behavior, a trend that was reinforced in 1975 by Edward O. Wilson's book, *Sociobiology*, a landmark work that included a suggestion that human social and mental life can be studied in terms of Darwinian evolution. It provoked ideological objections on the grounds that human sociobiology constitutes "biological determinism" lending itself to justifications of some of the least appealing features of the status quo. However objectionable sociobiology may seem to some, its advocates are innocent of the scathing political charges that were once leveled against them. Since the 1970s there has been an explosion of research in sociobiology and evolutionary psychology, fields which postulate that patterns of culture may be explained on Darwinian evolutionary principles. Although directed mainly at the social conduct of nonhuman animals, substantial work has concerned the sociobiological roots of human behavior and culture. Thus, for example, evolutionary accounts have been proposed for altruism, aggression, cooperation, incest taboos, alcoholism, gender differences, homosexuality, and attitudes toward children and strangers. In the 1960s and 1970s such ideas were attacked as offensive to liberal democratic ideals, for they seemed to limit the range of human freedom and social reform. The matter remains contested, notwithstanding the works of Richard Dawkins (1941–), author of *The Selfish Gene* (1976) and Professor for Public Understanding of Science at Oxford University from 1995 until 2008, and Wilson's groundbreaking *Sociobiology*. Increasingly this research is gaining acceptance from scientists and the informed public alike.

In the course of its final triumph Darwinism had to vanquish still another doctrine—Lamarckism, reduced to its essential claim that characteristics acquired in the course of a lifetime can be inherited. The doctrine's appeal is based on both its postulate of a direct influence of the environment on the process of evolution and the rapidity of the process in producing new forms. Darwin himself, as we saw, accepted the principle while casting about for an explanation of variation, and

Lamarckism continued to find favor in a few circles well into the twentieth century. The eminent psychologist Jean Piaget, for example, adopted it at one point, but the episode surrounding the Soviet agronomist Trofim Lysenko (1898–1976) had devastating consequences. Lysenko rejected Mendelian science in favor of Lamarck; he received political support and government sanction from Joseph Stalin from the mid-1930s through the late 1940s in the misguided hope that plant breeders could rapidly improve agriculture. The upshot was a criminal politicization of science, less food production, starvation, and the destruction of fledgling genetic research in the Soviet Union. Since then Lamarckism has found no support and is generally repudiated. What is sometimes referred to as the "Central Dogma of Genetics" asserts that the master molecule of DNA carries the codes for all hereditary traits, that DNA makes RNA, which makes proteins and the building blocks of life, and so no mechanism exists whereby the modification of a trait through life experience can effect the DNA that gave rise to it. The existence of retroviruses, viruses such as HIV that infect the DNA of host cells for their own replication, complicate this picture as does genomics or studies of the regulation of biochemical processes surrounding DNA and its expression, but in the end, the chemistry of the double helix effectively sounded the death knell of Lamarck's doctrine.

The uncovering of the double helical structure of DNA in 1953 by Watson and Crick provided a conceptional breakthrough for understanding reproduction, the nature of heredity, and the molecular basis of evolution. Also in 1953 a noted set of experiments that produced amino acids—the chemical building blocks of life—by electrifying a broth of more simple compounds supported the idea that life itself originated in the primeval conditions of the early Earth. Alternative views include the idea that life arose from the catalytic action of clays, while some scientists at one point proposed an extraterrestrial origin for life on Earth known as panspermia, whereby spores originating elsewhere in the universe were either sent or otherwise landed and flourished on a hospitable Earth. The rich organic chemistry of comets lends some probability to this view, but this theory begs the quesiton of how life arose elsewhere. Once again, in examining contemporary science we encounter not final answers but a vigorous process of inquiry.

Along with molecular biology have come refined accounts of the history of life on Earth. As research into the details of plant and animal evolution have mounted over the last decades, the fundamental perspective has remained the Darwinian one of evolution through natural selection. Understanding the molecular basis of inheritance has provided important new ways to analyze evolutionary history and to classify living plants and animals. New approaches, including "cladistics," "molecular clocks," and the study of mitochondrial DNA, permit scientists to measure the rate of evolutionary change and evolutionary distances between species. Does everyone agree? Hardly, and biological scientists today pursue lively debates over a range of topics, including, for example, whether or not birds are the direct descendants of dinosaurs and whether evolution proceeds at a constant, slow pace as Darwin claimed or whether periods of evolutionary stability are "punctuated" by periods of rapid change. Fundamentalist Christian groups and fundamentalist sects in other religions have seized on the latter debates as evidence of the failure of Darwinian evolutionary theory, but to do so misses the point that such debates constitute the norm in science. Darwinism today provides the paradigm for many fields of research. While interpretative differences divide scientists over some details, virtually universal consensus has crystallized around the fundamental tenets that species are the fruits of time and evolution and that natural selection represents its main method. In the past century and a half the world of the biological sciences has largely recapitulated the trajectory of Darwin's own biography—from a divinity student at Christ's College to the recognition that the diversity of life is an artifact of nature.

Sciences of the Mind

Scientists have not limited their inquiries to developing understandings of the physical world, however, or of life in some abstract sense. For centuries the human mind, consciousness, and the mysterious inner-self have been the subjects of inquiry by philosophers and psychologists driven to hold a mirror to reflect the spark within themselves. A well-recognized philosophical tradition devoted to these subjects stretches from Plato and Aristotle, through Descartes, and into the Enlightenment of the eighteenth century. But only in the nineteenth century did a distinct science of psychology emerge. Thereafter, psychology and the scientific study of mind and behavior grew in tandem with other natural scientific disciplines, and the contributions of psychology to scientific and natural philosophical understanding in our day have been enormous. But, unlike the physical sciences where researchers generally agree on the fundamentals of their field, the history of psychology in the twentieth and now twenty-first centuries is the story of diverse and conflicting schools of thought. Varieties of physics and chemistry do not compete for followers. Psychology, however, is more akin to Pre-Socratic natural philosophy, where different systems vie for adherents.

Rooted in the laboratory tradition of the physical sciences in nineteenth-century Germany, the first school of modern psychology was German experimental psychology. The great German physicist, Hermann von Helmholtz may be said to have started this tradition through his experimental work on the perception of colors and sounds. A student of Helmholtz's, the oft-hailed "father of modern psychology," Wilhelm Wundt (1832–1920), established the first laboratory dedicated to experimental psychology at the University of Leipzig in 1879, where, using introspective techniques he investigated sensation and perception. Edward Bradford Titchener (1867–1927) studied with Wundt in Leipzig, translated Wundt's *Principles of Physiological Psychology* into English, and carried Wundt's views to America and Cornell University in 1892. There, as professor of psychology, Titchener developed structuralist theory that looked to define the unit elements of consciousness, something like a periodic table of mental events. A few years earlier the philosopher and professor of psychology at Harvard, William James (1842–1910), supervised an early experimental psychology lab at Harvard that employed some of Wundt's students. James also wrote an influential text, *Principles of Psychology* (1890), that argued for functionalism in psychology, or the examination of the function of consciousness and looking to behaviors and the effect of behavior on feeling and consciousness. After studying with James and Wundt, G. Stanley Hall (1844–1924), organized an experimental psychology laboratory at Johns Hopkins University in 1883, and Hall was also responsible for the initial professionalization of the discipline in the United States by launching the *American Journal of Psychology* in 1887 and founding the American Psychological Association in 1892. Although interpretive differences separated these early investigators, they were all committed to the scientific, read experimental, study of the mind and mental events.

In subsequent decades experimental psychology came to place evermore emphasis on the objective study of behavior and to disregard subjective elements of human mental experience. The work of the great Russian experimenter and 1904 Nobel Prize winner Ivan Petrovitch Pavlov (1849–1936) contributed significantly to these developments. Pavlov is famous for his physiological research and the discovery of conditioned reflexes, whereby Pavlov's dogs learned to associate the sound of a metronome with the presentation of food, and would then salivate at the sound of the tone alone. In the United States J. B. Watson (1878–1958), for a while professor of experimental psychology at Johns Hopkins University, extended Pavlov's results. He developed the stimulus-response model and redefined the William James school of functionalism. The result was a brand of psychology known as behaviorism. In such works as *Behavior: An Introduction to Comparative Psychology* (1914) and *Psychology from the Standpoint of a Behaviorist* (1919), Watson pushed behaviorism to the extremes of abandoning introspection altogether, banishing subjectivity as unreliable and intrinsically unscientific, eliminating consciousness as an object of study, rejecting hereditary factors as explanatory in psychology, and distinguishing the human and the animal only in degree. He dismissed Freud's contemporaneous work as mysticism and nonsense. Watson's mantle was taken up by the famous Harvard psychologist B. F. Skinner

(1904–90), who likewise spurned the study of mental events in favor of focusing on behavior and factors that condition behavior. Skinner interested himself in behavior modification, and he discovered operant conditioning where positive and negative rewards variously shape behavior. Skinner popularized behaviorist psychology with his utopian novel, *Walden Two* (1948) and other works, and he achieved a certain notoriety by having his baby daughter play and sleep in a "baby box" of his invention.

Psychology has never been a wholly abstract science, and its utility in various applied guises helps explain its multifaceted character and lack of conceptual unity. Among the many domains of applied psychology, the development of intelligence testing stands out as an early area of application. Darwin's cousin Francis Galton (1822–1911) originated this line of applied research with his book *Hereditary Genius* (1869). Galton is important to the story, not least for the introduction of statistics into psychology, and the extent to which research in psychology drove the advance of applied statistics should be noted. Another student of Wundt's and for many years professor of psychology at Columbia University, James McKean Cattell (1860–1944) was a pioneer in the realm of quantitative intelligence testing, introducing the term "mental test" in 1890. Cattell's particular tests proved unreliable, but he was nonetheless instrumental in promoting such approaches and psychology in general, as publisher of several specialized psychology journals and reference works such as *American Men of Science* (later *American Men and Women of Science*). In 1905 Alfred Binet (1857–1911), director of the laboratory of experimental psychology at the Sorbonne in Paris, introduced the first IQ tests that putatively measured a child's intelligence relative to his or her age group. In 1916 the American psychologist and head of the psychology department at Stanford University, L. M. Terman (1877–1956), published his influential revision that became the Stanford-Binet IQ test, the current version of which is still in use a hundred years later. IQ tests have become standard throughout the world ever since, dicing populations along a single scale of intelligence. Tellingly, the *Journal of Applied Psychology* appeared in 1917 in the middle of World War I when the U.S. Army was administering aptitude tests to two million recruits to judge their suitability for soldiering and as potential officers. Similar sorts of tests were given in World War II and were used especially for identifying candidates to become pilots.

Educational testing represents a related area of applied psychology, as almost every parent and college student knows. The College Board was founded in 1900, in part to assure the suitability of applicants for higher education. It spawned the Educational Testing Service (ETS) in 1947 to handle the increasing demand for the College Board's Scholastic Aptitude Test (SATs, now called the Scholastic Assessment Test). Every year millions of students take tests offered by the Educational

Testing Service, including the SATs, the Graduate Record Examinations (GREs), the Test of English as a Foreign Language (TOEFL), and related examinations. In 2015 the ETS had 3,200 employees, many hundreds with advanced degrees in psychology, statistics, and psychometrics, the field devoted to the measurement of mental traits, abilities, and processes. ETS is a world center for research in testing, and its computerized testing today is of extreme scientific and technical sophistication. Educational testing is also big business. In 2012 ETS generated slightly over $1 billion in revenue, a figure it hopes to "grow aggressively" as the organization expands on a worldwide basis. The validity of these sorts of tests is stoutly debated. Correlation between good scores on the test and success in college is not clear, and a number of colleges and universities no longer require the SAT for admission.

Early on, the science of psychology found a home in advertising. Walter Dill Scott, professor of psychology at Northwestern University, published his pathbreaking *Psychology of Advertising* in 1908, and it is again telling that, after J. B. Watson was forced out of Johns Hopkins in 1920 for marital infidelity, he then spent more than three decades working in the advertising industry. Industrial and organizational psychology is another, related subdiscipline that studies and applies research on human behavior to business management, leadership, team building, training, productivity, and other workplace and human resource issues.

Whether psychology thus sold its soul to business is a question readers will judge for themselves. That various brands of psychology have been consistently, even maliciously misapplied over the years is not in doubt. Early psychological studies of "feeble mindedness," for example, lent scientific credibility to the eugenics movement and to racist attitudes in general. Lobotomies and the overuse of insulin shock and electrical shock therapies to treat mental illness fit into this category. The polygraph test, or lie detector, has an ambiguous status, but the application of psychology to political ends seems especially shameful, as in the development of "truth serums" or the creation in the United States in 1950 of a National Psychological Strategy Board to combat communism through propaganda. Robert Oppenheimer said that with the Manhattan project physics came to know sin. The sins of physics may be big, but those of psychology may be more insidious.

Then, what can be said of Sigmund Freud (1856–1939) and Freudian psychoanalysis? Freud's work embodies a categorically different school of psychological thought, one that broke new ground in understanding the human character and condition and that proved highly influential throughout most of the twentieth century. As a physician, Freud came to his views through an initial study of hypnosis and the medical condition then known as hysteria. But he soon left traditional medicine to chart his own course through the stormy seas of mental life. In 1900 he published his pathbreaking *Interpretation of Dreams*, followed by the *Psychopathology of Everyday Life* in 1901. These works and Freud's many subsequent publications opened new vistas on how we look upon ourselves and others. Freud brought the unconscious and unconscious motivations to the fore, thus undermining the view of humans as wholly rational creatures. Freud articulated several theoretical models for his psychology, but the most famous involved what he labeled the Id (unconscious passions and desires), the Ego (the psychic interface with reality), and the Superego (the internalized moral and behavioral censor).

The upshot of Freud's views was to envision human behavior and our subjective psychological experience as the outcome of a dynamic struggle among conflicting internal psychological pressures. Similarly, his emphasis on the libido and human sexuality likewise transformed understandings of psychological experience and what it is to be human. Especially influential were his emphases on childhood sexuality in particular and the trials and tribulations of childhood he saw as formative of the adult character. Central to Freud's views were stages of childhood psychosexual development—oral, anal, and phallic—and his Oedipus complex wherein boys seek to kill their fathers and possess their mothers. (For Freud, girls are beset with "penis envy," wanting to be like boys.) Freud introduced or repackaged a host of psychological concepts—repression, sublimation, projection, rationalization, denial, regression, identification, fixation, and so on that formed the bedrock of his psychology and that have become staples of popular culture.

From the beginning Freud treated patients through a new form of therapy, psychoanalysis. Believing that mental illness, particularly neurosis, could be traced to early childhood traumas, chiefly sexual ones, Freud developed new techniques for his brand of therapeutic practice that included putting patients in repose on a couch in his office, interpreting dreams, and using free association and slips of the tongue to probe the psyches of his patients. Cures were supposedly effected through the "transference" onto the psychoanalyst of unresolved conflict and affect from childhood. Freud himself was authoritarian in formulating the true faith of psychoanalytic theory and expelling apostate interpreters from his school. Freudian psychoanalysis became almost a caricature of itself, especially in America in the 1950s and 1960s, as certain social circles faddishly flocked to "headshrinkers." Psychoanalysis has also rightly been criticized for being sexist, circular in its logic, and unscientific in the sense that many of its fundamental tenets are untestable and abstracted from any bases in biology or anatomy. All that notwithstanding, Freudian theory was perhaps the most influential brand of psychology in the twentieth century. If Freud's legacy endures more in its general, almost poetic insights into the human condition than it does within the practice of science or medicine today, his influence is no less significant for all that.

The Swiss psychologist and early Freudian Carl Jung (1875–1961) broke with Freud in 1912 to establish his own school of analytic psychology. The distinctive trait of Jungian psychology is its emphasis on a collective unconscious shared by all humans and on archetypes in the collective unconscious that symbolize the deepest common elements of human experience, such as sex and death. While not directly related, Jung's psychology shares certain similarities with Gestalt psychology developed by the Czech psychologist Max Wertheimer (1880–1943) and his students Wolfgang Köhler (1887–1967) and Kurt Koffka (1886–1941). Put forward in opposition to structuralism and to behaviorism, the major claims of Gestalt psychology are that we perceive through patterns of organized wholes (*Gestalten*) and that breakthroughs of insight provide the keys to learning. The work of the Swiss child developmental psychologist Jean Piaget (1896–1980) can be included in this context. Piaget labeled his work genetic epistemology, and he understood children to pass through distinct conceptual and moral stages as they matured. The personality test emerged out of this constellation of Jungian and Gestalt psychology. Based on Jung, the Myers-Briggs Type Indicator was introduced in the 1940s to situate individuals along the axes of extroversion-introversion, sensing-intuiting, thinking-feeling, judging-perceiving. The most popular such test today, introduced in 1987, involves the "Big Five" measures of neuroticism, extraversion, openness to experience, agreeableness, and conscientiousness. Howard Gardner's popular treatment, *Theories of Multiple Intelligence* (1983) did wonders to expand popular conceptions of the diversity of human psychological experience and potential.

Clinical psychology is yet another notable subfield of modern psychology. It blends scientific training and research with clinical practice and treating patients suffering from addictions, depression, eating disorders, mental disabilities, and the like. Clinical psychology often overlaps with physiology and clinical medicine, as does neuropsychology which studies the relationship between the brain and mental life with its own scientific orientation and research methods, notably animal experimentation.

With roots in several of these antecedent areas, cognitive psychology is a major, but relatively new branch of psychology that arose in the 1960s and 1970s, partly as a response to behaviorism. With its emphasis on cognition and the active role of the intellect in shaping psychological experience, cognitive psychology has become increasingly popular among professionals and amateurs alike, and the subfield today exists at the intersection of linguistics, artificial intelligence, computer science, information processing, mathematical modeling, neuropsychology, neurophysiology, brain imaging techniques, and philosophy. The therapeutic dimension of cognitive psychology holds that irrational thinking is the source of most mental illness, and it seeks to substitute more realistic and less distorted ideas in patients as a means of changing feelings and behaviors. Drug therapy, based on sophisticated research on brain chemistry, often accompanies such treatment. Much less expensive than psychoanalysis or institutionalizing patients, cognitive therapy and drug treatments are now the favorite options of insurance companies and the mental health establishment.

To this already long list we might add developmental psychology, which investigates the cognitive and emotional trajectories of the individual from the womb to the tomb, and social psychology, which is concerned with such cultural and social issues as racism, gender, violence, and group behavior in general. In 2015 the American Psychological Association incorporated fifty-four different divisions into its organization, including sports psychology, gerontological psychology, and peace psychology. Research in each of these fields has brought insights to science and often solace to humanity, but, to repeat the main point, achievements in psychology in the last century have sprung from and embody a diversity of theoretical perspectives, not a single, agreed-upon conceptual frame guiding research that is the case with more mature scientific disciplines.

The science of psychology is immensely popular. Approximately 6 percent of all bachelor's degrees and fully 19 percent of science and engineering undergraduate degrees awarded in 2012 went to undergraduate majors in psychology. From 1968 to 2001 the number of undergraduate psychology majors increased over 300 percent and added another 42 percent from 2002 through 2012; the number of masters and Ph.D.s in psychology rose similarly over these periods. Psychologists find many employment opportunities in industry, as well as in counseling and academe. Despite the institutional and occupational success of psychology, for a variety of intellectual, social, and financial reasons a holistic vision of the human mind and human psychological experience is still lacking. There is no consensus.

And Then . . .

A defining characteristic of science in the modern era is its explosive growth, and as the volume of scientific activity mushroomed in the twentieth century, so, too, has theoretical innovation expanded across a number of scientific disciplines and sub-disciplines. In geology, for example, the idea that the continents may actually be "plates" riding on the earth's mantle is of considerable theoretical import. This idea, generally rejected after it was first put forward in 1915 by the German geologist Alfred Wegener (1880–1930), gained virtually universal acceptance in the 1960s for a variety of technical and social reasons. Understanding plate tectonics and continental drift has allowed scientists to "run the film backwards" and thus to recapitulate the geological history of the earth, with all the implications that holds for both geology and biology.

Another noteworthy example of theoretical novelty in recent decades concerns evidence uncovered in 1980 suggesting that a catastrophic meteorite or cometary impact caused a mass extinction, notably of the dinosaurs, at the end of the Cretaceous period 65 million years ago. The significance of this finding lies in what it says about the role of accident in the history of life, including the ultimate fate of humans. Similarly, planetary space science along with pictures sent back from space—including the glorious shot of the "pale blue dot" of the earth taken from the Voyager spacecraft in 1991 when it was 4 billion miles in space—have transformed how we visualize the earth and its planetary near neighbors.

Because of the exponential growth of scientific enterprise, theoretical research today is conducted in tiny, hyper-specialized areas of inquiry. Only specialists in these areas are aware of theoretical developments in their fields. Put another way, the scope of theoretical inquiry is so vast and yet so fragmented today that for a very long time now, no one person has been able to grasp the totality of science as natural philosophy in all its particulars. Nevertheless, fostered by science writers and science journalism, in a loose way elements from the highly specialized frontiers of research combine to forge today's scientific worldview.

Human inquiry into nature today has produced awe-inspiring accounts of the cosmos, majestic analyses of the world's constituent matter, breathtaking explanations of the history of life on this planet, and complex and multifaceted portraits of human society and what it is to be human. While the same might be said of the intellectual creations of any scientific civilization at any period, the point applies with unprecedented force today because of the scope, level of detail, and cogency of contemporary natural philosophy. Is what we think today in some final sense true? Clearly not. Still, science represents the best means we have for saying things about the world, and today's story is demonstrably better than any other account given to date. Of course, that realization does not mean that the story will not change again in the future, and undoubtedly it will.

The Bomb, the Internet, and the Genome

How, then, did science and technology achieve the merger and integration that we observe today? That was the question with which we began this book, and in the course of this historical survey we have collected more than a few clues to its answer. In the beginning there was only technology. Then, 6,000 years ago in the first civilizations, science originated in the form of written traditions of mathematical and astronomical knowledge. But this development occurred in the social context of state-level societies where the central government saw to it that the mathematical and astronomical sciences as well as expertise in general were tapped to serve the needs of a complex agricultural economy. This pattern of state support for useful knowledge repeated itself wherever and to whatever extent strong central states appeared. In classical Greece, where the state was weak, science in the form of natural philosophy and technology as craft remained estranged and unpatronized by the state. Later, centralized societies influenced by Greek philosophy merged pure and practical science traditions, but, again, without the great body of contemporary technology being affected by either. Along the way applications did emerge from the world of science that proved widely useful, as occurred in cartography and the development of gunpowder. But only in the nineteenth and twentieth centuries did a new connection emerge between science and industry and a recognition of the possibilities of what might be gained from applying research and theory from the world of science to practical problems and technology development. The result is a dramatic expansion in the applications of science to technology, this time under the rubric of R&D—research and development.

This chapter is about the advent of a particular, atypical kind of "applied science," one that involves the transformation of theoretical work from the forefront of scientific research directly into practical, useful technologies by government and industry. This marriage of the cutting edge of abstract research with technological application along with abundant support for creating new science-based technologies both from government and industry flowered in the twentieth century. It represents an important novelty and is the capstone of any study of science and technology in world history. Here, finally, along with everything else, science and technology achieved the integration we see around us today, a melding that has given rise to the cliché of technology as applied science.

In the nineteenth century and with mounting force in the twentieth, as we saw, industrialization was synonymous with the growth of powerful new technological systems, some of which, such as the production of electricity or radio broadcasting, derived from the world of contemporary science. Henry Ford's Model T was not directly the product of any applied science, but already the car embodied industrial and engineering components that were science-based in some loose sense. As noted previously, in the nineteenth century and thereafter many growing industries established research and development departments explicitly to tap scientific expertise in order to develop or improve products for the military or the marketplace. The use of chemical weapons in World War I is a sad reminder of just how much science and science-based technologies had proven themselves useful to governments and industries by the first decades of the twentieth century. For the most part, however, these examples represent "applied science" of a more empirical and trial-and-error sort, less sweepingly theoretical and with applications filtered down through engineering education and practice. The cases explored in this chapter are distinctively different.

As notable as were prior developments and industrialization generally, a new and fruitful paradigm of applied science emerged in the middle third of the twentieth century. Indeed, cutting-edge and advanced technologies and technological systems that exploit equally cutting-edge research science for practical and applied purposes are a distinguishing characteristic of our age. The atomic bomb, the computer, and antibiotics are powerful exemplars of the capacities of cutting-edge science transmuted into technology. This merger of science from the research front and technology institutionalized the pursuit of a new kind of applied science, and it garnered significant resources in support of exploring the research frontiers of science for practical benefits. Furthermore, the new and close connection between theoretical science and practical, industrial application produced new, high-tech industries, such as nuclear power or the computer industry, that have joined the ranks of the other economically significant industries and technological systems that characterize life in the twenty-first century. Indeed, these science-based creations of human ingenuity and engineering may be seen, for better or worse, as the highest achievements of industrial civilization.

Science-based technologies have proven to be powerful agents of social change and have revolutionized the way a significant portion of humanity now lives. Some of the major technologies where the forefront of theory provided the bases for innovation, application, and industrial development include atomic weapons, computers, medicine, biotechnology, and wireless communications. How cutting-edge science is incorporated into these industries and applications deserves our attention and poses questions about the concept of "applied science" itself, questions that are explored further in this chapter's concluding section.

The idea that inquiry into nature can and should be useful or applied goes back at least to the earliest civilizations where subventioned priests and astrologers found auguries in the heavens. With Francis Bacon in the seventeenth century we observed the emergence of the explicit ideology of the utility of science and the notion that science and scientists ought to exploit nature to benefit humanity. What is impressive about industrial civilization in the twentieth and now twenty-first centuries is the extraordinary success of governments and industry in tapping science and the research frontier for practical ends. These successes have not been without significant repercussions.

Science and the Bomb

The development and use of the atomic bomb by the United States during World War II marks a watershed in the history of modern science and technology. The reasons are twofold. One, the case dramatically revealed the practical potential of cutting-edge science or what could emerge from turning theory directly to useful ends. Secondly, it demonstrated what might be forthcoming when government supported large-scale scientific research and development (R&D) with abundant resources. The novelty of the atomic bomb case—and what it portended for the future of science and technology in the second half of the twentieth century—stemmed from the combination of these two factors: large-scale government initiatives to exploit high-level scientific theory for practical ends. It set a new pattern that changed not only traditional relations between science and government, but how we think about applied science in general.

The story of the atomic bomb is well known and can be quickly summarized. The scientific theory that made the bomb possible emerged only in 1938 and 1939. In 1938 the German physicist Otto Hahn (1879–1968) demonstrated that certain heavy elements (such as uranium) could fission or be split into simpler components, and then in 1939 Lise Meitner (1878–1968), an Austrian chemist who emigrated to Sweden from Nazi Germany, proposed a theoretical explanation for fission and calculated the immense amounts of energy that in principle could be released from an explosive nuclear chain reaction. With war under way in Europe, Allied physicists recognized the destructive potential of a nuclear bomb and the threat that Germany might develop one. In a historic letter dated August 2, 1939, Albert Einstein wrote to U.S. President Franklin Roosevelt about the matter, and, as a result, Roosevelt authorized a small exploratory project. In the fall of 1941, on the eve of the U.S. entry into World War II, Roosevelt gave the go-ahead for a major initiative to create atomic weapons. The result was the largest science-based R&D venture in history. The resulting Manhattan Project, under the overall command of U.S. general Leslie Groves, involved 125,000 people working in thirty-seven installations across the country, and it ended up costing $1.9 billion in 1940s dollars or more like $30 billion in today's dollars. In December 1942 beneath a football stadium at the University of Chicago, the Italian émigré scientist Enrico Fermi (1901–54) succeeded in creating the first controlled nuclear chain reaction. And, on July 16, 1945, the team directed by American physicist J. Robert Oppenheimer set off the world's first atomic explosion at the Trinity site near the Los Alamos lab in New Mexico. On August 6 the Enola Gay dropped a uranium-235 bomb on Hiroshima, Japan, instantly killing 70,000 people, and on August 9 a plutonium-239 bomb fell on Nagasaki, incinerating somewhat fewer. Japan surrendered five days later.

The atomic bomb brought World War II to a dramatic end. At the same time, the bomb launched the Cold War that followed. The political, scientific, and military-industrial establishments of the United States and the Soviet Union raced to develop larger atomic weapons and then, from 1952, even more powerful hydrogen or thermonuclear bombs that derived their energy not only from the fission of heavy elements but also from the fusion of hydrogen into helium.

World War II also saw a number of other government-funded, applied-science projects, such as radar, penicillin production, jet engines, and the earliest electronic computers. World War II established a new paradigm for science and government relations that endured to the end of the Cold War and continues in one guise or another today, that of comparatively large-scale government investment in pure and applied science in the hope of large-scale payoffs in industry, agriculture, medicine, and military technologies. The success of the Manhattan Project—in so closely linking novelties in theory to an immediately useful application—likewise forged a new image of connections between science and technology. These endeavors, which in so many ways—historically, institutionally, sociologically—had for so long been largely separate enterprises, became fused to a significant measure in reality and much more so in the public mind. Since World War II it has become common to think of technology as applied science.

In 1961 U.S. President Dwight Eisenhower warned of the unwarranted influence of a growing "military-industrial complex." Today, especially, and not only in the United States, support for scientific research and science-based technologies for use in military and national security applications has grown enormously. Although nuclear weap-

Fig. 19.1. The Atomic Age begins. J. Robert Oppenheimer and General Leslie R. Groves inspect the remains of the Trinity test site following the first explosion of a nuclear bomb (New Mexico, July 16, 1945). Inset: The "Little Boy" atomic bomb exploding over Hiroshima, Japan, on August 6, 1945.

ons development may have been scaled back since the fall of the Soviet Union in 1991, scientifico-military research, much of it secret, continues to be conducted at record-breaking levels by the government itself and through contracts with industry and universities. This research and development proceeds apace in a variety of esoteric areas of science pertaining to the most lethal and sophisticated weaponry the world has ever known—globe-encircling telecommunications, remote battlefield operations, "smart" weaponry, and elaborate equipment for soldiers in the field, among a myriad of other instances of applied science in the technological systems of armed forces around the world. Plainly, the military and military-related research spurred the modern marriage of theory and application in science and technology. Research in security-related and anti-terrorist matters today is driven by this same set of considerations.

The nuclear power industry was one offshoot of Cold War research that merged physics and electrification. Although the rate of growth of nuclear power plants has fallen in recent decades, some countries (notably France) remain dependent on nuclear power for electric power production. The exploration of space in the second half of the twentieth century and today was a singular outgrowth of the developing interconnection of science and the military. Late in World War II, Werner von Braun (1912–77, a Ph.D. in physics from the University of Berlin) and his German colleagues succeeded in launching armed V-2 rockets toward England. Intense research and development involving rockets and ballistic missiles became a defining part of Cold War research, especially when missiles could complement, then replace long-range bombers for delivering nuclear weapons. The USSR and the technological missile system of the USSR military sent the first artificial satellite, Sputnik I, into orbit on October 4, 1957; and the same system shot the first human, Yuri Gagarin, into earth orbit on April 12, 1961. These grand accomplishments fueled the race for space, and President Kennedy's announcement later in 1961 that the United States would "land a man on the moon before the decade is out" was a dramatic declaration inspired by competition with the Russians for supremacy in space. The resulting Apollo project and the actual landing of humans on the moon in 1969 was a historical triumph of human technological accomplishment. Most post-Apollo space endeavors—the Space Shuttle, the International Space Station, and projected manned missions to Mars—share a background in the Cold War. A number of nations including France, China, Japan, and India along with the European Union now have national space programs, and they seem to involve political and economic priorities more than scientific ones. The human presence in space seems especially driven by these considerations. Space programs have become commercialized, with commercial telecommunications satellites and commercial launch systems, and it goes without saying that space transport systems, satellites, space probes, and the complicated systems required for humans to live and work in space embody highly advanced technologies that are science-based and applied. Today, space science per se only piggybacks on national/military and commercial space industries. The scientific exploration of space by remote probes is cheaper and more effective than sending humans into space.

Chips and the Internet

Computers represent another major domain where applied science has radically transformed technology and society in recent decades. The historical roots of the modern electronic computer go back to the automated Jacquard loom (1801), which used punch cards to control weaving patterns. Another stream in this river of technological change flowed from the calculating needs of industrializing society and notably the mechanical "calculating engines" of Charles Babbage (1791–1871) who with Ada Lovelace (1815–52), the British mathematician and daughter of Lord Byron, conceived of a universal calculating machine or computer. Improved mechanical and electro-mechanical calculators followed through the 1930s; these were underwritten in part by the U.S. Census Bureau and used for producing numerical tables and processing data. Not unexpectedly, World War II pushed developments in computing that resulted in the Mark I calculator in 1944 and the Navy-sponsored ENIAC in 1946. After the war, a series of evermore sophisticated general purpose, stored-program digital electronic computers emerged, including the UNIVAC machine, the first commercially available and mass-produced central computer. It appeared on the market in 1951. This type of general purpose computer run by programs was based on principles articulated in 1937 by Alan Turing (1912–54) and amplified by Turing and by John von Neumann (1903–57) among others through the 1950s.

The key scientifico-technical innovation for the development of computers was the creation of the first solid-state transistor in 1947 by William Shockley, a mechanical engineer, and a team of scientists working at Bell Laboratories, an achievement for which Shockley won the

Fig. 19.2. Early computers. An electronic calculating behemoth with 18,000 vacuum tubes, the ENIAC machine (1946) filled a large room and consumed huge amounts of power. (Note women attendants.)

Nobel Prize in physics in 1956. Such solid-state devices supplanted vacuum tubes (which regularly failed), and they permitted the industrial and commercial development of the first practical large computers in the 1950s and 1960s, such as the early standard, the IBM 360. The miniaturization of the transistor and the creation of the semiconductor computer chip was a related tour de force by materials scientists and engineers. The exponential growth of the computer revolution was underway, a pattern of growth captured by Gordon Moore's "law" enunciated in 1965: it states that the number of electronic switches on semiconductors doubles every eighteen months. This prediction still holds true five decades later as physicists explore computing possibilities on the nanometer (10^{-9} meter) level. From the 1960s onward mainframe computers manufactured by the IBM Corporation, Digital Equipment Corporation, and other companies came into increasingly widespread use in the military, in the banking and emerging credit-card industries, in airline reservation systems, in and around stock markets, and on university campuses wired with terminals hooked up to central multiprocessing and timesharing machines and printers. The computer center became a new social space.

In a host of realms physical scientists, computer scientists, and computer engineers toiled to improve the amazing technology of the computer: software and programming languages; the silicon "chip" itself and evermore elaborate and faster integrated circuits; high-tech methods to manufacture the same; increasingly sophisticated magnetic media and data-storage technologies (the "floppy disk," CDs, and now the ubiquitous USB flash drives); improved means of displaying information; and industry standards in general. But nothing did more to fully unleash the computer and information revolution than the introduction of the personal computer in the 1970s and 1980s. Hobbyists began to build personal computers from kits and commercially available microprocessors and floppy disks in the early '70s. In 1976 college drop-outs Steve Jobs and Steve Wozniak brought to market the *Apple II*, and in 1981 IBM branded the name PC with its own machine running DOS (Disk Operating System) software licensed from Microsoft Corporation and a young Bill Gates. Refinements in the PC included the mouse and the graphical user interface with its windows and icons, developed at the Palo Alto Research Center of the Xerox Corporation. Improved operating systems and evermore sophisticated software packages for word processing, spreadsheets, and databases transformed the new toys into indispensable machines for the office, home, and modern life. The computer became a mighty and expensive new consumer item, and woe to businesses that were slow making investments in the technology. In 1981 the Commodore company's VIC-20 alone sold one million units. In 1989 Microsoft sales reached $1 billion. A radical transformation was underway.

No reader today needs to be informed of the explosive proliferation of computing devices. Computers are ubiquitous, and they exist at the core of industrial civilization to the point where industrial civilization as we know it would be impossible without them. Essentially no one anticipated the computer, however, or how it would transform society. Yet computing and personal computing have penetrated deeply into economies and societies, and the computer revolution has effected an enormous social transformation in barely two or three decades. The number of personal computers and computing systems of all sorts is substantial and growing exponentially, not to mention in more recent years a proliferation of personal devices connected to the Internet. But the computer revolution is a revolution in progress. In the year 2013, for example, the most intensely computerized country was the United States, with 310 million personal computers. This is a significant number, representing essentially one computer for every American. China ranked second in the world in 2013 with 195 million PCs, up spectacularly from fourth in the world in 2000 with only 20 million PCs in that country. Per capita penetration of the PC in China has exploded from 1.5 percent of the population to roughly 15 percent. On a world basis, in 2014 there were approximately 2 billion personal computers in use, or roughly one PC for every three or four persons, up from one computer for roughly every seven persons in 2000, a figure suggesting progress and also likely further social penetration of the computer. Be it noted that the computer industry in its manifold guises, with such brands as Apple, IBM, Lenovo, Sony, Dell, and so on, is big business. For example, IBM announced plans to spend $3 billion on chip development in the time frame 2014–19. Still, given the ways computing is evermore ubiquitous and accessible through so many different platforms, the dominance of the stand-alone PC seems to have stalled.

The extraordinary novelty and perhaps the key to understanding the computer as technological agent of social change was the idea of connecting computers together into networks. First it became possible to connect a remote terminal with mainframe computers over phone lines using telephone modems. Then, funded by DARPA (the Defense Advanced Research Projects Agency), the first true computer network, ARPANET, arose around a handful of nodes in 1969. ARPANET grew to link more and more universities and research labs, and the utility of the network began to demonstrate itself, not least in the new modality of email. The World Wide Web—WWW, that phantasmagorical set of web pages accessible over the Internet—emerged in 1991, and the Internet and the World Wide Web on top of it stood poised for commercial development. In the last two decades the world has witnessed the growth of the incredible and transformative global information system that the Internet is today.

The Internet is only partly about the routers or servers that make up the physical system, and it is only partly about the Internet Service Providers (ISPs) and the economics of Web-based sales or services.

The Internet is mostly about communication and access to information. The Internet and World Wide Web serve as the substrate and technical basis for personal communication and access to information. People around the globe now have instant access to one another via email and social media. But no less significant, in what amounts to being a universal library, the Internet and the Web also bring the world's information instantaneously to video screens in homes and offices and personal devices on the road everywhere across the globe. Commercial search engines comb the Web and fulfill every request for information, considered in its most general sense, and place it immediately at our disposal. The ancient Library at Alexandria is less than nothing compared to the libraries of information available on demand over the global Internet today. Computers and the Internet now give us information of unimaginable variety at the click of a mouse. Some of this information is idiosyncratic and unreliable, but much of it is credible and authoritative; some of it may be morally or otherwise offensive, but it is all there.

With computers the Internet has become a platform for entertainment of every variety . . . music, video gaming, and video clips of all kinds. The development of social media is a notable outgrowth of the Internet, as social life to a real degree has become transferred to the Web and such applications as Facebook, where family, friends, and communities can be sustained via their connections over the ether. Facebook dates only from 2004, and Facebook and its ilk (Weibo in China) have permeated an entire generation that is connected through social media. We keep in touch and redefine communities through Twitter. With Skype we have instantaneous face-to-face connections with most everyone anywhere in the world with an Internet connection. We now live our lives and share our lives with others in new ways, including social photography through Instagram or Snapchat, for example. Dating sites bring people together. Anyone and everyone can blog and compete for attention in a public social space that simply did not exist previously—a new town square. All this, not to mention what we can find posted on YouTube, Tumblr, Reddit, or like sites or by consulting Wikipedia or asking Google or Bing. In this connection one also thinks of gaming and the digital distribution of video games via Steam where literally millions of players are online simultaneously. Yes, the Internet even brings libraries and rich databases to your laptop or other device. Pornography is big, and there are now e-sports and e-sport competitions globally. Subtle generational and group differences sort out users and collectivities, but in ways that confer unprecedented choice and freedom. In many respects personal, social, and family life has been transformed by the Internet and social media, and the ultimate outcome of what these technologies will do to foster and maintain social connections remains to be seen.

The Internet has also emerged as a new global marketplace with virtually unlimited retailing, including such crossroads in cyberspace marketing as Amazon.com, iTunes for music, or Netflix for movies and streaming video. Online auction and shopping Websites such as eBay have become familiar landmarks in the e-world available to us. Brick and mortar stores, notably bookstores, have declined considerably from a retailing heyday of not that long ago and are sustained largely by civic pride, shopping necessity, or for entertainment excursions to the local mall.

While many caveats need to be introduced, the effect of the computer revolution and the Internet has been to liberate and democratize information. Those with access to computers have access to information and can make their own judgments; they are less in the grip of traditional experts and guardians of information. Those without computers and access to the Internet and the wonders of the World Wide Web, of course, are denied this capability. The powerfully global and globalizing technology of the Internet has the paradoxical effect, not of dominating people by limiting their choices, but of allowing individuals the freedom to exploit the Internet according to their own priorities and less according to any "establishment" or political authority. The Internet fosters a sense of community among users who share common interests to the point where cyberspace may be thought of as the jostling of splintered groups sharing shifting interests. By the same token, firewalls imposed by governments, notably China, restrict and control speech and access to the Internet that limit this democratic opening afforded by the technology.

The flip side of the Internet and Web that we need not forget is the constant information gathering that goes on based on the choices and keystrokes we users make. Advertising seems not to pay the bills for most Websites, so the business model has shifted to surveillance of users' habits and offering enticing services at the cost of being spied upon. The information collected is then sold to companies, advertisers, and political groups. Big Data is a tool of Big Brother and the big corporation, and it amplifies the issue of privacy mentioned above. The Internet is far more interactive and more intimate than television, a more passive medium long derided for the remote control and the couch potato.

Computers and digital recordings have completely transformed the music industry, which is now based on computer processing and digital storage and retrieval. Laser-read and -written disks have now given way to music downloaded from the cloud, and storage has moved to solid state flash drives. The pornography industry has boomed, thanks to the possibility of private access via computer to erotic pictures and videos distributed across the World Wide Web. And, a further word needs to be said about computer and video games. With ravenous appetites for speed, processing power, and storage, computer gaming has been a major driver of the computer revolution ever since the introduction of the primitive game PONG by the Atari company in 1976. Engrained in youth subcultures around the world, computer games have become exquisitely complex with ever improving capabilities for graphical representation. Today it costs $50 million to develop and introduce a new computer game, and a successful game is worth many millions more. The stakes for entire new gaming systems such as Xbox or PlayStation are even higher. Along these lines, developments initially for the military of supercomputers and supercomputing allowed for improved handling of graphical and visual representations of information. Today, larger and faster personal computers are capable of these intensive computational tasks, including gaming, digital photography, and digital home video. Assemblages of supercharged PCs can cope with processing needs met by the supercomputers of just yesterday, and these machines provide the computing basis for related developments in artificial intelligence and data mining. Finally in this connection, computers and computing power have crossed over to the movie industry. Digital projection systems are now common, and an entire subgenre of digital animation and digitally enhanced movies has arisen. Today, many films incorporate aspects of digital and computer-based images that were impossible prior to the advent of computers. Developments regarding the computer are recent, practically more current events than history, and they testify yet again to the place of high-tech industries and the rapid pace of change in industrialization civilization in the twenty-first century.

Telecommunications and Lifestyle

The integration of computers into the film and music industries illustrates how the technological systems of industrial civilization have synergistically combined to change society and effect entirely new modes of cultural existence for a handful of privileged nations and classes. Nowhere is this lesson more plain than in considering telecommunications technologies and the coming into being of the "wired" lifestyle.

Increased speed of communications is itself a hallmark of industrial civilization. With the advent of the telegraph, the steamboat, the telephone, radio communications, transoceanic cables, communication satellites, and now the Internet, human communities have been in increasingly closer touch over the last two hundred years, and the pace of life in industrial civilization has correspondingly increased.

The mobile cellular telephone represents a landmark science-based communications technology, and its effects have accelerated trends in industrial civilization long in the making. Through the 1980s telephones meant hardwired, usually landline connections controlled by monopolistic companies such as AT&T—the American Telephone and Telegraph Company ("Ma Bell"). Telephones linked places, not people. The first cell phone, the Motorola Dyna-Tac, dates from 1973.

The first trials of a "cell" system that could track and maintain multiple roaming calls took place in Chicago in 1977 with 2000 customers, and the Japanese built the first commercial cellular telephone system in Tokyo in 1979. Early cell phones were expensive items costing upwards of several thousand dollars, and the successful implantation of the new technology depended not only on lowered costs, but also on the deregulation of the telephone industry occurring in the United States in 1982, and on licensing changes introduced by the Federal Communications Commission in the same year. The cell phone quickly took off as an indispensable tool for modern living. The United States saw its first one million cell-phone subscribers in 1987; in 2004 1.3 billion mobile phones were in use around the world. By 2014 that number had risen to the point where the number of cell phones in the world was projected to equal the world's population—in other words, essentially everybody has or will soon have a cell phone. Not unexpectedly, China today is the country with the most cell phones, over 1.2 billion (quadrupling from 300 million in 2006). That this explosion of cell phone use occurred over a period of only two decades or so testifies to dramatic developments in infrastructure and business creation to put into place the complete technological system to support cell phone use. The cell phone industry is a multibillion-dollar industry, and there now are many more cellular phone subscribers in the world than customers using fixed telephone landlines. Cell phones and smart phones, but other devices as well, incorporate rare earth metals in their manufacture, and this need has spurred mining operations, merchant trading, and the supply chain around the world.

Cell phones themselves are already wonders of miniaturization. Coverage areas of cell phones systems are expanding and will soon encompass most areas of the world. Telephone numbers are now transportable, and they may assume the practical importance of Social Security or national identification numbers, except that "calling" no longer means "dialing," with biometric indicators a more likely wave of the future.

The effects of "cutting the cord" have been dramatic. Communications today are increasingly tied to persons, not places, and the concomitant lifestyle changes are evident everywhere. With the cell phone came instant messaging, which provides a different form of communication between people, one with behavioral and social consequences different from the traditional telephone call. Etiquette and mindsets have changed as people scramble to answer rings that now jingle insistently in meetings and other social settings. Drivers who use their cell phones while moving are a known hazard, with automobile accidents and deaths rising as a result. The cell phone has facilitated closer communications among circles of family, friends, and business associates, but the technology has also increased our collective alienation, as the cell phone turns us away from civil society and inward toward private cohorts. The more democratic pay phone is virtually extinct, but a distinct advantage of cell-phone technology is that it has allowed developing countries to "leap frog" over the stage of extending prohibitively expensive land lines in favor of implanting cell-phone systems that, while not cheap, have made it possible for less-developed countries to achieve something close to communications parity with the developed world.

In the smartphone the marriage of cell-phone technology with the Internet and the World Wide Web proved a dramatic convergence. The Internet and all that it offers in terms of information and functionality and what started out as telephone communications technology have merged in the extraordinary device of today's smartphone: a hand-held personal communications tool that literally puts the world at our fingertips. It combines the power and functions of a telephone, high-quality still and video cameras, television, a social media platform, a music player, a GPS device, a personal digital assistant (PDA), and a desktop computer. Applications (apps) of an astonishing variety can be downloaded onto one's phone, allowing the world to be tailored to individual tastes and experience. The concepts behind the smartphone and the kind of connectivity at issue here date back to the 1970s, if not to Dick Tracy's Two-Way Wrist Radio from 1952. Smartphones piggybacked on cell phone systems and the Internet, but mass utilization of smartphones awaited the maturation of mobile wireless technology and the technological systems underlying the Internet. Designed for business users, the first BlackBerry came on the market in 2003. Apple introduced the first iPhone in 2007 and came to dominate an expanding market along with sophisticated Android systems offered by other manufacturers and wireless providers. In 2014 nearly 60 percent of Americans owned a smartphone; the number of smartphones in the world was over one billion and counting.

The Genie of the Genome

The discovery of the double helical structure of DNA by James Watson and Francis Crick in 1953 marked a milestone in the scientific understanding of the nature of life and inheritance. Today, the burgeoning fields concerned with genetics and DNA, including applications in medicine, agriculture, and forensics, evidence the monumental social and economic significance of Watson's and Crick's discovery. DNA technologies represent the cutting edge of twenty-first-century applied science that has significant economic and other ramifications.

The transformation of this esoteric bit of knowledge concerning DNA from the research front, and the teasing out of its philosophical and practical implications, unfolded through a series of converging technical breakthroughs. The 1962 Nobel Prize was awarded for the discovery of the structure of DNA, and in that year scientists first "cracked" the DNA code and linked DNA sequences with the production of proteins, a step necessary for any practical work with genes. Recombinant DNA techniques that date from 1972 allowed researchers to "cut and paste" DNA from one strand to another, a powerful technique making it possible to splice DNA sequences from one life form to another. The concurrent development of gel electrophoresis technologies that permit the molecular comparison of genes and DNA samples provided another powerful new tool in this growing DNA toolkit. Then in 1985, working for the biotech firm Cetus Corporation, Kary B. Mullis invented the polymerase chain reaction (PCR). The technique allows for the massive amplification of any DNA sample; it can generate billions of copies of a gene sequence in a matter of hours, thus opening the door for applied science and genetics on an industrial scale. That Mullis won the 1993 Nobel Prize in Chemistry speaks volumes about how much genetics and biotechnology stemmed from the research front in science. More than 10,000 state and commercial laboratories in the United States specialize in these kinds of DNA analytical techniques.

The Human Genome Project is a glorious example of developments in this sphere. The Project began in 1990 with the ambitious goal of sequencing the three billion base pairs in human DNA and cataloguing the estimated 100,000 human genes. The thirteen-year effort was underwritten by the U.S. Congress and coordinated initially through the U.S. Department of Energy and the National Institutes of Health. In 1998 a private company, Celera Genomics headed by Dr. J. Craig Venter, entered the competition promising to sequence the entire human genome in three years using advanced, automated sequencing technology. After some clashes regarding ownership and use of the information being compiled, the public and private endeavors ended up cooperating. In 2001 scientists completed a first draft of the human genome, and they arrived at a final catalogue of the entire human genome, reduced down to around 20,000 genes, in April of 2003. This was an expensive, "Big Science" project that cost a total of $3 billion and that came in early and under budget because of quickly declining unit costs of DNA tests and because of the unusual public-private competition that spurred developments. The extraordinary potential of the study of human genes goes without saying, and the results of the Human Genome Project will provide the basis for research and application for decades to come.

The mapping of the DNA of other species has followed, in large measure because the research infrastructure was in place to study the genomes of other creatures, generally those of importance to humans, such as the cow, dog, mouse, honey bee, mosquito, and various viruses. Studies of primates contribute to the understanding of our evolutionary relationships with other organisms. Genomics, or the study of the chemistry and functioning of whole genomes and sets of genes, is a notable area of research in this domain with potentially important medical payoffs. With increasingly efficient research techniques, it will soon cost essentially nothing to decode the genome of any species or individual. Along these lines, in 1998, a private biotech company, de-CODE, in cooperation with the government of Iceland began building a database of genetic information of all 275,000 inhabitants of the island in order to isolate genes for specific illnesses. This kind of information is proving powerful, profitable, and problematic.

As the references to Celera Genomics and the deCODE corporation indicate, knowledge of genes and DNA has become heavily commercialized, and biotech industries have arisen to exploit and profit from proprietary knowledge. Companies have, indeed, patented human DNA sequences (genes). Around 1,500 biotech companies are active in the United States, and they employ thousands of scientists and technicians. Many are small and of an entrepreneurial, start-up variety, but many are huge companies worth billions of dollars. Genentech was the first of these, launched in 1981. In 2012 Genentech, having become a division of the Swiss pharmaceutical giant Roche, employed over 12,000 people (20% with advanced degrees); Roche itself had $50 billion in sales that year. The locus of these industries and their proximity to regional university and scientific centers in and around Silicon Valley, San Francisco, New York, Boston, and San Diego is to be noted. Increasingly, researchers and their supporting institutions, notably universities, are concerned with establishing and protecting intellectual property.

More than anywhere else to date, however, the impact of applied genetics is felt in agriculture. Genetically engineered crops and genetically modified foods have effected a historic agricultural revolution, one far surpassing the so-called green revolution of the 1960s that brought new varieties, particularly of rice, into cultivation. In this new revolution genetically modified organisms (GMOs)—corn, soybean, wheat, cotton, sugar beets, among many crops—have been introduced into widespread cultivation virtually around the world, excepting Europe. A genetically modified potato and a genetically modified salmon are in development. Fields tests are now underway to grow genetically modified plants that produce drugs. Agriculture—or, better, agribusiness—in countries like Argentina and the United States has been extensively impacted. The agribusiness giant, the Monsanto Company, for example, sells seeds that have been genetically modified to resist pests and herbicides. One result of genetically engineered crops is a lowered dependence on environmentally polluting pesticides, but the use of herbicides has increased tenfold over the last two decades. Monsanto sells not only the seeds, but also the herbicides that are doused on farmers' fields, and it contractually prohibits farmers from setting aside their own seed and instead requires that new patented seed be purchased annually.

The foods produced from genetically engineered seed and crops now enter most of the processed foods that humans consume. Most commercial breakfast cereals, chips, and protein bars, for example, are confected out of GMOs, and the list goes on and on. Sixty countries worldwide now require GMO labeling, but the United States does not. A related, but silent, revolution presently underway concerns chemically enhanced and chemically fortified foods and their additives, such as folate, now routinely added to cereals to prevent birth defects. The long-term ecological and health impacts of such human tinkering is not clear, and a decided backlash against GMOs has developed with demand for non-GMO foods skyrocketing.

Genetic testing for disease is another component of the application of genetic research. Tay-Sachs disease, for example, a fatal genetic disorder that causes progressive destruction of the central nervous system in children, is caused by abnormalities on human chromosome 15, and the disease has now been virtually eliminated by testing and counseling of carriers. Tests exist that can check for Down syndrome and other genetic abnormalities in utero allowing parents to prepare themselves for the challenges of raising a child with a genetic abnormality or deciding to terminate the pregnancy. Genetic testing of this sort also allows parents to select the gender of their offspring, and a new eugenics movement is underway. This "neo-eugenics," unlike quasi-state-run efforts of the first half of the twentieth century, is consumer demanded, technology driven, and physician enabled. As much benefit as DNA testing may bring, such testing threatens discrimination in medical and insurance coverage.

Cloning is a related technology that produces genetically identical copies of an individual. In 1996 the sheep Dolly became the first cloned mammal. In general, although of great economic potential for animal husbandry, cloning has not proven itself. One area where cloning may prove important is in preserving endangered species. The issue of cloning humans has received some attention. It is technically possible to clone a human, but whether this possibility ever turns into something more than a wild fantasy remains to be seen.

What is not an oddity is the creation of DNA databanks for use in criminal forensics and law enforcement. The technology of polymerase chain reaction allows comparison of any two pieces of DNA, say one taken from a crime scene and another from a possible suspect. In the United States, DNA data banks now store the DNA profiles of all convicted felons in the criminal justice system and everyone in the military. These data banks will soon supplant fingerprints as the major (and more reliable) tool for use in the pursuit of criminals and terrorists. Naturally, these developments raise serious concerns regarding privacy, civil liberties, and the freedom of citizens to control their own genetic information. Indeed, the technology and applied science of DNA have resulted in a "brave new world" of remarkable powers and scary possibilities. On the dark side of DNA and the witches' brew of knowledge that science and technology have stirred up are the possibilities of bioterrorism.

Medical Technologies

Throughout the developed world, today's medicine is an important arena of high-tech applied science. Particularly in recent decades, the exploitation of science for medical applications has been relentlessly successful, and stunning developments have taken place regarding the application of fundamental research in biology, chemistry, and physics to pharmaceuticals, medical technologies, and the practice of medicine. Highly sophisticated, science-based medical technologies have emerged that have transformed medicine and that have improved the identification, understanding, and treatment of disease. These applied-science technologies have changed what it means to be a patient and a physician.

Many different technologies of medical care have soothed humanity over the centuries. In Islam and the West, in China, and in India elaborate theoretical frameworks arose both to explain disease and to point the way to treatments. Remedies offered were of varying degrees of efficacy, but the dream was always to link knowledge to practice, natural philosophy to health. In seventeenth-century Europe René Descartes articulated the modern vision of scientific medicine, that is, medical practice continually advanced and perfected through research and improved scientific understanding. In the nineteenth century, as we saw, the ideology of scientific medicine began to become a reality with anesthesia, antisepsis, the germ theory of disease, and modernized hospitals. But only in more recent periods does Descartes' vision seem at least partially fulfilled, with scientific medicine and science-based medical technologies in full flower in industrial civilization today.

Medical practice has always been a technology or, better, a set of technologies, and contemporary scientific medicine has become thoroughly colonized by technology. This tradition is only nominally Western, for today organized and institutionalized medicine of this sort is found in hospitals and medical centers around the world. Indeed, contemporary scientific medicine is one grand science-based technological system, as anyone knows who has ever been a hospital patient or visited a doctor's office. The technologies applied to monitor and improve the health of individuals and societies are the results of craft practice, to be sure, but only in the sense of being highly specialized, science-based, and like none other in the history of medicine prior to industrialization.

The traditional model of bedside practice wherein a doctor treats and counsels a patient who is experiencing illness is no longer the norm, and one observes increasingly depersonalized treatment of individuals within the technological machine of modern medicine. Today, patients are evaluated in light of clinical data obtained from diagnostic tests pertaining to specific ailments, and disease often defined as test results that deviate from some norm. (And don't overlook the science-based and exquisitely sophisticated battery of blood tests and other routine screening tests available to patients and physicians today.) The subjective experience of patients is less relevant, often completely so. Meanwhile, physicians undergo years of special training, in part to develop the knowledge and expertise necessary to use the scientific technologies that provide the results required for diagnosis and treatment. Physicians are thus necessarily removed from treating the whole patient.

The increasing specialization of doctors is also integrally connected to industrialization. Factory-like hospitals insure a steady pool of "clinical material" for physicians. With sufficient patient populations, doctors can develop specializations, and institutions can offer physicians access to technological equipment that is either too cumbersome or too costly for a local hospital or practitioner to maintain personally. Authority for the delivery and administration of medical treatment have been largely taken away from physicians themselves (despite some persistence of "doctor's orders") and given over to detailed protocols that follow an intricate and cascading decision tree of treatments dictated and overseen by government social service agencies and insurance companies. The fact that in many instances health care delivery is given on a for-profit basis puts an unconscionable moral burden on us all.

In the developed world today the hospital has become an advanced center of science-based technologies. The ordinary hospital operating room in particular is a miracle of modern technology. All over the world in this amazing technological setting surgeons save lives and increase human happiness. Esoteric and high-tech surgeries push the norm. In the 1950s surgeons at the vanguard perfected open heart and cardio-pulmonary bypass surgery employing artificial heart-lung devices, and today what was once extraordinary has become routine heart surgery to repair birth defects, replace heart valves, and less invasively, but no less technologically sophisticated, to insert stents and perform other feats. In 1967 the South African surgeon Dr. Christiaan Barnard performed the first heart transplant, and today heart, lung, kidney, and other organ transplants are not uncommon. An artificial heart meant to replace the original was first implanted in a human in 1982, but at present the technology is still experimental and under continuing development; artificial pumps and assists, however, have shown great promise for persons with weak or injured hearts. The trend in surgery today is toward minimally invasive procedures using laparoscopic techniques that embody their own advanced technologies. Now it is possible to perform delicate surgeries using fully robotic devices that allow for operations to be conducted remotely. And, there are truly intricate surgeries of staggering technological complexity, such as in utero surgery or the separation of conjoined twins. These technological marvels raise social and ethical questions about what is being achieved and about the allocation of limited resources.

Today's medicine is widely dependent on science-based technology and machines. The medical application of X-rays followed almost immediately upon their discovery in 1895, and, once appropriated by the medical profession, the X-ray became the first of a number of powerful diagnostic and imaging tools whose origins are rooted in scientific research. The medical use of X-rays is common today in a variety of circumstances from mammograms, to dental offices, to advanced diagnostic procedures. Digital X-ray technology is now the norm, and in testimony to the role of the computer and computer technology in medicine today, the CAT scan (computerized axial tomography) takes imaging technology to new levels by using X-rays and the computer to generate 3-D images. Along the same lines and springing from the field of nuclear physics, the MRI (magnetic resonance imaging) and the PET scan (positron emission tomography) exploit properties of atomic nuclei and of radioactive substances to produce images and information for research and clinical use. (See figure 19.3.) The 2003 Nobel Prize in Medicine or Physiology was awarded for fundamental scientific research that went into creating MRI technology. And, let us not forget the army of other science-based machines grinding away in medical facilities around the world: automatic pulse, blood pressure, and oxygenation-measuring instruments, ECGs (or EKGs, electrocardiograms), EEGs (electroencephalograms), the ultrasound, bone scanners, and even the humble thermometer, now an electronic device that takes core temperature by sensing infrared radiation.

The development of pharmaceuticals and the chemistry underlying them represent another area where knowledge from the realm of science found medically and economically valuable applications that have contributed significantly to human health and welfare. The British physician Alexander Fleming (1881–1955) discovered penicillin somewhat accidentally in 1928; he was knighted in 1944 and won the Nobel Prize in medicine in 1945. Penicillin was an antibiotic "wonder drug" developed by medical and scientific research, and it is hard to imagine life and medicine before the age of antibiotics that began less than a century ago. World War II hastened the mass production of penicillin and other drugs, and the pharmaceutical industry has grown explosively since then, with global sales in 2013 reaching $839 billion. The pharmaceutical industry is now dominated by a handful of multinational corporations that are responsible for virtually all new drugs that come into widespread use. To bring a new drug to market costs hundreds of millions of dollars, and in 2014 the global pharmaceutical industry invested 135 billion dollars in drug research and development.

Fig. 19.3. High-tech, science-based medicine. A SPECT-CT imaging machine such as this one combines a 3-D CT (computed tomography) image from a 360° X-ray scan and an image derived from a nuclear medicine SPECT (single photon emission computed tomography) scan. In the latter, patients are injected with a radioactive isotope and a nuclear medicine gamma camera rotates 360° around the patient measuring the ionizing gamma ray radiation and creating another image in three dimensions. At special facilities, nuclear medicine technologists operate the machine and combine the images at a separate station, images that are then "read" by nuclear medicine specialists.

In this intense context of investment and research many new antibiotics and drugs are developed through systematic experimentation, "designer" chemistry, and elaborate trials. The successful introduction of oral contraceptives in the 1960s was a landmark instance of this process. "The Pill" gave women greater control over their reproductive lives, and it proved a key ingredient in the sexual revolution of the '60s. Chemical/hormonal birth control continues to affect the lives of millions of women (and men), and the case admirably illustrates the social impact that can flow from this nexus of applied science and pharmaceutical commerce. But that impact can be sharply negative, too, as when pharmaceuticals already on the market are found to have dangerous side effects. The science-based and high-tech pharmaceutical industry makes available an impressive variety of manufactured chemical and biological compounds that are prescribed and consumed for an equally impressive range of ailments. With the introduction of Prozac in 1986, the pharmaceutical industry produced and marketed new antidepressants, and today tens of millions of people stave off depression chemically. Anti-inflammatory and anti-allergy drugs are other classes of effective and profitable products introduced via the pharmaceutical industry. How many millions of people today take allergy medications and baste themselves with steroids? Millions of people today take cholesterol-lowering drugs; "attention deficit disorder" in children is treated by other drugs; heartburn, now elevated to "acid reflux disease," is alleviated with several sorts of stomach acid suppressants, such as H2 antagonists and proton pump inhibitors (PPI). The medical establishment and the pharmaceutical industry sometimes seem to create the "disease" for which they offer chemicals as treatments. The widespread advertising of prescription drugs has prompted millions to ask their doctors if a given drug "is right" for them. Along these lines, tens of millions of men worldwide have used Viagra, the drug brought to market in 1998 by Pfizer, Inc., for treating so-called sexual dysfunction, with U.S. sales reaching over $300 million and world sales a tad under $2 billion in 2013.

But all is not pharmacy. In another chapter in the history of modern scientific medicine, the discovery of vitamins in 1911 led to the prevention and treatment of vitamin deficiency diseases. And, building on the work of Louis Pasteur, the development of vaccines in the twentieth century against typhoid, tetanus, diphtheria, and measles greatly impacted mortality and improved public health. The vaccines developed in the 1950s by Jonas Salk and A. B. Sabin against poliomyelitis created tremendous public health benefits and attention.

Techniques surrounding in vitro fertilization likewise exemplify high-tech medical science and practice today. Human in vitro fertilization dates from 1978 and embodies a very sophisticated and expensive set of scientifico-medical techniques and procedures that regularly brings happiness to those looking to have children. The number of children born through in vitro fertilization and assisted reproduction techniques nearly doubled between 1996 and 2001, with 41,000 newborns that year. The total number passed 5 million such babies born through 2014. Improved treatments for infertility have raised a host of unprecedented ethical and legal issues concerning the rights of children, surrogate mothers, and of biological and nonbiological parents.

Materials science and bio-medical engineering are other disciplines where scientific research has been turned to fruitful medical applications by scientists, engineers, physicians, and surgeons. The development of artificial joints, artificial limbs, dialysis equipment, pacemakers, hearing aids (now digital), cochlear implants, and similar bionic parts or extensions have further transformed medicine and helped millions. And let us not forget dentistry and the phenomenal range of dental technologies that have their origin in this same base of knowledge and industry. Today's dental crown, a marvel of contemporary technology and materials science, is ringing testimony to human scientifico-technological achievement.

And the future? Evermore sophisticated and expensive high-tech technologies are in the offing in medical research. Practical artificial blood is on the horizon. Stem cell research holds the promise of microscopic organ transplants, and stem cell therapy offers the prospect of nerve regeneration to reverse spinal cord paralysis. Today new vaccines stand on the forefront of research and development in medical science. From the vanguard of physics and materials science comes nanotechnology, an applied-science discipline that envisions the construction of medical devices one hundred-thousandth the size of a human hair. The prospect of molecular medicine powered by nano-scale machines for tailored drug delivery or even internal microsurgery is not outlandish. Also in the offing is pharmocogenomics, an applied-science field that blends traditional drug development with separate research in genetics to deliver personalized medicine and drugs tailored to the profile of individuals derived from their genetic information. Another notable "cross-over" technology of this sort increasingly used in drug-related research involves biologically sensitive materials embedded in microarrays of computer chips that permit broad and quick tests of chemicals as potential drug and medicinal agents.

The successes of modern scientific medicine are undeniable. The virtual eradication of polio and the technical elimination of smallpox may be added to the list of these phenomenal achievements that have alleviated human suffering. Science and medicine today offer unprecedented vistas for health and longevity not possible before industrialization. A female born in Japan in 2014 has a life expectancy of 87.3 years! Life expectancy at birth in the United States in 1901 was only 49 years; in 1930 it rose to 59.7 years, and in 2014 the figure stood at 79.8 years (77.2 for men and 82.4 for women). In 2014 almost 26 percent of the population of Japan was over 65; 33 percent over the age of 60, and projections are that by 2025 there will be only two Japanese under the age of 65 for every Japanese senior over that age. Another consequence of increasing longevity is the explosion of populations of the truly old, with the number of centenarians in developed countries, and in Japan in particular, skyrocketing in recent decades. That number is expected to increase by a factor of eighteen by 2050, mostly in the developed world. The social consequences of these demographic trends are enormous in terms of providing social services, funding national social security and health plans, and representing another divide between the rich and the poor and the developed and developing worlds. These consequences remain to be fully felt.

Nevertheless, institutionalized scientific medicine and medical practice of the sort under discussion here offers a mixed legacy, and many fundamental premises have been questioned. The vision of an all-powerful scientific medicine that can vanquish disease with a pill has been challenged, certainly since the thalidomide drug disaster of the late 1950s and early 1960s. Thalidomide was introduced in 1956 as a sedative and to counteract morning sickness in pregnant women. Recognition that thalidomide caused serious birth defects was slow to emerge, and was even resisted in certain quarters. The product was not taken off the market until 1961, a disaster that left 5,000 thalidomide-deformed survivors in forty-six countries. The regular recall of drugs or after-the-fact recognition that one treatment or another is ill-advised, such as universally prescribed hormone replacement therapy for menopausal women, does little to alleviate the suspicion that we do not adequately evaluate the long-term effects of new drugs, much less consider non-pharmaceutical alternatives to treating disease.

In fact, industrial civilization itself is responsible for some of the medical challenges faced by the world today. The problem of resistant bacteria or herbicide-resistant super-weeds results directly from the overuse of antibiotics and pesticides. International travel permits the rapid spread of disease, as in the SARS or Ebola epidemics, and the opening up of previously remote places has facilitated the spread of exotic new disease agents such as HIV and AIDS. Researchers have not had the same success with antiviral drugs that they have had with antibacterial ones, and medical gene therapy has not worked well to date. Epidemics of diabetes and obesity stem directly from unhealthy lifestyles and consumerism encouraged by mass advertising.

Compounding these shortcomings, the dominance of pharmaceutical companies and the search for profit have skewed research and the provision of health care. The scientifico-pharmaco-medico establishment (Big Pharma) is firmly organized and institutionalized around the globe; and representing powerful financial interests, it is not about to go away. The fear is that alternative views of what constitutes health and how to promote personal and collective health are not generally pursued because no profit is seen in the endeavor. Socially oriented investments in preventative medicine or public health measures, for example, or research avenues that do not seem lucrative, such as in vitamins and vaccines, are not explored with anything like the level of support that goes into the pursuit of more financially promising drugs. Increasing skepticism exists over high-tech, science-based "Western" medicine. The limitations, not the glories of modern medicine, explain why people, even in developed countries, are turning to alternative medical traditions and treatments to live saner, healthier lives.

Most alarmingly, the growing expense of science- and technology-based medical procedures divides rich and poor nations, and this gap likewise divides industrial societies into subgroups of people who can and cannot afford advanced medicine. The overwhelming majority of people on the planet today have little or no access to medical services that are commonplace in the developed world. This divergence raises serious political and ethical questions concerning the allocation, distribution, and rationing of limited resources. Such is the paradox of medicine in industrial civilization today.

Varieties of Applied Science

One overriding conclusion stands out: tapping science to create new technologies is a modern phenomenon. Although applied science first manifested itself with government support of science and scientific experts in the pristine civilizations beginning in the fourth millennium BCE, until the nineteenth century the worlds of learning and the crafts remained largely segregated, as we have seen. Similarly, the ideology that knowledge of nature should be drawn on to promote the general welfare did not emerge before the seventeenth century, and, as we have now seen, that ideal became realized only in the nineteenth century and substantially so only in the twentieth century. The support for science and applied knowledge from government and industry has grown to match the long-standing rhetoric of the utility of science. Technology as applied science is a defining element of industrial civilization in the twenty-first century and gives us the cliché we started out to examine. As characteristic and significant as the intellectual, technical, and social rapprochement of technology with science has been, the precise bonds between contemporary science and technology need to be unraveled.

Plainly, science and technology today evidence a spectrum of connections. As this chapter has shown, theoretical developments at the forefront of scientific research can be impressively applied to practical technologies. But even here, we can discern a variety of connections. In the case of the atomic bomb, for example, there was still a delay, albeit a short one, separating theoretical innovation and practical invention. Another consideration arises when we consider that often scientists and engineers, sometimes in different institutional settings, are separately involved in research on the one hand and development and application on the other. This partition preserves the classical distinction between pure scientific research and the separate application of knowledge. By contrast, the case of the laser, invented by Gordon Gould in 1957, blurs the distinction between discovery and application, in that the underlying physics and the practical tool emerged simultaneously "in a flash," as Gould put it. The fact that Gould, then a graduate student at Columbia University, dropped his studies and himself undertook to develop and commercialize the laser erodes the neat categories of scientist, engineer, and entrepreneur and adds further nuance to the concept of applied science.

In many other areas, however, even today it is misleading to envision technology as simply or directly "applied science." For example, the education and on-the-job training of engineers and technologists entail significant doses of science, yet rarely does that component consist of advanced theoretical science from the research front. Rather, "boiled down" science usually suffices for the practicing engineer or science-based technologist. Even NASA scientists and engineers had no need of and did not apply cutting-edge research in relativity or quantum mechanics in launching the Apollo moon missions or any of the other successes of the U.S. space program. The old celestial mechanics proffered by Newton in the seventeenth century and perfected by Laplace at the turn of the nineteenth century provides an entirely sufficient basis for calculating trajectories of rockets and space probes.

Even where scientific knowledge forms the basis of a device or a technology, so much else is ordinarily involved that it is misleading to characterize a new technology as simply applied science. For example, in 1938 Chester A. Carlson, a patent attorney with a chemistry background, using his "boiled-down" knowledge of optics and photochemistry, invented xerography, a dry-process photocopy method. Necessity was hardly the mother of invention in this case, as Carlson spent fruitless decades trying to persuade various backers, including a skeptical International Business Machines Corporation (IBM), of the utility of his substitute for carbon paper. In the end, it took engineering and design improvements—experience, intuition, and inspiration—that had little to do with science to perfect a practical photocopying machine; and, also, a marketing decision to lease rather than sell the machines proved the key to the stunning success of the first Xerox copiers in the 1960s. Only at that point did a perceived need for photocopies multiply and the photocopy machine become a technological commonplace. In this instance invention was once again the mother of necessity.

The fact that many different styles and brands of printer/copy machine or TV or computer exist on the market today reveals a key difference between contemporary science and technology. That is, the scientific community generally recognizes only one solution to a given scientific puzzle or problem. A particular protein, for example, will have only one agreed-upon chemical makeup and structure. In technology, by contrast, even in science-based technologies, multiple design and engineering solutions are commonplace. Sometimes different designs are intended for different uses, such as personal copiers versus large office copiers. Other times, in the variety of music players or personal devices presently available, multiple designs exist merely for marketing purposes. Women get pink razors, while men get multi-bladed macho ones in blue or black.

Science and technology have indeed forged historically significant new types of interaction. These range from what has been labeled a "strong" direct application of scientific theory to an applied-science product or process, such as the laser or the atomic bomb, to "weak" or "boiled-down" interactions of science and technology of the sort illustrated by Chester Carlson and his copy machine. By the same token, the traditional independence of technology from science and natural philosophy also continues today. Technical innovations continue to crop up that have little or nothing to do with the world of science or theory and that represent continuations of independent traditions in technology, engineering, and the crafts characteristic of technology since its roots in prehistory. In 1994, for example, Navy veteran Theodore M. Kiebke received patent 5,361,719 from the U.S. patent office for a new variety of cat litter made out of wheat. His invention had potentially significant consequences for an industry worth over $2 billion annually in 2015, yet it seems unreasonable to claim that Mr. Kiebke's cat litter represented any kind of applied science. The same might be said of the development of the charcoal briquette by Henry

Ford back in the 1920s and '30s to deal with scrap timber from automobile manufacture, or "The World's Most Comfortable Pillow," invented and manufactured by Michael J. Lindell of Canhassen, Minnesota, as advertised in 2015.

The historian of technology Thomas Parke Hughes summed up from a systems point of view how best to think of modern applied science and the transformation of natural knowledge into practical technologies:

> The relationships between engineers and scientists and between technology and science have long held the attention of historians, especially historians of science. From the systems point of view the distinctions tend to fade. . . . Persons committed emotionally and intellectually to problem solving associated with system creation and development rarely take note of disciplinary boundaries.

Across a spectrum of connections, then, modern applied science in its manifold forms has transformed human existence and brought industrial civilization to its present state. As never before, science and science-based technologies have integrated themselves into human societies and the world economy.

The history of technology shows how hard it is to predict the course of technological change, much less its social effects. We can only be confident in saying that many of the grand technologies, particularly the science-based ones, that had roots in the nineteenth and twentieth centuries will continue to blend together and shape the supersystem of industrial civilization as the twenty-first century unfolds. Whether today's technologies make people happier or wiser is another question, but regardless, to conclude our study, we need to address the state of science as a social institution in the world today.

Under Today's Pharaohs

The promise of socially useful knowledge is an old theme that harks back to the origins of civilization. Throughout this study we have seen that governments, rulers, states, and political entities, while often indifferent to abstract knowledge, always and everywhere valued useful knowledge and patronized experts who possessed (or claimed to possess) such knowledge. Utility in the service of the state was the goal, and, to a degree, ancient and medieval monarchs got their money's worth in cadres of specialists who could keep lists, tally the harvest, calculate the days and seasons, follow the sun and moon, collect taxes, construct altars, forecast the future, heal the sick, tell time, engineer and direct public works, maintain inventories, make maps, and indicate the direction of Mecca, to list some of the activities undertaken by specialists over the millennia. As noteworthy as such activities were for the development of science and society, up until the nineteenth century the overall effect of science and medicine on the organization and functioning of states and societies remained small. That is, expert activities such as reckoning, healing, and forecasting were important and perhaps even necessary, but civilizations were nonetheless overwhelmingly traditional societies where agriculture and customary crafts provided the mainsprings of civilized life. Although "science" may have played a real part in practical affairs, until recently that part remained a small one, and state support for any scientific expertise remained at a comparatively low level.

Consider, as an example, France during the era of its Sun King, Louis XIV, whose personal reign extended from 1661 to 1715. At the time, France was technologically and culturally the most advanced civilization in Europe, if not in the world, and no other contemporary state supported the natural sciences to the degree that France did. The Royal Academy of Sciences in Paris stood as the premier scientific institution in the world. It was explicitly dedicated to the sciences;

its list of members reads like a Who's Who in late seventeenth- and eighteenth-century science; its expeditions and scientific work were unparalleled, its publications unrivaled. Nevertheless, the Academy was seriously underfunded in the last decades of the seventeenth century, its members paid late, if at all. Fine arts and literary academies received much higher subventions than did the science academy, and scientific academicians held a lower social status than their literary and fine arts counterparts. Indeed, a smaller and ephemeral academy of technology received relatively more government support, and the Academy of Sciences was constantly pressured to turn its attentions to useful ends. The Academy controlled the award of patents, so in this particular the scientists of the Academy oversaw technology even if they themselves contributed little to innovation, and otherwise the Academy had but little impact on the economic life of Old-Regime France. While we need to locate the modern origins of government support for science in seventeenth- and eighteenth-century France, we also need to recognize the comparatively limited impact of science on practical affairs until more recent times.

The marriage of science and industry in the nineteenth century was the great novelty, and only in the twentieth century did the enterprise of science fully break out of its historically limited social role as handmaiden to governments. As documented in previous chapters, science in our day has transformed itself into an essential element for the functioning of industrial civilization. While far from shedding its connections to government, science has become wedded to technology and to industry, and it has assumed a fundamental role as a force of production in the modern world.

Science as a Mode of Production

Science and technology have become indisputably significant forces shaping the modern world, and, indeed, a defining characteristic of today's world is the full integration of science with technology and with the engines of economic production. Marxist thinkers in the former Soviet Union recognized the novelty of this development, labeling it the twentieth century's "scientific-technological revolution." The term has not been taken up generally, but it does capture much of what is at issue here: the effective merger of science and technology in the twentieth century and the increasing importance of science-based technologies and technology-based sciences in the twenty-first century, not merely as cultural embellishments to civilization, but as powerful forces operating at the heart of the economies and societies of the developed world and, increasingly, the whole world.

We have previously identified three different modes for conceptualizing science in society: the Babylonian, where the state employs trained

Fig. 20.1. World spending on science and technology. Led by the United States, industrialized nations now spend more than 1.6 *trillion* dollars annually on scientific research and the development of new technologies. The overall percentage of U.S. R&D expenditures has been declining steadily, and relative to its economy, the United States now ranks eighth in the world in such expenditures. China's R&D spending is rising rapidly, although these investments go more toward applied research and development.

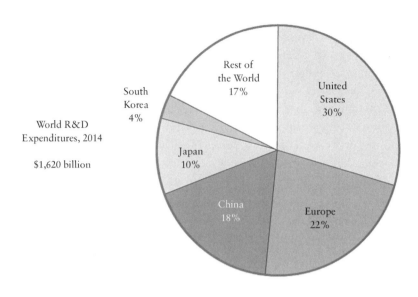

World R&D
Expenditures, 2014

$1,620 billion

Fig. 20.2. Growing support for science and technology. U.S. spending on research and development jumped dramatically with the Cold War and especially after the Soviet launch of *Sputnik* in 1957. Through the mid-1970s the U.S. government provided the bulk of these funds, but then private industry funding began to skyrocket.

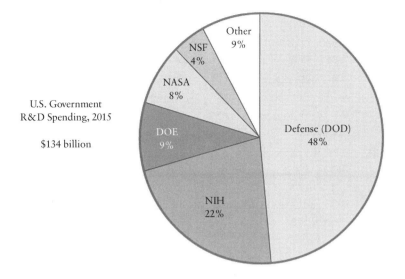

U.S. Government
R&D Spending, 2015

$134 billion

Fig. 20.3. Hellenistic priorities. Not quite half of U.S. government spending on research on science and technology is allocated to military defense. Almost all of defense-related spending supports applied science and technology. Only a small fraction of federal funds goes to non-applied research, notably in the National Science Foundation (NSF).

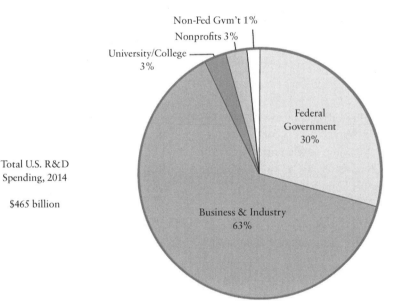

Total U.S. R&D
Spending, 2014

$465 billion

Fig. 20.4. Industry driven. Private industry now outspends the federal government on R&D by more than a two-to-one ratio, with funds going overwhelmingly to product development and applied research, especially in the life sciences. Research to advance knowledge in the Hellenic tradition takes place only here and there in universities, government, and industry.

experts in the service of governance; the Hellenic, where disinterested inquiry into nature takes place on the social periphery; and, the Hellenistic, where government patronage is aimed toward the utilitarian, as in the Babylonian model, but allows a certain latitude for the theoretical and not immediately practical, as in Hellenistic natural philosophy. A new term is required for what we see has happened in recent history. The modern marriage of science and industry took place in the nineteenth century, as we saw, all the while as the Hellenistic model persisted in the West into the twentieth century. The larger and more dramatic merger of science, government, technology, and industry that we are familiar with today arose as the Industrial Revolution made itself felt, particularly with the industrialization of war. This new pattern of government and industry support for science becomes especially apparent after World War II. The scale of the patronage for pure and applied science is unprecedented; the lines between and among science and technology, development, discovery, invention, and innovation have become blurred virtually beyond recognition. Something new has presented itself in the historical record. That something is not peripheral to the economic heart of societies, as was Hellenic natural philosophy; it is not something essential, but limited to keep things going, as in Babylonian or Hellenistic support for cadres of experts. No, this amalgamation of science and technology and forging a new order of things with government and industry was new, a key force in modern economies and a defining feature of industrial civilization. And so we discover another model and another way to place science and technology in social context, a fourth model associated with industrial civilization.

The resources allocated to research and development reveal this fundamental new role for science in society. In 2014, for example, governments and industry were spending a world total of 1.618 trillion U.S. dollars on research and development (R&D). (See figure 20.1.) The United States still led the world with $465 billion or roughly 30 percent of world R&D spending, but the U.S. percentage is down sharply from 41 percent in 2000 and 38 percent in 2003. Collectively, Europe follows at 22 percent. China comes third at 18 percent, but China is the leading single nation after the United States to invest in R&D, and China is rising in this category, too. In 2003 China with 9 percent stood behind Japan, but China and South Korea are increasing their R&D expenditures more rapidly than anyone else, and projections are that China could surpass the United States in total R&D spending by 2020. Japan has fallen to third place after China at 10 percent of the world total with South Korea next at 4 percent. Considered as a percentage of Gross Domestic Product (GDP), Israel led the way in 2014, spending 4.2 percent of its GDP on R&D, followed by South Korea (3.6%), Finland (3.5%), and Japan and Sweden (3.4%). Not coincidentally, these nations have a higher proportion of scientists and engineers. The United States stood in eighth position with 2.8 percent of its GDP invested in research and development.

Considered over time, the present level of government and industry funding for science and scientific research is even more impressive. For example, in 1930 the government of the United States supported science to the tune of a mere $160 million or 0.2 percent of what was then called the Gross National Product (GNP). By 1945 those figures stood at $1.52 billion and 0.7 percent of GNP. With the Cold War in full swing in the 1950s, U.S. public and private R&D funding began to mount steadily, as indicated in the accompanying figure. Adjusted for inflation, total spending in the United States on research and development has quadrupled since 1965. A sea change has taken place in the last half century as government and industry have placed historically new and higher priorities on underwriting scientific research and development.

Considered as a proportion of the total federal budget, after hitting a high of almost 12 percent in the post-Sputnik era of the early 1960s, the percentage of government spending set aside for R&D has hovered in the range of 3–5 percent. Considered as a percentage of the discretionary funds available to the federal government, its R&D spending is even higher, varying in the 11–14 percent range since the mid-1970s.

Through the 1980s military and perceived defense needs overwhelmingly drove government patronage for science in the United States. The R&D budget of the Department of Defense (DOD) is down by almost 30 percent since a post-9/11 peak in 2010, but still, for 2015 the DOD R&D budget ($64 billion) was two and a quarter times that of the next largest consumer of federal science dollars, the National Institutes of Health (the NIH, a unit of the Department of Health and Human Services), at $29 billion. Across all agencies, defense-related science expenditures continue to consume more than half of all federal R&D science dollars. The descending order of funding from the DOD to the NIH, the Department of Energy, NASA, and down to the National Science Foundation (NSF) is revealing in this regard. Funding for defense, health, energy, and space projects (which have always entailed significant political and employment goals) comes well before support of the agency nominally charged with the promotion of scientific research, the National Science Foundation (founded in 1950). For 2015 a mere 4 percent of the federal R&D budget was allocated to the NSF. The budgetary philosophy at the NSF increasingly directs research dollars into so-called strategic areas, such as the National Nanotechnology Initiative, linked to national needs and harnessing science to economic growth.

Notable perspectives emerge when one considers trends in U.S. government funding for R&D by discipline over time. Through the 1970s, government subventions to nondefense research in the physical sciences, the life sciences, and engineering were approximately equal. Beginning in the 1970s, however, R&D funds for biology and the life sciences began to outpace support for other disciplines, and from the mid-1990s life-science funding mushroomed, while support for other disciplines remained essentially stagnant. By 2015 the budget for the National Institutes of Health more than doubled in absolute dollars since 1990, and today not quite half of non-defense government R&D spending is devoted to research in health and the life sciences. This level of support is over twice that for research in any other discipline with engineering following in second place with 11 percent of federal

R&D funding. Mathematics and computer science, for understandable national security reasons, display a pattern of exponential growth, quintupling since the 1980s, but in 2012 these fields received just over one tenth the funding of the life sciences. Similarly, federal R&D devoted to counterterrorism increased six times in the two years following September 11, 2001. Spending on research and development in the opaque world of national security and state surveillance is hard to track down, but the figure would seem to be increasing and likely to increase even more in coming years.

Two conclusions concerning federal support for research and development are to be drawn from these statistics. One, the fall of the Soviet Union and the end of the Cold War in 1991 represent a watershed for U.S. science and government. Prior to that date the story centers on the physical sciences and on military hardware. After that date, emphasis shifts to the life sciences and the revolution in biology and genetics discussed in the previous chapter. The profound impact of that revolution is the second conclusion to be noted. Today, because of their health and economic potentials the life and biomedical sciences have supplanted the physical sciences as the main beneficiary of the government's purse. As notable a development as this shift to health and the life sciences is, we shouldn't forget the vast sums that still go to R&D for the military.

As important as government support for scientific research has been and continues to be, the overall figures reveal an even larger seismic shift in the topology of patronage for science since the end of the Cold War. In the 1990s a new era began with business and industry displacing government as the primary source of R&D funding. In virtually every developed country today, industry now supports scientific research and development to a considerably greater degree than does government, usually by at least a factor of two! In the United States, for example, in 2014 business and industry were responsible for 63 percent of R&D *funding*, and the figures for Japan and South Korea topped 70 percent. The untold story is that, although industry funding for R&D mirrored that of government spending through the 1980s, industry R&D then began to take off rapidly. In the period 1970–2014 industry R&D budgets increased by six times to something like $300 billion.

We need to ask what kind of research is at play here and who does it. "Development" involves prototyping, scaling, marketing, and other things necessary for transforming discoveries or innovations into practical realities. Not surprisingly, development receives the bulk of funds spent in the United States on R&D, some 60 percent overall with the remainder divided between applied research (22%) and basic research (18%). Federal support allocates moneys almost evenly between research and development. As for the research component, it is evenly divided, too, between applied research and basic research, and federal support for basic research amounts to 23 percent of total federal R&D spending. When one looks at the sectors (government, industry, university) that actually carry out research and development work, industry predominated, responsible for 70 percent of R&D in the United States. Paralleling industrial development worldwide, North America currently contributes 35 percent of the world total of industry-funded R&D, followed by Asia (31%) and then Europe (28%). The investments of business and industry in R&D go overwhelmingly (71%) into development itself; only 21 percent of industry's own funds are given over to applied research, and industry's overall share of support for basic research is consistently less than 4 percent of total industrial R&D expenditures in the United States. In other words, although business and industry in the United States have now eclipsed the government as the main underwriters of research in science and technology, their orientation is almost entirely toward the development of immediately useful and profitable applications of knowledge. As much as is claimed for the private sector, government is the incubator of innovation. This new role for industry has had dramatic implications concerning human capital and the employment of scientists in positions away from what were traditional careers in universities and government.

The U.S. government itself does comparatively little R&D work (9% of the U.S. total in 2012). That figure would be 3 percent higher if federally funded research centers were included. The DOD does by far most (almost 90%) of government R&D work. The federal government, although its contributions in dollars are comparatively small, remains the major player supporting basic research. At roughly 50 percent of the total, the NIH is the major performer of both basic research and applied research. Scientists and other researchers in colleges and universities conduct 13 percent of R&D in the United States, and the majority of basic research is conducted by research universities. Colleges and universities remain the locus of many fundamental discoveries in science with great practical potential. In other words, even if business and industry are now responsible for most R&D expenditures, the basic research on which their profits often depend takes place in universities and federal research centers. Basic research is still conveniently and largely unknowingly underwritten by the taxpayer.

These same patterns and pressures are also apparent in other countries that support scientific research through government and industry subventions. In China over 80 percent of R&D funds goes toward development with smaller slivers allocated toward applied or basic research. With slightly more emphasis on applied research, Europe parallels the United States in terms of government and industry R&D funding. Although science is currently funded at historically unprecedented levels by government and industry, the rationale for such support essentially has not changed since ancient Babylonia—the hoped-for practical and, these days, profitable benefits that may be derived from science and expert knowledge.

Technical Careers

An elaborate system of institutions and social practices surrounds the activities of scientists and engineers today. The human capital involved with science and engineering in the United States now approaches 6 million people. In 1950 that number was less than 200,000, and since then the science and engineering labor force (S&E) has outpaced over-all employment growth in the United States by a factor of four. This figure is expected to continue to grow faster than other occupations over the next decade but at a somewhat slower rate. Not surprisingly, mathematics and computer sciences outpaced other areas of science and engineering growth since the 1980s, and the life sciences and mathematical/computer scientists will be the leaders in future growth. Education in science, technology, engineering, and mathematics (the STEM fields) is a top priority worldwide.

According to the U.S. Bureau of Labor Statistics, in 2013 1,135,000 people were employed specifically as scientists and another 632,000 in complementary scientific research and development services. Engineers numbered 1,547,580 with another 900,590 employed in engineering services. In addition, computer and mathematical occupations (which overlap with employment categories of scientists and engineers) added a whopping 3,700,000 individuals to these totals. Of the personnel involved specifically with research and development (R&D again), engineering degree holders comprise more than one-third of that population with those with degrees in computer science and mathematics constituting another 17 percent.

In the early nineteenth century scientific careers were haphazard and concerned only a handful of people. Although wide latitude exists in scientific employment, the social role of the scientist and the social pathways available to become one are quite narrow and rigidly defined. In many respects, whether someone will be a scientist is already determined in his or her middle school years. Almost universally, one has to graduate from high school with the requisite science courses, attend college, and complete an undergraduate major in a field of science. Then, one usually has to pursue graduate work and, on the basis of original research, obtain a Ph.D. Research scientists normally continue training for a period of years after the Ph.D. in temporary, postdoctoral positions known as postdocs. Postdoc positions are increasingly common, especially as tenure-track jobs in academe are of limited availability. The Ph.D. constitutes the scientist's license, and from there career paths for scientists diverge. Traditionally, the norm has been to pursue academic careers of research and teaching in colleges or universities. But increasingly, young men and women in science find productive lives for themselves in private industry and in government service.

Although they require much the same initial undergraduate training, engineering careers take a different path, and the contrasts between the world of science and the training and practice of engineers and technologists remain profound and revealing. In the matter of education, for example, university training of engineers arose only in the nineteenth century, and today an undergraduate degree in engineering (the B.E., bachelor of engineering) is a terminal degree satisfactory for entering professional ranks and employment as a working engineer. A master's degree in a field of engineering or technology is the advanced degree that practitioners most often seek to round out their educations, but few go further to obtain a Ph.D. in an engineering field. Those who do often return to academe as professors of engineering; fewer Ph.D. engineers are employed in industry and by government. Licensing and public practice as a professional engineer (P.E.) is another professional step for a handful of engineers.

Business and industry employ the bulk of scientists and engineers and depend on their expertise. Private for-profit companies account for almost 70 percent of the total S&E labor force, but notably they employ only 35 percent of those holding S&E doctorates. (The non-profit sector exists at a few percent level.) As centers for basic and applied research, colleges and universities are homes for 18 percent of the overall number of scientists and engineers and 41 percent of S&E doctoral holders. All levels of government employ 12 percent of the S&E workforce. Science and engineering employment in the U.S. federal government reflects what we saw earlier regarding government funding for R&D. In this case the DOD dominated with 43 percent of government employed scientists and engineers in 2012, with the NIH, NASA, the Nuclear Regulatory Commission (NRC), NSF, and the DOE following. Scientists and engineers and those in related occupations are generally well paid and earn more than comparable positions, and they have low unemployment rates; the mean salary among life scientists, physical scientists, and computer and mathematical scientists in 2010 was $78,000; for engineers the figure was $87,000.

What scientists do today spans a broad range of practice: from disinterested pure research in universities or specialized research facilities to applied science work in industry or public service. Generally speaking, whether in academe, industry, or government, younger scientists tend to be the more active producers of scientific knowledge. The more one moves up the career ladder in science today, the more one tends to leave active research and turn toward scientific administration and directing the research efforts of others.

The status of women in science changed dramatically in the second half of the twentieth century. Women have never been completely absent from the social or intellectual histories of science, and as they gained admission to universities in the nineteenth century, gradually increasing numbers of women have become scientists. Heroic accounts often, and rightly, single out, among others, Marie Curie (1867–1934), Rosalind Franklin (1920–1958), and Barbara McClintock (1902–1992) as iconic figures. Their careers do exemplify the increasing presence and professionalism of women in modern science. But these women faced trials unknown to their male colleagues. Despite the international fame that followed her 1903 Nobel Prize, for example, Marie Curie was rejected as a member of the French Academy of Sciences, largely on account of her gender, and she became caught up in a vicious public scandal that slandered her as the other woman in the breakup of the marriage of the physicist Paul Langevin. British physical chemist Rosalind Franklin was instrumental in the discovery of the double helix of DNA in 1953, but her story presents cautionary lessons as well, because the social system of British science in the 1950s did little to welcome her or give her credit for her work; she was not permitted to take tea with her male colleagues and had to do so with the janitorial staff. The case of Barbara McClintock, who won the 1983 Nobel Prize for her discovery of mobile genetic elements, offers a more positive, but not unmixed example, as McClintock's primary employment for over two decades was in a staff position at the Cold Spring Harbor research laboratory on Long Island in New York State. It remains difficult for women to pursue research and applied-science careers in academe and industry, not least because of childbearing. But mirroring other shifts in Western societies over the last half century, opportunities have improved for women, and the idea of a female scientist is no longer exceptional.

Yet, women remain underrepresented in science and engineering. The number of women working in S&E fields has increased from 12 percent in 1980 to 21 percent in 1993 to 28 percent in 2013. Women earned 40 percent of the Ph.D.s in science and engineering, but are in the workplace at only that lower figure of 28 percent, which implies that women tend to drift away from science and engineering careers. The median age of scientists and engineers in the United States is 41 years old. Women in science and engineering tend to be younger than the age profile of their male counterparts, and the retirement of baby boom scientists and engineers is shifting the gender balance further in a positive direction. The distribution of women across fields in science and engineering is notably uneven also: Women constitute the majority in the social sciences (52%) and are commanding in the field of psychology in particular (67%). They are close to the majority in the biological and medical sciences (now 48%, down slightly from 2006). On the other hand, only 23% of job holders in computer or mathematical sciences are women, and that figure has declined over the last two decades. Women hold only 13% of jobs in engineering, but that figure is increasing. Depending on education levels, women in S&E are paid 25–35 percent less than their male counterparts.

The world of science and engineering is becoming more diverse, but it remains a bastion populated largely by white males. Specifically, white males made up 50 percent of the S&E labor force in the United States in 2013. In terms of race and ethnicity whites (male and female) dominate with 70 percent, but that percentage has declined from 80 percent in 2001 and 84 percent in 1993. Asians are the next largest group among scientists and engineers in the United States at 18 percent, both as American-born and as immigrants. Asians are overrepresented in science and engineering compared to their proportion in the general U.S. population of 4.7 percent, and they are more highly concentrated in computer, computer engineering, and information science occupations (variously from 22 to 40% of employees). Blacks or African Americans compose just under 5 percent, and Hispanic and Latino scientists and engineers just over 5 percent of the S&E workforce. Both blacks and Hispanics are underrepresented in science and engineering compared to the general population, which is 12 percent black and 17 percent Hispanic. The numbers of black and Hispanic engineers and scientists are rising, but in engineering blacks number only 4 percent and Hispanics only 6 percent of engineers. A handful of Native Americans and mixed race professionals populate the worlds of science and engineering. More so than the rest of America, as a social institution science and engineering have a long way to go to be inclusive in terms of gender or race.

Foreign-born scientists and engineers constitute a notable subset of the S&E workers in the United States. They represent about 25 percent of the total and fully 40 percent of S&E Ph.D. holders. Foreign-born scientists and engineers may have received some or all of their technical and scientific education abroad or in the United States. Approximately half of foreign students who obtain a Ph.D. in the United States end up remaining in the country and joining its S&E cadres. U.S. H-1B visas allow companies to hire foreign workers with specialized expertise, and of all H-1B visas issued in 2010, most—39 percent—went to citizens of India followed by China at 10 percent; if one counts only H-1B visas issued to doctorate holders, the places are reversed with China at 29 percent and India second with 16 percent.

Work that involves science and engineering occurs throughout the world, and it occupies perhaps 6.5 million people in total. The number of scientists, engineers, related cadres, and support staff has ballooned worldwide from 2.8 million in 1995, and as governments and business continue to make investments in science and engineering, the figures will continue to rise. By all measures the United States and Europe remain the world's leaders and are largely comparable, although Europe produces almost three times as many science Ph.D.s as does the United States. But as other indicators reveal, China is not far behind in these categories either. Chinese universities and technical institutes produce more engineering Ph.D.s than any other country, and China far outnumbers any other nation in the numbers of students who receive undergraduate degrees in science and engineering, something on the order of seven to one for engineering degrees over Japan and the United States. Figures indicate that China, Japan, and the rest of Asia account for 45 percent of all undergraduate degrees in sciences and 56 percent of undergraduate degrees in engineering. The United States lags behind at 10 percent and 4 percent at fourth and fifth place respectively. Publication patterns are shifting on a world scale, too. Europe is the largest producer of scientific and engineering journal articles followed by the United States, both at steady levels over the last two decades. Collectively Asia has moved into a close third position, with rapidly rising China having moved ahead of Japan in the publication of original articles in 2007.

Mainstream scientists are expected to publish research results in scientific journals, and the old adage of "publish or perish" applies to life in contemporary science, although publication is felt to be more important in academe than in industry. Journals are rated by their "impact factor," and authors accumulate what is labeled an h-index measuring their overall scientific productivity and gravitas. Indicative of other changes in science, the number of multi-authored papers has risen dramatically since the 1950s, and today multi-authored papers are the norm in scientific publication. The premier American scientific periodical, *Science*, for example, publishes only 10 percent of the reports and articles submitted to it, a remarkably low figure given the self-selection and high qualifications of authors who seek to be published in this journal. In a similar vein, today the overall success rate of applications to the National Science Foundation stands at about 20 percent, although more competitive programs often fund only 15 percent of proposals. These figures illustrate just how competitive the world of science is and they help explain the presence of monumental egos at the top of the social pyramid of science.

Participation in scientific societies and at scientific meetings comprises another requisite part of scientific life today. Normally, scientists belong to several societies that represent their specialized interests and their general professional status. For example, in addition to joining smaller organizations concerned with specialized research areas, a physicist will undoubtedly belong to the American Physical Society (1899), a chemist to the American Chemical Society (1876), an astronomer to the American Astronomical Society (1899), and they will all belong to the American Association for the Advancement of Science (1848). Such organizations usually hold annual conventions where scientists gather to present their research, to participate in the organizational life of the society, and to party.

Most scientific research today is a costly enterprise, and obtaining grants is another essential aspect of normal scientific practice. Academic scientists in particular spend a great deal of time on a grant treadmill that requires them to write proposals to get funds to do research to support more proposal writing and more research. Grants typically include funds for principal investigators' salaries, graduate and postdoc support, and equipment. In what amounts to government and foundation subsidies of institutions, a significant element of grant budgets are so-called overhead or indirect costs, on the order of one-half to one-third of a total grant budget, paid not to support research but to underwrite the grantee's institution. Private and public agencies, such as the National Science Foundation (NSF), solicit proposals and judge them through intricate procedures that involve outside referees, internal study groups and panels, and financial decisions by program officers and governing boards. As competition has increased, the average age at which scientists usually receive a first major grant has risen to 45 years old. For the sociology of science the implication of this trend is that career paths now extend from the Ph.D. through to a series of postdocs and junior positions and a trajectory of junior author of papers prior to becoming a senior scientist and a principal investigator (PI) on a major research grant.

Honorary organizations such as the National Academy of Sciences (1863) in the United States and like institutions in other countries cap the elaborate social organization of contemporary science. A number of international organizations and commissions, such as the International Council of Scientific Unions (1931), govern science on a world level. In the public mind at least, at the very top of the social and institutional hierarchies of contemporary science stand the Nobel Prizes, given annually from 1901 in physics, chemistry, and physiology or medicine. (Other Nobel Prizes are given for literature and peace and from 1968 for work in economics.) A prolific inventor like Edison, the Swede Alfred Nobel (1833–1896) made his fortune as a chemist, engineer, and industrialist. He is most famous as the inventor of dynamite, and in his will he bequeathed almost $200 million in today's dollars to fund the prizes. A Nobel Prize brings crowning prestige to awardees and a cash award worth almost $1.5 million. The likewise prestigious but less-well-remunerated Fields Medal (at $15,000), sometimes called the Nobel Prize of mathematics, is awarded only every four years to mathematicians under the age of 40. Other prizes and rewards are there as landmarks in the professional lives of men and women in science.

Science philanthropy is a notable new aspect of the sociology of science and scientific research. The largely American super-rich are now funneling billions of dollars into scientific research. The prime example is the Bill and Melinda Gates Foundation, which, based on the tremendous wealth Gates accumulated as co-founder of the Microsoft Corporation, has spent $10 billion on global health initiatives.

The other co-founder, Paul Allen, donated $500 million to establish an institute for brain research in Washington State. Private donors now account for 30 percent of research money in the top research universities. Such philanthropists and the philanthropic institutions they have spawned fund grants, prizes, and whole research institutes. The particular interests of donors, in a specific disease, say, such as diabetes or melanoma, would seem to skew the direction that disinterested research would have otherwise taken, but with a certain ambivalence researchers have welcomed such funding.

Engineering is now also a fully professional occupation that shares similarities with the world of science. Both engineering and science are highly competitive enterprises involving research, and it usually takes sophisticated technical education and training to enter the engineering profession. Engineering, like science, is organized into a host of specialized and professional societies, such as the American Society of Mechanical Engineers (1880) and the National Academy of Engineering (1964), while any number of specialized periodicals also serve the needs and interests of practicing engineers.

Both engineer/technologists and scientists pursue research, but here, too, fundamental differences distinguish the enterprises of science and technology even today. Scientific research, for example, most often focuses on very narrowly defined problems, such as the structure of a particular protein or the intensity of magnetic fields at the southern pole of the sun. Such research generally takes into account only a limited number of other recently published scientific papers, and it is most often directed at a restricted "invisible college" of fellow research practitioners. Engineering and technical research, by contrast, ordinarily encompasses a broader problem set (for example, creating practical technologies for video conferencing or electric cars). Solutions very often involve a more diverse range of elements—choice of materials, the aesthetics of design, manufacturing considerations, financing, and marketing. And the consumers of engineering solutions are not usually other engineers or scientists but, more often, governments, corporations, and the general public.

Science and technology drew closer together in the twentieth century through increased understanding and exploitation of science for applied ends. That said, however, it may be more analytically useful to distinguish pure science on the one hand from a merged applied science and technology on the other. That is, a social and institutional separation exists today not, as in the past, between science and technology but between theoretical scientific inquiries and scientifico-technological applied science. The product of scientific research, for example, is new knowledge. Applied scientists, technologists, and engineers, on the other hand, strive to produce useful material things or processes. The results of scientific research are generally not of immediate economic value. Because they get professional credit when other people cite and use their work, university scientists are inclined to "give away" the results of their research in the form of freely accessible published papers in the hopes of nonpecuniary social and professional rewards. In contrast, the products of engineers and applied scientists possess real economic value. Thus engineers and the companies that employ them tend to keep work secret until they receive a patent that secures their economic rights. This difference in work product—papers versus patents—is telling of differences between the world of pure science and the worlds of applied science and technology.

The merger of science and technology becomes evident in cutting-edge industries like biotechnology and computers where, these days, new scientific discoveries are regularly patented rather than published and made freely available to other researchers. As a result, a cultural shift is underway in the world of science as the disinterested pursuit of knowledge for knowledge's sake is increasingly giving way to concerns for owning and protecting intellectual property, patenting discoveries, and spawning entrepreneurial activities based on the results of scientific research. One indication of this trend is the number of patents issued by the U. S. Patent Office, a number that has grown with remarkable rapidity in recent years. In 2013 the number of patent applications stood at over 570,000, a figure that doubled since 2000 and quadrupled since 1990. In the period 2000–2013 the number of patents granted jumped by 76 percent to almost 278,000. The number of patents issued to foreigners has remained steady at about 50 percent of the total since the mid-1980s. In 2010 nearly 90 percent of grants to non-U.S. inventors went to citizens from Japan, Europe, South Korea, and elsewhere in Asia. The vast majority of patents (92% in 2013) are assigned to private corporations; individuals accounted for not quite 7 percent of the total number of patents assigned in 2013, with U.S. and foreign governments responsible for the remaining 1 percent. The percentage of patents for inventions assigned to colleges and universities essentially doubled in the late 1990s, and today 11 percent of patents are held by colleges and universities as concerns over intellectual property and licensing have taken hold. These and like statistics evidence an effective shift away from traditional patterns of scientific discovery and the public dissemination of science toward the privatization of knowledge and its appropriation for commercial purposes. This conclusion is in harmony with what we have already noted relative to scientific employment and the application of scientific discovery in technology.

Explosive Growth

In assessing contemporary science, we need to recognize that more is involved than simply the linear evolution of scientific ideas or successive stages in the social and professional development of science. The

Fig. 20.5. Exponential growth. The scientific enterprise has grown exponentially over the three centuries following the Scientific Revolution, as indicated in this classic logarithmic graph of the growth of scientific journals since the seventeenth century. The quantitative study of parameters affecting science can shed useful light on the nature of the scientific enterprise.

exponential growth of science represents another characteristic feature of the history of modern science. By every indication, science has grown this way since the seventeenth century, outpacing other social indicators such as population. In the accompanying classic figure from 1961, for example, one sees that the scale of the scientific enterprise has increased a millionfold since the seventeenth century, doubling in size about every 15 years. More recent work confirms the continuing exponential growth of the scientific enterprise through the year 2000 at a doubling rate of 8 to 9 years, although by 2012 that rate may have finally plateaued.

Several paradoxical consequences follow from the exponential growth of science in the modern era. For example, a high proportion of all the scientists who ever lived—some 80 to 90 percent—are said to be alive today. And, given exponential growth, the scientific enterprise clearly cannot continue to expand indefinitely, for such growth would sooner or later consume all resources, human, financial, and otherwise. Indeed, as predicted, an exponential growth rate for science has not been maintained since the 1960s and 1970s, and while the enterprise of science continues to enlarge, that growth has leveled off. In any particular area of science, however—especially a hot new one such as genomics or brain science—exponential growth to an ultimate plateau remains the typical pattern.

Other metrics, notably citation studies, add to these basic understandings of the character of the contemporary sciences. Scientists back up their results by referencing other papers, and much can be learned from studying citation patterns. Sociometric studies reveal, for example, that a significant percentage of published work is never cited or actively used. In essence, much of the production of contemporary science disappears into "black holes" in the literature. (In the humanities an even larger fraction goes uncited.) Studies also show a drastically skewed inequality of scientific productivity among researchers. Thus, a handful will be big producers while the vast majority of any sample will produce little work, if any. Typically, for 100 scientific authors, two will write 25 percent of the papers. In all, the top 10 will be responsible for 50 percent of the papers, while the other 90 will produce the remaining 50 percent. Most scientists produce only one or two papers in a career. A corollary of this differential productivity affects institutions, too, in that a handful of elite research universities attract top producers and dominate the production of science.

The compounding geometrical growth of the scientific enterprise has had a predictable impact on national scientific establishments. It is more difficult to expand an existing large scientific establishment than it is to grow smaller national cadres. This prediction has proven out. In particular, the dominance of the United States as the world's leading national community of scientists has begun to be eclipsed by other national traditions. Statistics regarding Nobel Prizes, numbers of doctorates, patents, and the like make the same point. The era of U.S. dominance in science and technology has ended, and the United States faces a scientific brain drain of its own. This globalizing trend is doubtless going to continue as natural sciences become increasingly part of world culture, not only intellectually, but socially and institutionally as well.

Citation studies also show a peculiar "half-life" of scientific information. That is, new work in science tends to evolve out of other recent work, and therefore the citation and presumably the utility of scientific knowledge fall off with time. Older scientific results prove less useful to practicing scientists than more recent science, and so, in contrast with the traditional humanities, the scientific enterprise displays a characteristic present-mindedness. The works of Shakespeare or of Homer, for example, still speak to readers, writers, and literary critics today, whereas Newton's works, much less Aristotle's, are simply of no value to the practicing scientist. As a result, old science is consistently ignored in the teaching of science. Science textbooks serve to convey the content—and not the history—of science to students. Science teachers and textbooks often associate some historical figure with a scientific law, such as the analysis of the action of a spring that goes under the name of Hooke's law, but the putative historical connections between Hooke and his law are never explored. Taken together, the half-life of scientific information revealed in citation studies, the fact of scientific revolution, the demographic preponderance of living scientists, and pedagogical practices, all make plain the distinctive fact that science repudiates its past.

Big Science and Technoscience

The conduct of scientific research using large, expensive instruments operated by specialized workers in industrial-scale installations, or what has been called Big Science, did not originate in the twentieth century. The Imperial Astronomical Bureau of classical China, the grand observatories created by Tycho Brahe in the sixteenth century, and the international expeditions to observe the transits of Venus in the eighteenth century may be understood as Big Science to some extent. Nevertheless, these examples pale before extraordinary developments in the twentieth and twenty-first centuries that can only be characterized as the industrialization of scientific research.

In the nineteenth century, the individual scientist working alone or with a few coworkers in a small laboratory represented the dominant mode for the production of scientific knowledge. With the development of nuclear physics in the twentieth century, however, that age-old pattern changed. "Atom smashers" represent the industrialization of research, and the Tevatron particle accelerator at the Fermi National Accelerator Laboratory (Fermilab) in Batavia, Illinois, is a prime example of state-supported Big Science. Inaugurated in 1983 and running under the aegis of the U.S. Department of Energy until it was shut down in 2011, this giant machine used 1,000 superconducting magnets cooled to −268 degrees C (−450 degrees F) and spaced along a ring four miles in circumference, and it attained stupendous energy levels of 1 trillion electron volts. Elsewhere as in physics, twentieth- and now twenty-first-century research in many disciplines began to necessitate huge installations and costly capital equipment, increasingly beyond the resources of individual experimenters or even universities or private research facilities. Teams of scientific researchers began to replace the labors of individual scientists. Each team member became a specialized science worker in charge of one aspect of a complicated research endeavor. Scientific papers issuing from such team-based science came to be authored sometimes by hundreds of individuals. For example, the discovery of the "top quark" in 1995 was produced at Fermilab by two separate teams, each numbering 450 scientists and technicians staffing two detectors costing $100 million apiece.

The Large Hadron Collider (LHC), part of the European Center for Nuclear Research (CERN) on the French-Swiss border near Geneva, succeeded the Tevatron in Illinois as the world's largest and most powerful particle accelerator in 2008. Thirty countries funded the LHC at a final cost of almost $2 billion. Four thousand scientists, engineers, and support staff work at CERN itself; the CERN annual budget in 2013 was over $1.3 billion (U.S.). Ten thousand scientists and engineers from over 100 countries became involved in the construction of the LHC. The instrument itself consists of a 27-kilometer (17-mile) ring built 100 meters underground. Using thousands of superconducting electromagnets, themselves connected to the electric grid through superconducting electric cables, the LHC accelerates two beams of heavy particles (protons or lead nuclei) in opposite directions to near the speed of light before guiding them to smash together. The superconducting magnets have to be cooled to near absolute zero; the tubes that carry the beams are kept at ultra-high vacuum. Four massive particle detectors the size of six-story buildings capture the spectacular results of the collisions. The electric power requirements of the installation are gargantuan, rivaling that of the nearby city of Geneva. From 2008 through 2012 the LHC ran at a power level of 4 trillion electron volts, four times the power of the Tevatron accelerator in the United States, and using this machine and the unrivaled distributed computing power of the LHC facility, a team of 700 scientists and technicians from 44 different institutions confirmed the discovery of the elusive particle known at the Higgs boson in 2012. In so doing these researchers have extended the standard model in particle physics in important ways and opened the door to further research, and for this two theorists, Peter Higgs and François Englert, won the 2013 Nobel Prize in physics. The LHC then shut down for maintenance, to reboot in 2015 at near its designed power of 7 trillion electron volts. The LHC is the world's largest machine, its biggest piece of technology. It is inspiring to think

Fig. 20.6. (opposite) Natural philosophy and Big Science. The Large Hadron Collider (LHC) at CERN, the European Center for Nuclear Research, near Geneva, Switzerland, is the world's largest machine and its most powerful high-energy particle accelerator. Pictured is a portion of the 27-kilometer (17-mile) ring buried deep underground that uses superconducting magnets to accelerate particle beams to almost the speed of light before smashing them together in spectacular collisions that help physicists probe the ultimate constituents of the universe. In 2012 scientists at the LHC produced data confirming the existence of the Higgs boson, an elementary particle that plays a role in creating mass. Scientific research that uses the LHC is industrial-scale Big Science, but it would seem to have little practical application.

that the primary purpose of this triumph of human technology and human ingenuity is simply to advance our knowledge of nature. Hellenic natural philosophy finds a remarkable contemporary successor in the Big Science installation of the LHC. Individual or small-group research continues in some fields, such as botany, mathematics, and paleontology, but in other areas, such as particle physics, biomedicine, and space exploration, Big Science represents a material and characteristic element of twentieth- and now twenty-first-century science.

The divergent fates of two government-sponsored Big Science projects highlight the forces shaping such research today. The first concerns the cancellation in 1994 of the Superconducting Supercollider, a gigantic, publicly funded particle accelerator then under construction in Texas designed to pack twenty times the energy of the Tevatron at the Fermi Laboratory in Illinois. Although scientists made vague and sometimes shameless claims about its potential utility (to "cure cancer," for example), most recognized that such a fantastic machine would be of primary importance for natural philosophy, that it would help produce exciting new knowledge about the history and the fundamental constituents of the universe. When the costs of the Supercollider escalated to $11 billion in public dollars, politicians could no longer justify its expense, even as a public works project, and they shut it down. In contrast, a second Big-Science effort, the multi-billion-dollar Human Genome Project, discussed in the previous chapter, was never under threat of cancellation, but not because it was less expensive than the Superconducting Supercollider. The reason for its success was that the effort to map the panoply of human DNA promised to be of great practical utility for society at large, in discovering and possibly leading to cures for such genetic diseases as Huntington's disease, cystic fibrosis, and muscular dystrophy, as well as for its obvious possibilities in the fields of agricultural science. This politicians could line up behind.

Not all Big Science projects succeed or fail simply because of their immediate social or economic utility. Here one thinks of the great and expensive Big Science efforts at space exploration. Launched in 1990 the Hubble Space Telescope, for example, cost approximately $3 billion to build, service, and operate over the course of the period 1990–2014. The much-delayed James Webb Space Telescope (JWST), designed to replace the Hubble Space Telescope in 2018, will have a life-cycle cost of $8.7 billion. The Mars Exploration Rover Mission project that lifted off in 2003 had an initial cost of $820 million; five mission extensions have pushed the total to near $1 billion. (Sending back remarkable pictures and data, the rover *Spirit* kept going until 2010; the rover *Opportunity* still chugs along in 2015 having covered over 25 miles (40 k) on the Martian surface.) The more advanced rover *Curiosity* landed on Mars in 2012 and continues the exploration of the Red Planet with a price tag of $2.5 billion. NASA has plans for another rover to launch in 2020 that is budgeted at $1.9 billion. The multinational Cassini-Huygens Mission to Saturn and Titan, launched in 1997, was designed to orbit and study Saturn and its rings and moons from 2004 to 2008; the mission has been twice extended, and the spacecraft will soldier on as a working scientific instrument through 2017. The total expense for this project is $3.3 billion. The scientific and natural philosophical value of the Hubble telescope, the Mars rovers, and the Cassini mission is incalculable; but the practical utility of these endeavors is virtually nil. More plainly than small-scale research, Big Science projects underscore the tension between the disinterested pursuit of scientific inquiry and pressures for useful outcomes for research, especially when that research requires public funding, which is the case almost by definition for Big Science. That tension has been a defining feature of the social history of science from the era of Alexander the Great and Hellenistic Greek antiquity, and it continues today, most especially in Big Science projects.

National and international space programs deserve to be mentioned here. Fifty countries have such civil and scientific programs, most notably including NASA, the European Space Agency, and the space agencies of Russia, France, Japan, and China. The collective nonmilitary space budgets of these agencies runs to $45 billion per annum. The end of the U.S. space shuttle program has left Russia the only nation capable of servicing the International Space Station, the multinational cooperative endeavor launched in 1998 and now having cost $150 billion. Space has been privatized, and many satellite and communications companies are involved. For the super-rich the reality of space tourism looms. And then, space has become heavily militarized. Spy satellites of various nations look down on us from space; military world communications pass through space; and missiles can shoot through space. None of this, or the science and technology involved, comes cheap. Along these lines, too, we should not overlook ground-based telescopes that dot mountaintops, particularly in Chile and Hawaii. Larger and ever-more-sophisticated of these telescopes are coming online, and they, too, are costly Big Science endeavors whose primary goal is to extend the boundaries of our knowledge of nature.

The dependence of Big Science on sophisticated equipment and cadres of technicians to run machines brings up a related aspect of the modern scientific enterprise that deserves attention. This is sometimes labeled "technoscience." The term is a loose one with shades of meaning that include both scientized technologies of the sort examined in the last chapter (atomic weapons, genetic engineering, computers, etc.) and technologized science of a sort intrinsic to Big Science, where stupendous instruments are essential for research. Regarding the latter face of technoscience, the argument is that the technologies of scientific instrumentation are not just necessary implements for the conduct of research, but that they have transformed the very nature of scientific practice and that in a certain sense they create the phenomena being investigated. Of course, experimental science has been a crucial kind of scientific pursuit since the seventeenth century, as evermore numerous and sophisticated instruments have become central to the research enterprise. In this sense, for example, from Galileo's primitive spyglass of 1609 down to the upcoming James Webb Space Telescope research in astronomy has depended on the telescope as an investigative tool. For technoscience, however, at a certain point the increase in the quantity and complexity of instruments produced a qualitative change in the nature of research itself. It is no longer apt, in other words, to think of a laboratory as a place where someone called a scientist straightforwardly deploys an instrument to investigate and report self-evident facts concerning nature. Today, the research laboratory is a complicated knowledge-producing factory that begs many questions concerning the sociology and technology of knowledge.

Technology now drives research, and laboratories today marshal virtual arsenals of instruments of extraordinary complexity—confocal microscopes, mass spectrometers, lasers, and so on and so forth. It is impossible to list the bewildering array of analyzers, controllers, counters, emulators, indicators, meters, modulators, regulators, samplers, sensors, simulators, synthesizers, and testers available and required for research. Many of these instruments cost hundreds of thousands of dollars, and whole industries exist to provide equipment and supplies to the research community. The American Institute of Physics even publishes a journal, *Review of Scientific Instruments*, that tries to keep track of developments in scientific instrumentation. In the world of genetic research alone entire categories of instruments are devoted to apoptosis, bioinformatics, immunoblotting, signal transduction, and like areas of specialized research. The Stratagene Company, for example, proudly advertises its two-step Prolytica™ ^{18}O kit, which "quantitates [sic] proteins efficiently and economically [and] unlike traditional ^{18}O labeling methods gives you ≥98% labeling efficiency." Such piece of apparatus is useless and meaningless to all but the person, himself or herself usually a specialist in the lab, whose job is to operate the same.

Scholars who have thought about technoscience have rightly construed scientific instruments as "inscription machines," meaning that, based on certain inputs, instruments produce certain outputs of "data." Furthermore, scientific instruments are "blackboxed," which means that their inner workings are no longer subject to question or controversy in the scientific community. Everyone in a research community accepts that the output of standard instruments unproblematically corresponds to whatever was fed into them and that they produce reliable information, for otherwise they would simply not be scientific instruments. Yet, to students of technoscience, an elaborate and nonobvious set of social processes surrounds the operation of scientific instruments and the extraction of useful information from them. Inputs are never quite identical, and raw data spilling out of machines has to be massaged before it can be submitted for publication and consumption by the rest of the scientific community; there is nothing sinister or unethical about this activity because the squiggles and other inscriptions forthcoming from the apparatus are meaningless in and of themselves. They have to be filtered through the mind and intelligence of the investigator and spurious or extraneous bits scrubbed before data can take on meaning in the context of a community of researchers.

From these points of view, a different picture emerges of what is going on in laboratory research. The traditional, lone scientist is replaced by the industrialization of research and a hierarchical division of labor within the lab. Scientific worker bees service instruments and pass instrumental excreta up the social ladder of the laboratory to mid-level and senior scientists who digest the same and who in turn redirect workers and instruments on the one hand and who on the other hand produce manuscripts and grant applications. From this point of view, the research lab is itself an entity and the key functional unit of analysis. The lab is a real physical place that has to be paid for and maintained. Laboratories are partially composed of people and partially of machines, and they are animated by complicated social and technical practices. Laboratories have inputs in the form of what walks in the door, what comes in the mail, and what is received electronically. These inputs are people, supplies, reports, papers, and ideas. Laboratories also have outputs, notably the scientific papers that are sent to various journals and grants submitted to funding agencies. Laboratories associated with particular invisible colleges are thus seen as units interacting with one another, scientific equipment manufacturers, scientific journals, government offices, foundations, and so forth outward through the complex social and institutional system that constitutes technoscience today. This conception of the laboratory and of scientific research takes some getting used to, but it does capture important features of the nature of the scientific enterprise today.

Industrial Civilization and Its Discontents

In his 1961 farewell address President Eisenhower famously cautioned the citizens of the United States to be wary of the "military-industrial complex" and the dangers it posed to American democracy. But Eisenhower also issued another warning in that speech, one less-well recalled, against "a scientific technological elite." His remarks encapsulate much of what has been said in this chapter:

> Akin to, and largely responsible for the sweeping changes in our industrial-military posture, has been the technological revolution during recent decades. In this revolution, research has become central; it also becomes more formalized, complex, and costly. A steadily increasing share is conducted for, by, or at the direction of, the Federal government. Today, the solitary

inventor, tinkering in his shop, has been overshadowed by task forces of scientists in laboratories and testing fields. In the same fashion, the free university, historically the fountainhead of free ideas and scientific discovery, has experienced a revolution in the conduct of research. Partly because of the huge costs involved, a government contract becomes virtually a substitute for intellectual curiosity. For every old blackboard there are now hundreds of new electronic computers. The prospect of domination of the nation's scholars by Federal employment, project allocations, and the power of money is ever present and is gravely to be regarded. Yet, in holding scientific research and discovery in respect, as we should, we must also be alert to the equal and opposite danger that public policy could itself become the captive of a scientific technological elite.

Eisenhower was concerned about the distorting influence of government funding on unfettered scientific research and about scientists possibly overreaching themselves, and in 1961 he did not foresee the melding of technoscience with commerce and industry that would come into full flower in our day. His anxieties over this scientific technological elite also seem quaint in being limited to effects on "public policy." Yet, his views concerning a scientific technological elite were nonetheless prescient, and they foreshadowed significant social and intellectual critiques of science and technology articulated in the succeeding years.

In the period following World War II science enjoyed an unquestioned moral, intellectual, and technical authority. Through the operations of its apparently unique "scientific method," theoretical science seemed to offer an infallible path to knowledge, while applied science in its various forms promised to better human existence. Paradoxically, or perhaps not, just as the effects of a fused science-technology culture began to refashion advanced societies—with the bomb, television, interstate highways, computers, the contraceptive pill—so, too, beginning in the 1960s did a wave of social reactions lead many people to question science and technology as the triumphant bearers of human progress. Jacques Ellul (1912–1994), the influential midcentury critic, coined the term "the technological society" in 1964 to capture the mixed character of modern technology, particularly science-based technologies. In Ellul's view we have made a Faustian bargain with technology; technology caters to our every whim, yet it has enslaved us. The environmental movement represents a related critical stream emerging out of concerns over acid rain, industrial pollution, global warming, depletion of the ozone layer, and the loss of biodiversity. Other critics have voiced reservations regarding consumerism as a dominant value in the developed world and the increasingly hectic pace of life. Still other express moral revulsion at the increasing disparity between scientifically and technologically rich and powerful nations and impoverished countries today who cannot afford the luxury of a scientific culture. The back-to-nature movement of the hippie counter-culture of the 1960s and 1970s likewise embodied these trends, and historical parallels are strong between such movements and romantic, anti-science reactions in the early nineteenth century. Technological failures of nuclear power plants such as the core meltdowns at the Fukushima nuclear power plant in 2011 or the explosion of the nuclear reactor at Chernobyl in 1986 make many people suspicious of the material benefits of science and technology. The spread of new diseases induces the same anxieties. Increasing societal concerns with ecology, recycling, "appropriate technologies," and "green" politics derive from and drive such suspicions.

Intellectual critiques of science that emerged in the 1960s have to be understood in the antiscience context of those times. Detailed work then and since in the philosophy of science and the sociology of science and scientific knowledge have challenged any uniquely privileged position for science as a means of knowing or staking claims about any ultimate truth. Many thinkers now recognize that scientific knowledge claims are in some measure relative, fallible, human creations and not final statements about an objective nature. Although some would seize on this conclusion to promote intellectual anarchy or the primacy of theology, paradoxically no better mechanism exists for understanding the natural world around us than the human enterprise of science and natural philosophy.

From most historical perspectives, then, the social and intellectual circumstances of science and technology today would seem unique. Although the patronage and institutionalization of science as useful knowledge originated at the dawn of civilization itself, government and business today funnel unprecedented resources into the support of pure and applied science at a categorically higher level, and science as a social institution plays a more fundamental role than ever. Science as natural philosophy had its origins with the Greeks, and, although that enterprise also continues today, the content of today's science is radically different not only from what the ancient Greeks thought but, more to the point, from fundamental scientific conceptions universally held as recently as the beginning of the twentieth century. Technology has even deeper roots in our biological heritage and in prehistory, as we have seen, yet the Industrial Revolution of the eighteenth and nineteenth centuries created an utterly new mode of human existence in the form of industrial civilization. The future of industrial civilization is the most pressing and uncertain issue facing humanity today.

The Medium of History

Our study has shown that technology has functioned as a fundamental driving force in human history. Technology proved decisive in fashioning and maintaining human societies in the Paleolithic and Neolithic eras and in every human culture since, and technology completely underpins industrial civilization today. Undoubtedly, then, as long as humankind exists and inhabits planet Earth, humans will continue to shape their world using their technologies.

We likewise saw that since the first civilizations science-based technologies operated as governments recruited experts and expert knowledge in the service of running the state—in mathematics, astronomy/astrology, engineering, alchemy, medicine, and, later, cartography. But the stronger link between science and industry that we are familiar with is a comparatively new connection forged since the Industrial Revolution. As each of us experiences on a daily basis, this science-industry connection represents a potent historical novelty with both positive and negative effects. The application of knowledge from the biomedical sciences, for example, has undoubtedly improved the human condition; the long-term effects of the chemical or science-based armament industries are more problematic. In the short term, the pace of "progress" in these domains is likely to continue. The long-term future of applied science in industry and in the military and the apparatus of national security is cloudy, and to some extent worrisome.

What of the future of science itself? In discussions oddly reminiscent of those that arose at the end of the nineteenth century in the context of the Classical World View, renewed talk has been heard of the "end of science." Usually what is meant is that science will shortly figure everything out, that it will reach a final conceptual conclusion of some sort. The "end of science" scenario derives from asking established scientists about problems remaining for their disciplines to solve. But that approach does not reveal the intellectual future of science. We do not now know which problem areas in biology or physics or cosmology or other yet-undreamed-of disciplines will lead to radically new formulations, and, besides, can science today tell us what gravity is? If science is the reasoned story humans tell about nature, then, as long as natural philosophy exists as a social and intellectual enterprise, that story would seem bound to continue to change. Indeed, one secure lesson of the history of science is that virtually every scientific theory to date has failed—to be replaced by a better one.

Because of its utility, science is highly valued and socially well supported, and it is hard to imagine industrial civilization without a highly developed scientific enterprise to help sustain it. Barring a cataclysmic collapse of world civilization, science as a social endeavor should continue. But what governments and industry and the world at large have always wanted from science was expertise and power, not values, which is why science has flourished in so many different political, ideological, and cultural contexts over time. There would seem to be nothing inevitable about the continuation of secular natural philosophy. Indeed, on a global plane, the vision of the world and humanity articulated by the natural sciences today is probably a minority view and a fragile one. As so much of the popular mentality turns away from the traditions of science and valuing the human enterprise of trying to decode the world around us, science could well lose its central position in high culture.

We have largely left out of this account any explicit consideration of the discipline of history and the research enterprise that leads us to tell this story and not another. We have made this omission reluctantly but deliberately, in order to simplify the presentation for a particular audience, but the serious reader who has come this far needs to recognize that what we have said regarding science and technology in history comes to us not through some objective or final appeal to unquestioned realities of the past, but through the living interpretative medium of history as practiced by historians. The reader who consults works in the accompanying guide to resources will discover not truth and unanimity but a vibrant craft practice, lively debates concerning explanations of historical change, and sophisticated research undertaken to support or undermine one argument or another. We cannot therefore offer timeless conclusions for the edification of readers. Instead, we need to acknowledge the historiographical bases and biases of this account and the limitations inherent in all narrative.

As the study of the past, the discipline of history provides an unreliable guide to thinking about the future. The often-quoted aphorism of the philosopher George Santayana (1863–1952), "Those who cannot remember the past are condemned to repeat it," has its appeal both in suggesting that historical study can be useful and in warding off the notion that history is pursued by historians only for the pleasure of pure understanding. History is essential to understanding the present, of that we can be assured, but in the final analysis Santayana's commandment rings hollow because it posits that what happened in the past is applicable to changed circumstances in the future. The tendency is for commentators today to cherry pick examples from the past to support positions having little to do with understanding the past and much to do with current events. Does history itself give us lessons? Clio, the muse of history, is silent. History certainly does not cyclically repeat itself—the present differs from the past, the future will differ from the present, and, therefore, what we may learn of the past can be of only limited utility in understanding, much less affecting, the future.

Enlightenment thinkers of the eighteenth century first popularized the idea that there was a secular, progressive direction to history. The Enlightenment's view of human progress and social betterment predominated well into the second half of the twentieth century, and histories of science and technology were called on to demonstrate that progress. From today's critical perspectives, however, and from the long view encompassed in this study, the notion of progress is considerably less clear-cut or meaningful. Undoubtedly great change has lifted humanity to unprecedented heights. But how are we to define progress exactly? Is progress as we know it inevitable? Is it reversible? Is it sustainable? In particular, the Industrial Revolution and its consequences over the last three centuries have transformed historical circumstances so rapidly and in such profound ways that the current modes of intensified industrial existence seem unlikely to endure.

Guide to Resources

General Orientations

Given the nature of this book as an introduction to the histories of science and technology for the general reader, it was thought best not to clutter the pages with citations and references. Instead, we are providing this supplementary guide for those readers who may wish to delve further into the subject or who may seek clarification of some episode of the story. All Internet addresses were functional at the time of printing, but the Internet and the World Wide Web today are continuing to develop at a rapid pace, and information available over the Web is notoriously transient as well as variable in quality and reliability. (Roughly 50 percent of the sites listed in the 2006 edition of this book were still viable in 2015.) In addition, we think that these days, readers' first instincts would be to Google whatever might interest them from this book rather than typing in one of the URLs listed below. That circumstance, if true, is revealing of the current state of affairs. We offer these suggestions nonetheless in the hopes of their being useful, and in listing Web resources we can make a virtue out of our dilemma and underscore an analytical point by noting that future readers will find here a telling snapshot of the state of this technology in the second decade of the twenty-first century. Michael N. Geselowitz at the end of volume 7 of W. Bernard Carlson, ed., *Technology in World History* (New York: Oxford University Press, 2005) provides a similar list of recommended Websites, and the present guide to resources extracts from and updates this list as well.

The best general entree into the history of science remains David C. Lindberg and Ronald L. Numbers, general editors, *The Cambridge History of Science*, 8 vols. (Cambridge: Cambridge University Press, 2002–). An excellent resource, available in many libraries, that covers the history of science (and in some measure technology) through the 1970s is Charles C. Gillispie, ed., *Dictionary of Scientific Biography*, 16 vols. (New York: Scribner's, 1970–80). Other authoritative general sources include J. L. Heilbron, ed., *The Oxford Companion to the History of Modern Science* (Oxford: Oxford University Press, 2003); Helge Kragh, *An Introduction to the Historiography of Science* (Cambridge: Cambridge University Press, 1987); R. C. Olby, G. N. Cantor,

J.R.R. Christie, and A.M.S. Hodge, eds., *Companion to the History of Modern Science* (London: Routledge, 1990); William F. Bynum, ed., *Dictionary of the History of Science* (Princeton: Princeton University Press, 1985); Ivor Grattan-Guinness, ed., *Companion Encyclopedia of the History and Philosophy of the Mathematical Sciences*, 2 vols. (Baltimore: Johns Hopkins University Press, 2003); and William H. Brock, *The Fontana History of Chemistry* (London: HarperCollins, 1992). Two books written and edited by Andrew Ede and Lesley B. Cormack are noteworthy, *A History of Science in Society: A Reader* (Peterborough, Ont.: Broadview Press, 2007) and *A History of Science in Society: From Philosophy to Utility*, 2nd ed. (Toronto: University of Toronto Press, 2012). See also *ISIS, Official Journal of the History of Science Society* (USA) and its annual *Critical Bibliography*.

For science studies see Sheila Jasanoff, Gerald E. Markle, James C. Petersen, and Trevor Pinch, eds., *Handbook of Science and Technology Studies* (Thousand Oaks, Calif.: Sage Publications, 1995); Mario Biagioli, ed., *The Science Studies Reader* (New York and London: Routledge, 1999); and Steve Fuller; *The Philosophy of Science and Technology Studies* (New York: Routledge, 2006).

For the Western scientific tradition and more modern science, see Frederick Gregory, *Natural Science in Western History* (Boston: Houghton Mifflin, 2008); Todd Simmons, *Makers of Western Science: The Works and Words of 24 Visionaries from Copernicus to Watson and Crick* (Jefferson, N.C.: McFarland & Co., 2012); John Henry, *A Short History of Scientific Thought* (New York: Palgrave Macmillan, 2012); David C. Lindberg, *The Beginnings of Western Science*, 2nd ed. (Chicago: University of Chicago Press, 2007); and Edward Grant, *A History of Natural Philosophy from the Ancient World to the Nineteenth Century* (Cambridge: Cambridge University Press, 2007).

Helaine Selin, *Science across Cultures: An Annotated Bibliography of Books on Non-Western Science, Technology, and Medicine* (New York: Garland Publishing, 1992), is valuable for its global perspective; her *Encyclopedia of the History of Science, Technology, and Medicine in Non-Western Cultures* (Dordrecht: Kluwer Academic Publishers, 1997) remains highly recommended. John North, *The Norton History of Astronomy and Cosmology* (New York and London: W.W. Norton & Co., 1995), and Marcia Ascher, *Mathematics Elsewhere: An Exploration of Ideas Across Cultures* (Princeton: Princeton University Press, 2002), are similarly multicultural in perspective. Related sources include H. Floris Cohen, *How Modern Science Came into the World: Four Civilizations, One 17th-Century Breakthrough* (Amsterdam: Amsterdam University Press, 2010) and Patricia Fara, *Science: A Four Thousand Year History* (Oxford: Oxford University Press, 2010).

For technology and its history, see Donald Cardwell, *The Norton History of Technology* (New York: Norton, 1995); Thomas J. Misa, *Leonardo to the Internet: Technology and Culture from the Renaissance to the Present*, 2nd ed. (Baltimore: Johns Hopkins University Press, 2011); Daniel Headrick, *Technology: A World History* (Oxford and New York: Oxford University Press, 2009); Jolyon Goddard, *National Geographic Concise History of Science and Invention: An Illustrated Time Line* (Washington, D.C.: National Geographic, 2010); and Neil MacGregor, *A History of the World in 100 Objects* (New York: Viking, 2011). Providing a ready entry to technology in world cultures, the aforementioned W. Bernard Carlson, ed., *Technology in World History,* 7 vols. (New York: Oxford University Press, 2005) is a handsomely illustrated encyclopedia with explanatory chapters written by experts. See also the scholarly journal, *Technology and Culture: The International Quarterly of the Society for the History of Technology*; and Merritt Roe Smith and Leo Marx, eds., *Does Technology Drive History: The Dilemma of Technological Determinism* (Cambridge, Mass.: The MIT Press, 1994). A 2006 series of videos produced by A&E Television Networks, *Engineering an Empire*, brings to life ancient engineering in practice in a number of civilizations.

Two books by Jared Diamond, *Guns, Germs, and Steel: The Fates of Human Societies* (New York: W.W. Norton & Co, 1997) and *Collapse: How Societies Choose to Fail or Succeed* (New York: Viking, 2005), will be of interest to those taken with the themes of this book. Readers are encouraged to encounter the writings of Thomas S. Kuhn whose seminal work, *The Structure of Scientific Revolutions* (1962), redirected the understanding of science and its history; his work may also be approached through Kuhn, "The History of Science," in *The Essential Tension: Selected Studies in Scientific Tradition and Change* (Chicago: University of Chicago Press, 1977). Diederick Raven, Wolfgang Krohn, and Robert S. Cohen, eds., *Edgar Zilsel. The Social Origins of Modern Science* (Dordrecht/Boston: Kluwer Academic Publishers, 2000) opens up further dimensions to the historiography of science. The full development of the geographical thesis that recurs throughout this book will be found in Harold Dorn, *The Geography of Science* (Baltimore: Johns Hopkins University Press, 1991).

A not insubstantial "juvenile" literature treats the history of science and technology. The author Charlie Samuels, for example, has written a dozen short books for younger readers that explore the subjects, e.g., *Ancient Science: Prehistory—A.D. 500* or *Revolutions in Science: 1500–1700* (New York: Gareth Stevens Publishers, 2011–14). His books are multicultural and global in outlook, accurate, informative, and nicely illustrated, and they and like works deserve to be mentioned here for readers young and old.

Web Resources

Virtual Library. History: Science, Technology and Medicine:
    vlib.iue.it/history/topical/science.html
Society for Social Studies of Science (4S):
    www.4sonline.org
History of Science Society:
    www.hssonline.org
Society for the History of Technology (SHOT):
    www.shot.jhu.edu
STS Wiki: Department of Social Studies of Science, University of Vienna:
    http://www.stswiki.org
4000 Years of Women in Science:
    https://www.astr.ua.edu/4000WS
Center for History and New Media:
    http://chnm.gmu.edu
National Library of Medicine, History of Medicine:
    www.nlm.nih.gov/hmd

History of Mathematics:
    alepho.clarku.edu/~djoyce/mathhist/mathhist.html
    http://www.math.tamu.edu/~dallen/masters/hist_frame.htm
History of Early Mathematics:
    http://www.history.mcs.st-and.ac.uk/Indexes/HistoryTopics.html
HPSTM: A Selection of Web and Other Resources (Thomas Settle):
    www.imss.fi.it/~tsettle
Knowledge Web: A Project of the James Burke Institute
    www.k-web.org
Library of Congress:
    http:// www.loc.gov/rr/scitech/tracer-bullets/historyoftechtb.html
Museum of the History of Science, Oxford, U.K.:
    http://www.mhs.ox.ac.uk
National Museum of Industry and Science, London, U.K. (interactive):
    http://www.nmsi.ac.uk
    http://www.sciencemuseum.org.uk/online_science.aspx
Time Measurement (National Institute of Standards and Technology):
    http://physics.nist.gov/GenInt/Time/time.html

CHAPTER 1. Humankind Emerges

The evolutionary origin of human beings is inherently fascinating and is now
the subject of several well-written, semipopular books. The field of study has
made great strides over the past fifty years based on both a series of remark-
able fossil discoveries and developments in molecular biology. Ian Tattersall's
The Human Odyssey: Four Million Years of Human Evolution (New York:
Prentice Hall, 1993) and Jean Guilaine, ed., *Prehistory: The World of Early
Man* (New York: Facts on File, 1991), provide solid introductions. An account
of one of the most sensational discoveries is Donald Johanson and Maitland
Edey, *Lucy: The Beginnings of Humankind* (New York: Warner Books, 1981).
Prehistory—the Old and New Stone Ages—receives classic treatment in
V. Gordon Childe, *Man Makes Himself* (New York: NAL/Dutton, 1983). Al-
exander Marshack, *The Roots of Civilization* (Wakefield: Moyer Bell, 1992),
is a pioneering work on the earliest records of astronomical interest among
Old Stone Age populations. And a reliable survey of the entire field across the
world is Robert J. Wenke, *Patterns in Prehistory: Mankind's First Three Million Years*, 5th ed. (New York: Oxford University Press, 2007). The journal
Archaeoastronomy is a valuable, if specialized, source. Lydia V. Pyne and Stephen J. Pyne, *The Last Lost World: Ice Ages, Human Origins, and the Invention of the Pleistocene* (New York: Penguin Books, 2012) provides an updated
account of the science and the history of science of the Upper Paleolithic. See
also Brian Fagan, *People of the Earth: An Introduction to World Prehistory*,
14th ed. (Boston: Pearson, 2014) and Michael Chazan, *World Prehistory and
Archaeology: Pathways Through Time* (Boston: Pearson, 2014).

Web Resources

ArchNet: WWW Archaeology:
    http://ari.asu.edu/archnet
Human Evolution:
    www.archaeologyinfo.com/index.html

3-D Gallery of Human Ancestors:
   www.anth.ucsb.edu/projects/human
Stone Tool Technologies:
   http://humanorigins.si.edu/evidence/behavior/tools
Paleolithic art:
   www.paleolithicartmagazine.org
   http://www.britishmuseum.org/whats_on/past_exhibitions/2013/ice_age
      _art.aspx
The Caves of Lascaux:
   http://www.culture.fr/Actualites/Architecture-Patrimoine/Lascaux
   http://www.bradshawfoundation.com/lascaux/

CHAPTER 2. The Reign of the Farmer

The Neolithic era is mainly the province of the archeologist. An authoritative overview is Colin Renfrew and Paul Bahn, *Archaeology: Theories, Methods, and Practice* (New York: Thames & Hudson, 1991). Astronomy in the preliterate Neolithic is accessible only through its architectural monuments. Anthony Aveni, *World Archaeoastronomy* (Cambridge: Cambridge University Press, 1989), covers the subject extensively. On Stonehenge, see Jean-Pierre Mohen, *The World Megaliths* (New York: Facts on File, 1990) and Christopher Chippindale, *Stonehenge Complete* (New York: Thames & Hudson, 1983); Gerald S. Hawkins and John B. White, *Stonehenge Decoded* (New York: Delta, 1963), remains a valuable study. And Jo Anne Van Tilburg, *Easter Island: Archaeology, Ecology, and Culture* (Washington, D.C.: Smithsonian Institution Press, 1994), provides an authoritative introduction.

Web Resources

Easter Island:
   http://www.bradshawfoundation.com/easter
Stone Pages: A Guide to European Megaliths:
   www.stonepages.com
How the Shaman Stole the Moon by William H. Calvin:
   faculty.washington.edu/wcalvin/bk6
The Agricultural Revolution (Washington State University):
   http://richard-hooker.com/sites/worldcultures/AGRI/INDEX.HTM
Virtual Archaeology (links by John W. Hoopes):
   http://people.ku.edu/~hoopes
Marija Gimbutas—Life and Work (Pacifica Graduate Institute):
   http://pacifica.edu/innercontent-m.aspx?id=1762
The Origins of Agriculture:
   http://www.adbio.com/science/agri-history.htm
Stonehenge:
   http://www.christiaan.com/stonehenge/

CHAPTER 3. Pharaohs and Engineers

The advent of civilization saw the origin of the centralized state, large cities, and higher learning. It was the beginning of the "historical" era—the period that, for the first time, allows the historian to examine documents produced by human civilizations. An informative survey of early civilizations across the world is C. C. Lamberg-Karlovsky and Jeremy A. Sabloff, *Ancient Civilizations: The Near East and Mesoamerica* (Prospect Heights, Ill.: Waveland Press, 1987). A presentation of various theories that attempt to account for the origin of the earliest states is Ronald Cohen and Elman R. Service, eds., *Origins of the State* (Philadelphia: Institute for the Study of Human Issues, 1978). In a major study, Haicheng Wang, *Writing and the Ancient State: Early China in Comparative Perspective* (New York: Cambridge University Press, 2014), unpacks the nexus of writing, expertise, and centralizing authority in the history of all of the pristine civilizations. Karl Wittfogel, *Oriental Despotism* (New Haven: Yale University Press, 1957), is a learned and provocative, albeit idiosyncratic, approach to the subject. Monumental building in the centralized state was, in complexity, an order of magnitude beyond Neolithic structures: on the Egyptian case, see J.-P. Lepre, *The Egyptian Pyramids: A Comprehensive, Illustrated Reference* (Jefferson, N.C.: McFarland and Co., 1990); Zahi A. Hawass, *The Pyramids of Ancient Egypt* (Pittsburgh: Carnegie Museum of Natural History, 1990); and for an attempt by a physicist to explain the purposes of the ancient pharaohs in undertaking their great building projects, see Kurt Mendelssohn, *Riddle of the Pyramids* (New York: Thames & Hudson, 1986).

Regarding science, Otto Neugebauer, *The Exact Sciences in Antiquity*, 2nd ed. (New York: Dover Publications, 1969), provides an expert standard treatment of the subject. For a more detailed view of the Egyptian case, see Marshall Clagget, *Ancient Egyptian Science*, 2 vols. (Philadelphia: American Philosophical Society, 1989–95). And the reader who wishes to see original ancient documents colorfully reproduced should consult Gay Robins and Charles Shute, *The Rhind Mathematical Papyrus: An Ancient Egyptian Text* (New York: Dover Publications, 1987), and O.A.W. Dilke, *Mathematics and Measurement* (London: British Museum Publications, 1987); see also John Steele, ed., *Calendars and Years II: Astronomy and Time in the Ancient and Medieval World* (Oxford: Oxbow, 2011).

Web Resources

The Egyptian Pyramids (NOVA):
   www.pbs.org/wgbh/pages/nova/pyramid/
The Oriental Institute:
   www.oi.uchicago.edu
The Ancient Indus Valley:
   www.harappa.com/har/haro.html
Exploring Ancient World Cultures:
   eawc.evansville.edu/index.htm
Mesopotamia: Washington State University, World Civilizations:
   http://public.wsu.edu/~brians/world_civ/worldcivreader
Egyptology Resources (Newton Institute):
   www.newton.cam.ac.uk/egypt
The Ancient World Web:
   www.julen.net/ancient
Near-East and Middle East Archaeology:
   www.cyberpursuits.com/archeo/ne-arch.asp

Smith College Museum of Ancient Inventions:
   www.smith.edu/hsc/museum/ancient_inventions/hsclist.htm

CHAPTER 4. Greeks Bearing Gifts
CHAPTER 5. Alexandria and After

Greek science is the subject of many scholarly studies. G.E.R. Lloyd, *Early Greek Science: Thales to Aristotle* (New York: Norton, 1970) and his *Greek Science after Aristotle* (New York: Norton, 1973) provide a standard popular introduction. His *Adversaries and Authorities: Investigations into Ancient Greek and Chinese Science* (Cambridge: Cambridge University Press, 1996) is a landmark comparative study, as is Geoffrey Lloyd and Nathan Sivin, *The Way and the Word: Science and Medicine in Early China and Greece* (New Haven: Yale University Press, 2003). Still valuable is Marshall Clagett, *Greek Science in Antiquity*, rev. ed. (Mineola, N.Y.: Dover Publications, 2001; original edition, 1955). The reader who needs a more detailed survey will find it in George Sarton, *Ancient Science through the Golden Age of Greece* (New York: Dover, 1993) and *Hellenistic Science and Culture in the Last Three Centuries B.C.* (New York: Dover, 1993). James Evans, *The History and Practice of Ancient Astronomy* (New York and Oxford: Oxford University Press, 1998), presents a valuable, technical survey of what ancient astronomers actually did.

Joseph Ben-David, *The Scientist's Role in Society: A Comparative Study* (Chicago: University of Chicago Press, 1984), details a sociological approach to situating knowledge in antiquity and since. Lois N. Magner, *A History of the Life Sciences* (New York: Marcel Dekker, 1994), provides a summary of ancient views of matters biological. On technology in antiquity, see John W. Humphrey, John P. Oleson, and Andrew N. Sherwood, eds., *Greek and Roman Technology: A Sourcebook* (New York: Routledge, 2002); J. G. Landels, *Engineering in the Ancient World* (Berkeley: University of California Press, 1978); and L. Sprague de Camp, *The Ancient Engineers* (New York: Ballantine Books, 1988).

The volume edited by Jason König et al., *Ancient Libraries* (New York: Cambridge University Press, 2013), provides an illuminating, comprehensive, and up-to-date scholarly overview of libraries in the Greco-Roman world.

Web Resources

Ancient Greece Web Sites:
   http://www.bbc.co.uk/schools/primaryhistory/ancient_greeks
The Perseus Project:
   www.perseus.tufts.edu
The Internet Classics Archive:
   classics.mit.edu
Diotima: Women and Gender in the Ancient World:
   http://mcl.as.uky.edu/classics
   www.stoa.org/diotima
Hypatia of Alexandria:
   cosmopolis.com/people/hypatia.html
Classics and Mediterranean Archaeology Page:
   http://homes.chass.utoronto.ca/~kloppen/Classics.htm

Technology Museum of Thessaloniki: Ancient Greek Scientists:
  http://www.tmth.edu.gr/en/aet.html
Archimedes:
  http://www.mcs.drexel.edu/~crorres/Archimedes/contents.html
The Antikythera Machine:
  www.math.sunysb.edu/~tony/whatsnew/column/antikytheraI-0400/kyth1
  .html
Roman Catapult (The Discovery Channel):
  http://www.sciencechannel.com/tv-shows/what-the-ancients-knew/videos
  /what-the-ancients-knew-catapult-balista.htm
History of the Roman Empire:
  http://www.roman-empire.net/index.html

CHAPTER 6. The Enduring East

The reader who wishes to dip more deeply into Islamic science and civiliza-tion will find the following helpful: Jim Al-Khalili, *The House of Wisdom: How Arabic Science Saved Ancient Knowledge and Gave Us the Renaissance* (New York: Penguin Books, 2010); John Freely, *Aladdin's Lamp: How Greek Science Came to Europe Through the Islamic World* (New York: Alfred A. Knopf, 2009); Seyyed Hossein Nasr, *Science and Civilization in Islam* (New York: Barnes and Noble Books, 1992); George Saliba, *A History of Arabic Astronomy* (New York: New York University Press, 1994), *Islamic Science and the Making of the European Renaissance* (Cambridge, Mass.: MIT Press, 2011); Aydin Sayili, *The Observatory in Islam* (New York: Arno Press, 1981); Michael Adas, ed., *Islamic and European Expansion: The Forging of a Global Order* (Philadelphia: Temple University Press, 1993); and, for technology, Ahmad Y. al-Hassan and Donald R. Hill, *Islamic Technology: An Illustrated History* (Lanham: UNIPUB, 1992), and Donald R. Hill, *Islamic Science and Engineering* (Chicago: Kazi Publications, 1996); A. Mark Smith, *From Sight to Light: The Passage from Ancient to Modern Optics* (Chicago: University of Chicago Press, 2014). Rushdi Rashid is publishing a five-volume work, *A History of Arabic Sciences and Mathematics* (London and New York: Routledge, 2012–). Specialists will also appreciate Muzaffar Iqbal, *New Perspectives on the History of Islamic Science* (Burlington, Vt.: Ashgate, 2012). See also Fuat Sezgin, *Science and Technology in Islam* (Frankfurt am Main: Institute für Geschichte der Arabisch-Islamischen Wissenschaften an der Johann Wolf-gang Goethe-Universität, 2011); Michael Hamilton Morgan, *Lost History: The Enduring Legacy of Muslim Scientists, Thinkers, and Artists* (Washing-ton, D.C.: National Geographic, 2007). Of particular value in placing Islamic science in a comparative context is Toby E. Huff, *The Rise of Early Modern Science: Islam, China, and the West*, 2nd ed. (New York: Cambridge Univer-sity Press, 2003). Also of note is Ekmeleddin Ihsanoglu, *Science, Technology, and Learning in the Ottoman Empire* (Aldershot, Hampshire, UK: Ashgate, 2004).

Web Resources

Middle East Studies Internet Resources:
  http://files.eric.ed.gov/fulltext/EJ854295.pdf

Muslim Scientists and Scholars:
  www.ummah.net/history/scholars/
Early Islam:
  eawc.evansville.edu/ispage.htm
The Alchemy Website:
  www.levity.com/alchemy
Muslim Heritage:
  www.muslimheritage.com
Sassanid Empire:
  www.iranchamber.com/history/sassanids/sassanids.php
Byzantine Studies on the Internet:
  www.fordham.edu/halsall/byzantium

CHAPTER 7. The Middle Kingdom

No student of the history of science should neglect Joseph Needham's many-splendored volumes, *Science and Civilization in China*, 24 vols. to date (Cambridge: Cambridge University Press, 1954–). Needham's magisterial work is accessible in the abridgement by Colin A. Ronan, *The Shorter Science and Civilization in China: An Abridgement of Joseph Needham's Original Text*, 5 vols. (Cambridge: Cambridge University Press, 1978–95). Less formidable, but no less valuable, is Derk Bodde, *Chinese Thought, Society, and Science: The Intellectual and Social Background of Science and Technology in Pre-Modern China* (Honolulu: University of Hawaii Press, 1991). And for insightful comparative approaches see Huff, *Rise of Early Modern Science* (listed in chap. 6 above), G.E.R. Lloyd, *Adversaries and Authorities: Investigations into Ancient Greek and Chinese Science*, and Geoffrey Lloyd and Nathan Sivin, *The Way and the Word: Science and Medicine in Early China and Greece* (listed in chaps. 4–5 above). Dagmar Schäfer, *Cultures of Knowledge: Technology in Chinese History* (Leiden and Boston: Brill, 2012) is an important collection exploring the sociology and transmission of technical knowledge in traditional Chinese society. Also to be noted is R. Po-chia Hsia, *A Jesuit in the Forbidden City: Matteo Ricci, 1552–1610* (Oxford: Oxford University Press, 2010).

Web Resources

The Joseph Needham Research Institute:
  www.nri.org.uk/
Ancient China Web Sites:
  eawc.evansville.edu/www/chpage.htm
The China Page:
  www.chinapage.com/china-rm.html
History of China:
  www-chaos.umd.edu/history/toc.html
China and East Asian Chronology:
  http://www.metmuseum.org/toah/ht/index-east-asia.html
Condensed China:
  http://condensedchina.com
Chinese Philosophy Page (Links):
  http://www.iep.utm.edu/mod-chin

Oriental Medicine:
    http://www.orientalmedicine.com

CHAPTER 8. Indus, Ganges, and Beyond

Readings in the histories of science and technology in precolonial India are limited. Among the few surveys of the precolonial period are S. Balaachandra Rao, *Indian Mathematics and Astronomy* (Bangalore: Jnana Deep Publications, 1994); Debiprasad Chattopadhyaya, ed., *Studies in the History of Science in India*, 2 vols. (New Delhi: Editorial Enterprises, 1982); Irfan Habib, *Technology in Medieval India: c. 650–1750* (New Delhi: Tulika Books, 2008); and, in various editions with various titles and subtitles, O. P. Jaggi, *History of Science, Technology, and Medicine in India*, 15 vols. (Delhi: Atma Ram and Sons, 1969–86).

David Pingree, "History of Mathematical Astronomy in India" in *Dictionary of Scientific Biography*, ed. C. C. Gillispie (New York: Scribner's, 1978), 15:533–633, is specialized, but still essential reading, as is Toke Lindegaard Knudsen, *The Siddhantasundara of Jnanaraja: An English Translation with Commentary* (Baltimore: Johns Hopkins University Press, 2014), and B. S. Yadav and Man Mohan, eds., *Ancient Indian Leaps into Mathematics* (Birkhäuser, 2011). *The Indian Journal of History of Science* is another source. On the Khmer empire, see Eleanor Mannikka, *Angkor Wat: Time, Space, and Kingship* (Honolulu: University of Hawaii Press, 1996); Charles Higham, *The Civilization of Angkor* (London: Phoenix, 2003); Michael D. Coe, *Angkor and the Khmer Civilization* (London: Thames & Hudson, 2003).

Web Resources

Links Relating to Ancient India:
    eawc.evansville.edu/www/inpage.htm
India's History:
    www.geographia.com/india/india02.htm
History of India: Philosophy, Science, and Technology:
    http://www.indianscience.org
Ancient India:
    http://www.ancientindia.co.uk
Indian (Wootz) Steel:
    http://www.tms.org/pubs/journals/JOM/9809/Verhoeven-9809.html
What is Ayurveda?:
    www.ayur.com/about.html
Khmer Civilization:
    www.cambodia-travel.com/khmer-civilization.htm

CHAPTER 9. The New World

Ancient American civilizations have become active fields of research. See especially Jeremy A. Sabloff, *The Cities of Ancient Mexico* (London: Thames & Hudson, 1989) and his *The New Archaeology and the Ancient Maya* (New York: W. H. Freeman, 1990); Michael D. Coe, *The Maya*, 8th ed. (New York: Thames & Hudson, 2011); Brian M. Pagan, *Kingdoms of Gold, Kingdoms of*

Jade: The Americas Before Columbus (London: Thames & Hudson, 1991); Craig Morris and Adriana von Hagen, *The Inka Empire and Its Andean Origins* (New York: American Museum of Natural History/Abbeville Press, 1993); Michael E. Moseley, *The Incas and Their Ancestors* (London: Thames & Hudson, 1993); Craig Morris and Adriana von Hagen, *The Incas: Lords of the Four Quarters* (New York: Thames & Hudson, 2011). Linda Schele and Mary Ellen Miller, *The Blood of Kings* (New York: George Braziller, 1986), have contributed to a reinterpretation that now highlights aggression and violence in these civilizations that were formerly thought to be pacific.

On the Anasazi, see David Muench and Donald G. Pike, *Anasazi: Ancient People of the Rock* (New York: Harmony Books, 1974); Lynne Sebastian et al., eds., *The Chaco Anasazi: Sociopolitical Evolution in the Prehistoric Southwest* (Cambridge and New York: Cambridge University Press, 1996); and David E. Stuart, *Anasazi America: Seventeen Centuries on the Road from Center Place*, 2nd ed. Albuquerque: University of New Mexico Press, 2014).

On science and expertise in pre-Columbian America: Anthony F. Aveni, *Empires of Time: Calendars, Clocks, and Cultures* (New York: Basic Books, 1989); Bernard R. Ortiz de Montellano, *Aztec Medicine, Health, and Nutrition* (New Brunswick, N.J.: Rutgers University Press, 1990); Anna Sofaer and contributors to The Solstice Project, *Chaco Astronomy: An Ancient American Cosmology* (Santa Fe, New Mexico, 2014); William L. Fash, *Scribes, Warriors, and Kings: The City of Copan and the Ancient Maya* (London: Thames & Hudson, 1991); Maria Longhena, *Mayan Script: A Civilization and Its Writing* (New York: Abbeville Press, 2000); Michael D. Coe and Mark Van Stone, *Reading the Maya Glyphs* (London: Thames & Hudson, 2001).

Web Resources

Maya Links:
   www.ruf.rice.edu/~jchance/link.html
Mesoamerican Archaeology:
   http://www.angelfire.com/zine/meso
   http://pages.ucsd.edu/~dkjordan/arch/mexchron.html
The Maya Astronomy Page:
   www.michielb.nl/maya/astro.html
The Mayan Epigraphic Database Project:
   www.iath.virginia.edu/med/
NOVA Online: Ice Mummies of the Inca:
   http://www.pbs.org/wgbh/nova/icemummies
Chaco Canyon National Monument:
   www.nps.gov/chcu/home.htm
Cahokia Mounds Links:
   http://cahokiamounds.org
   http://whc.unesco.org/en/list/198

CHAPTER 10. Plows, Stirrups, Guns, and Plagues

Over the past fifty years the history of medieval and early modern European technology has been a flourishing field of research with the result that the reader who wishes to explore the subject further may choose from a wealth of publications. A classic work, Lynn White, Jr., *Medieval Technology and Social Change* (Oxford: Oxford University Press, 1966), interprets the rise of feudalism in terms of technological innovation; Arnold Pacey, *The Maze of Ingenuity: Ideas and Idealism in the Development of Technology*, 2nd ed. (Cambridge, Mass.: The MIT Press, 1992), focuses on the early modern period; Jean Gimpel, *The Medieval Machine: The Industrial Revolution of the Middle Ages* (New York: Penguin Books, 1983), remains valuable. And a readable survey is Frances and Joseph Gies, *Cathedral, Forge, and Waterwheel: Technology and Invention in the Middle Ages* (New York: HarperCollins, 1994); see also Roberta J. Magnusson, *Water Technology in the Middle Ages: Cities, Monasteries, and Waterworks after the Roman Empire* (Baltimore: Johns Hopkins University Press, 2001); Vibeke Olson, ed., *Working with Limestone: The Science, Technology, and Art of Medieval Limestone Monuments* (Burlington, Vt.: Ashgate, 2011).

Michael Roberts, *The Military Revolution, 1560–1660* (Belfast: Queen's University Press, 1956), opened a field of research on the military technologies of early modern Europe and their social consequences. This route has been followed by other historians including Geoffrey Parker, *The Military Revolution: Military Innovation and the Rise of the West, 1000–1800* (New York: Cambridge University Press, 1988); Carlo M. Cipolla, *Guns, Sails, and Empires: The Technological Innovation and the Early Phases of European Expansion, 1400–1700* (New York: Pantheon Books, 1966); and William H. McNeill, *The Pursuit of Power: Technology, Armed Force, and Society since A.D. 1000* (Chicago: University of Chicago Press, 1984). More recent works on this subject include Bert S. Hall, *Weapons and Warfare in Renaissance Europe: Gunpowder, Technology, and Tactics* (Baltimore: Johns Hopkins University Press, 2001); Alfred W. Crosby, *Throwing Fire: Projectile Technology through History* (Cambridge: Cambridge University Press, 2002); and Clifford J. Rogers, ed., *The Oxford Encyclopedia of Medieval Warfare and Military Technology* (New York: Oxford University Press, 2010).

On medieval science itself, see David C. Lindberg, ed., *Science in the Middle Ages* (Chicago: University of Chicago Press, 1978); Edward Grant, *The Nature of Natural Philosophy in the Late Middle Ages* (Washington, D.C.: Catholic University of America Press, 2010), *The Foundations of Modern Science in the Middle Ages* (New York: Cambridge University Press, 1996), and his earlier *Physical Science in the Middle Ages* (New York: Cambridge University Press, 1977); Hunt Janin, *The University in Medieval Life* (Jefferson, N.C.: McFarland & Co., 2008); Jeffrey R. Wigelsworth, *Science and Technology in Medieval European Life* (Westport, Conn.: Greenwood Press, 2006). John Freely, *Before Galileo: The Birth of Modern Science in Medieval Europe* (New York, London: Overlook Duckworth, 2012), is a more recent contribution.

Web Resources

Medieval Europe Web Sites:
   eawc.evansville.edu/www/mepage.htm
Medieval Crusades:
   www.medievalcrusades.com
The Black Death:
   http://www.eyewitnesstohistory.com/plague.htm

Medieval Science:

    https://explorable.com/middle-ages-science

Medieval Technology Pages:

    scholar.chem.nyu.edu/tekpages/Technology.html

    people.clemson.edu/~pammack/lec122/medag.htm

Gothic Architecture:

    http://www.elore.com/gothic.htm

Medieval Science:

    http://physics.ucr.edu/~wudka/Physics7/Notes_www/node40.html

    https://explorable.com/middle-ages-science

CHAPTERS 11–13. The Scientific Revolution—General

The story of the Scientific Revolution is one of the centerpieces of the history of science. H. Floris Cohen, *The Scientific Revolution: A Historiographical Inquiry* (Chicago: University of Chicago Press, 1994), provides a sweeping entree into the well-developed literature surrounding the Scientific Revolution. I. B. Cohen, *Puritanism and the Rise of Modern Science: The Merton Thesis* (New Brunswick, N.J.: Rutgers University Press, 1990), is another essential piece of historiography.

Among the more recent retellings and reinterpretations, see Wilbur Applebaum, ed., *Encyclopedia of the Scientific Revolution: From Copernicus to Newton* (New York: Garland Publishing, 2008); Margaret C. Jabob, *The Scientific Revolution: A Brief History with Documents* (Boston: Bedford/St. Martin's, 2010); Steven Shapin, *The Scientific Revolution* (Chicago: University of Chicago Press, 1996); David Knight, *Voyaging in Strange Seas: The Great Revolution in Science* (New Haven, Conn.: Yale University Press, 2014); John Henry, *The Scientific Revolution and the Origins of Modern Science*, 2nd ed. (Houndmills, Hampshire, UK: Palgrave Macmillan, 2001); David C. Lindberg and Robert S. Westman, eds., *Reappraisals of the Scientific Revolution* (Cambridge: Cambridge University Press, 1990). Roy Porter and Mikuláš Teich, eds., *The Scientific Revolution in National Context* (Cambridge: Cambridge University Press, 1992). Also useful is Norriss S. Hetherington, *Cosmology: Historical, Literary, Philosophical, Religious, and Scientific Perspectives* (New York: Garland Publishing, 1993). Older, but still excellent, accounts are Arthur Koestler, *The Sleepwalkers: A History of Man's Changing Vision of the Universe* (New York: Viking Penguin, 1990); Herbert Butterfield, *The Origins of Modern Science, 1300–1800*, rev. ed. (New York: Free Press, 1965); I. Bernard Cohen, *The Birth of a New Physics, revised and updated* (New York: Norton, 1985); and Richard S. Westfall, *The Construction of Modern Science: Mechanism and Mechanics*, 2nd ed. (Cambridge: Cambridge University Press, 1977).

Other works that have recast the interpretative context for the Scientific Revolution include William Eamon, *Science and the Secrets of Nature: Books of Secrets in Medieval and Early Modern Culture* (Princeton, N.J.: Princeton University Press, 1994); Pamela H. Smith, *The Business of Alchemy: Science and Culture in the Holy Roman Empire* (Princeton, N.J.: Princeton University Press, 1994); Pamela O. Long, *Artisan/Practitioners and the Rise of the New Sciences, 1400–1600* (Corvallis, Ore.: Oregon State University Press, 2011); Ofer Gal and Raz Chen-Morris, *Baroque Science* (Chicago and London: University of Chicago Press, 2013); David C. Goodman, *Power and Penury: Government, Technology, and Science in Philip II's Spain* (Cambridge: Cambridge University Press, 1988); Frank J. Swetz, *Capitalism and Arithmetic: The New Math of the 15th Century* (LaSalle, Ill.: Open Court, 1987); Carolyn Merchant, *The Death of Nature: Women, Ecology, and the Scientific Revolution* (San Francisco: HarperSanFrancisco, 1990).

See also Toby E. Huff, *Intellectual Curiosity and the Scientific Revolution: A Global Perspective* (Cambridge and New York: Cambridge University Press, 2011); Dan Falk, *The Science of Shakespeare: A New Look at the Playwright's Universe* (New York: St. Martin's Press, 2014); Mordechai Feingold and Victor Navarro-Brotons, *Universities and Science in the Early Modern Period* (Dordrecht: Springer, 2006).

CHAPTER 11. Copernicus Incites a Revolution

On Copernicus, Tycho, and Kepler, see Owen Gingerich, *The Book That Nobody Read: Chasing the Revolutions of Nicolaus Copernicus* (New York: Penguin, 2005); Bruce Stephenson, *Kepler's Physical Astronomy* (Princeton, N.J.: Princeton University Press, 1994), and his companion *The Music of the Heavens: Kepler's Harmonic Astronomy* (Princeton, N.J.: Princeton University Press, 1994); Max Caspar, *Kepler*, trans. and ed. C. Doris Hellman with a new introduction and references by Owen Gingerich (New York: Dover Publications, 1993); James A. Connor, *Kepler's Witch: An Astronomer's Discovery of Cosmic Order Amid Religious War, Political Intrigue, and the Heresy Trial of His Mother* (New York: HarperCollins, 2004); Victor E. Thoren, *The Lord of Uraniborg: A Biography of Tycho Brahe* (Cambridge: Cambridge University Press, 1990). Thomas S. Kuhn, *The Copernican Revolution* (Cambridge: Harvard University Press, 1957), remains insightful.

More recent works include André Goddu, *Copernicus and the Aristotelian Tradition: Education, Reading, and Philosophy in Copernicus's Path to Heliocentrism* (Leiden and Boston: Brill, 2010); Robert S. Westman, *The Copernican Question: Prognostication, Skepticism, and Celestial Order* (Berkeley: University of California Press, 2011); Arthur Mazer, *Shifting the Earth: The Mathematical Quest to Understand the Motion of the Universe* (Hoboken, N.J.: Wiley, 2011); Dava Sobel, *A More Perfect Heaven: How Copernicus Revolutionized the Cosmos* (New York: Walker, 2011).

Web Resources

The Scientific Revolution (Links):
    www.historyteacher.net/APEuroCourse/WebLinks/WebLinks
      -ScientificRevolution.htm
Museum of the History of Science (Oxford):
    www.mhs.ox.ac.uk
De Revolutionibus Orbium Coelestium:
    webexhibits.org/calendars/year-text-Copernicus.html
Tycho Brahe:
    www.tychobrahe.com/uk/om_museet.html
Museum of Science: Leonardo:
    www.mos.org/leonardo

CHAPTER 12. The Crime and Punishment of Galileo Galilei

Galileo's scientific works are among the few in the history of science that can be read and enjoyed by the general reader. Stillman Drake, *Discoveries and Opinions of Galileo* (Garden City, N.Y.: Doubleday Anchor, 1990), provides a sampler. Also available are Galileo's major publications: *Galileo on the World Systems: A New Abridged Translation and Guide*, trans. and ed. Maurice A. Finocchiaro (Berkeley: University of California Press, 1997) and *Two New Sciences*, trans. S. Drake (Madison: University of Wisconsin Press, 1992); the former is also available as *Dialogue Concerning the Two Chief World Systems*, trans. S. Drake (Berkeley: University of California Press, 1967). For a discussion of Galileo's scientific work, Stillman Drake, *Galileo at Work: His Scientific Biography* (New York: Dover Publications, 1995) is the standard source. See also his *Galileo: A Very Short Introduction* (New York: Sterling Pub., 2009). A recent biography is David Wootton, *Galileo: Watcher of the Skies* (New Haven: Yale University Press, 2010).

Maurice Finocchiaro has produced an impressive body of work regarding Galileo. *The Galileo Affair: A Documentary History* (Berkeley: University of California Press, 1989), provides an essential starting point for Galileo's persecution by the Inquisition; see also Finocchiaro's *Retrying Galileo, 1633–1992* (Berkeley and Los Angeles: University of California Press, 2005); *Defending Copernicus and Galileo: Critical Reasoning in the Two Affairs* (Dordrecht and New York: Springer, 2010); and *The Routledge Guidebook to Galileo's Dialogue* (New York: Routledge–Taylor and Francis, 2014). Eileen Reeves, *Galileo's Glassworks: The Telescope and the Mirror* (Cambridge, Mass.: Harvard University Press, 2008) is to be consulted.

Thomas F. Mayer, ed., *The Trial of Galileo, 1612–1633* (North York, Ont.: University of Toronto Press, 2012) is a recent update. Georgio de Santillana, *The Crime of Galileo* (Alexandria, Va.: Time-Life Books, 1981), remains a valuable account, as do Mario Biagioli, *Galileo Courtier: The Practice of Science in the Culture of Absolutism* (Chicago: University of Chicago Press, 1993); Pietro Redondi, *Galileo Heretic* (Princeton, N.J.: Princeton University Press, 1987); and Michael Segré, *In the Wake of Galileo* (New Brunswick, N.J.: Rutgers University Press, 1991). Dava Sobel, *Galileo's Daughter: A Historical Memoir of Science, Faith, and Love* (New York: Penguin, 2000) is a notable addition to the Galileo literature. Bertolt Brecht's play, *Life of Galileo* (New York: Penguin Books, 2008) remains a classic in the Galileo literature.

Web Resources

The Galileo Project:
   es.rice.edu/ES/humsoc/Galileo
Museum of the History of Science, Florence:
   http://www.museogalileo.it/en/index.html
"Galileo's Battle for the Heavens":
   www.pbs.org/nova/galileo
Jesuit Scientists:
   http://www.faculty.fairfield.edu/jmac/sjscient.htm
   http://innovation.ucdavis.edu/people/publications/Biagioli_Texts
      _Contexts_Jesuits.pdf

Descartes, *Discourse on Method*:
   www.literature.org/authors/descartes-rene/reason-discourse
   http://www.earlymoderntexts.com/pdfs/descartes1637.pdf

CHAPTER 13. "God said, '*Let Newton be!*'"

The science and career of Isaac Newton have been the subjects of a major scholarly industry. Richard S. Westfall, *Never at Rest: A Biography of Isaac Newton* (Cambridge: Cambridge University Press, 1983), a masterpiece of scientific biography, remains the essential starting point; see also its abridgement, *The Life of Isaac Newton* (Cambridge: Cambridge University Press, 1993). A. Rupert Hall, *Isaac Newton, Adventurer in Thought* (Oxford: Blackwell, 1992); Rob Iliffe, *Newton: A Very Short Introduction* (Oxford and New York: Oxford University Press, 2007); and James Gleick, *Isaac Newton* (New York: Vintage, 2004) also provide standard accounts. I. Bernard Cohen and Richard S. Westfall, eds., *Newton: Texts, Backgrounds, Commentaries: A Norton Critical Edition* (New York: W. W. Norton, 1995), is a similarly important introductory source. Mordechai Feingold, *The Newtonian Moment: Isaac Newton and the Making of Modern Culture* (New York and Oxford: The New York Public Library/Oxford University Press, 2004) is a handsome, accessible presentation of Newton's life, work, and influence. John Fauvel et al., *Let Newton Be!* (Oxford: Oxford University Press, 1988), is another engaging, popular work to the same ends. Patricia Fara, *Newton: The Making of Genius* (New York: Columbia University Press, 2002) inventively traces Newton's changing reputation over time. B.J.T. Dobbs, *The Janus Faces of Genius: The Role of Alchemy in Newton's Thought* (Cambridge: Cambridge University Press, 1991), provides a pathbreaking scholarly study of Newton's private science. I. B. Cohen and Anne Whitman's *Isaac Newton: The Principia: Mathematical Principles of Natural Philosophy* presents a new translation of Newton's *Principia* (Berkeley, Los Angeles, and London: University of California Press, 1999); this landmark work is now the standard in English, and it includes Cohen's 370-page "Guide to Newton's Principia." Colin Pask, *Magnificent Principia: Exploring Isaac Newton's Masterpiece* (Amherst, N.Y.: Prometheus Books, 2013) likewise provides an in-depth study of this great work. See also Jed Z. Buchwald and I. Bernard Cohen, eds., *Isaac Newton's Natural Philosophy* (Cambridge, Mass.: MIT Press, 2001); Jed Z. Buchwald and Mordechai Feingold, *Newton and the Origin of Civilization* (Princeton: Princeton University Press, 2013).

For the general historical context, see Margaret C. Jacob, *The Newtonians and the English Revolution* (New York: Gordon and Breach, 1990). Betty Jo Teeter Dobbs and Margaret C. Jacob, *Newton and the Culture of Newtonianism* (Atlantic Highlands, N.J.: Humanities Press, 1995), provide a ready introduction to a broad range of themes. The study by Sarah Dry, *The Newton Papers: The Strange and True Odyssey of Isaac Newton's Manuscripts* (Oxford and New York: Oxford University Press, 2014) tells an important story. For more background and for one of Newton's chief contemporaries and competitors, see Lisa Jardine, *The Curious Life of Robert Hooke: The Man Who Measured London* (New York: HarperCollins, 2003). See also Jardine's *On a Grander Scale: The Outstanding Career of Sir Christopher Wren* (London:

HarperCollins, 2002) and Michael Hunter, *Boyle: Between God and Science* (New Haven, Conn.: Yale University Press, 2009). For Newton's immediate social and cultural impact, see Dorinda Outram, *The Enlightenment*, 3rd ed. (Cambridge and New York: Cambridge University Press, 2013) and sources cited below, chapter 15.

Web Resources

The Newton Project:
   www.newtonproject.sussex.ac.uk
The Newton Project Canada:
   www.isaacnewton.ca
Newton Biographies:
   www.newton.cam.ac.uk/newtlife.html
   galileoandeinstein.physics.virginia.edu/lectures/newton.html
The Alchemy Virtual Library:
   www.levity.com/alchemy/home.html

CHAPTER 14. Textiles, Timber, Coal, and Steam

Studies of the Industrial Revolution and its global impact have become more sophisticated over the years. See George Basalla, *The Evolution of Technology* (Cambridge: Cambridge University Press, 1989); Arnold Pacey, *Technology in World Civilization* (Cambridge: The MIT Press, 1992.); Peter N. Stearns, *The Industrial Revolution in World History*, 4th ed. (Boulder, Colo.: Westview Press, 2013); *The ABC-CLIO World History Companion to the Industrial Revolution* (Santa Barbara, Calif.: ABC-CLIO, 1996); Vaclav Smil, *Energy in World History* (Boulder, Colo.: Westview Press, 1994); Alfred Crosby, *Ecological Imperialism: The Biological Expansion of Europe, 900–1900* (New York: Cambridge University Press, 1993); Daniel Headrick, *Tools of Empire: Technology and European Imperialism in the Nineteenth Century* (New York: Oxford University Press, 1981). See also Wolfgang Schivelbusch, *The Railway Journey: The Industrialization of Time and Space in the Nineteenth Century*, with a new preface (Berkeley, Calif.: University of California Press, 2014); T. S. Ashton, *The Industrial Revolution, 1760–1830*, with a New Preface and Bibliography by Pat Hudson (Oxford: Oxford University Press, 1998); E. J. Hobsbawm, *Industry and Empire: The Birth of the Industrial Revolution* (New York: The New Press, 1999). E. P. Thompson, *The Making of the English Working Class* (New York: Pantheon Books, 1964) remains a landmark in the literature of the Industrial Revolution. David Landes, *The Unbound Prometheus: Technological Change and Industrial Development in Western Europe, 1750 to the Present*, 2nd ed. (Cambridge: Cambridge University Press, 2003) is an updated edition of a classic account. On the cultural impact of science on the processes of industrialization, see Margaret C. Jacob, *Scientific Culture and the Making of the Industrial West* (New York: Oxford University Press, 1997); see also her *The First Knowledge Economy: Human Capital and the European Economy, 1750–1850* (Cambridge: Cambridge University Press, 2014), and Joel Mokyr, *The Gifts of Athena: Historical Origins of the Knowledge Economy* (Princeton: Princeton University Press, 2002).

Among the newer literature, see Christine Rider, ed., *Encyclopedia of the*

Age of the Industrial Revolution, 1700–1920 (Westport, Conn.: Greenwood Press, 2007); Jeff Horn, Leonard N. Rosenband, and Merritt Roe Smith, eds., *Reconceptualizing the Industrial Revolution* (Cambridge, Mass.: MIT Press, 2010); Emma Griffin, *Liberty's Dawn: A People's History of the Industrial Revolution* (New Haven: Yale University Press, 2013); Robert C. Allen, *The British Industrial Revolution in Global Perspective* (Cambridge and New York: Cambridge University Press, 2009); David Philip Miller, *James Watt, Chemist: Understanding the Origins of the Steam Age* (London: Pickering & Chatto, 2009); Peter M. Jones, *Industrial Enlightenment: Science, Technology, and Culture in Birmingham and the West Midlands, 1760–1820* (Manchester and New York: Manchester University Press, 2008); Carroll W. Pursell, *The Machine in America: A Social History of Technology*, 2nd ed. (Baltimore: Johns Hopkins University Press, 2007); Michael Andrew Zmolek, *Rethinking the Industrial Revolution: Five Centuries of Transition from Agrarian to Industrial Capitalism in England* (Leiden; Boston: Brill, 2013).

Web Resources

Library of Congress: Economic History:
   http://www.loc.gov/teachers/additionalresources/relatedresources/ushist
      /special/economic.html
Railroad History:
   www.rrhistorical.com/index.html
National Railway Museum:
   http://www.nrm.org.uk
Isambard Kingdom Brunel (1806–1859):
   http://www.brunel.ac.uk/about/history/isambard-kingdom-brunel
Internet Modern History Source Book:
   www.fordham.edu/halsall/mod/modsbook.html
Timeline of Modern Agriculture:
   http://www.ars.usda.gov/is/timeline/comp.htm
American Textile History Museum:
   http://www.athm.org

CHAPTER 15. Legacies of Revolution: Newton to Einstein

The interpretation put forth in this chapter takes off from Thomas S. Kuhn, "Mathematical versus Experimental Traditions in the Development of Physical Science," in *The Essential Tension: Selected Studies in Scientific Tradition and Change* (Chicago: University of Chicago Press, 1977). A related view of post-Newtonian science is presented in I. Bernard Cohen, *Revolution in Science* (Cambridge: Belknap Press of Harvard University Press, 1985); see also his earlier *Franklin and Newton: An Inquiry into Speculative Newtonian Experimental Science* (Philadelphia: American Philosophical Society, 1956), and Mary Terrall, *The Man Who Flattened the Earth: Maupertuis and the Sciences in the Enlightenment* (Chicago: University of Chicago Press, 2002). For an entree into the intricacies of nineteenth-century science, see Christa Jungnickel and Russell McCormmach, *Intellectual Mastery of Nature*, 2. vols. (Chicago: University of Chicago Press, 1990); Crosbie Smith and M. Norton Wise, *Energy and Empire: William Thomson, Lord Kelvin, 1824–1907* (Cambridge: Cambridge University Press, 1989); Ursula DeYoung, *A Vision of Modern Science: John Tyndall and the Role of the Scientist in Victorian Culture* (New York: Palgrave Macmillan, 2011); Robert H. Kargon, *Science in Victorian Manchester: Enterprise and Expertise* (New Brunswick, N.J.: Transaction Publishers, 2010); Bruce J. Hunt, *Pursuing Power and Light: Technology and Physics from James Watt to Albert Einstein* (Baltimore: Johns Hopkins University Press, 2010); Stephen G. Brush, *A History of Modern Science: A Guide to the Second Scientific Revolution, 1800–1950* (Ames: Iowa State University Press, 1988); still valuable is Edmund Whittaker, *A History of the Theories of Aether and Electricity* (New York: Dover Publications, 1989). Lewis Pyenson and Susan Sheets-Pyenson, *Servants of Nature: A History of Scientific Institutions, Enterprises, and Sensibilities* (New York and London: W. W. Norton, 2000) traces the social history of science in the period. See also David M. Knight, *The Making of Modern Science: Science, Technology, Medicine, and Modernity, 1789–1914* (Cambridge, UK; Malden, Mass.: Polity Press, 2009).

Among other related works see George N. Vlahakis et al., *Imperialism and Science: Social Impact and Interaction* (Santa Barbara, Calif.: ABC-CLIO, 2006); John Tresch, *The Romantic Machine: Utopian Science and Technology after Napoleon* (Chicago and London: University of Chicago Press, 2012); Christine MacLeod, *Heros of Invention: Technology, Liberalism, and British Identity, 1750–1914* (Cambridge and New York: Cambridge University Press, 2007); Ross Thomson, *Structures of Change in the Mechanical Age: Technological Innovation in the United States, 1790–1865* (Baltimore: Johns Hopkins University Press, 2009); James Essinger, *A Female Genius: How Ada Lovelace, Lord Byron's Daughter, Started the Computer Age* (London: Gibson Square, 2014); Renée L. Bergland, *Maria Mitchell and the Sexing of Science: An Astronomer Among the American Romantics* (Boston: Beacon Press, 2008). David N. Livingstone and Charles W. J. Withers, eds., *Geographies of Nineteenth-Century Science* (Chicago and London: University of Chicago Press, 2011).

More recent works touching on science and technology across the eighteenth and nineteenth centuries include Tom Shachtman, *Gentlemen Scientists and Revolutionaries: The Founding Fathers in the Age of Enlightenment* (New York: Palgrave Macmillan, 2014); Keith Steward Thomson, *Jefferson's Shadow: The Story of His Science* (New Haven: Yale University Press, 2012); Jim A. Bennett and Sofia Talas, *Cabinets of Experimental Philosophy in Eighteenth-Century Europe* (Leiben: Brill, 2013); Stephen Gaukroger, *The Collapse of Mechanism and the Rise of Sensibility: Science and the Shaping of Modernity, 1680–1760* (Oxford: Clarendon Press, 2010); Richard Holmes, *The Age of Wonder: How the Romantic Generation Discovered the Beauty and Terror of Science* (New York: Vintage Books, 2010); Larrie D. Ferreiro, *Ships and Science: The Birth of Naval Architecture in the Scientific Revolution, 1600–1800* (Cambridge, Mass.: MIT Press, 2007); Christoph Irmscher, *Louis Agassiz: Creator of American Science* (Boston: Houghton Mifflin Harcourt, 2013); Robert Fox, *The Savant and the State: Science and Cultural Politics in Nineteenth-Century France* (Baltimore: Johns Hopkins University Press, 2012).

Web Resources

Chemical Heritage Foundation:
  www.chemheritage.org
  http://www.chemheritage.org/discover/online-resources/chemistry-in
    -history
Panopticon Lavoisier:
  moro.imss.fi.it/lavoisier/main.asp
Victorian Science (Links):
  http://www.victorianweb.org/science/index.html
Famous Physicists:
  cnr2.kent.edu/~manley/physicists.html
The Royal Institution, Heritage:
  http://www.rigb.org/about/heritage-and-collections
A Concise History of Thermodynamics:
  www.thermohistory.com

CHAPTER 16. Life Itself

As in the case of Galileo some of Darwin's writings are eminently readable, especially *The Origin of Species*, which is widely available in various imprints. Janet Browne treats the book itself in her *Darwin's Origin of Species: A Biography* (New York: Grove Press, 2007); see also Michael Ruse and Robert J. Richards, eds., *The Cambridge Companion to the "Origin of Species"* (Cambridge: Cambridge University Press, 2009). Janet Browne is the accomplished author of the now standard biography of Darwin in two volumes: *Charles Darwin: Voyaging* (Princeton, N.J.: Princeton University Press, 1996) and *Charles Darwin: The Power of Place* (London: Jonathan Cape, 2002). Other of Darwin's writings are accessible in Philip Appleman, ed., *Darwin: A Norton Critical Edition*, 3rd ed. (New York: Norton, 2001) and Thomas E. Glick and David Kohn, eds., *Darwin on Evolution* (Indianapolis, Ind.: Hackett Publishing, 1996). Other starting points include Michael Ruse, ed., *The Cambridge Encyclopedia of Darwin and Evolutionary Thought* (Cambridge: Cambridge University Press, 2013); Ronald K. Wetherington, *Readings in the History of Evolutionary Theory: Selections from Primary Sources* (New York: Oxford University Press, 2012); and Russell Re Manning, ed., *The Oxford Handbook of Natural Theology* (Oxford: Oxford University Press, 2013). Peter J. Bowler's work, especially his *Charles Darwin: The Man and His Influence* (Cambridge: Cambridge University Press, 1996) and *Evolution: The History of an Idea*, rev. ed. (Berkeley: University of California Press, 1989), provides solid historical overviews to Darwin and the development of evolutionary theory as well as another entree into the prodigious Darwin scholarship. The reader who requires assistance with the biological principles and their development in the nineteenth and twentieth centuries will find it in Maitland A. Edey and Donald C. Johanson, *Blueprints: Solving the Mystery of Evolution* (New York: Penguin Books, 1989). David Kohn, ed., *The Darwinian Heritage* (Princeton, N.J.: Princeton University Press, 1985), is a major collection of essays on the subject. The principle of natural selection receives a spirited presentation in Richard Dawkins, *The Blind Watchmaker: Why the Evidence of Evolution Reveals a Universe Without Design* (New York: Norton, 1996); similar but more philosophical is Daniel C. Dennet's *Darwin's Dangerous Idea* (New York: Simon and Schuster, 1995). Jonathan Weiner, *The Beak of the Finch: A Story of Evolution in Our Time* (New York: Knopf, 1994), provides a gracefully written account of evolution in action. Carl N. Degler, *In Search of Human Nature: The Decline and Revival of Darwinism in American Social Thought* (New York: Oxford University Press, 1991), presents a thorough narrative of Darwinian controversies in the social sciences. And Edward O. Wilson, *On Human Nature* (Cambridge, Mass.: Harvard University Press, 1978), is a popular discussion of sociobiology, one of the controversial outgrowths of evolutionary theory.

Web Resources

Darwin and Darwinianism (Links):
  www.human-nature.com/darwin
  darwin.baruch.cuny.edu/index.html
  www.aboutdarwin.com
Darwin Online:
  http://darwin-online.org.uk
Darwin Manuscripts Project:
  http://www.amnh.org/our-research/darwin-manuscripts-project
On the Origin of Species:
  www.bartleby.com/11
Voyage of the Beagle:
  www.bartleby.com/29
Enter Evolution: Theory and History:
  www.ucmp.berkeley.edu/history/evolution.html
Darwin—American Museum of Natural History:
  http://www.amnh.org/exhibitions/past-exhibitions/darwin
Evolution: PBS/NOVA:
  http://www.pbs.org/wgbh/evolution
Darwin's Tree of Life (Natural History Museum, U.K.):
  http://www.nhm.ac.uk/nature-online/evolution/tree-of-life/darwin-tree

CHAPTER 17. Toolmakers Take Command

The literature on 20th- and 21st-century technologies treated in this chapter is substantial. Studies include David Edgerton, *The Shock of the Old: Technology and Global History since 1900* (Oxford and New York: Oxford University Press, 2007) and the key works by Thomas P. Hughes, *American Genesis: A Century of Invention and Technological Enthusiasm, 1870–1970* (New York: Viking, 1989) and his *Networks of Power: Electrification in Western Society, 1880–1930* (Baltimore: Johns Hopkins University Press, 1983); Christopher Cumo, *Science and Technology in 20th-Century America* (Westport, Conn.: Greenwood Press, 2007); Mark H. Rose, *Cities of Light and Heat: Domesticating Gas and Electricity in Urban America* (University Park, Penn.: Penn State University Press, 1995); David E. Brown, *Inventing Modern America: From the Microwave to the Mouse* (Cambridge, Mass., and London: MIT Press, 2003); Douglas Brinkley, *Wheels for the World: Henry Ford, His Company, and a Century of Progress, 1903–2003* (New York: Penguin, 2004);

works by Paul Israel, including his *Edison: A Life of Invention* (John Wiley & Sons, 1998); Robert Friedel and Paul Israel with Bernard S. Finn, *Edison's Electric Light: The Art of Invention*, revised and updated edition (Baltimore: Johns Hopkins University Press, 2010); W. Bernard Carlson, *Tesla: Inventor of the Electrical Age* (Princeton, N.J.: Princeton University Press, 2013); Matthew Josephson, *Edison: A Biography* (New York: Wiley, 1992); Jonathan Metcalf, ed., *Flight: 100 Years of Aviation* (London: DK ADULT, 2002); Robert E. Gallamore and John R. Meyer, *American Railroads: Decline and Renaissance in the Twentieth Century* (Cambridge, Mass.: Harvard University Press, 2014); James J. Flink, *The Automobile Age* (Cambridge, Mass.: MIT Press, 1990); Susan Strasser, *Waste and Want: A Social History of Trash* (New York: Metropolitan Books/Henry Holt and Co., 1999); Michelle Hilmes and James Loviglio, eds., *Radio Reader: Essays in the Cultural History of Radio* (New York and London: Routledge, 2001); Geoffrey Nowell-Smith, ed., *The Oxford History of World Cinema* (Oxford: Oxford University Press, 1999); James Lastra, *Sound Technology and the American Cinema* (New York: Columbia University Press, 2000); Kenneth Bilby, *The General: David Sarnoff and the Rise of the Communications Industry* (New York: HarperCollins, 1986); Daniel Stashower, *The Boy Genius and the Mogul: The Untold Story of Television* (New York: Broadway Books, 2002). On globalization, see Thomas L. Friedman, *The Lexus and the Olive Tree: Understanding Globalization*, updated and expanded ed. (New York: Anchor Books, 2000) and his *The World Is Flat: A Brief History of the Twenty-First Century* (New York: Farrar, Straus and Giroux, 2005). Eric Chaline, *Fifty Machines That Changed the Course of History* (Buffalo, N.Y.: Firefly Books, 2012), confines its treatment to notable nineteenth- and twentieth-century machines.

See also the pathbreaking works by Ruth Schwartz Cowan, *A Social History of American Technology* (New York: Oxford University Press, 1997) and *More Work for Mother: The Ironies of Household Technology from the Open Hearth to the Microwave* (New York: Basic Books, 1985); Carolyn M. Goldstein, *Creating Consumers: Home Economists in Twentieth-Century America* (Chapel Hill, N.C.: University of North Carolina Press, 2012); Nellie Oudshoorn and Trevor Pinch, eds., *How Users Matter: The Co-Construction of Users and Technology* (Cambridge, Mass.: MIT Press, 2005); Nina E. Lehrman, Ruth Oldenziel, and Arwen P. Mohun, *Gender and Technology: A Reader* (Baltimore: Johns Hopkins University Press, 2003); James Essinger, *Jacquard's Web: How a Hand-loom Led to the Birth of the Information Age* (New York: Oxford University Press, 2004).

Web Resources

Smithsonian Institution, Transportation History:
   http://amhistory.si.edu/onthemove
History of Radio:
   http://transition.fcc.gov/omd/history/radio/documents/short_history.pdf
Spark Museum of Electrical Invention:
   http://www.sparkmuseum.org/collections/dawn-of-the-electrical-age
   -(1600-1800)
IEEE Virtual Museum:
   www.ieee-virtual-museum.org

IEEE Global History Network:
 http://www.ieeeghn.org/wiki/index.php/Special:Home
IEEE History Center:
 www.ieee.org/organizations/history_center
Henry Ford Museum and Greenfield Village:
 http://www.hfmgv.org
Ford Motor Company Heritage:
 www.ford.com/en/heritage
Household Technologies (Library of Congress):
 http://www.loc.gov/rr/scitech/tracer-bullets/householdtb.html
Modern Transportation (National Museum of American History):
 http://amhistory.si.edu/onthemove/themes/story_47_1.html
The Aviation History Online Museum:
 www.aviation-history.com
History of Flight Timeline (American Institute of Aeronautics and Astronautics—AIAA):
 https://www.aiaa.org/HistoryTimeline/
Military Technologies:
 www.aeragon.com/01/index.html
National Inventors Hall of Fame:
 www.invent.org
The History of Sound Recording Technology:
 http://www.recording-history.org
 http://www.recording-history.org/HTML/tech.php
Television History: The First 75 Years:
 www.tvhistory.tv
The Globalization Website:
 www.sociology.emory.edu/globalization
U.S. Energy Information Administration:
 www.eia.gov
International Energy Agency:
 www.iea.org
World Bank: World Development Indicators Database
 www.worldbank.org/data/countrydata/countrydata.html
World Fact Book:
 https://www.cia.gov/library/publications/the-world-factbook/index.html

CHAPTER 18. The New Aristotelians

The physical sciences in the twentieth century are presented in many popular and semipopular books. The following is a small sample: Gerard Piel, *The Age of Science: What Scientists Learned in the Twentieth Century* (New York: Basic Books, 2001); Russell McCormmach, *Night Thoughts of a Classical Physicist* (Cambridge: Harvard University Press, 1982); Stephen W. Hawking, *The Illustrated A Brief History of Time*, updated and expanded (New York: Bantam Books, 1996); Steven Weinberg, *The First Three Minutes: A Modern View of the Origin of the Universe*, updated ed. (New York: Basic Books, 1993); George Gamow, *Thirty Years That Shook Physics: The Story of Quantum Theory* (New York: Dover, 1985); David Lindley, *The End of Physics:*

The Myth of a Unified Theory (New York: Basic Books, 1993). More scholarly and detailed is Jed Z. Buchwald and Robert Fox, eds., *The Oxford Handbook of the History of Physics* (Oxford and New York: Oxford University Press, 2014), which covers the subject from the seventeenth century. See also Helge Kragh, *Quantum Generations: A History of Physics in the Twentieth Century* (Princeton: Princeton University Press, 1999); Simon Singh, *Big Bang: The Origin of the Universe* (New York: HarperCollins, 2004); Cornelia Dean, ed., *"The New York Times" Book of Physics and Astronomy: More Than 100 Years of Covering the Expanding Universe* (New York: Sterling, 2013); James Kakalios, *The Amazing Story of Quantum Mechanics* (New York: Gotham Books, 2011). On the end of science see John Horgan, *The End of Science: Facing the Limits of Knowledge in the Twilight of the Scientific Age* (Reading, Mass.: Helix Books, 1996); Russell Stannard, *The End of Discovery* (Oxford and New York: Oxford University Press, 2010). See also Siegmund Brandt, *The Harvest of a Century: Discoveries of Modern Physics in 100 Episodes* (Oxford and New York: Oxford University Press, 2009).

On the substantial literature dealing with Albert Einstein, see Michel Janssen and Christoph Lehner, *The Cambridge Companion to Einstein* (New York: Cambridge University Press, 2014); David C. Cassidy, *Einstein and Our World* (Atlantic Highlands, N.J.: Humanities Press, 1995); Ronald William Clark, *Einstein: The Life and Times* (New York: Wings Books, 1995); Abraham Païs, *"Subtle is the Lord . . .": The Science and Life of Albert Einstein* (Oxford: Oxford University Press, 1982); Jeremy Bernstein, *Albert Einstein and the Frontiers of Physics* (New York: Oxford University Press, 1996), and his *Einstein* (New York: Penguin, 1976); Richard Staley, *Einstein's Generation: The Origins of the Relativity Revolution* (Chicago: University of Chicago Press, 2008). See also Richard Reeves, *A Force of Nature: The Frontier Genius of Ernest Rutherford* (New York: W. W. Norton, 2008). Alan Lightman, *Einstein's Dreams: A Novel* (New York: Warner Books, 1994) provides poetic insight into Einstein's worldview.

James D. Watson's story of the discovery of the structure of DNA details not only that landmark of twentieth-century science but also the realities of contemporary scientific practice; it is best accessible in James Watson, *The Double Helix: A Personal Account of the Discovery of the Structure of DNA*, A Norton Critical Edition, ed. Gunther S. Stent (New York: Norton, 1980). See also James D. Watson and Andrew Berry, *DNA: The Secret of Life* (New York: Knopf, 2003); Brenda Maddox, *Rosalind Franklin: The Dark Lady of DNA* (New York: HarperCollins, 2002); and Evelyn Fox Keller, *A Feeling for the Organism: The Life and Work of Barbara McClintock* (New York: Owl Books, 2003). David Joravsky, *The Lysenko Affair* (Chicago: University of Chicago Press, 1970), details a notable episode in the history of the life sciences in the twentieth century. Edward O. Wilson, *Consilience: The Unity of Knowledge* (New York: Alfred A. Knopf, 1998) is a work to be consulted relative to contemporary natural philosophy.

On the history of psychology, see Duane P. Schultz and Sydney Ellen Schultz, *A History of Modern Psychology*, 8th ed. (Belmont, Calif.: Wadsworth Publishing, 2004); David Hothersall, *History of Psychology*, 4th ed. (Columbus, Ohio: McGraw Hill, 2004); C. James Goodwin, *A History of Modern Psychology*, 5th ed. (Hoboken, N.J.: John Wiley & Sons, 2015); Bernard J. Baars,

The Cognitive Revolution in Psychology (New York: Guilford Press, 1986); Roger E. Backhouse and Philippe Fontaine, *The History of the Social Sciences since 1945* (Cambridge and New York: Cambridge University Press, 2010).

Web Resources

Albert Einstein Online (Links):
   www.westegg.com/einstein
American Institute of Physics, Center for the History of Physics:
   www.aip.org/history-programs/physics-history
Modern Science:
   http://undsci.berkeley.edu/article/modern_science
Nobel Laureates:
   http://ddd.uab.cat/pub/ppascual/ppascualcor/1997/ppascualcor_07_234@
      benasque.pdf
Nobel Prizes:
   nobelprize.org
   www.almaz.com
Marie Curie:
   http://www.nobelprize.org/nobel_prizes/physics/laureates/1903/marie
      -curie-facts.html
History of the Lawrence Livermore National Lab:
   https://www.llnl.gov/str/Hacker.html
Center for Evolutionary Psychology:
   www.psych.ucsb.edu/research/cep
Linus Pauling and the Race for DNA:
   osulibrary.orst.edu/specialcollections/coll/pauling/dna/index.html
History of the Heredity Molecule:
   www.mun.ca/biology/scarr/2250_History.html
Edwin Hubble:
   www.edwinhubble.com
   http://www.pbs.org/wgbh/aso/databank/entries/bahubb.html
The Official String Theory Website:
   superstringtheory.com/index.html
Black Holes and Neutron Stars:
   antwrp.gsfc.nasa.gov/htmltest/rjn_bht.html
Sources for the History of Quantum Physics:
   http://www.amphilsoc.org/guides/ahqp

CHAPTER 19. The Bomb, the Internet, and the Genome

The twentieth-century merger of science and technology has been approached from many directions. The paradigmatic event, the invention and building of the atomic bomb, is described in Richard Rhodes, *The Making of the Atomic Bomb* (New York: Simon and Schuster, 1986); see also his *Dark Sun: The Making of the Hydrogen Bomb* (New York: Touchstone, 1996). See also Ray Monk, *Robert Oppenheimer: A Life inside the Center* (New York: Doubleday, 2013); Kai Bird, *American Prometheus: The Triumph and Tragedy of J. Robert Oppenheimer* (New York: Vintage Books, 2006), and for more on science and technology in World War II, see Ad Maas and Hans Hooijmaijers, *Scientific*

Research in World War II: What Scientists Did in the War (New York: Rout-ledge/Taylor & Francis: 2009); Brian J. Ford, Secret Weapons: Technology, Science, and the Race to Win World War II (Oxford and Long Island City, N.Y.: Osprey, 2011); Jordynn Jack, Science on the Home Front: American Women Scientists in World War II (Urbana: University of Illinois Press, 2009).

On science and industry, the works of the business historian Alfred D. Chandler, Jr., are to be consulted. See his Shaping the Industrial Century: The Remarkable Story of the Evolution of the Modern Chemical and Pharmaceu-tical Industries (Cambridge, Mass.: Harvard University Press, 2005); Invent-ing the Electronic Century: The Epic Story of the Consumer Electronics and Computer Science Industries (New York: Free Press, 2001); and editor with James W. Cortada, A Nation Transformed by Information: How Information Has Shaped the United States from Colonial Times to the Present (Oxford and New York: Oxford University Press, 2000).

On genetics, see A. H. Sturtevant, A History of Genetics (Cold Spring Har-bor, N.Y.: Cold Spring Harbor Laboratory Press, 2001); Henry Gee, Jacob's Ladder: The History of the Human Genome (New York: W.W. Norton, 2004); Philip R. Reilly, Is It in Your Genes? How Genes Influence Common Disorders and Diseases That Affect You and Your Family (Cold Spring Harbor, N.Y.: Cold Spring Harbor Laboratory Press, 2004); and his Orphan: The Quest to Save Children with Rare Genetic Disorders (Cold Spring Harbor, NY: Cold Spring Harbor Laboratory Press, 2015); Paul Lawrence Farber, Mixing Races: From Scientific Racism to Modern Evolutionary Ideas (Baltimore: Johns Hop-kins University Press, 2011).

On computers and computing, see Andrew L. Russell, Open Standards and the Digital Age: History, Ideology, and Networks (New York: Cambridge Uni-versity Press, 2014); Martin Campbell-Kelly, William Aspray, Nathan Ens-menger, and Jeffrey R. Yost, Computer: A History of the Information Ma-chine, 3rd ed. (Boulder, Colo.: Westview Press, 2014); Thomas J. Misa, Digital State: The Story of Minnesota's Computing Industry (Minneapolis: University of Minnesota Press, 2013); Michael R. Williams, A History of Computing, 2nd ed. (New York: Wiley-IEEE Computer Society Press, 1997); Paul E. Ceruzzi, A History of Modern Computing, 2nd ed. (Cambridge, Mass.: MIT Press, 2003); James Wallace and Jim Erickson, Hard Drive: Bill Gates and the Making of the Microsoft Empire (New York: HarperCollins, 1993); Michael A. Hiltzik, Dealers of Lightning: Xerox PARC and the Dawn of the Computer Age (New York: HarperBusiness, 2000); Subrata Dasgupta, It Began with Babbage: The Genesis of Computer Science (Oxford: Oxford University Press, 2014).

On other topics treated in this chapter, see Joan Lisa Bromberg, NASA and the Space Industry (Baltimore: Johns Hopkins University Press, 2000); Richard Ling, The Mobile Connection: The Cell Phone's Impact on Society (San Fran-cisco: Morgan Kaufmann, 2004); Nick Taylor, Laser: The Inventor, the Nobel Laureate, and the Thirty-Year Patent War (New York: Citadel Press, 2000).

Web Resources

A-Bomb WWW Museum:
   http://atomicbombmuseum.org
The Nuclear Weapon Archive:
   nuclearweaponarchive.org

The NUKEMAP by Alex Wellerstein:
  nuclearsecrecy.com/nukemap
Leo Szilard Online:
  www.dannen.com/szilard.html
Human Genome Project History:
  www.ornl.gov/sci/techresources/Human_Genome/project/hgp.shtml
The Internet Society (History and Links):
  www.isoc.org/internet/history
Telecom History:
  www.privateline.com/history.html

CHAPTER 20. Under Today's Pharaohs

Sociological analyses of today's science are presented in Ben-David, *The Scientist's Role in Society* (listed above in chap. 4), and Derek J. da Solla Price, *Little Science, Big Science . . . and Beyond* (New York: Columbia University Press, 1986). See also Daniel S. Greenberg, *The Politics of Pure Science*, 2nd ed. (Chicago: University of Chicago Press, 1999), and David Dickson, *The New Politics of Science*, with a new preface (Chicago: University of Chicago Press, 1988). John Ziman, *The Force of Knowledge* (Cambridge: Cambridge University Press, 1976), is still of value as an introduction. Peter L. Galison and Bruce Hevly, eds., *Big Science: The Growth of Large-Scale Research* (Stanford: Stanford University Press, 1992); Mario Biagioli and Peter Galison, eds., *Scientific Authorship: Credit and Intellectual Property in Science* (New York and London: Routledge, 2003); Peter Galison and Emily Thompson, *The Architecture of Science* (Cambridge, Mass.: MIT Press, 1999) are also to be consulted.

For innovative approaches to the study of contemporary science and technology, see Bruno Latour and Steve Woolgar, *Laboratory Life: The Construction of Scientific Facts* (Princeton, N.J.: Princeton University Press, 1986); Bruno Latour, *Science in Action* (Cambridge, Mass.: Harvard University Press, 1987); Wiebe E. Bijker, Thomas P. Hughes, and Trevor Pinch, *The Social Construction of Technological Systems* (Cambridge, Mass.: MIT Press, 1989); and Weibe E. Bijker and John Law, *Shaping Technology/Building Society: Studies in Sociotechnical Change* (Cambridge, Mass.: MIT Press, 1991); Donald MacKenzie and Judy Wajcman, eds., *The Social Shaping of Technology*, 2nd ed. (Buckingham, U.K. and Philadelphia: Open University Press, 1999).

On women in science, see Barbara Betsch-McGreyne, *Nobel Women in Science: Their Lives, Struggles, and Momentous Discoveries*, 2nd ed. (Washington, D.C.: National Academies Press, 2001); Yu Xei and Kimberlee A. Shauman, *Women in Science: Career Processes and Outcomes* (Cambridge, Mass.: Harvard University Press, 2003); Sue V. Rosser, *The Science Glass Ceiling: Academic Women Scientists and the Struggle to Succeed* (New York: Routledge, 2004); Monique Frize, *The Bold and the Brave: A History of Women in Science and Engineering* (Ottawa: University of Ottawa Press, 2009); Tiffany K. Wayne, *American Women of Science since 1900* (Santa Barbara, Calif.: ABC-CLIO, 2011); Margaret W. Rossiter, *Women Scientists in America: Forging a New World since 1972* (Baltimore: Johns Hopkins University Press, 2012).

For other topics treated in this chapter see Martin Beech, *The Large Hadron Collider: Unraveling the Mysteries of the Universe* (New York: Springer,

2010); Benjamin A. Elman, *A Cultural History of Modern Science in China* (Cambridge, Mass.: Harvard University Press, 2009); Audra J. Wolfe, *Competing with the Soviets: Science, Technology, and the State in Cold War America* (Baltimore: Johns Hopkins University Press, 2013); Philip Mirowski, *Science-Mart: Privatizing American Science* (Cambridge, Mass.: Harvard University Press, 2011); Mark Solovey, *Shaky Foundations: The Politics-Patronage-Social Science Nexus in Cold War America* (New Brunswick, N.J.: Rutgers University Press, 2013).

See also Bureau of Labor Statistics, U.S. Department of Labor, *Occupational Outlook Handbook, 2014–15 Edition* (New York: Skyhorse Publishing, 2014); National Science Foundation / National Center for Science and Engineering Statistics, *Science and Engineering Indicators 2012* (Arlington, Va.: NSF/NCSES, 2012).

Web Resources

National Science Foundation, Division of Science Resource Statistics:
www.nsf.gov/sbe/srs/stats.htm
U.S. Patent and Trademark Office, Patent Activity:
www.uspto.gov/web/offices/ac/ido/oeip/taf/h_counts.htm
R&D Funding Data (American Association for the Advancement of Science):
http://www.aaas.org/page/historical-trends-federal-rd
http://www.aaas.org/program/rd-budget-and-policy-program
Bureau of Labor Statistics:
http://www.bls.gov
World Bank data:
data.worldbank.org/data-catalog
OECD (Organisation for Economic Co-operation and Development) Main Science and Technology Indicators 2013:
http://www.oecd.org/sti/msti.htm
Batelle Corporation forecasts:
http://www.battelle.org/docs/tpp/2014_global_rd_funding_forecast
.pdf?sfvrsn=4
Center for the Study of Technology and Society:
www.tecsoc.org
American Physical Society, Careers and Employment:
www.aps.org/jobs/index.cfm
Science and Technology in the People's Republic of China:
http://www.loc.gov/rr/scitech/tracer-bullets/scitechchinatb.html
The Virtual Nuclear Tourist:
www.nucleartourist.com
The Public Library of Science:
http://www.plos.org
American Institute of Physics:
www.aip.org
Association for Women in Science:
www.awis.org
Committee on Women in Science and Engineering:
http://sites.nationalacademies.org/pga/cwsem

Illustration Credits

Maps, figures, and illustrations not included in the list below are original art or from the collections of the authors.

Fig. 1.2. © 1982 Jay H. Matternes. Reprinted with permission.

Fig. 1.3. Archive Photos.

Fig. 1.4. Gerald S. Hawkins, *Beyond Stonehenge* (1973). Reprinted with the kind permission of the author.

Fig. 2.2. Kathleen Mary Kenyon, *Digging Up Jericho: The Results of the Jericho Excavations, 1952–1956.* © 1957 by Frederick A. Praeger, Inc. Reprinted by permission of Henry Holt and Co., Inc.

Fig. 2.3. J. Easton, *Conjectures on That Mysterious Monument of Ancient Art, Stonehenge, on Salisbury Plain* (Salisbury: J. Easton, 1815).

Fig. 2.5. Archive Photos.

Fig. 3.1. Archive Photos.

Fig. 5.4. Image © Estate of Mary E. and Dan Todd, reprinted with permission.

Fig. 6.1. © Museum of the History of Science, Oxford, U.K. Reprinted with permission.

Fig. 7.2. Joseph Needham, *Science and Civilization in China*, vol. 2 (Cambridge: Cambridge University Press, 1956), facing p. 363. Reprinted with permission from Cambridge University Press.

Fig. 7.3. Arnold Pacey, *Technology in World Civilization* (Cambridge, Mass.: MIT Press, 1992), p. 65. Reprinted with permission.

Fig. 7.6. Joseph Needham, *Science and Civilization in China*, vol. 4, pt. 2 (Cambridge: Cambridge University Press, 1965), 449. Reprinted with permission from Cambridge University Press.

Fig. 7.7. Joseph Needham, *Science and Civilization in China*, vol. 3 (Cambridge: Cambridge University Press, 1959), 629. Reprinted with permission from Cambridge University Press.

Fig. 9.1. Reprinted with permission from J. Justeson and T. Kaufman, "A Decipherment of Epi-Olmec Hieroglyphic Writing," *Science* 259 (19 March 1993). © 1993 American Association for the Advancement of Science.

Fig. 9.5. Archive Photos.

Fig. 9.6. Leslie Leland Locke, *The Ancient Quipu or Peruvian Knot Record* (New York: The American Museum of Natural History, 1923).

Fig. 9.7. Photo by Paul Logsdon, reprinted with the kind permission of Marcia L. Logsdon.

Fig. 10.2. Paul Lacroix, Ferdinand Séré and A. Rivaux, *Le Moyen Âge et la Renaissance,* vol. 1 (Paris: Administration, 1848).

Fig. 10.5. Original data of H. Dorn.

Fig. 11.1. Vesalius, *De humani corporis fabrica* (Basel, 1555). Reprinted with permission from the Berg Collection, the New York Public Library, Astor, Lenox, and Tilden Foundations.

Fig. 11.4. Jean Bleau, *Geographia* (Amsterdam, 1662). Reprinted with permission from the Spencer Collection, the New York Public Library, Astor, Lenox, and Tilden Foundations.

Fig. 13.4. Robert Boyle, *Nova experimenta physico-mecanica* (Rotterdam, 1669). Reprinted with permission from the Wheeler Collection, Rare Book Division, the New York Public Library, Astor, Lenox, and Tilden Foundations.

Fig. 14.1. Engravings of Industries Collection, Baker Library, Harvard Business School (Box 1, Mss 442), reprinted with permission.

Fig. 14.4. Archive Photos.

Fig. 15.1. Abbé Nollet, *Leçons de physique expérimentale* (Paris, 1745–48). Reprinted with permission from the Wheeler Collection, Rare Book Division, the New York Public Library, Astor, Lenox, and Tilden Foundations.

Fig. 15.3. Michael Faraday, *Experimental Researches in Chemistry and Physics* (London and New York: Taylor & Francis, 1859).

Fig. 15.4. Archive Photos.

Fig. 15.5. U.S. Department of the Navy, Naval Historical Center photograph.

Fig. 15.6. Archive Photos.

Map 16.1. Charles Darwin, *Journal of Researches . . . during the Voyage of the H.M.S. Beagle* (London: J. Murray, 1873).

Fig. 16.1. Archive Photos.

Fig. 17.2. Archive Photos.

Fig. 17.3. Images courtesy of Wikipedia (France) and the Wright Brothers Aeroplane Company (West Milton, Ohio).

Fig. 17.4. Image courtesy of the Advertising Archives.

Fig. 18.2. Courtesy of Institut Curie, Association Curie et Joliot-Curie, Paris.

Fig. 18.3. Photo courtesy of NASA/STScl.

Fig. 19.1. Collage of images from http://nuclearweaponsarchive.com and www.hiroshima-is.ac.jp.

Fig. 19.2. Reproduced by permission of the Computer History Museum; http://www.computerhistory.org.

Fig. 19.3. Photo courtesy of Philips Healthcare. All rights reserved.

Fig. 20.5. D. Price, *Science Since Babylon* (New Haven: Yale University Press, 1961), 166. Reprinted by permission from Yale University Press.

Fig. 20.6. Image courtesy of CERN (Image HI-1307174-13, July 2013).

Index

288, 294; in Greece, 70–94, 96, 99, 103, 105–6, 109–11; and Islam, 119, 123, 125; in medieval Europe, 202, 204, 207–9, 212, 221; and modern period, 401, 406, 409, 419, 427, 444, 474, 479; in 19th century, 328, 330, 333–34, 337, 344; in Renaissance, 226, 243, 246–48, 250, 254, 263, 265–67

Natural selection, 24, 350, 357, 359–64, 413

Natural theology, 279, 283–84, 289, 293, 316, 350

Navies, 145–46, 215, 217–20, 255, 285, 287, 294, 307, 340–42

Nazis, 415, 429

Neanderthals, 20–26, 362, 416

Nebulous bodies, 410

Needles, 25

Neolithic era, 13–14, 17–19, 21, 30–47, 49–51, 53, 59, 73, 186–88, 192–93, 195, 197, 202, 297, 299, 307, 312, 337, 366, 481, 489

Neoplatonism, 226, 231, 266, 282

Neptune, 322

Nestorian Christians, 120

Netherlands, 12, 218, 220, 222, 244, 255, 263–65, 274, 338, 404

Neumann, John von, 433

Neutrino, 408

Neutron, 408

New Age beliefs, 2

Newcomen, Thomas, 309–10, 314, 319

New Deal, 369

Newton, Isaac, 8, 12, 91, 133, 210–11, 228–29, 246, 256, 261–62, 265, 267–72, 274–85, 288–90, 292–94, 298, 306, 317, 320–23, 327, 329–30, 332–33, 364, 401, 405–6, 451, 471, 500

New York City, viii, 341–42, 345, 366–70, 389

Nile River, 49–51, 55, 57–59, 61, 65, 67, 76–77, 96, 120, 126, 218

Nineteenth Amendment, 383

Ninety-Five Theses (Luther), 228

Nobel, Alfred, 467

Nobel Prize, 334, 347, 378, 403, 408, 411, 414, 417, 420, 434, 440–41, 446, 464, 467, 471

Nollet, Jean-Antoine, 324

Nomadism, 25, 31–34, 121, 135, 187, 192, 302, 338

Norfolk system, 307

North America, early civilizations in, 186–93

North Star (Polaris), 191, 232, 286

Nuclear energy, 366, 369, 371, 408, 428–29, 431, 479

Nuclear weapons, 7, 399, 410, 428–32, 452, 478

Nutrifying spirits (Galenic physiology), 108–9

Oats, 198–99

Observatories, astronomical, 126–29, 131, 152–53, 178, 180, 186, 190–92, 265, 274, 281, 286–87, 472

Obsidian, 54, 56–57

Oersted, Hans Christian, 328, 330–31

Oil (petroleum), 313, 366, 369, 371, 376, 379, 383, 398–99

Olives, 73

Olmecs, 44, 53, 174

Oppenheimer, J. Robert, 422, 430–31

Optics, 101, 130, 209, 244, 260, 263–65, 268, 271–72, 282–83, 287–89, 321–22, 329, 332

Oresme, Nicole, 208, 211–12, 258

Organ transplants, 445

Oribasius, 110

Origin of Species (Darwin), 360–64

Orreries, 43, 152–53

Osiander, Andreas, 234, 249

Ottoman Empire, 117, 120, 214–15, 219

Oven, 382

Oxen, 46, 57, 198

Oxygen, 325–26

Ozone layer, 399

Palenque (Mayan city), 179

Paleolithic era, 1, 4, 13–14, 17, 19, 21–33, 35–38, 43–44, 49, 53, 87, 186–87, 192–93, 378, 416, 481, 488

Paleontology, 17, 24, 30, 416

Paley, William, 350–51

Pangenesis, 363

Pantheism, 70

Papacy, 200, 213, 219, 228–30, 234, 250, 252–54, 262

Paper, 127, 141–44

Papyrus, 57, 64, 96

Paracelsus, 226–27

Parallax, stellar, 99, 233, 236